Student Solutions Manual

College Algebra

EIGHTH EDITION

Jerome E. Kaufmann
Western Illinois University (Retired)

Karen L. Schwitters
Seminole State College of Florida

Prepared by

Ross Rueger
College of the Sequoias

BROOKS/COLE
CENGAGE Learning™

Australia • Brazil • Japan • Korea • Mexico • Singapore • Spain • United Kingdom • United States

BROOKS/COLE
CENGAGE Learning™

For product information and technology assistance, contact us at **Cengage Learning Customer & Sales Support, 1-800-354-9706**

For permission to use material from this text or product, submit all requests online at **www.cengage.com/permissions** Further permissions questions can be emailed to **permissionrequest@cengage.com**

ISBN-13: 978-1-111-99045-9

ISBN-10: 1-111-99045-X

Brooks/Cole
20 Davis Drive
Belmont, CA 94002
USA

Cengage Learning is a leading provider of customized learning solutions with office locations around the globe, including Singapore, the United Kingdom, Australia, Mexico, Brazil, and Japan. Locate your local office at **www.cengage.com/global**

Cengage Learning products are represented in Canada by Nelson Education, Ltd.

To learn more about Brooks/Cole, visit **www.cengage.com/brookscole**

Purchase any of our products at your local college store or at our preferred online store **www.cengagebrain.com**

Printed in the United States of America
2 3 4 5 6   18 17 16 15 14

Contents

Preface

This *Student's Solutions Manual* contains complete solutions to every odd-numbered problem in the exercise sets and all end-of-chapter exercises for Kaufmann/Schwitters' *College Algebra* (Eighth Edition). I have attempted to format solutions for readability and accuracy, and apologize to you for any errors that you may encounter. If you have any comments, suggestions, error corrections, or alternative solutions please feel free to drop me a note. If you prefer, you can send me e-mail with corrections or comments (address at the bottom of this page).

I would like to thank Cynthia Ashton of Cengage Learning for her guidance throughout this project, as well as Jerome Kaufmann and Karen Schwitters for asking me to be involved. This book provides a thorough treatment of college algebra topics, which should help you to prepare for your future mathematics courses. Remember to attempt all problems before looking up the solution in this manual.

I would especially like to thank Matt Bourez of College of the Sequoias for his complete error-checking of this manuscript. Any errors you encounter are undoubtedly mine, and not his.

Ross Rueger
College of the Sequoias
915 South Mooney Boulevard
Visalia, CA 93277
rossrueger@gmail.com

February, 2012

Chapter 0
Some Basic Concepts of Algebra: A Review

0.1 Some Basic Ideas

1. This statement is true.

3. This statement is false ($\sqrt{3}$ is a real number which is not a rational number).

5. This statement is false.

7. This statement is true.

9. This statement is false.

11. The natural numbers set is $\{46\}$.

13. The integer set is $\{0, -14, 46\}$.

15. The irrational number set is $\left\{\sqrt{5}, -\sqrt{2}, -\pi\right\}$.

17. The nonpositive integer set is $\{0, -14\}$.

19. The statement is $N \subseteq R$.

21. The statement is $N \subseteq I$.

23. The statement is $H \not\subseteq Q$.

25. The statement is $W \subseteq I$.

27. The statement is $I \not\subseteq W$.

29. The statement is $\{0, 2, 4, ...\} \subseteq W$.

31. The statement is $\{-2, -1, 0, 1, 2\} \not\subseteq W$.

33. The set is $\{1\}$.

35. The set is $\{0, 1, 2, 3\}$.

37. The set is $\{..., -2, -1, 0, 1\}$.

39. The set is $\varnothing$ (empty set).

41. The set is $\{0, 1, 2\}$.

43.
   a. The distance is $35 - 17 = 18$.
   b. The distance is $12 - (-14) = 26$.
   c. The distance is $18 - (-21) = 39$.
   d. The distance is $-17 - (-42) = 25$.
   e. The distance is $-21 - (-56) = 35$.
   f. The distance is $0 - (-37) = 37$.

45. This is the commutative property of multiplication.

47. This is the identity property of multiplication.

49. This is the multiplication property of -1.

51. This is the distributive property.

53. This is the commutative property of multiplication.

55. This is the distributive property.

57. This is the associative property of multiplication.

59. Evaluating: $5x + 3y = 5(-2) + 3(-4) = -10 - 12 = -22$

61. Evaluating: $-3ab - 2c = -3(-4)(7) - 2(-8) = 84 + 16 = 100$

63. Evaluating: $(a - 2b) + (3c - 4) = \left[6 - 2(-5)\right] + \left[3(-11) - 4\right] = (6 + 10) + (-33 - 4) = 16 - 37 = -21$

65. Evaluating: $\dfrac{-2x + 7y}{x - y} = \dfrac{-2(-3) + 7(-2)}{-3 - (-2)} = \dfrac{6 - 14}{-3 + 2} = \dfrac{-8}{-1} = 8$

67. Evaluating: $(5x - 2y)(-3x + 4y) = \left[5(-3) - 2(-7)\right]\left[-3(-3) + 4(-7)\right] = (-15 + 14)(9 - 28) = (-1)(-19) = 19$

1

69. Evaluating: $5x + 4y - 9y - 2y = 5(2) + 4(-8) - 9(-8) - 2(-8) = 10 - 32 + 72 + 16 = 66$

71. Evaluating: $-5x + 8y + 7y + 8x = -5(5) + 8(-6) + 7(-6) + 8(5) = -25 - 48 - 42 + 40 = -75$

73. Evaluating: $|3x + y| + |2x - 4y| = |3(5) - 3| + |2(5) - 4(-3)| = |15 - 3| + |10 + 12| = 12 + 22 = 34$

75. Evaluating: $\left|\dfrac{2a - 3b}{3b - 2a}\right| = \left|\dfrac{2(-4) - 3(-8)}{3(-8) - 2(-4)}\right| = \left|\dfrac{-8 + 24}{-24 + 8}\right| = \left|\dfrac{16}{-16}\right| = |-1| = 1$

77. Evaluating: $2(3x + 4) - 3(2x - 1) = 2(3(-2) + 4) - 3(2(-2) - 1) = 2(-6 + 4) - 3(-4 - 1) = -4 + 15 = 11$

79. Evaluating:
$$5(a - 3) - 4(2a + 1) - 2(a - 4) = 5(-3 - 3) - 4(2(-3) + 1) - 2(-3 - 4)$$
$$= 5(-6) - 4(-5) - 2(-7)$$
$$= -30 + 20 + 14$$
$$= 4$$

81. Plotting the point $(-4, 1)$:

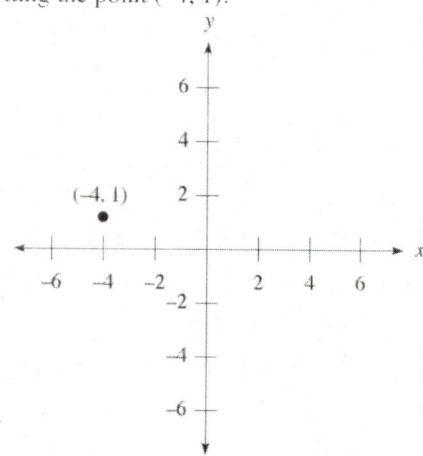

83. Plotting the point $(0, -3)$:

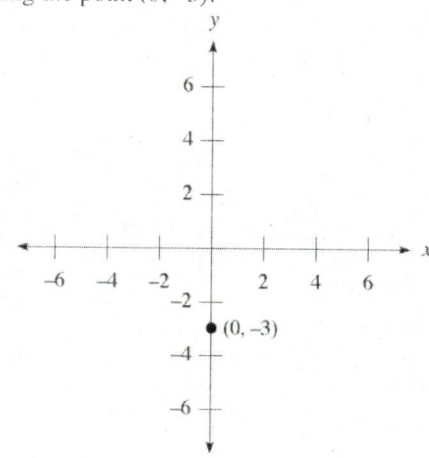

85. Plotting the point $(5, -1)$:

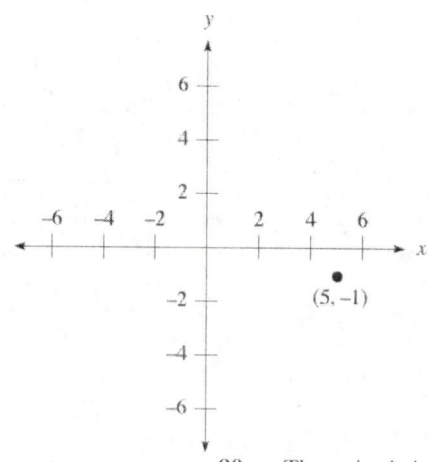

87. The point is in quadrant IV.
91. The point is in quadrant I.

89. The point is in quadrant III.

0.2 Exponents

1. Evaluating the expression: $2^{-3} = \dfrac{1}{2^3} = \dfrac{1}{8}$

3. Evaluating the expression: $-10^{-3} = -\dfrac{1}{10^3} = -\dfrac{1}{1000}$

5. Evaluating the expression: $\dfrac{1}{3^{-3}} = 3^3 = 27$

7. Evaluating the expression: $\left(\dfrac{1}{2}\right)^{-2} = 2^2 = 4$

9. Evaluating the expression: $\left(-\dfrac{2}{3}\right)^{-3} = \left(-\dfrac{3}{2}\right)^3 = -\dfrac{27}{8}$

11. Evaluating the expression: $\left(-\dfrac{1}{5}\right)^0 = 1$

13. Evaluating the expression: $\dfrac{1}{\left(\dfrac{4}{5}\right)^{-2}} = \dfrac{1}{\left(\dfrac{5}{4}\right)^2} = \dfrac{1}{\dfrac{25}{16}} = \dfrac{16}{25}$

15. Evaluating the expression: $2^5 \cdot 2^{-3} = 2^2 = 4$

17. Evaluating the expression: $10^{-6} \cdot 10^4 = 10^{-2} = \dfrac{1}{10^2} = \dfrac{1}{100} = 0.01$

19. Evaluating the expression: $10^{-2} \cdot 10^{-3} = 10^{-5} = \dfrac{1}{10^5} = \dfrac{1}{100,000} = 0.00001$

21. Evaluating the expression: $\left(3^{-2}\right)^{-2} = 3^4 = 81$

23. Evaluating the expression: $\left(4^2\right)^{-1} = 4^{-2} = \dfrac{1}{4^2} = \dfrac{1}{16}$

25. Evaluating the expression: $\left(3^{-1} \cdot 2^2\right)^{-1} = 3 \cdot 2^{-2} = \dfrac{3}{2^2} = \dfrac{3}{4}$

27. Evaluating the expression: $\left(4^2 \cdot 5^{-1}\right)^2 = 4^4 \cdot 5^{-2} = \dfrac{4^4}{5^2} = \dfrac{256}{25}$

29. Evaluating the expression: $\left(\dfrac{2^{-2}}{5^{-1}}\right)^{-2} = \dfrac{2^4}{5^2} = \dfrac{16}{25}$

31. Evaluating the expression: $\left(\dfrac{3^{-2}}{8^{-1}}\right)^2 = \dfrac{3^{-4}}{8^{-2}} = \dfrac{8^2}{3^4} = \dfrac{64}{81}$

33. Evaluating the expression: $\dfrac{2^3}{2^{-3}} = 2^6 = 64$

35. Evaluating the expression: $\dfrac{10^{-1}}{10^4} = 10^{-5} = \dfrac{1}{10^5} = \dfrac{1}{100,000} = 0.00001$

37. Evaluating the expression: $3^{-2} + 2^{-3} = \dfrac{1}{3^2} + \dfrac{1}{2^3} = \dfrac{1}{9} + \dfrac{1}{8} = \dfrac{8}{72} + \dfrac{9}{72} = \dfrac{17}{72}$

39. Evaluating the expression: $\left(\dfrac{2}{3}\right)^{-1} - \left(\dfrac{3}{4}\right)^{-1} = \dfrac{3}{2} - \dfrac{4}{3} = \dfrac{9}{6} - \dfrac{8}{6} = \dfrac{1}{6}$

41. Evaluating the expression: $\left(2^{-4} + 3^{-1}\right)^{-1} = \left(\dfrac{1}{2^4} + \dfrac{1}{3}\right)^{-1} = \left(\dfrac{1}{16} + \dfrac{1}{3}\right)^{-1} = \left(\dfrac{3}{48} + \dfrac{16}{48}\right)^{-1} = \left(\dfrac{19}{48}\right)^{-1} = \dfrac{48}{19}$

43. Simplifying the expression: $x^3 \cdot x^{-7} = x^{-4} = \dfrac{1}{x^4}$

45. Simplifying the expression: $a^2 \cdot a^{-3} \cdot a^{-1} = a^{-2} = \dfrac{1}{a^2}$

47. Simplifying the expression: $\left(a^{-3}\right)^2 = a^{-6} = \dfrac{1}{a^6}$

49. Simplifying the expression: $\left(x^3 y^{-4}\right)^{-1} = x^{-3} y^4 = \dfrac{y^4}{x^3}$

51. Simplifying the expression: $\left(ab^2 c^{-1}\right)^{-3} = a^{-3} b^{-6} c^3 = \dfrac{c^3}{a^3 b^6}$

53. Simplifying the expression: $\left(2x^2 y^{-1}\right)^{-2} = 2^{-2} x^{-4} y^2 = \dfrac{y^2}{4x^4}$

55. Simplifying the expression: $\left(\dfrac{x^{-2}}{y^{-3}}\right)^{-2} = \dfrac{x^4}{y^6}$

57. Simplifying the expression: $\left(\dfrac{2a^{-1}}{3b^{-2}}\right)^{-2} = \dfrac{2^{-2}a^2}{3^{-2}b^4} = \dfrac{3^2 a^2}{2^2 b^4} = \dfrac{9a^2}{4b^4}$

59. Simplifying the expression: $\dfrac{x^{-5}}{x^{-2}} = x^{-3} = \dfrac{1}{x^3}$

61. Simplifying the expression: $\dfrac{a^2 b^{-3}}{a^{-1}b^{-2}} = a^3 b^{-1} = \dfrac{a^3}{b}$

63. Simplifying the expression: $\left(4x^3 y^2\right)\left(-5xy^3\right) = -20x^4 y^5$

65. Simplifying the expression: $\left(-3xy^3\right)^3 = -27x^3 y^9$

67. Simplifying the expression: $\left(\dfrac{2x^2}{3y^3}\right)^3 = \dfrac{8x^6}{27y^9}$

69. Simplifying the expression: $\dfrac{72x^8}{-9x^2} = -8x^6$

71. Simplifying the expression: $\left(2x^{-1}y^2\right)\left(3x^{-2}y^{-3}\right) = 6x^{-3}y^{-1} = \dfrac{6}{x^3 y}$

73. Simplifying the expression: $\left(-6a^5 y^{-4}\right)\left(-a^{-7}y\right) = 6a^{-2}y^{-3} = \dfrac{6}{a^2 y^3}$

75. Simplifying the expression: $\dfrac{24x^{-1}y^{-2}}{6x^{-4}y^3} = 4x^3 y^{-5} = \dfrac{4x^3}{y^5}$

77. Simplifying the expression: $\dfrac{-35a^3 b^{-2}}{7a^5 b^{-1}} = -5a^{-2}b^{-1} = -\dfrac{5}{a^2 b}$

79. Simplifying the expression: $\left(\dfrac{14x^{-2}y^{-4}}{7x^{-3}y^{-6}}\right)^{-2} = \left(2xy^2\right)^{-2} = 2^{-2}x^{-2}y^{-4} = \dfrac{1}{4x^2 y^4}$

81. Writing as a single fraction: $x^{-1} + x^{-2} = \dfrac{1}{x} + \dfrac{1}{x^2} = \dfrac{x}{x^2} + \dfrac{1}{x^2} = \dfrac{x+1}{x^2}$

83. Writing as a single fraction: $x^{-2} - y^{-1} = \dfrac{1}{x^2} - \dfrac{1}{y} = \dfrac{y}{x^2 y} - \dfrac{x^2}{x^2 y} = \dfrac{y - x^2}{x^2 y}$

85. Writing as a single fraction: $3a^{-2} + 2b^{-3} = \dfrac{3}{a^2} + \dfrac{2}{b^3} = \dfrac{3b^3}{a^2 b^3} + \dfrac{2a^2}{a^2 b^3} = \dfrac{3b^3 + 2a^2}{a^2 b^3}$

87. Writing as a single fraction: $x^{-1}y - xy^{-1} = \dfrac{y}{x} - \dfrac{x}{y} = \dfrac{y^2}{xy} - \dfrac{x^2}{xy} = \dfrac{y^2 - x^2}{xy}$

89. Simplifying the expression: $\left(3x^a\right)\left(4x^{2a+1}\right) = 12x^{a+2a+1} = 12x^{3a+1}$

91. Simplifying the expression: $\left(x^a\right)\left(x^{-a}\right) = x^{a-a} = x^0 = 1$

93. Simplifying the expression: $\dfrac{x^{3a}}{x^a} = x^{3a-a} = x^{2a}$

95. Simplifying the expression: $\dfrac{-24y^{5b+1}}{6y^{-b-1}} = -4y^{5b+1+b+1} = -4y^{6b+2}$

97. Simplifying the expression: $\dfrac{(xy)^b}{y^b} = \dfrac{x^b y^b}{y^b} = x^b$

99. Writing in scientific notation: $62,000,000 = (6.2)(10)^7$

101. Writing in scientific notation: $0.000412 = (4.12)(10)^{-4}$

103. Writing in decimal form: $(1.8)(10)^5 = 180,000$

105. Writing in decimal form: $(2.3)(10)^{-6} = 0.0000023$

107. Performing the operations: $\dfrac{0.00052}{0.013} = \dfrac{(5.2)(10)^{-4}}{(1.3)(10)^{-2}} = (4)(10)^{-2} = 0.04$

109. Performing the operations: $\sqrt{900,000,000} = \sqrt{(9)(10)^8} = (3)(10)^4 = 30,000$

111. Performing the operations: $\sqrt{0.0009} = \sqrt{(9)(10)^{-4}} = (3)(10)^{-2} = 0.03$

117.
   a. Calculating: $(4576)^4 \approx (4.385)(10)^{14}$

   b. Calculating: $(719)^{10} \approx (3.692)(10)^{28}$

   c. Calculating: $(28)^{12} \approx (2.322)(10)^{17}$

   d. Calculating: $(8619)^6 \approx (4.100)(10)^{23}$

   e. Calculating: $(314)^5 \approx (3.052)(10)^{12}$

   f. Calculating: $(145,723)^2 \approx (2.124)(10)^{10}$

0.3 Polynomials

1. Performing the operations: $(5x^2 - 7x - 2) + (9x^2 + 8x - 4) = 14x^2 + x - 6$

3. Performing the operations: $(14x^2 - x - 1) - (15x^2 + 3x + 8) = 14x^2 - x - 1 - 15x^2 - 3x - 8 = -x^2 - 4x - 9$

5. Performing the operations: $(3x - 4) - (6x + 3) + (9x - 4) = 3x - 4 - 6x - 3 + 9x - 4 = 6x - 11$

7. Performing the operations:
$$\left(8x^2 - 6x - 2\right) + \left(x^2 - x - 1\right) - \left(3x^2 - 2x + 4\right) = 8x^2 - 6x - 2 + x^2 - x - 1 - 3x^2 + 2x - 4 = 6x^2 - 5x - 7$$

9. Performing the operations: $5(x - 2) - 4(x + 3) - 2(x + 6) = 5x - 10 - 4x - 12 - 2x - 12 = -x - 34$

11. Finding the product: $3xy\left(4x^2 y + 5xy^2\right) = 12x^3 y^2 + 15x^2 y^3$

13. Finding the product: $6a^3 b^2 \left(5ab - 4a^2 b + 3ab^2\right) = 30a^4 b^3 - 24a^5 b^3 + 18a^4 b^4$

15. Finding the product: $(x + 8)(x + 12) = x^2 + 8x + 12x + 96 = x^2 + 20x + 96$

17. Finding the product: $(n - 4)(n - 12) = n^2 - 4n - 12n + 48 = n^2 - 16n + 48$

19. Finding the product: $(s - t)(x + y) = sx + sy - tx - ty$

21. Finding the product: $(3x - 1)(2x + 3) = 6x^2 - 2x + 9x - 3 = 6x^2 + 7x - 3$

23. Finding the product: $(4x - 3)(3x - 7) = 12x^2 - 28x - 9x + 21 = 12x^2 - 37x + 21$

25. Finding the product: $(x + 4)^2 = (x + 4)(x + 4) = x^2 + 4x + 4x + 16 = x^2 + 8x + 16$

27. Finding the product: $(2n + 3)^2 = (2n + 3)(2n + 3) = 4n^2 + 6n + 6n + 9 = 4n^2 + 12n + 9$

29. Finding the product:
$$(x + 2)(x - 4)(x + 3) = (x + 2)\left(x^2 - x - 12\right) = x^3 - x^2 - 12x + 2x^2 - 2x - 24 = x^3 + x^2 - 14x - 24$$

31. Finding the product:
$$(x - 1)(2x + 3)(3x - 2) = (x - 1)\left(6x^2 + 5x - 6\right) = 6x^3 + 5x^2 - 6x - 6x^2 - 5x + 6 = 6x^3 - x^2 - 11x + 6$$

33. Finding the product: $(x - 1)\left(x^2 + 3x - 4\right) = x^3 + 3x^2 - 4x - x^2 - 3x + 4 = x^3 + 2x^2 - 7x + 4$

35. Finding the product: $(t - 1)\left(t^2 + t + 1\right) = t^3 + t^2 + t - t^2 - t - 1 = t^3 - 1$

37. Finding the product: $(3x+2)(2x^2-x-1)=6x^3-3x^2-3x+4x^2-2x-2=6x^3+x^2-5x-2$

39. Finding the product:

$$(x^2+2x-1)(x^2+6x+4)=x^4+6x^3+4x^2+2x^3+12x^2+8x-x^2-6x-4=x^4+8x^3+15x^2+2x-4$$

41. Finding the product: $(5x-2)(5x+2)=(5x)^2-(2)^2=25x^2-4$

43. Finding the product:

$$\begin{aligned}\left(x^2-5x-2\right)^2 &= \left(x^2-5x-2\right)\left(x^2-5x-2\right)\\ &= x^4-5x^3-2x^2-5x^3+25x^2+10x-2x^2+10x+4\\ &= x^4-10x^3+21x^2+20x+4\end{aligned}$$

45. Finding the product: $(2x+3y)(2x-3y)=(2x)^2-(3y)^2=4x^2-9y^2$

47. Finding the product:

$$\begin{aligned}(x+5)^3 &= (x+5)(x+5)(x+5)\\ &= (x+5)\left(x^2+10x+25\right)\\ &= x^3+10x^2+25x+5x^2+50x+125\\ &= x^3+15x^2+75x+125\end{aligned}$$

49. Finding the product:

$$\begin{aligned}(2x+1)^3 &= (2x+1)(2x+1)(2x+1)\\ &= (2x+1)\left(4x^2+4x+1\right)\\ &= 8x^3+8x^2+2x+4x^2+4x+1\\ &= 8x^3+12x^2+6x+1\end{aligned}$$

51. Finding the product:

$$\begin{aligned}(4x-3)^3 &= (4x-3)(4x-3)(4x-3)\\ &= (4x-3)\left(16x^2-24x+9\right)\\ &= 64x^3-96x^2+36x-48x^2+72x-27\\ &= 64x^3-144x^2+108x-27\end{aligned}$$

53. Finding the product:

$$\begin{aligned}(5x-2y)^3 &= (5x-2y)(5x-2y)(5x-2y)\\ &= (5x-2y)\left(25x^2-20xy+4y^2\right)\\ &= 125x^3-100x^2y+20xy^2-50x^2y+40xy^2-8y^3\\ &= 125x^3-150x^2y+60xy^2-8y^3\end{aligned}$$

55. Using Pascal's triangle: $(a+b)^7=a^7+7a^6b+21a^5b^2+35a^4b^3+35a^3b^4+21a^2b^5+7ab^6+b^7$

57. Using Pascal's triangle:

$$\begin{aligned}(x-y)^5 &= x^5+5x^4(-y)+10x^3(-y)^2+10x^2(-y)^3+5x(-y)^4+(-y)^5\\ &= x^5-5x^4y+10x^3y^2-10x^2y^3+5xy^4-y^5\end{aligned}$$

59. Using Pascal's triangle:

$$\begin{aligned}(x+2y)^4 &= x^4+4x^3(2y)+6x^2(2y)^2+4x(2y)^3+(2y)^4\\ &= x^4+4x^3\bullet 2y+6x^2\bullet 4y^2+4x\bullet 8y^3+16y^4\\ &= x^4+8x^3y+24x^2y^2+32xy^3+16y^4\end{aligned}$$

61. Using Pascal's triangle:

$$\begin{aligned}(2a-b)^6 &= (2a)^6+6(2a)^5(-b)+15(2a)^4(-b)^2+20(2a)^3(-b)^3+15(2a)^2(-b)^4+6(2a)(-b)^5+(-b)^6\\ &= 64a^6-6\bullet 32a^5b+15\bullet 16a^4b^2-20\bullet 8a^3b^3+15\bullet 4a^2b^4-6\bullet 2ab^5+b^6\\ &= 64a^6-192a^5b+240a^4b^2-160a^3b^3+60a^2b^4-12ab^5+b^6\end{aligned}$$

63. Using Pascal's triangle:

$$\begin{aligned}\left(x^2+y\right)^7 &= \left(x^2\right)^7+7\left(x^2\right)^6y+21\left(x^2\right)^5y^2+35\left(x^2\right)^4y^3+35\left(x^2\right)^3y^4+21\left(x^2\right)^2y^5+7\left(x^2\right)y^6+y^7\\ &= x^{14}+7x^{12}y+21x^{10}y^2+35x^8y^3+35x^6y^4+21x^4y^5+7x^2y^6+y^7\end{aligned}$$

65. Using Pascal's triangle:

$$(2a-3b)^5 = (2a)^5 + 5(2a)^4(-3b) + 10(2a)^3(-3b)^2 + 10(2a)^2(-3b)^3 + 5(2a)(-3b)^4 + (-3b)^5$$

$$= 32a^5 + 5(16a^4)(-3b) + 10(8a^3)(9b^2) + 10(4a^2)(-27b^3) + 5(2a)(81b^4) - 243b^5$$

$$= 32a^5 - 240a^4b + 720a^3b^2 - 1080a^2b^3 + 810ab^4 - 243b^5$$

67. Performing the division: $\dfrac{15x^4 - 25x^3}{5x^2} = \dfrac{15x^4}{5x^2} - \dfrac{25x^3}{5x^2} = 3x^2 - 5x$

69. Performing the division: $\dfrac{30a^5 - 24a^3 + 54a^2}{-6a} = \dfrac{30a^5}{-6a} - \dfrac{24a^3}{-6a} + \dfrac{54a^2}{-6a} = -5a^4 + 4a^2 - 9a$

71. Performing the division: $\dfrac{-20a^3b^2 - 44a^4b^5}{-4a^2b} = \dfrac{-20a^3b^2}{-4a^2b} - \dfrac{44a^4b^5}{-4a^2b} = 5ab + 11a^2b^4$

73. Finding the product: $\left(x^a + y^b\right)\left(x^a - y^b\right) = \left(x^a\right)^2 - \left(y^b\right)^2 = x^{2a} - y^{2b}$

75. Finding the product: $\left(x^b + 4\right)\left(x^b - 7\right) = x^{2b} + 4x^b - 7x^b - 28 = x^{2b} - 3x^b - 28$

77. Finding the product: $\left(2x^b - 1\right)\left(3x^b + 2\right) = 6x^{2b} + 4x^b - 3x^b - 2 = 6x^{2b} + x^b - 2$

79. Finding the product: $\left(x^{2a} - 1\right)^2 = \left(x^{2a} - 1\right)\left(x^{2a} - 1\right) = x^{4a} - 2x^{2a} + 1$

81. Finding the product:

$$\left(x^a - 2\right)^3 = \left(x^a - 2\right)\left(x^a - 2\right)\left(x^a - 2\right)$$

$$= \left(x^a - 2\right)\left(x^{2a} - 4x^a + 4\right)$$

$$= x^{3a} - 4x^{2a} + 4x^a - 2x^{2a} + 8x^a - 8$$

$$= x^{3a} - 6x^{2a} + 12x^a - 8$$

89.
  a. Computing: $19^2 = (20-1)^2 = 20^2 - 2(20)(1) + 1^2 = 400 - 40 + 1 = 361$

  b. Computing: $29^2 = (30-1)^2 = 30^2 - 2(30)(1) + 1^2 = 900 - 60 + 1 = 841$

  c. Computing: $49^2 = (50-1)^2 = 50^2 - 2(50)(1) + 1^2 = 2500 - 100 + 1 = 2401$

  d. Computing: $79^2 = (80-1)^2 = 80^2 - 2(80)(1) + 1^2 = 6400 - 160 + 1 = 6241$

  e. Computing: $38^2 = (40-2)^2 = 40^2 - 2(40)(2) + 2^2 = 1600 - 160 + 4 = 1444$

  f. Computing: $58^2 = (60-2)^2 = 60^2 - 2(60)(2) + 2^2 = 3600 - 240 + 4 = 3364$

0.4 Factoring Polynomials

1. Factoring: $6xy - 8xy^2 = 2xy(3 - 4y)$

3. Factoring: $12x^2y^3z^4 - 6x^4y^3z^3 + 6x^2y^3z^2 = 6x^2y^3z^2\left(2z^2 - x^2z + 1\right)$

5. Factoring: $x(z+3) + y(z+3) = (z+3)(x+y)$

7. Factoring: $3x + 3y + ax + ay = 3(x+y) + a(x+y) = (x+y)(3+a)$

9. Factoring: $ax - ay - bx + by = a(x-y) - b(x-y) = (x-y)(a-b)$

11. Factoring: $9x^2 - 25 = (3x+5)(3x-5)$

13. Factoring: $1 - 81n^2 = (1+9n)(1-9n)$

15. Factoring: $(x+4)^2 - y^2 = (x+4+y)(x+4-y)$

17. Factoring: $9s^2 - (2t-1)^2 = (3s+2t-1)(3s-2t+1)$

19. Factoring: $x^2 - 5x - 14 = (x-7)(x+2)$

21. Factoring: $15 - 2x - x^2 = (5+x)(3-x)$

23. The polynomial $x^2 + 7x - 36$ does not factor.

25. Factoring: $3x^2 - 11x + 10 = (3x - 5)(x - 2)$

27. Factoring: $10x^2 + 17x + 7 = (10x + 7)(x + 1)$

29. Factoring: $10x^2 + 39x - 27 = (5x - 3)(2x + 9)$

31. Factoring: $36a^2 - 12a + 1 = (6a - 1)(6a - 1) = (6a - 1)^2$

33. Factoring: $8x^2 + 2xy - y^2 = (4x - y)(2x + y)$

35. The polynomial $2n^2 - n - 5$ does not factor.

37. Factoring: $x^3 - 8 = (x - 2)(x^2 + 2x + 4)$

39. Factoring: $64x^3 + 27y^3 = (4x + 3y)(16x^2 - 12xy + 9y^2)$

41. Factoring: $4x^2 + 16 = 4(x^2 + 4)$

43. Factoring: $x^3 - 9x = x(x^2 - 9) = x(x + 3)(x - 3)$

45. Factoring: $9a^2 - 42a + 49 = (3a - 7)(3a - 7) = (3a - 7)^2$

47. Factoring: $2n^3 + 6n^2 + 10n = 2n(n^2 + 3n + 5)$

49. Factoring: $2n^3 + 14n^2 - 20n = 2n(n^2 + 7n - 10)$

51. Factoring: $4x^3 + 32 = 4(x^3 + 8) = 4(x + 2)(x^2 - 2x + 4)$

53. Factoring: $x^4 - 4x^2 - 45 = (x^2 + 5)(x^2 - 9) = (x^2 + 5)(x + 3)(x - 3)$

55. Factoring: $2x^4y - 26x^2y - 96y = 2y(x^4 - 13x^2 - 48) = 2y(x^2 + 3)(x^2 - 16) = 2y(x^2 + 3)(x + 4)(x - 4)$

57. Factoring: $(a + b)^2 - (c + d)^2 = (a + b + c + d)(a + b - c - d)$

59. Factoring: $x^2 + 8x + 16 - y^2 = (x + 4)^2 - y^2 = (x + 4 + y)(x + 4 - y)$

61. Factoring: $x^2 - y^2 - 10y - 25 = x^2 - (y^2 + 10y + 25) = x^2 - (y + 5)^2 = (x + y + 5)(x - y - 5)$

63. Factoring: $60x^2 - 32x - 15 = (10x + 3)(6x - 5)$

65. Factoring: $84x^3 + 57x^2 - 60x = 3x(28x^2 + 19x - 20) = 3x(7x - 4)(4x + 5)$

67. Factoring: $x^{2a} - 16 = (x^a + 4)(x^a - 4)$

69. Factoring: $x^{3n} - y^{3n} = (x^n - y^n)(x^{2n} + x^n y^n + y^{2n})$

71. Factoring: $x^{2a} - 3x^a - 28 = (x^a + 4)(x^a - 7)$

73. Factoring: $2x^{2n} + 7x^n - 30 = (2x^n - 5)(x^n + 6)$

75. Factoring: $x^{4n} - y^{4n} = (x^{2n} + y^{2n})(x^{2n} - y^{2n}) = (x^{2n} + y^{2n})(x^n + y^n)(x^n - y^n)$

77. a. Factoring: $x^2 + 35x + 96 = (x + 32)(x + 3)$ b. Factoring: $x^2 + 27x + 176 = (x + 11)(x + 16)$

c. Factoring: $x^2 - 45x + 504 = (x - 21)(x - 24)$ d. Factoring: $x^2 - 26x + 168 = (x - 14)(x - 12)$

e. Factoring: $x^2 + 60x + 896 = (x + 28)(x + 32)$ f. Factoring: $x^2 - 84x + 1728 = (x - 36)(x - 48)$

0.5 Rational Expressions

1. Simplifying: $\dfrac{14x^2y}{21xy} = \dfrac{7xy \cdot 2x}{7xy \cdot 3} = \dfrac{2x}{3}$

3. Simplifying: $\dfrac{-63xy^4}{-81x^2y} = \dfrac{9xy \cdot 7y^3}{9xy \cdot 9x} = \dfrac{7y^3}{9x}$

5. Simplifying: $\dfrac{(2x^2y^2)^3}{(3xy)^2} = \dfrac{8x^6y^6}{9x^2y^2} = \dfrac{x^2y^2 \cdot 8x^4y^4}{x^2y^2 \cdot 9} = \dfrac{8x^4y^4}{9}$

7. Simplifying: $\dfrac{a^2 + 7a + 12}{a^2 - 6a - 27} = \dfrac{(a + 4)(a + 3)}{(a - 9)(a + 3)} = \dfrac{a + 4}{a - 9}$

9. Simplifying: $\dfrac{2x^3 + 3x^2 - 14x}{x^2 y + 7xy - 18y} = \dfrac{x\left(2x^2 + 3x - 14\right)}{y\left(x^2 + 7x - 18\right)} = \dfrac{x\left(2x + 7\right)\left(x - 2\right)}{y\left(x + 9\right)\left(x - 2\right)} = \dfrac{x\left(2x + 7\right)}{y\left(x + 9\right)}$

11. Simplifying: $\dfrac{x^3 - y^3}{x^2 + xy - 2y^2} = \dfrac{\left(x - y\right)\left(x^2 + xy + y^2\right)}{\left(x - y\right)\left(x + 2y\right)} = \dfrac{x^2 + xy + y^2}{x + 2y}$

13. Simplifying: $\dfrac{2y - 2xy}{x^2 y - y} = \dfrac{2y\left(1 - x\right)}{y\left(x^2 - 1\right)} = \dfrac{-2y\left(x - 1\right)}{y\left(x + 1\right)\left(x - 1\right)} = -\dfrac{2}{x + 1}$

15. Simplifying: $\dfrac{8x^2 + 4xy - 2x - y}{4x^2 - 4xy - x + y} = \dfrac{4x\left(2x + y\right) - 1\left(2x + y\right)}{4x\left(x - y\right) - 1\left(x - y\right)} = \dfrac{\left(2x + y\right)\left(4x - 1\right)}{\left(x - y\right)\left(4x - 1\right)} = \dfrac{2x + y}{x - y}$

17. Simplifying:

$$\dfrac{27x^3 + 8y^3}{3x^2 - 15x + 2xy - 10y} = \dfrac{\left(3x + 2y\right)\left(9x^2 - 6xy + 4y^2\right)}{3x\left(x - 5\right) + 2y\left(x - 5\right)} = \dfrac{\left(3x + 2y\right)\left(9x^2 - 6xy + 4y^2\right)}{\left(x - 5\right)\left(3x + 2y\right)} = \dfrac{9x^2 - 6xy + 4y^2}{x - 5}$$

19. Performing the operations: $\dfrac{4x^2}{5y^2} \bullet \dfrac{15xy}{24x^2 y^2} = \dfrac{60x^3 y}{120x^2 y^4} = \dfrac{x}{2y^3}$

21. Performing the operations: $\dfrac{-14xy^4}{18y^2} \bullet \dfrac{24x^2 y^3}{35y^2} = -\dfrac{14 \bullet 24x^3 y^7}{35 \bullet 18y^4} = -\dfrac{8x^3 y^3}{15}$

23. Performing the operations: $\dfrac{7a^2 b}{9ab^3} \div \dfrac{3a^4}{2a^2 b^2} = \dfrac{7a^2 b}{9ab^3} \bullet \dfrac{2a^2 b^2}{3a^4} = \dfrac{14a^4 b^3}{27a^5 b^3} = \dfrac{14}{27a}$

25. Performing the operations: $\dfrac{5xy}{x + 6} \bullet \dfrac{x^2 - 36}{x^2 - 6x} = \dfrac{5xy}{x + 6} \bullet \dfrac{\left(x + 6\right)\left(x - 6\right)}{x\left(x - 6\right)} = 5y$

27. Performing the operations: $\dfrac{5a^2 + 20a}{a^3 - 2a^2} \bullet \dfrac{a^2 - a - 12}{a^2 - 16} = \dfrac{5a\left(a + 4\right)}{a^2\left(a - 2\right)} \bullet \dfrac{\left(a + 3\right)\left(a - 4\right)}{\left(a + 4\right)\left(a - 4\right)} = \dfrac{5\left(a + 3\right)}{a\left(a - 2\right)}$

29. Performing the operations:

$$\dfrac{x^2 + 5xy - 6y^2}{xy^2 - y^3} \bullet \dfrac{2x^2 + 15xy + 18y^2}{xy + 4y^2} = \dfrac{\left(x + 6y\right)\left(x - y\right)}{y^2\left(x - y\right)} \bullet \dfrac{\left(2x + 3y\right)\left(x + 6y\right)}{y\left(x + 4y\right)} = \dfrac{\left(x + 6y\right)^2\left(2x + 3y\right)}{y^3\left(x + 4y\right)}$$

31. Performing the operations: $\dfrac{9y^2}{x^2 + 12x + 36} \div \dfrac{12y}{x^2 + 6x} = \dfrac{9y^2}{x^2 + 12x + 36} \bullet \dfrac{x^2 + 6x}{12y} = \dfrac{9y^2}{\left(x + 6\right)^2} \bullet \dfrac{x\left(x + 6\right)}{12y} = \dfrac{3xy}{4\left(x + 6\right)}$

33. Performing the operations:

$$\dfrac{2x^2 + 3x}{2x^3 - 10x^2} \bullet \dfrac{x^2 - 8x + 15}{3x^3 - 27x} \div \dfrac{14x + 21}{x^2 - 6x - 27} = \dfrac{2x^2 + 3x}{2x^3 - 10x^2} \bullet \dfrac{x^2 - 8x + 15}{3x^3 - 27x} \bullet \dfrac{x^2 - 6x - 27}{14x + 21}$$

$$= \dfrac{x\left(2x + 3\right)}{2x^2\left(x - 5\right)} \bullet \dfrac{\left(x - 5\right)\left(x - 3\right)}{3x\left(x + 3\right)\left(x - 3\right)} \bullet \dfrac{\left(x - 9\right)\left(x + 3\right)}{7\left(2x + 3\right)}$$

$$= \dfrac{x - 9}{42x^2}$$

35. Performing the operations: $\dfrac{x + 4}{6} + \dfrac{2x - 1}{4} = \dfrac{2x + 8}{12} + \dfrac{6x - 3}{12} = \dfrac{8x + 5}{12}$

37. Performing the operations: $\dfrac{x + 1}{4} + \dfrac{x - 3}{6} - \dfrac{x - 2}{8} = \dfrac{6x + 6}{24} + \dfrac{4x - 12}{24} - \dfrac{3x - 6}{24} = \dfrac{6x + 6 + 4x - 12 - 3x + 6}{24} = \dfrac{7x}{24}$

39. Performing the operations: $\dfrac{7}{16a^2 b} + \dfrac{3a}{20b^2} = \dfrac{7}{16a^2 b} \bullet \dfrac{5b}{5b} + \dfrac{3a}{20b^2} \bullet \dfrac{4a^2}{4a^2} = \dfrac{35b}{80a^2 b^2} + \dfrac{12a^3}{80a^2 b^2} = \dfrac{12a^3 + 35b}{80a^2 b^2}$

41.   Performing the operations: $\dfrac{1}{n^2}+\dfrac{3}{4n}-\dfrac{5}{6}=\dfrac{12}{12n^2}+\dfrac{9n}{12n^2}-\dfrac{10n^2}{12n^2}=\dfrac{12+9n-10n^2}{12n^2}$

43.   Performing the operations: $\dfrac{3}{4x}+\dfrac{2}{3y}-1=\dfrac{3}{4x}\cdot\dfrac{3y}{3y}+\dfrac{2}{3y}\cdot\dfrac{4x}{4x}-1\cdot\dfrac{12xy}{12xy}=\dfrac{9y+8x-12xy}{12xy}$

45.   Performing the operations: $\dfrac{3}{2x+1}+\dfrac{2}{3x+4}=\dfrac{3}{2x+1}\cdot\dfrac{3x+4}{3x+4}+\dfrac{2}{3x+4}\cdot\dfrac{2x+1}{2x+1}=\dfrac{9x+12+4x+2}{(3x+4)(2x+1)}=\dfrac{13x+14}{(3x+4)(2x+1)}$

47.   Performing the operations: $\dfrac{4x}{x^2+7x}+\dfrac{3}{x}=\dfrac{4x}{x(x+7)}+\dfrac{3}{x}\cdot\dfrac{x+7}{x+7}=\dfrac{4x+3x+21}{x(x+7)}=\dfrac{7x+21}{x(x+7)}$

49.   Performing the operations:

$$\dfrac{4a-4}{a^2-4}-\dfrac{3}{a+2}=\dfrac{4a-4}{(a+2)(a-2)}-\dfrac{3}{a+2}\cdot\dfrac{a-2}{a-2}=\dfrac{4a-4-3a+6}{(a+2)(a-2)}=\dfrac{a+2}{(a+2)(a-2)}=\dfrac{1}{a-2}$$

51.   Performing the operations: $\dfrac{3}{x-1}-\dfrac{2}{4x-4}=\dfrac{3}{x-1}\cdot\dfrac{4}{4}-\dfrac{2}{4(x-1)}=\dfrac{12}{4(x-1)}-\dfrac{2}{4(x-1)}=\dfrac{10}{4(x-1)}=\dfrac{5}{2(x-1)}$

53.   Performing the operations: $\dfrac{4}{n^2-1}+\dfrac{2}{3n+3}=\dfrac{4}{(n+1)(n-1)}\cdot\dfrac{3}{3}+\dfrac{2}{3(n+1)}\cdot\dfrac{n-1}{n-1}=\dfrac{12+2n-2}{3(n+1)(n-1)}=\dfrac{2n+10}{3(n+1)(n-1)}$

55.   Performing the operations:

$$\dfrac{3}{x+1}+\dfrac{x+5}{x^2-1}-\dfrac{3}{x-1}=\dfrac{3}{x+1}\cdot\dfrac{x-1}{x-1}+\dfrac{x+5}{(x+1)(x-1)}-\dfrac{3}{x-1}\cdot\dfrac{x+1}{x+1}$$

$$=\dfrac{3x-3+x+5-3x-3}{(x+1)(x-1)}$$

$$=\dfrac{x-1}{(x+1)(x-1)}$$

$$=\dfrac{1}{x+1}$$

57.   Performing the operations:

$$\dfrac{5}{x^2+10x+21}+\dfrac{4}{x^2+12x+27}=\dfrac{5}{(x+7)(x+3)}\cdot\dfrac{x+9}{x+9}+\dfrac{4}{(x+9)(x+3)}\cdot\dfrac{x+7}{x+7}$$

$$=\dfrac{5x+45+4x+28}{(x+7)(x+3)(x+9)}$$

$$=\dfrac{9x+73}{(x+3)(x+7)(x+9)}$$

59.   Performing the operations:

$$\dfrac{5}{x^2-1}-\dfrac{2}{x^2+6x-16}=\dfrac{5}{(x+1)(x-1)}\cdot\dfrac{(x+8)(x-2)}{(x+8)(x-2)}-\dfrac{2}{(x+8)(x-2)}\cdot\dfrac{(x+1)(x-1)}{(x+1)(x-1)}$$

$$=\dfrac{5x^2+30x-80-2x^2+2}{(x+1)(x-1)(x+8)(x-2)}$$

$$=\dfrac{3x^2+30x-78}{(x+1)(x-1)(x+8)(x-2)}$$

61.   Performing the operations: $\dfrac{3x}{x^2-6x+9}-\dfrac{2}{x-3}=\dfrac{3x}{(x-3)^2}-\dfrac{2}{x-3}\cdot\dfrac{x-3}{x-3}=\dfrac{3x-2x+6}{(x-3)^2}=\dfrac{x+6}{(x-3)^2}$

63.   Performing the operations:

$$x-\dfrac{x^2}{x-1}+\dfrac{1}{x^2-1}=x\cdot\dfrac{(x+1)(x-1)}{(x+1)(x-1)}-\dfrac{x^2}{x-1}\cdot\dfrac{x+1}{x+1}+\dfrac{1}{(x+1)(x-1)}=\dfrac{x^3-x-x^3-x^2+1}{(x+1)(x-1)}=\dfrac{-x^2-x+1}{(x+1)(x-1)}$$

65. Performing the operations:

$$\frac{2n^2}{n^4-16}-\frac{n}{n^2-4}+\frac{1}{n+2}=\frac{2n^2}{\left(n^2+4\right)\left(n+2\right)\left(n-2\right)}-\frac{n}{\left(n+2\right)\left(n-2\right)}\cdot\frac{n^2+4}{n^2+4}+\frac{1}{n+2}\cdot\frac{\left(n^2+4\right)\left(n-2\right)}{\left(n^2+4\right)\left(n-2\right)}$$

$$=\frac{2n^2-n^3-4n+n^3-2n^2+4n-8}{\left(n^2+4\right)\left(n+2\right)\left(n-2\right)}$$

$$=\frac{-8}{\left(n^2+4\right)\left(n+2\right)\left(n-2\right)}$$

67. Performing the operations:

$$\frac{2x+1}{x^2-3x-4}+\frac{3x-2}{x^2+3x-28}=\frac{2x+1}{\left(x-4\right)\left(x+1\right)}\cdot\frac{x+7}{x+7}+\frac{3x-2}{\left(x+7\right)\left(x-4\right)}\cdot\frac{x+1}{x+1}$$

$$=\frac{2x^2+15x+7+3x^2+x-2}{\left(x-4\right)\left(x+1\right)\left(x+7\right)}$$

$$=\frac{5x^2+16x+5}{\left(x-4\right)\left(x+1\right)\left(x+7\right)}$$

69. **a.** Simplifying: $\dfrac{7}{x-1}+\dfrac{2}{1-x}=\dfrac{7}{x-1}-\dfrac{2}{x-1}=\dfrac{5}{x-1}$

b. Simplifying: $\dfrac{5}{2x-1}+\dfrac{8}{1-2x}=\dfrac{5}{2x-1}-\dfrac{8}{2x-1}=-\dfrac{3}{2x-1}$

c. Simplifying: $\dfrac{4}{a-3}-\dfrac{1}{3-a}=\dfrac{4}{a-3}+\dfrac{1}{a-3}=\dfrac{5}{a-3}$

d. Simplifying: $\dfrac{10}{a-9}-\dfrac{5}{9-a}=\dfrac{10}{a-9}+\dfrac{5}{a-9}=\dfrac{15}{a-9}$

e. Simplifying: $\dfrac{x^2}{x-1}-\dfrac{2x-3}{1-x}=\dfrac{x^2}{x-1}+\dfrac{2x-3}{x-1}=\dfrac{x^2+2x-3}{x-1}=\dfrac{\left(x+3\right)\left(x-1\right)}{x-1}=x+3$

f. Simplifying: $\dfrac{x^2}{x-4}-\dfrac{3x-28}{4-x}=\dfrac{x^2}{x-4}+\dfrac{3x-28}{x-4}=\dfrac{x^2+3x-28}{x-4}=\dfrac{\left(x+7\right)\left(x-4\right)}{x-4}=x+7$

71. Simplifying: $\dfrac{\dfrac{5}{x^2}-\dfrac{3}{x}}{\dfrac{1}{y}+\dfrac{2}{y^2}}=\dfrac{\dfrac{5}{x^2}-\dfrac{3}{x}}{\dfrac{1}{y}+\dfrac{2}{y^2}}\cdot\dfrac{x^2y^2}{x^2y^2}=\dfrac{5y^2-3xy^2}{x^2y+2x^2}$

73. Simplifying: $\dfrac{1+\dfrac{1}{x}}{1-\dfrac{1}{x}}=\dfrac{1+\dfrac{1}{x}}{1-\dfrac{1}{x}}\cdot\dfrac{x}{x}=\dfrac{x+1}{x-1}$

75. Simplifying: $\dfrac{1-\dfrac{1}{n+1}}{1+\dfrac{1}{n-1}}=\dfrac{1-\dfrac{1}{n+1}}{1+\dfrac{1}{n-1}}\cdot\dfrac{\left(n+1\right)\left(n-1\right)}{\left(n+1\right)\left(n-1\right)}=\dfrac{n^2-1-n+1}{n^2-1+n+1}=\dfrac{n^2-n}{n^2+n}=\dfrac{n\left(n-1\right)}{n\left(n+1\right)}=\dfrac{n-1}{n+1}$

77. Simplifying: $\dfrac{\dfrac{-2}{x}-\dfrac{4}{x+2}}{\dfrac{3}{x^2+2x}+\dfrac{3}{x}}=\dfrac{\dfrac{-2}{x}-\dfrac{4}{x+2}}{\dfrac{3}{x\left(x+2\right)}+\dfrac{3}{x}}\cdot\dfrac{x\left(x+2\right)}{1}=\dfrac{-2x-4-4x}{3+3x+6}=\dfrac{-6x-4}{3x+9}$

79. Simplifying: $1+\dfrac{x}{1+\dfrac{1}{x}}=1+\dfrac{x}{1+\dfrac{1}{x}}\cdot\dfrac{x}{x}=1+\dfrac{x^2}{x+1}=1\cdot\dfrac{x+1}{x+1}+\dfrac{x^2}{x+1}=\dfrac{x^2+x+1}{x+1}$

81. Simplifying: $\dfrac{a}{\dfrac{1}{a}+4}+1 = \dfrac{a}{\dfrac{1}{a}+4}\cdot\dfrac{a}{a}+1 = \dfrac{a^2}{1+4a}+1 = \dfrac{a^2}{1+4a}+1\cdot\dfrac{1+4a}{1+4a} = \dfrac{a^2+4a+1}{4a+1}$

83. Simplifying:
$$\dfrac{\dfrac{1}{(x+h)^2}-\dfrac{1}{x^2}}{h} = \dfrac{\dfrac{1}{(x+h)^2}-\dfrac{1}{x^2}}{h}\cdot\dfrac{x^2(x+h)^2}{x^2(x+h)^2}$$
$$= \dfrac{x^2-(x+h)^2}{hx^2(x+h)^2}$$
$$= \dfrac{x^2-x^2-2xh-h^2}{hx^2(x+h)^2}$$
$$= \dfrac{-h(2x+h)}{hx^2(x+h)^2}$$
$$= -\dfrac{2x+h}{x^2(x+h)^2}$$

85. Simplifying:
$$\dfrac{\dfrac{1}{x+h+1}-\dfrac{1}{x+1}}{h} = \dfrac{\dfrac{1}{x+h+1}-\dfrac{1}{x+1}}{h}\cdot\dfrac{(x+1)(x+h+1)}{(x+1)(x+h+1)}$$
$$= \dfrac{x+1-x-h-1}{h(x+1)(x+h+1)}$$
$$= \dfrac{-h}{h(x+1)(x+h+1)}$$
$$= -\dfrac{1}{(x+1)(x+h+1)}$$

87. Simplifying:
$$\dfrac{\dfrac{2}{2x+2h-1}-\dfrac{2}{2x-1}}{h} = \dfrac{\dfrac{2}{2x+2h-1}-\dfrac{2}{2x-1}}{h}\cdot\dfrac{(2x-1)(2x+2h-1)}{(2x-1)(2x+2h-1)}$$
$$= \dfrac{4x-2-4x-4h+2}{h(2x-1)(2x+2h-1)}$$
$$= \dfrac{-4h}{h(2x-1)(2x+2h-1)}$$
$$= \dfrac{-4}{(2x-1)(2x+2h-1)}$$

89. Simplifying: $\dfrac{x^{-1}+2y^{-1}}{x-y} = \dfrac{\dfrac{1}{x}+\dfrac{2}{y}}{x-y}\cdot\dfrac{xy}{xy} = \dfrac{y+2x}{xy(x-y)} = \dfrac{y+2x}{x^2y-xy^2}$

91. Simplifying: $\dfrac{x+2x^{-1}y^{-2}}{4x^{-1}-3y^{-2}} = \dfrac{x+\dfrac{2}{xy^2}}{\dfrac{4}{x}-\dfrac{3}{y^2}}\cdot\dfrac{xy^2}{xy^2} = \dfrac{x^2y^2+2}{4y^2-3x}$

0.6 Radicals

1. Evaluating: $\sqrt{81} = 9$

3. Evaluating: $\sqrt[3]{125} = 5$

5. Evaluating: $\sqrt{\dfrac{36}{49}} = \dfrac{6}{7}$

7. Evaluating: $\sqrt[3]{-\dfrac{27}{8}} = -\dfrac{3}{2}$

9. Simplifying: $\sqrt{24} = \sqrt{4 \cdot 6} = 2\sqrt{6}$

11. Simplifying: $\sqrt{112} = \sqrt{16 \cdot 7} = 4\sqrt{7}$

13. Simplifying: $-3\sqrt{44} = -3\sqrt{4 \cdot 11} = -6\sqrt{11}$

15. Simplifying: $\dfrac{3}{4}\sqrt{20} = \dfrac{3}{4}\sqrt{4 \cdot 5} = \dfrac{3\sqrt{5}}{2}$

17. Simplifying: $\sqrt{12x^2} = \sqrt{4x^2 \cdot 3} = 2x\sqrt{3}$

19. Simplifying: $\sqrt{64x^4 y^7} = \sqrt{64x^4 y^6 \cdot y} = 8x^2 y^3 \sqrt{y}$

21. Simplifying: $\dfrac{3}{7}\sqrt{45xy^6} = \dfrac{3}{7}\sqrt{9y^6 \cdot 5x} = \dfrac{9y^3 \sqrt{5x}}{7}$

23. Simplifying: $\sqrt[3]{128} = \sqrt[3]{64 \cdot 2} = 4\sqrt[3]{2}$

25. Simplifying: $\sqrt[3]{16x^4} = \sqrt[3]{8x^3 \cdot 2x} = 2x\sqrt[3]{2x}$

27. Simplifying: $\sqrt[4]{48x^5} = \sqrt[4]{16x^4 \cdot 3x} = 2x\sqrt[4]{3x}$

29. Simplifying: $\sqrt{\dfrac{12}{25}} = \dfrac{\sqrt{12}}{\sqrt{25}} = \dfrac{2\sqrt{3}}{5}$

31. Simplifying: $\sqrt{\dfrac{7}{8}} = \dfrac{\sqrt{7}}{\sqrt{8}} \cdot \dfrac{\sqrt{2}}{\sqrt{2}} = \dfrac{\sqrt{14}}{\sqrt{16}} = \dfrac{\sqrt{14}}{4}$

33. Simplifying: $\dfrac{4\sqrt{6}}{\sqrt{10}} = \dfrac{4\sqrt{6}}{\sqrt{10}} \cdot \dfrac{\sqrt{10}}{\sqrt{10}} = \dfrac{4\sqrt{60}}{10} = \dfrac{4 \cdot 2\sqrt{15}}{10} = \dfrac{4\sqrt{15}}{5}$

35. Simplifying: $\dfrac{6\sqrt{3}}{7\sqrt{6}} = \dfrac{6\sqrt{3}}{7\sqrt{6}} \cdot \dfrac{\sqrt{6}}{\sqrt{6}} = \dfrac{6\sqrt{18}}{7\sqrt{36}} = \dfrac{18\sqrt{2}}{42} = \dfrac{3\sqrt{2}}{7}$

37. Simplifying: $\dfrac{\sqrt{5}}{\sqrt{12x^4}} = \dfrac{\sqrt{5}}{\sqrt{12x^4}} \cdot \dfrac{\sqrt{3}}{\sqrt{3}} = \dfrac{\sqrt{15}}{\sqrt{36x^4}} = \dfrac{\sqrt{15}}{6x^2}$

39. Simplifying: $\dfrac{\sqrt{12a^2 b}}{\sqrt{5a^3 b^3}} = \dfrac{\sqrt{12a^2 b}}{\sqrt{5a^3 b^3}} \cdot \dfrac{\sqrt{5ab}}{\sqrt{5ab}} = \dfrac{\sqrt{60a^3 b^2}}{\sqrt{25a^4 b^4}} = \dfrac{2ab\sqrt{15a}}{5a^2 b^2} = \dfrac{2\sqrt{15a}}{5ab}$

41. Simplifying: $\dfrac{\sqrt[3]{27}}{\sqrt[3]{4}} = \dfrac{3}{\sqrt[3]{4}} \cdot \dfrac{\sqrt[3]{2}}{\sqrt[3]{2}} = \dfrac{3\sqrt[3]{2}}{\sqrt[3]{8}} = \dfrac{3\sqrt[3]{2}}{2}$

43. Simplifying: $\dfrac{\sqrt[3]{2y}}{\sqrt[3]{3x}} = \dfrac{\sqrt[3]{2y}}{\sqrt[3]{3x}} \cdot \dfrac{\sqrt[3]{9x^2}}{\sqrt[3]{9x^2}} = \dfrac{\sqrt[3]{18x^2 y}}{\sqrt[3]{27x^3}} = \dfrac{\sqrt[3]{18x^2 y}}{3x}$

45. Simplifying: $5\sqrt{12} + 2\sqrt{3} = 5 \cdot 2\sqrt{3} + 2\sqrt{3} = 10\sqrt{3} + 2\sqrt{3} = 12\sqrt{3}$

47. Simplifying: $2\sqrt{28} - 3\sqrt{63} + 8\sqrt{7} = 2 \cdot 2\sqrt{7} - 3 \cdot 3\sqrt{7} + 8\sqrt{7} = 4\sqrt{7} - 9\sqrt{7} + 8\sqrt{7} = 3\sqrt{7}$

49. Simplifying: $\dfrac{5}{6}\sqrt{48} - \dfrac{3}{4}\sqrt{12} = \dfrac{5}{6} \cdot 4\sqrt{3} - \dfrac{3}{4} \cdot 2\sqrt{3} = \dfrac{10}{3}\sqrt{3} - \dfrac{3}{2}\sqrt{3} = \dfrac{20\sqrt{3} - 9\sqrt{3}}{6} = \dfrac{11\sqrt{3}}{6}$

51. Simplifying: $\dfrac{2\sqrt{8}}{3} - \dfrac{3\sqrt{18}}{5} - \dfrac{\sqrt{50}}{2} = \dfrac{4\sqrt{2}}{3} - \dfrac{9\sqrt{2}}{5} - \dfrac{5\sqrt{2}}{2} = \dfrac{40\sqrt{2}}{30} - \dfrac{54\sqrt{2}}{30} - \dfrac{75\sqrt{2}}{30} = -\dfrac{89\sqrt{2}}{30}$

53. Multiplying: $\left(4\sqrt{3}\right)\left(6\sqrt{8}\right) = 24\sqrt{24} = 24 \cdot 2\sqrt{6} = 48\sqrt{6}$

55. Multiplying: $2\sqrt{3}\left(5\sqrt{2} + 4\sqrt{10}\right) = 10\sqrt{6} + 8\sqrt{30}$

57. Multiplying: $3\sqrt{x}\left(\sqrt{6xy} - \sqrt{8y}\right) = 3\sqrt{6x^2 y} - 3\sqrt{8xy} = 3x\sqrt{6y} - 6\sqrt{2xy}$

59. Multiplying: $\left(\sqrt{3} + 2\right)\left(\sqrt{3} + 5\right) = 3 + 2\sqrt{3} + 5\sqrt{3} + 10 = 13 + 7\sqrt{3}$

61. Multiplying: $\left(4\sqrt{2} + \sqrt{3}\right)\left(3\sqrt{2} + 2\sqrt{3}\right) = 24 + 3\sqrt{6} + 8\sqrt{6} + 6 = 30 + 11\sqrt{6}$

63. Multiplying: $\left(6 + 2\sqrt{5}\right)\left(6 - 2\sqrt{5}\right) = (6)^2 - \left(2\sqrt{5}\right)^2 = 36 - 20 = 16$

65. Multiplying: $\left(\sqrt{x}+\sqrt{y}\right)^2 = \left(\sqrt{x}+\sqrt{y}\right)\left(\sqrt{x}+\sqrt{y}\right) = x+2\sqrt{xy}+y$

67. Multiplying: $\left(\sqrt{a}+\sqrt{b}\right)\left(\sqrt{a}-\sqrt{b}\right) = \left(\sqrt{a}\right)^2 - \left(\sqrt{b}\right)^2 = a-b$

69. Rationalizing the denominator: $\dfrac{3}{\sqrt{5}+2} = \dfrac{3}{\sqrt{5}+2}\cdot\dfrac{\sqrt{5}-2}{\sqrt{5}-2} = \dfrac{3\sqrt{5}-6}{5-4} = 3\sqrt{5}-6$

71. Rationalizing the denominator: $\dfrac{4}{\sqrt{7}-\sqrt{3}} = \dfrac{4}{\sqrt{7}-\sqrt{3}}\cdot\dfrac{\sqrt{7}+\sqrt{3}}{\sqrt{7}+\sqrt{3}} = \dfrac{4\sqrt{7}+4\sqrt{3}}{7-3} = \dfrac{4\sqrt{7}+4\sqrt{3}}{4} = \sqrt{7}+\sqrt{3}$

73. Rationalizing the denominator:

$$\frac{\sqrt{2}}{2\sqrt{5}+3\sqrt{7}} = \frac{\sqrt{2}}{2\sqrt{5}+3\sqrt{7}}\cdot\frac{2\sqrt{5}-3\sqrt{7}}{2\sqrt{5}-3\sqrt{7}} = \frac{2\sqrt{10}-3\sqrt{14}}{20-63} = \frac{2\sqrt{10}-3\sqrt{14}}{-43} = \frac{-2\sqrt{10}+3\sqrt{14}}{43}$$

75. Rationalizing the denominator: $\dfrac{\sqrt{x}}{\sqrt{x}-1} = \dfrac{\sqrt{x}}{\sqrt{x}-1}\cdot\dfrac{\sqrt{x}+1}{\sqrt{x}+1} = \dfrac{x+\sqrt{x}}{x-1}$

77. Rationalizing the denominator: $\dfrac{\sqrt{x}}{\sqrt{x}+\sqrt{y}} = \dfrac{\sqrt{x}}{\sqrt{x}+\sqrt{y}}\cdot\dfrac{\sqrt{x}-\sqrt{y}}{\sqrt{x}-\sqrt{y}} = \dfrac{x-\sqrt{xy}}{x-y}$

79. Rationalizing the denominator: $\dfrac{2\sqrt{x}+\sqrt{y}}{3\sqrt{x}-2\sqrt{y}} = \dfrac{2\sqrt{x}+\sqrt{y}}{3\sqrt{x}-2\sqrt{y}}\cdot\dfrac{3\sqrt{x}+2\sqrt{y}}{3\sqrt{x}+2\sqrt{y}} = \dfrac{6x+7\sqrt{xy}+2y}{9x-4y}$

81. Rationalizing the numerator:

$$\frac{\sqrt{2x+2h}-\sqrt{2x}}{h} = \frac{\sqrt{2x+2h}-\sqrt{2x}}{h}\cdot\frac{\sqrt{2x+2h}+\sqrt{2x}}{\sqrt{2x+2h}+\sqrt{2x}} = \frac{2x+2h-2x}{h\left(\sqrt{2x+2h}+\sqrt{2x}\right)} = \frac{2}{\sqrt{2x+2h}+\sqrt{2x}}$$

83. Rationalizing the numerator:

$$\frac{\sqrt{x+h-3}-\sqrt{x-3}}{h} = \frac{\sqrt{x+h-3}-\sqrt{x-3}}{h}\cdot\frac{\sqrt{x+h-3}+\sqrt{x-3}}{\sqrt{x+h-3}+\sqrt{x-3}}$$

$$= \frac{x+h-3-x+3}{h\left(\sqrt{x+h-3}+\sqrt{x-3}\right)}$$

$$= \frac{1}{\sqrt{x+h-3}+\sqrt{x-3}}$$

91. Simplifying: $\sqrt{16x^4} = 4x^2$ (since $x^2 \geq 0$)   **93.** Simplifying: $\sqrt{3y^5} = \sqrt{y^4\cdot 3y} = y^2\sqrt{3y}$ (since $y \geq 0$)

95. Simplifying: $\sqrt{28m^8} = \sqrt{4m^8\cdot 7} = 2m^4\sqrt{7}$ (since $m^4 \geq 0$)

97. Simplifying: $\sqrt{18d^7} = \sqrt{9d^6\cdot 2d} = 3d^3\sqrt{2d}$ (since $d \geq 0$)

99. Simplifying: $\sqrt{80n^{20}} = \sqrt{16n^{20}\cdot 5} = 4n^{10}\sqrt{5}$ (since $n^{10} \geq 0$)

101. **a.** Computing: $\sqrt{49,000,000} = \sqrt{(49)\left(10^6\right)} = 7(10)^3 = 7,000$

   b. Computing: $\sqrt{0.0025} = \sqrt{(25)\left(10^{-4}\right)} = 5(10)^{-2} = 0.05$

   c. Computing: $\sqrt{14,400} = \sqrt{(144)\left(10^2\right)} = 12(10)^1 = 120$

   d. Computing: $\sqrt{0.000121} = \sqrt{(121)\left(10^{-6}\right)} = 11(10)^{-3} = 0.011$

   e. Computing: $\sqrt[3]{27,000} = \sqrt[3]{(27)\left(10^3\right)} = 3(10)^1 = 30$

   f. Computing: $\sqrt[3]{0.000064} = \sqrt[3]{(64)\left(10^{-6}\right)} = 4(10)^{-2} = 0.04$

103. **a.** Approximating: $\sqrt[3]{24} \approx 2.9$ **b.** Approximating: $\sqrt[3]{32} \approx 3.2$
    c. Approximating: $\sqrt[3]{150} \approx 5.3$ **d.** Approximating: $\sqrt[3]{200} \approx 5.8$
    e. Approximating: $\sqrt[4]{50} \approx 2.7$ **f.** Approximating: $\sqrt[4]{250} \approx 4.0$

0.7 Relationship Between Exponents and Roots

1. Evaluating: $49^{1/2} = \sqrt{49} = 7$

3. Evaluating: $32^{3/5} = \left(\sqrt[5]{32}\right)^3 = (2)^3 = 8$

5. Evaluating: $-8^{2/3} = -\left(\sqrt[3]{8}\right)^2 = -2^2 = -4$

7. Evaluating: $\left(\dfrac{1}{4}\right)^{-1/2} = \left(\sqrt{\dfrac{1}{4}}\right)^{-1} = \left(\dfrac{1}{2}\right)^{-1} = 2$

9. Evaluating: $16^{3/2} = \left(\sqrt{16}\right)^3 = 4^3 = 64$

11. Evaluating: $(0.01)^{3/2} = \left(\sqrt{0.01}\right)^3 = (0.1)^3 = 0.001$

13. Evaluating: $64^{-5/6} = \left(\sqrt[6]{64}\right)^{-5} = 2^{-5} = \dfrac{1}{2^5} = \dfrac{1}{32}$

15. Evaluating: $\left(\dfrac{1}{8}\right)^{-1/3} = 8^{1/3} = \sqrt[3]{8} = 2$

17. Performing the operations: $\left(3x^{1/4}\right)\left(5x^{1/3}\right) = 15x^{1/4+1/3} = 15x^{7/12}$

19. Performing the operations: $\left(y^{2/3}\right)\left(y^{-1/4}\right) = y^{2/3-1/4} = y^{5/12}$

21. Performing the operations: $\left(4x^{1/4}y^{1/2}\right)^3 = 4^3 x^{3/4} y^{3/2} = 64x^{3/4} y^{3/2}$

23. Performing the operations: $\dfrac{24x^{3/5}}{6x^{1/3}} = 4x^{3/5-1/3} = 4x^{4/15}$

25. Performing the operations: $\dfrac{56a^{1/6}}{8a^{1/4}} = 7a^{1/6-1/4} = 7a^{-1/12} = \dfrac{7}{a^{1/12}}$

27. Performing the operations: $\left(\dfrac{2x^{1/3}}{3y^{1/4}}\right)^4 = \dfrac{2^4 x^{4/3}}{3^4 y} = \dfrac{16x^{4/3}}{81y}$

29. Performing the operations: $\left(\dfrac{x^2}{y^3}\right)^{-1/2} = \dfrac{x^{-1}}{y^{-3/2}} = \dfrac{y^{3/2}}{x}$

31. Performing the operations: $\left(\dfrac{4a^2 x}{2a^{1/2} x^{1/3}}\right)^3 = \dfrac{4^3 a^6 x^3}{2^3 a^{3/2} x} = \dfrac{64a^6 x^3}{8a^{3/2} x} = 8a^{6-3/2} x^{3-1} = 8a^{9/2} x^2$

33. Performing the operations: $\sqrt{2}\sqrt[4]{2} = 2^{1/2} 2^{1/4} = 2^{1/2+1/4} = 2^{3/4} = \sqrt[4]{2^3} = \sqrt[4]{8}$

35. Performing the operations: $\sqrt[3]{x}\sqrt[4]{x} = x^{1/3} x^{1/4} = x^{1/3+1/4} = x^{7/12} = \sqrt[12]{x^7}$

37. Performing the operations: $\sqrt{xy}\sqrt[4]{x^3 y^5} = x^{1/2} y^{1/2} x^{3/4} y^{5/4} = x^{1/2+3/4} y^{1/2+5/4} = x^{5/4} y^{7/4} = \sqrt[4]{x^5 y^7} = xy\sqrt[4]{xy^3}$

39. Performing the operations:
$$\sqrt[3]{a^2 b^2}\sqrt[4]{a^3 b} = a^{2/3} b^{2/3} a^{3/4} b^{1/4} = a^{2/3+3/4} b^{2/3+1/4} = a^{17/12} b^{11/12} = \sqrt[12]{a^{17} b^{11}} = a\sqrt[12]{a^5 b^{11}}$$

41. Performing the operations: $\sqrt[3]{4}\sqrt{8} = \sqrt[3]{2^2}\sqrt{2^3} = 2^{2/3} 2^{3/2} = 2^{2/3+3/2} = 2^{13/6} = \sqrt[6]{2^{13}} = 4\sqrt[6]{2}$

43. Performing the operations: $\dfrac{\sqrt{2}}{\sqrt[3]{2}} = \dfrac{2^{1/2}}{2^{1/3}} = 2^{1/2-1/3} = 2^{1/6} = \sqrt[6]{2}$

45. Performing the operations: $\dfrac{\sqrt[3]{8}}{\sqrt[4]{4}} = \dfrac{2}{\sqrt[4]{2^2}} = \dfrac{2}{2^{1/2}} = 2^{1-1/2} = 2^{1/2} = \sqrt{2}$

47. Performing the operations: $\dfrac{\sqrt[4]{x^9}}{\sqrt[3]{x^2}} = \dfrac{x^{9/4}}{x^{2/3}} = x^{9/4-2/3} = x^{19/12} = \sqrt[12]{x^{19}} = x\sqrt[12]{x^7}$

49. Rationalizing the denominator: $\dfrac{5}{\sqrt[3]{x}} = \dfrac{5}{\sqrt[3]{x}} \cdot \dfrac{\sqrt[3]{x^2}}{\sqrt[3]{x^2}} = \dfrac{5\sqrt[3]{x^2}}{x}$

51. Rationalizing the denominator: $\dfrac{\sqrt{x}}{\sqrt[3]{y}} = \dfrac{\sqrt{x}}{\sqrt[3]{y}} \cdot \dfrac{\sqrt[3]{y^2}}{\sqrt[3]{y^2}} = \dfrac{x^{1/2}y^{2/3}}{y} = \dfrac{x^{3/6}y^{4/6}}{y} = \dfrac{\sqrt[6]{x^3y^4}}{y}$

53. Rationalizing the denominator: $\dfrac{\sqrt[4]{x^3}}{\sqrt[5]{y^3}} = \dfrac{\sqrt[4]{x^3}}{\sqrt[5]{y^3}} \cdot \dfrac{\sqrt[5]{y^2}}{\sqrt[5]{y^2}} = \dfrac{x^{3/4}y^{2/5}}{y} = \dfrac{x^{15/20}y^{8/20}}{y} = \dfrac{\sqrt[20]{x^{15}y^8}}{y}$

55. Rationalizing the denominator: $\dfrac{5\sqrt[3]{y^2}}{4\sqrt[4]{x}} = \dfrac{5\sqrt[3]{y^2}}{4\sqrt[4]{x}} \cdot \dfrac{\sqrt[4]{x^3}}{\sqrt[4]{x^3}} = \dfrac{5y^{2/3}x^{3/4}}{4x} = \dfrac{5x^{9/12}y^{8/12}}{4x} = \dfrac{5\sqrt[12]{x^9y^8}}{4x}$

57. **a.** Simplifying: $\sqrt[3]{\sqrt{2}} = \left(2^{1/2}\right)^{1/3} = 2^{1/6} = \sqrt[6]{2}$ **b.** Simplifying: $\sqrt[3]{\sqrt[4]{3}} = \left(3^{1/4}\right)^{1/3} = 3^{1/12} = \sqrt[12]{3}$

   c. Simplifying: $\sqrt[3]{\sqrt{x^3}} = \left(x^{3/2}\right)^{1/3} = x^{1/2} = \sqrt{x}$ **d.** Simplifying: $\sqrt{\sqrt[3]{x^4}} = \left(x^{4/3}\right)^{1/2} = x^{2/3} = \sqrt[3]{x^2}$

61. Simplifying:

$$\dfrac{2(2x-1)^{1/2} - 2x(2x-1)^{-1/2}}{\left[(2x-1)^{1/2}\right]^2} = \dfrac{2(2x-1)^{1/2} - 2x(2x-1)^{-1/2}}{(2x-1)} \cdot \dfrac{(2x-1)^{1/2}}{(2x-1)^{1/2}} = \dfrac{2(2x-1) - 2x}{(2x-1)^{3/2}} = \dfrac{2x-2}{(2x-1)^{3/2}}$$

63. Simplifying:

$$\dfrac{\left(x^2+2x\right)^{1/2} - x(x+1)\left(x^2+2x\right)^{-1/2}}{\left[\left(x^2+2x\right)^{1/2}\right]^2} = \dfrac{\left(x^2+2x\right)^{1/2} - x(x+1)\left(x^2+2x\right)^{-1/2}}{\left(x^2+2x\right)} \cdot \dfrac{\left(x^2+2x\right)^{1/2}}{\left(x^2+2x\right)^{1/2}}$$

$$= \dfrac{x^2+2x-x^2-x}{\left(x^2+2x\right)^{3/2}}$$

$$= \dfrac{x}{\left(x^2+2x\right)^{3/2}}$$

65. Simplifying: $\dfrac{3(2x)^{1/3} - 2x(2x)^{-2/3}}{\left[(2x)^{1/3}\right]^2} = \dfrac{3(2x)^{1/3} - 2x(2x)^{-2/3}}{(2x)^{2/3}} \cdot \dfrac{(2x)^{2/3}}{(2x)^{2/3}} = \dfrac{3(2x) - 2x}{(2x)^{4/3}} = \dfrac{4x}{(2x)^{4/3}}$

67. **a.** Both values are 9.   **b.** Both values are 32.
   c. Both values are 8.   **d.** Both values are approximately 6.35.
   e. Both values are approximately 5.80.   **f.** Both values are approximately 27.47.

69. **a.** Evaluating: $7^{4/3} \approx 13.391$ **b.** Evaluating: $10^{4/5} \approx 6.310$
   c. Evaluating: $12^{2/5} \approx 2.702$ **d.** Evaluating: $19^{2/5} \approx 3.247$
   e. Evaluating: $7^{3/4} \approx 4.304$ **f.** Evaluating: $10^{5/4} \approx 17.783$

0.8 Complex Numbers

1. Computing: $(5+2i)+(8+6i)=13+8i$

3. Computing: $(8+6i)-(5+2i)=8+6i-5-2i=3+4i$

5. Computing: $(-7-3i)+(-4+4i)=-11+i$

7. Computing: $(-2-3i)-(-1-i)=-2-3i+1+i=-1-2i$

9. Computing: $\left(-\dfrac{3}{4}-\dfrac{1}{4}i\right)+\left(\dfrac{3}{5}+\dfrac{2}{3}i\right)=\left(-\dfrac{3}{4}+\dfrac{3}{5}\right)+\left(-\dfrac{1}{4}+\dfrac{2}{3}\right)i=-\dfrac{3}{20}+\dfrac{5}{12}i$

11. Computing: $\left(\dfrac{3}{10}-\dfrac{3}{4}i\right)-\left(-\dfrac{2}{5}+\dfrac{1}{6}i\right)=\left(\dfrac{3}{10}+\dfrac{2}{5}\right)+\left(-\dfrac{3}{4}-\dfrac{1}{6}\right)i=\dfrac{7}{10}-\dfrac{11}{12}i$

13. Computing: $(5+3i)+(7-2i)+(-8-i)=(5+7-8)+(3-2-1)i=4=4+0i$

15. Simplifying: $\sqrt{-9}=3i$ 17. Simplifying: $\sqrt{-19}=i\sqrt{19}$

19. Simplifying: $\sqrt{-\dfrac{4}{9}}=\dfrac{2}{3}i$ 21. Simplifying: $\sqrt{-8}=i\sqrt{4\cdot2}=2i\sqrt{2}$

23. Simplifying: $\sqrt{-27}=i\sqrt{9\cdot3}=3i\sqrt{3}$ 25. Simplifying: $\sqrt{-54}=i\sqrt{9\cdot6}=3i\sqrt{6}$

27. Simplifying: $3\sqrt{-36}=3\cdot6i=18i$ 29. Simplifying: $4\sqrt{-18}=4\cdot3i\sqrt{2}=12i\sqrt{2}$

31. Simplifying the complex number: $\dfrac{-4-\sqrt{-12}}{2}=\dfrac{-4-i\sqrt{12}}{2}=\dfrac{-4-2i\sqrt{3}}{2}=\dfrac{2\left(-2-i\sqrt{3}\right)}{2}=-2-i\sqrt{3}$

33. Simplifying the complex number: $\dfrac{-3-\sqrt{-18}}{3}=\dfrac{-3-i\sqrt{18}}{3}=\dfrac{-3-3i\sqrt{2}}{3}=-1-i\sqrt{2}$

35. Simplifying the complex number: $\dfrac{12+\sqrt{-45}}{6}=\dfrac{12+i\sqrt{45}}{6}=\dfrac{12+3i\sqrt{5}}{6}=\dfrac{4+i\sqrt{5}}{2}$

37. Simplifying: $\sqrt{-4}\sqrt{-16}=2i\cdot4i=8i^2=-8$ 39. Simplifying: $\sqrt{-2}\sqrt{-3}=i\sqrt{2}\cdot i\sqrt{3}=i^2\sqrt{6}=-\sqrt{6}$

41. Simplifying: $\sqrt{-5}\sqrt{-4}=i\sqrt{5}\cdot i\sqrt{4}=i^2\sqrt{20}=-2\sqrt{5}$

43. Simplifying: $\sqrt{-6}\sqrt{-10}=i\sqrt{6}\cdot i\sqrt{10}=i^2\sqrt{60}=-2\sqrt{15}$

45. Simplifying: $\sqrt{-8}\sqrt{-7}=i\sqrt{8}\cdot i\sqrt{7}=i^2\sqrt{56}=-2\sqrt{14}$

47. Simplifying: $\dfrac{\sqrt{-36}}{\sqrt{-4}}=\dfrac{6i}{2i}=3$ 49. Simplifying: $\dfrac{\sqrt{-54}}{\sqrt{-9}}=\dfrac{3i\sqrt{6}}{3i}=\sqrt{6}$

51. Finding the product: $(3i)(7i)=21i^2=-21=-21+0i$

53. Finding the product: $(4i)(3-2i)=12i-8i^2=8+12i$

55. Finding the product: $(3+2i)(4+6i)=12+8i+18i+12i^2=12+26i-12=0+26i$

57. Finding the product: $(4+5i)(2-9i)=8+10i-36i-45i^2=8-26i+45=53-26i$

59. Finding the product: $(-2-3i)(4+6i)=-8-12i-12i-18i^2=-8-24i+18=10-24i$

61. Finding the product: $(6-4i)(-1-2i)=-6+4i-12i+8i^2=-6-8i-8=-14-8i$

63. Finding the product: $(3+4i)^2=(3+4i)(3+4i)=9+12i+12i+16i^2=9+24i-16=-7+24i$

65. Finding the product: $(-1-2i)^2=(-1-2i)(-1-2i)=1+2i+2i+4i^2=1+4i-4=-3+4i$

67. Finding the product: $(8-7i)(8+7i)=64-56i+56i-49i^2=64+49=113=113+0i$

69. Finding the product: $(-2+3i)(-2-3i)=4-6i+6i-9i^2=4+9=13=13+0i$

71. Finding the quotient: $\dfrac{4i}{3-2i}=\dfrac{4i}{3-2i}\cdot\dfrac{3+2i}{3+2i}=\dfrac{12i+8i^2}{9-4i^2}=\dfrac{-8+12i}{9+4}=-\dfrac{8}{13}+\dfrac{12}{13}i$

73.   Finding the quotient: $\dfrac{2+3i}{3i} = \dfrac{2+3i}{3i} \cdot \dfrac{i}{i} = \dfrac{2i+3i^2}{3i^2} = \dfrac{2i-3}{-3} = 1 - \dfrac{2}{3}i$

75.   Finding the quotient: $\dfrac{3}{2i} = \dfrac{3}{2i} \cdot \dfrac{i}{i} = \dfrac{3i}{2i^2} = 0 - \dfrac{3}{2}i$

77.   Finding the quotient: $\dfrac{3+2i}{4+5i} = \dfrac{3+2i}{4+5i} \cdot \dfrac{4-5i}{4-5i} = \dfrac{12+8i-15i-10i^2}{16-25i^2} = \dfrac{12-7i+10}{16+25} = \dfrac{22}{41} - \dfrac{7}{41}i$

79.   Finding the quotient: $\dfrac{4+7i}{2-3i} = \dfrac{4+7i}{2-3i} \cdot \dfrac{2+3i}{2+3i} = \dfrac{8+14i+12i+21i^2}{4-9i^2} = \dfrac{8+26i-21}{4+9} = -\dfrac{13}{13} + \dfrac{26}{13}i = -1+2i$

81.   Finding the quotient: $\dfrac{3-7i}{-2+4i} = \dfrac{3-7i}{-2+4i} \cdot \dfrac{-2-4i}{-2-4i} = \dfrac{-6+14i-12i+28i^2}{4-16i^2} = \dfrac{-6+2i-28}{4+16} = -\dfrac{34}{20} + \dfrac{2}{20}i = -\dfrac{17}{10} + \dfrac{1}{10}i$

83.   Finding the quotient: $\dfrac{-1-i}{-2-3i} = \dfrac{-1-i}{-2-3i} \cdot \dfrac{-2+3i}{-2+3i} = \dfrac{2+2i-3i-3i^2}{4-9i^2} = \dfrac{2-i+3}{4+9} = \dfrac{5}{13} - \dfrac{1}{13}i$

85.   a.   The sum of the two numbers is $(a+bi)+(c+di) = (a+c)+(b+d)i$, so the conjugate of the sum is $(a+c)-(b+d)i$. The conjugates of the two numbers are $a-bi$ and $c-di$, so the sum of the conjugates is $(a-bi)+(c-di) = (a+c)-(b+d)i$, which is the same.

   b.   The product of the two numbers is $(a+bi)(c+di) = (ac-bd)+(bc+ad)i$, so the conjugate of the product is $(ac-bd)-(bc+ad)i$. The conjugates of the two numbers are $a-bi$ and $c-di$, so the product of the conjugates is $(a-bi)(c-di) = (ac-bd)-(bc+ad)i$, which is the same.

89.   a.   Finding the power: $(2+i)^3 = 2^3 + 3(2)^2 i + 3(2)i^2 + i^3 = 8 + 12i - 6 - i = 2 + 11i$

   b.   Finding the power: $(1-i)^3 = 1^3 + 3(1)^2(-i) + 3(1)(-i)^2 + (-i)^3 = 1 - 3i - 3 + i = -2 - 2i$

   c.   Finding the power: $(1-2i)^3 = 1^3 + 3(1)^2(-2i) + 3(1)(-2i)^2 + (-2i)^3 = 1 - 6i - 12 + 8i = -11 + 2i$

   d.   Finding the power: $(1+i)^4 = 1^4 + 4(1)^3 i + 6(1)^2 i^2 + 4(1)i^3 + i^4 = 1 + 4i - 6 - 4i + 1 = -4 + 0i$

   e.   Finding the power:
         $$(2-i)^4 = 2^4 + 4(2)^3(-i) + 6(2)^2(-i)^2 + 4(2)(-i)^3 + (-i)^4 = 16 - 32i - 24 + 8i + 1 = -7 - 24i$$

   f.   Finding the power:
         $$(-1+i)^5 = (-1)^5 + 5(-1)^4 i + 10(-1)^3 i^2 + 10(-1)^2 i^3 + 5(-1)i^4 + i^5 = -1 + 5i + 10 - 10i - 5 + i = 4 - 4i$$

Chapter 0 Review Problem Set

1.   Evaluating: $5^{-3} = \dfrac{1}{5^3} = \dfrac{1}{125}$

2.   Evaluating: $-3^{-4} = -\dfrac{1}{3^4} = -\dfrac{1}{81}$

3.   Evaluating: $\left(\dfrac{3}{4}\right)^{-2} = \left(\dfrac{4}{3}\right)^2 = \dfrac{16}{9}$

4.   Evaluating: $\dfrac{1}{\left(\frac{1}{3}\right)^{-2}} = \dfrac{1}{3^2} = \dfrac{1}{9}$

5.   Evaluating: $-\sqrt{64} = -8$

6.   Evaluating: $\sqrt[3]{\dfrac{27}{8}} = \dfrac{3}{2}$

7.   Evaluating: $\sqrt[5]{-\dfrac{1}{32}} = -\dfrac{1}{2}$

8.   Evaluating: $36^{-1/2} = \left(\sqrt{36}\right)^{-1} = 6^{-1} = \dfrac{1}{6}$

9.   Evaluating: $\left(\dfrac{1}{8}\right)^{-2/3} = 8^{2/3} = \left(\sqrt[3]{8}\right)^2 = 2^2 = 4$

10.   Evaluating: $-32^{3/5} = -\left(\sqrt[5]{32}\right)^3 = -2^3 = -8$

11.   Simplifying: $\left(3x^{-2}y^{-1}\right)\left(4x^4 y^2\right) = 12x^{-2+4}y^{-1+2} = 12x^2 y$

12. Simplifying: $\left(5x^{2/3}\right)\left(-6x^{1/2}\right) = -30x^{2/3+1/2} = -30x^{7/6}$

13. Simplifying: $\left(-8a^{-1/2}\right)\left(-6a^{1/3}\right) = 48a^{-1/2+1/3} = 48a^{-1/6} = \dfrac{48}{a^{1/6}}$

14. Simplifying: $\left(3x^{-2/3}y^{1/5}\right)^3 = 27x^{-2}y^{3/5} = \dfrac{27y^{3/5}}{x^2}$

15. Simplifying: $\dfrac{64x^{-2}y^3}{16x^3y^{-2}} = 4x^{-5}y^5 = \dfrac{4y^5}{x^5}$ 16. Simplifying: $\dfrac{56x^{-1/3}y^{2/5}}{7x^{1/4}y^{-3/5}} = 8x^{-7/12}y = \dfrac{8y}{x^{7/12}}$

17. Simplifying: $\left(\dfrac{-8x^2y^{-1}}{2x^{-1}y^2}\right)^2 = \left(-4x^3y^{-3}\right)^2 = 16x^6y^{-6} = \dfrac{16x^6}{y^6}$

18. Simplifying: $\left(\dfrac{36a^{-1}b^4}{-12a^2b^5}\right)^{-1} = \dfrac{-12a^2b^5}{36a^{-1}b^4} = -\dfrac{1}{3}a^3b = -\dfrac{a^3b}{3}$

19. Simplifying: $(-7x-3)+(5x-2)+(6x+4) = 4x-1$

20. Simplifying: $(12x+5)-(7x-4)-(8x+1) = 12x+5-7x+4-8x-1 = -3x+8$

21. Simplifying: $3(a-2)-2(3a+5)+3(5a-1) = 3a-6-6a-10+15a-3 = 12a-19$

22. Simplifying: $(4x-7)(5x+6) = 20x^2-35x+24x-42 = 20x^2-11x-42$

23. Simplifying: $(-3x+2)(4x-3) = -12x^2+8x+9x-6 = -12x^2+17x-6$

24. Simplifying: $(7x-3)(-5x+1) = -35x^2+15x+7x-3 = -35x^2+22x-3$

25. Simplifying: $(x+4)(x^2-3x-7) = x^3-3x^2-7x+4x^2-12x-28 = x^3+x^2-19x-28$

26. Simplifying: $(2x+1)(3x^2-2x+6) = 6x^3-4x^2+12x+3x^2-2x+6 = 6x^3-x^2+10x+6$

27. Simplifying: $(5x-3)^2 = (5x-3)(5x-3) = 25x^2-15x-15x+9 = 25x^2-30x+9$

28. Simplifying: $(3x+7)^2 = (3x+7)(3x+7) = 9x^2+21x+21x+49 = 9x^2+42x+49$

29. Simplifying:
$$(2x-1)^3 = (2x-1)(2x-1)^2 = (2x-1)\left(4x^2-4x+1\right) = 8x^3-8x^2+2x-4x^2+4x-1 = 8x^3-12x^2+6x-1$$

30. Simplifying:
$$\begin{aligned}(3x+5)^3 &= (3x+5)(3x+5)^2 \\ &= (3x+5)\left(9x^2+30x+25\right) \\ &= 27x^3+90x^2+75x+45x^2+150x+125 \\ &= 27x^3+135x^2+225x+125\end{aligned}$$

31. Simplifying:
$$\left(x^2-2x-3\right)\left(x^2+4x+5\right) = x^4+4x^3+5x^2-2x^3-8x^2-10x-3x^2-12x-15 = x^4+2x^3-6x^2-22x-15$$

32. Simplifying:
$$\left(2x^2-x-2\right)\left(x^2+6x-4\right) = 2x^4+12x^3-8x^2-x^3-6x^2+4x-2x^2-12x+8 = 2x^4+11x^3-16x^2-8x+8$$

33. Simplifying: $\dfrac{24x^3y^4-48x^2y^3}{-6xy} = \dfrac{24x^3y^4}{-6xy}-\dfrac{48x^2y^3}{-6xy} = -4x^2y^3+8xy^2$

34. Simplifying: $\dfrac{-56x^2y+72x^3y^2}{8x^2} = \dfrac{-56x^2y}{8x^2}+\dfrac{72x^3y^2}{8x^2} = -7y+9xy^2$

35. Factoring completely: $9x^2-4y^2 = (3x+2y)(3x-2y)$

36. Factoring completely: $3x^3 - 9x^2 - 120x = 3x(x^2 - 3x - 40) = 3x(x-8)(x+5)$

37. Factoring completely: $4x^2 + 20x + 25 = (2x+5)(2x+5) = (2x+5)^2$

38. Factoring completely: $(x-y)^2 - 9 = (x-y+3)(x-y-3)$

39. Factoring completely: $x^2 - 2x - xy + 2y = x(x-2) - y(x-2) = (x-2)(x-y)$

40. Factoring completely: $64x^3 - 27y^3 = (4x-3y)(16x^2 + 12xy + 9y^2)$

41. Factoring completely: $15x^2 - 14x - 8 = (3x-4)(5x+2)$

42. Factoring completely: $3x^3 + 36 = 3(x^3 + 12)$

43. The polynomial $2x^2 - x - 8$ is not factorable.

44. Factoring completely: $3x^3 + 24 = 3(x^3 + 8) = 3(x+2)(x^2 - 2x + 4)$

45. Factoring completely: $x^4 - 13x^2 + 36 = (x^2 - 9)(x^2 - 4) = (x+3)(x-3)(x+2)(x-2)$

46. Factoring completely: $4x^2 - 4x + 1 - y^2 = (2x-1)^2 - y^2 = (2x-1+y)(2x-1-y)$

47. Performing the operations: $\dfrac{8xy}{18x^2y} \cdot \dfrac{24xy^2}{16y^3} = \dfrac{2x^2y^3}{3x^2y^4} = \dfrac{2}{3y}$

48. Performing the operations: $\dfrac{-14a^2b^2}{6b^3} \div \dfrac{21a}{15ab} = \dfrac{-14a^2b^2}{6b^3} \cdot \dfrac{15ab}{21a} = -\dfrac{5a^3b^3}{3ab^3} = -\dfrac{5a^2}{3}$

49. Performing the operations: $\dfrac{x^2 + 3x - 4}{x^2 - 1} \cdot \dfrac{3x^2 + 8x + 5}{x^2 + 4x} = \dfrac{(x+4)(x-1)}{(x+1)(x-1)} \cdot \dfrac{(3x+5)(x+1)}{x(x+4)} = \dfrac{3x+5}{x}$

50. Performing the operations: $\dfrac{9x^2 - 6x + 1}{2x^2 + 8} \cdot \dfrac{8x + 20}{6x^2 + 13x - 5} = \dfrac{(3x-1)^2}{2(x^2+4)} \cdot \dfrac{4(2x+5)}{(2x+5)(3x-1)} = \dfrac{2(3x-1)}{x^2 + 4}$

51. Performing the operations: $\dfrac{3x-2}{4} + \dfrac{5x-1}{3} = \dfrac{3x-2}{4} \cdot \dfrac{3}{3} + \dfrac{5x-1}{3} \cdot \dfrac{4}{4} = \dfrac{9x-6}{12} + \dfrac{20x-4}{12} = \dfrac{29x-10}{12}$

52. Performing the operations:

$$\dfrac{2x-6}{5} - \dfrac{x+4}{3} = \dfrac{2x-6}{5} \cdot \dfrac{3}{3} - \dfrac{x+4}{3} \cdot \dfrac{5}{5} = \dfrac{6x-18}{15} - \dfrac{5x+20}{15} = \dfrac{6x-18-5x-20}{15} = \dfrac{x-38}{15}$$

53. Performing the operations: $\dfrac{3}{n^2} + \dfrac{4}{5n} - \dfrac{2}{n} = \dfrac{3}{n^2} \cdot \dfrac{5}{5} + \dfrac{4}{5n} \cdot \dfrac{n}{n} - \dfrac{2}{n} \cdot \dfrac{5n}{5n} = \dfrac{15}{5n^2} + \dfrac{4n}{5n^2} - \dfrac{10n}{5n^2} = \dfrac{-6n+15}{5n^2}$

54. Performing the operations: $\dfrac{5}{x^2 + 7x} - \dfrac{3}{x} = \dfrac{5}{x(x+7)} - \dfrac{3}{x} \cdot \dfrac{x+7}{x+7} = \dfrac{5}{x(x+7)} - \dfrac{3x+21}{x(x+7)} = \dfrac{5 - 3x - 21}{x(x+7)} = \dfrac{-3x - 16}{x(x+7)}$

55. Performing the operations:

$$\dfrac{3x}{x^2 - 6x - 40} + \dfrac{4}{x^2 - 16} = \dfrac{3x}{(x-10)(x+4)} \cdot \dfrac{x-4}{x-4} + \dfrac{4}{(x+4)(x-4)} \cdot \dfrac{x-10}{x-10}$$

$$= \dfrac{3x^2 - 12x}{(x-10)(x+4)(x-4)} + \dfrac{4x - 40}{(x-10)(x+4)(x-4)}$$

$$= \dfrac{3x^2 - 12x + 4x - 40}{(x-10)(x+4)(x-4)}$$

$$= \dfrac{3x^2 - 8x - 40}{(x-10)(x+4)(x-4)}$$

56. Performing the operations:

$$\frac{2}{x-2} - \frac{2}{x+2} - \frac{4}{x^3-4x} = \frac{2}{x-2} \cdot \frac{x(x+2)}{x(x+2)} - \frac{2}{x+2} \cdot \frac{x(x-2)}{x(x-2)} - \frac{4}{x(x+2)(x-2)}$$

$$= \frac{2x^2+4x}{x(x+2)(x-2)} - \frac{2x^2-4x}{x(x+2)(x-2)} - \frac{4}{x(x+2)(x-2)}$$

$$= \frac{2x^2+4x-2x^2+4x-4}{x(x+2)(x-2)}$$

$$= \frac{8x-4}{x(x+2)(x-2)}$$

57. Simplifying: $\dfrac{\dfrac{3}{x}-\dfrac{2}{y}}{\dfrac{5}{x^2}+\dfrac{7}{y}} = \dfrac{\dfrac{3}{x}-\dfrac{2}{y}}{\dfrac{5}{x^2}+\dfrac{7}{y}} \cdot \dfrac{x^2y}{x^2y} = \dfrac{3xy-2x^2}{5y+7x^2}$

58. Simplifying: $\dfrac{3-\dfrac{2}{x}}{4+\dfrac{3}{x}} = \dfrac{3-\dfrac{2}{x}}{4+\dfrac{3}{x}} \cdot \dfrac{x}{x} = \dfrac{3x-2}{4x+3}$

59. Simplifying:

$$\frac{\dfrac{3}{(x+h)^2}-\dfrac{3}{x^2}}{h} = \frac{\dfrac{3}{(x+h)^2}-\dfrac{3}{x^2}}{h} \cdot \frac{x^2(x+h)^2}{x^2(x+h)^2}$$

$$= \frac{3x^2-3(x+h)^2}{hx^2(x+h)^2}$$

$$= \frac{3x^2-3x^2-6xh-3h^2}{hx^2(x+h)^2}$$

$$= \frac{-h(6x+3h)}{hx^2(x+h)^2}$$

$$= -\frac{6x+3h}{x^2(x+h)^2}$$

60. Simplifying:

$$\frac{6(x^2+2)^{1/2}-6x^2(x^2+2)^{-1/2}}{\left[(x^2+2)^{1/2}\right]^2} = \frac{6(x^2+2)^{1/2}-6x^2(x^2+2)^{-1/2}}{x^2+2} \cdot \frac{(x^2+2)^{1/2}}{(x^2+2)^{1/2}}$$

$$= \frac{6(x^2+2)-6x^2}{(x^2+2)^{3/2}}$$

$$= \frac{6x^2+12-6x^2}{(x^2+2)^{3/2}}$$

$$= \frac{12}{(x^2+2)^{3/2}}$$

61. Simplifying: $5\sqrt{48} = 5\sqrt{16 \cdot 3} = 5 \cdot 4\sqrt{3} = 20\sqrt{3}$

62. Simplifying: $3\sqrt{24x^3} = 3\sqrt{4x^2 \cdot 6x} = 3 \cdot 2x\sqrt{6x} = 6x\sqrt{6x}$

63.   Simplifying: $\sqrt[3]{32x^4y^5} = \sqrt[3]{8x^3y^3 \bullet 4xy^2} = 2xy\sqrt[3]{4xy^2}$

64.   Simplifying: $\dfrac{3\sqrt{8}}{2\sqrt{6}} = \dfrac{6\sqrt{2}}{2\sqrt{6}} \bullet \dfrac{\sqrt{6}}{\sqrt{6}} = \dfrac{6\sqrt{12}}{12} = \dfrac{12\sqrt{3}}{12} = \sqrt{3}$

65.   Simplifying: $\sqrt{\dfrac{5x}{2y^2}} = \dfrac{\sqrt{5x}}{y\sqrt{2}} \bullet \dfrac{\sqrt{2}}{\sqrt{2}} = \dfrac{\sqrt{10x}}{2y}$

66.   Simplifying: $\dfrac{3}{\sqrt{2}+5} = \dfrac{3}{\sqrt{2}+5} \bullet \dfrac{\sqrt{2}-5}{\sqrt{2}-5} = \dfrac{3\sqrt{2}-15}{2-25} = \dfrac{3\sqrt{2}-15}{-23} = \dfrac{15-3\sqrt{2}}{23}$

67.   Simplifying: $\dfrac{4\sqrt{2}}{3\sqrt{2}+\sqrt{3}} = \dfrac{4\sqrt{2}}{3\sqrt{2}+\sqrt{3}} \bullet \dfrac{3\sqrt{2}-\sqrt{3}}{3\sqrt{2}-\sqrt{3}} = \dfrac{24-4\sqrt{6}}{18-3} = \dfrac{24-4\sqrt{6}}{15}$

68.   Simplifying: $\dfrac{3\sqrt{x}}{\sqrt{x}-2\sqrt{y}} = \dfrac{3\sqrt{x}}{\sqrt{x}-2\sqrt{y}} \bullet \dfrac{\sqrt{x}+2\sqrt{y}}{\sqrt{x}+2\sqrt{y}} = \dfrac{3x+6\sqrt{xy}}{x-4y}$

69.   Simplifying: $\sqrt{5}\sqrt[3]{5} = 5^{1/2} \bullet 5^{1/3} = 5^{1/2+1/3} = 5^{5/6} = \sqrt[6]{5^5}$

70.   Simplifying: $\sqrt[3]{x^2}\sqrt[4]{x} = x^{2/3} \bullet x^{1/4} = x^{2/3+1/4} = x^{11/12} = \sqrt[12]{x^{11}}$

71.   Simplifying: $\sqrt{x^3}\sqrt[3]{x^4} = x^{3/2} \bullet x^{4/3} = x^{3/2+4/3} = x^{17/6} = \sqrt[6]{x^{17}} = x^2\sqrt[6]{x^5}$

72.   Simplifying: $\sqrt{xy}\sqrt[5]{x^3y^2} = x^{1/2}y^{1/2} \bullet x^{3/5}y^{2/5} = x^{1/2+3/5}y^{1/2+2/5} = x^{11/10}y^{9/10} = \sqrt[10]{x^{11}y^9} = x\sqrt[10]{xy^9}$

73.   Simplifying: $\dfrac{\sqrt{5}}{\sqrt[3]{5}} = \dfrac{5^{1/2}}{5^{1/3}} = 5^{1/2-1/3} = 5^{1/6} = \sqrt[6]{5}$

74.   Simplifying: $\dfrac{\sqrt[3]{x^2}}{\sqrt[4]{x^3}} = \dfrac{x^{2/3}}{x^{3/4}} = x^{2/3-3/4} = x^{-1/12} = \dfrac{1}{x^{1/12}} = \dfrac{1}{\sqrt[12]{x}} \bullet \dfrac{\sqrt[12]{x^{11}}}{\sqrt[12]{x^{11}}} = \dfrac{\sqrt[12]{x^{11}}}{x}$

75.   Performing the operations: $(-7+3i)+(-4-9i) = -11-6i$

76.   Performing the operations: $(2-10i)-(3-8i) = 2-10i-3+8i = -1-2i$

77.   Performing the operations: $(-1+4i)-(-2+6i) = -1+4i+2-6i = 1-2i$

78.   Performing the operations: $(3i)(-7i) = -21i^2 = 21+0i$

79.   Performing the operations: $(2-5i)(3+4i) = 6-15i+8i-20i^2 = 6-7i+20 = 26-7i$

80.   Performing the operations: $(-3-i)(6-7i) = -18-6i+21i+7i^2 = -18+15i-7 = -25+15i$

81.   Performing the operations: $(4+2i)(-4-i) = -16-8i-4i-2i^2 = -16-12i+2 = -14-12i$

82.   Performing the operations: $(5-2i)(5+2i) = 25-4i^2 = 25+4 = 29+0i$

83.   Performing the operations: $\dfrac{5}{3i} = \dfrac{5}{3i} \bullet \dfrac{i}{i} = \dfrac{5i}{3i^2} = \dfrac{5i}{-3} = 0-\dfrac{5}{3}i$

84.   Performing the operations: $\dfrac{2+3i}{3-4i} = \dfrac{2+3i}{3-4i} \bullet \dfrac{3+4i}{3+4i} = \dfrac{6+9i+8i+12i^2}{9-16i^2} = \dfrac{6+17i-12}{9+16} = \dfrac{-6+17i}{25} = -\dfrac{6}{25}+\dfrac{17}{25}i$

85.   Performing the operations: $\dfrac{-1-2i}{-2+i} = \dfrac{-1-2i}{-2+i} \bullet \dfrac{-2-i}{-2-i} = \dfrac{2+4i+i+2i^2}{4-i^2} = \dfrac{2+5i-2}{4+1} = \dfrac{5i}{5i} = 0+i$

86.   Performing the operations: $\dfrac{-6i}{5+2i} = \dfrac{-6i}{5+2i} \bullet \dfrac{5-2i}{5-2i} = \dfrac{-30i+12i^2}{25-4i^2} = \dfrac{-30i-12}{29} = -\dfrac{12}{29}-\dfrac{30}{29}i$

87.   Simplifying: $\sqrt{-100} = 10i$ 88.   Simplifying: $\sqrt{-40} = i\sqrt{4 \bullet 10} = 2i\sqrt{10}$

89.   Simplifying: $4\sqrt{-80} = 4i\sqrt{16 \bullet 5} = 16i\sqrt{5}$ 90.   Simplifying: $\sqrt{-9}\sqrt{-16} = (3i)(4i) = 12i^2 = -12$

91. Simplifying: $\sqrt{-6}\sqrt{-8} = \left(i\sqrt{6}\right)\left(i\sqrt{8}\right) = i^2\sqrt{48} = -\sqrt{16\cdot 3} = -4\sqrt{3}$

92. Simplifying: $\dfrac{\sqrt{-24}}{\sqrt{-3}} = \dfrac{i\sqrt{24}}{i\sqrt{3}} = \dfrac{2\sqrt{6}}{\sqrt{3}}\cdot\dfrac{\sqrt{3}}{\sqrt{3}} = \dfrac{2\sqrt{18}}{3} = \dfrac{6\sqrt{2}}{3} = 2\sqrt{2}$

93. Using scientific notation: $\dfrac{(0.0064)(420,000)}{(0.00014)(0.032)} = \dfrac{(6.4)(10)^{-3}(4.2)(10)^5}{(1.4)(10)^{-4}(3.2)(10)^{-2}} = \dfrac{(6)(10)^2}{(10)^{-6}} = (6)(10)^8 = 600,000,000$

94. Using scientific notation: $\dfrac{(8600)(0.0000064)}{(0.0016)(0.000043)} = \dfrac{(8.6)(10)^3(6.4)(10)^{-6}}{(1.6)(10)^{-3}(4.3)(10)^{-5}} = \dfrac{(8)(10)^{-3}}{(10)^{-8}} = (8)(10)^5 = 800,000$

Chapter 0 Test

1.   **a.**   Evaluating: $-7^{-2} = -\dfrac{1}{7^2} = -\dfrac{1}{49}$ **b.**   Evaluating: $\left(\dfrac{3}{2}\right)^{-3} = \left(\dfrac{2}{3}\right)^3 = \dfrac{8}{27}$

　　　c.   Evaluating: $\left(\dfrac{4}{9}\right)^{3/2} = \left(\sqrt{\dfrac{4}{9}}\right)^3 = \left(\dfrac{2}{3}\right)^3 = \dfrac{8}{27}$ **d.**   Evaluating: $\sqrt[3]{\dfrac{27}{64}} = \dfrac{3}{4}$

2.   Finding the product: $\left(-3x^{-1}y^2\right)\left(5x^{-3}y^{-4}\right) = -15x^{-4}y^{-2} = -\dfrac{15}{x^4y^2}$

3.   Simplifying: $(-3x-4)-(7x-5)+(-2x-9) = -3x-4-7x+5-2x-9 = -12x-8$

4.   Simplifying: $(5x-2)(-6x+4) = -30x^2+12x+20x-8 = -30x^2+32x-8$

5.   Simplifying: $(x+2)\left(3x^2-2x-7\right) = 3x^3-2x^2-7x+6x^2-4x-14 = 3x^3+4x^2-11x-14$

6.   Simplifying:

$$\begin{aligned}
(4x-1)^3 &= (4x-1)(4x-1)^2 \\
&= (4x-1)\left(16x^2-8x+1\right) \\
&= 64x^3-32x^2+4x-16x^2+8x-1 \\
&= 64x^3-48x^2+12x-1
\end{aligned}$$

7.   Simplifying: $\dfrac{-18x^4y^3-24x^5y^4}{-2xy^2} = \dfrac{-18x^4y^3}{-2xy^2} - \dfrac{24x^5y^4}{-2xy^2} = 9x^3y+12x^4y^2$

8.   Factoring completely: $18x^3-15x^2-12x = 3x\left(6x^2-5x-4\right) = 3x(3x-4)(2x+1)$

9.   Factoring completely: $30x^2-13x-10 = (6x-5)(5x+2)$

10.   Factoring completely: $8x^3+64 = 8\left(x^3+8\right) = 8(x+2)\left(x^2-2x+4\right)$

11.   Factoring completely: $x^2+xy-2y-2x = x(x+y)-2(x+y) = (x+y)(x-2)$

12.   Simplifying: $\dfrac{6x^3y^2}{5xy} \div \dfrac{8y}{7x^3} = \dfrac{6x^3y^2}{5xy}\cdot\dfrac{7x^3}{8y} = \dfrac{42x^6y^2}{40xy^2} = \dfrac{21x^5}{20}$

13.   Simplifying: $\dfrac{x^2-4}{2x^2+5x+2}\cdot\dfrac{2x^2+7x+3}{x^3-8} = \dfrac{(x+2)(x-2)}{(2x+1)(x+2)}\cdot\dfrac{(2x+1)(x+3)}{(x-2)\left(x^2+2x+4\right)} = \dfrac{x+3}{x^2+2x+4}$

14.   Simplifying: $\dfrac{3n-2}{4}-\dfrac{4n+1}{6} = \dfrac{3n-2}{4}\cdot\dfrac{3}{3}-\dfrac{4n+1}{6}\cdot\dfrac{2}{2} = \dfrac{9n-6}{12}-\dfrac{8n+2}{12} = \dfrac{9n-6-8n-2}{12} = \dfrac{n-8}{12}$

15. Simplifying:

$$\frac{5}{2x^2 - 6x} + \frac{4}{3x^2 + 6x} = \frac{5}{2x(x-3)} \cdot \frac{3(x+2)}{3(x+2)} + \frac{4}{3x(x+2)} \cdot \frac{2(x-3)}{2(x-3)}$$

$$= \frac{15x + 30}{6x(x-3)(x+2)} + \frac{8x - 24}{6x(x-3)(x+2)}$$

$$= \frac{23x + 6}{6x(x-3)(x+2)}$$

16. Simplifying: $\dfrac{4}{n^2} - \dfrac{3}{2n} - \dfrac{5}{n} = \dfrac{4}{n^2} \cdot \dfrac{2}{2} - \dfrac{3}{2n} \cdot \dfrac{n}{n} - \dfrac{5}{n} \cdot \dfrac{2n}{2n} = \dfrac{8}{2n^2} - \dfrac{3n}{2n^2} - \dfrac{10n}{2n^2} = \dfrac{8 - 13n}{2n^2}$

17. Simplifying: $\dfrac{\dfrac{2}{x} - \dfrac{5}{y}}{\dfrac{3}{x} + \dfrac{4}{y^2}} = \dfrac{\dfrac{2}{x} - \dfrac{5}{y}}{\dfrac{3}{x} + \dfrac{4}{y^2}} \cdot \dfrac{xy^2}{xy^2} = \dfrac{2y^2 - 5xy}{3y^2 + 4x}$

18. Simplifying: $6\sqrt{28x^5} = 6\sqrt{4x^4 \cdot 7x} = 6 \cdot 2x^2\sqrt{7x} = 12x^2\sqrt{7x}$

19. Simplifying: $\dfrac{5\sqrt{6}}{3\sqrt{12}} = \dfrac{5\sqrt{6}}{3\sqrt{12}} \cdot \dfrac{\sqrt{3}}{\sqrt{3}} = \dfrac{5\sqrt{18}}{3\sqrt{36}} = \dfrac{15\sqrt{2}}{18} = \dfrac{5\sqrt{2}}{6}$

20. Simplifying: $\dfrac{\sqrt{6}}{2\sqrt{2} - \sqrt{3}} = \dfrac{\sqrt{6}}{2\sqrt{2} - \sqrt{3}} \cdot \dfrac{2\sqrt{2} + \sqrt{3}}{2\sqrt{2} + \sqrt{3}} = \dfrac{2\sqrt{12} + \sqrt{18}}{8 - 3} = \dfrac{4\sqrt{3} + 3\sqrt{2}}{5}$

21. Simplifying: $\sqrt[3]{48x^4y^5} = \sqrt[3]{8x^3y^3 \cdot 6xy^2} = 2xy\sqrt[3]{6xy^2}$

22. Simplifying: $(-2 - 4i) - (-1 + 6i) + (-3 + 7i) = -2 - 4i + 1 - 6i - 3 + 7i = -4 - 3i$

23. Simplifying: $(5 - 7i)(4 + 2i) = 20 - 28i + 10i - 14i^2 = 20 - 18i + 14 = 34 - 18i$

24. Simplifying: $(7 - 6i)(7 + 6i) = 49 - 42i + 42i - 36i^2 = 49 + 36 = 85 + 0i$

25. Simplifying: $\dfrac{1 + 2i}{3 - i} = \dfrac{1 + 2i}{3 - i} \cdot \dfrac{3 + i}{3 + i} = \dfrac{3 + 6i + i + 2i^2}{9 - i^2} = \dfrac{3 + 7i - 2}{10} = \dfrac{1}{10} + \dfrac{7}{10}i$

Chapter 1
Equations, Inequalities, and Problem Solving

1.1 Linear Equations and Problem Solving

1. Solving the equation:
$$9x - 3 = -21$$
$$9x = -18$$
$$x = -2$$

The solution set is $\{-2\}$.

3. Solving the equation:
$$13 - 2x = 14$$
$$-2x = 1$$
$$x = -\frac{1}{2}$$

The solution set is $\left\{-\frac{1}{2}\right\}$.

5. Solving the equation:
$$3n - 2 = 2n + 5$$
$$n - 2 = 5$$
$$n = 7$$

The solution set is $\{7\}$.

7. Solving the equation:
$$-5a + 3 = -3a + 6$$
$$-2a + 3 = 6$$
$$-2a = 3$$
$$a = -\frac{3}{2}$$

The solution set is $\left\{-\frac{3}{2}\right\}$.

9. Solving the equation:
$$-3(x + 1) = 7$$
$$-3x - 3 = 7$$
$$-3x = 10$$
$$x = -\frac{10}{3}$$

The solution set is $\left\{-\frac{10}{3}\right\}$.

11. Solving the equation:
$$4(2x - 1) = 3(3x + 2)$$
$$8x - 4 = 9x + 6$$
$$-x - 4 = 6$$
$$-x = 10$$
$$x = -10$$

The solution set is $\{-10\}$.

13.   Solving the equation:
$$3(n-1) = -2(n+4) + 6(n-3)$$
$$3n-3 = -2n-8+6n-18$$
$$3n-3 = 4n-26$$
$$-n-3 = -26$$
$$-n = -23$$
$$n = 23$$

The solution set is $\{23\}$.

15.   Solving the equation:
$$3(2t-1) - 2(5t+1) = 4(3t+4)$$
$$6t-3-10t-2 = 12t+16$$
$$-4t-5 = 12t+16$$
$$-16t = 21$$
$$t = -\frac{21}{16}$$

The solution set is $\left\{-\frac{21}{16}\right\}$.

17.   Solving the equation:
$$-2(y-4) - (3y-1) = -2 + 5(y+1)$$
$$-2y+8-3y+1 = -2+5y+5$$
$$-5y+9 = 5y+3$$
$$-10y = -6$$
$$y = \frac{3}{5}$$

The solution set is $\left\{\frac{3}{5}\right\}$.

19.   Solving the equation:
$$\frac{-6x}{7} = 12$$
$$7 \cdot \frac{-6x}{7} = 7 \cdot 12$$
$$-6x = 84$$
$$x = -14$$

The solution set is $\{-14\}$.

21.   Solving the equation:
$$\frac{3}{4}n - \frac{1}{12}n = 6$$
$$12\left(\frac{3}{4}n - \frac{1}{12}n\right) = 12 \cdot 6$$
$$9n - n = 72$$
$$8n = 72$$
$$n = 9$$

The solution set is $\{9\}$.

23.   Solving the equation:
$$\frac{h}{2} + \frac{h}{5} = 1$$
$$10\left(\frac{h}{2} + \frac{h}{5}\right) = 10 \cdot 1$$
$$5h + 2h = 10$$
$$7h = 10$$
$$h = \frac{10}{7}$$

The solution set is $\left\{\frac{10}{7}\right\}$.

25.   Solving the equation:
$$\frac{y}{5} - 2 = \frac{y}{2} + 1$$
$$10\left(\frac{y}{5} - 2\right) = 10\left(\frac{y}{2} + 1\right)$$
$$2y - 20 = 5y + 10$$
$$-3y = 30$$
$$y = -10$$

The solution set is $\{-10\}$.

27.   Solving the equation:
$$\frac{c+5}{7} + \frac{c-3}{4} = \frac{5}{14}$$
$$28\left(\frac{c+5}{7} + \frac{c-3}{4}\right) = 28 \cdot \frac{5}{14}$$
$$4c+20+7c-21 = 10$$
$$11c-1 = 10$$
$$11c = 11$$
$$c = 1$$

The solution set is $\{1\}$.

29.   Solving the equation:
$$\frac{n-3}{2} - \frac{4n-1}{6} = \frac{2}{3}$$
$$6\left(\frac{n-3}{2} - \frac{4n-1}{6}\right) = 6 \cdot \frac{2}{3}$$
$$3n-9-4n+1 = 4$$
$$-n-8 = 4$$
$$-n = 12$$
$$n = -12$$

The solution set is $\{-12\}$.

31.   Solving the equation:
$$\frac{2t+3}{6} - \frac{t-9}{4} = 5$$
$$12\left(\frac{2t+3}{6} - \frac{t-9}{4}\right) = 12 \cdot 5$$
$$4t+6-3t+27 = 60$$
$$t+33 = 60$$
$$t = 27$$

The solution set is $\{27\}$.

33. Solving the equation:

$$\frac{3n-1}{8} - 2 = \frac{2n+5}{7}$$

$$56\left(\frac{3n-1}{8} - 2\right) = 56\left(\frac{2n+5}{7}\right)$$

$$21n - 7 - 112 = 16n + 40$$

$$21n - 119 = 16n + 40$$

$$5n = 159$$

$$n = \frac{159}{5}$$

The solution set is $\left\{\frac{159}{5}\right\}$.

35. Solving the equation:

$$\frac{2t-3}{6} + \frac{3t-2}{4} + \frac{5t+6}{12} = 4$$

$$12\left(\frac{2t-3}{6} + \frac{3t-2}{4} + \frac{5t+6}{12}\right) = 12 \cdot 4$$

$$4t - 6 + 9t - 6 + 5t + 6 = 48$$

$$18t - 6 = 48$$

$$18t = 54$$

$$t = 3$$

The solution set is $\{3\}$.

37. Solving the equation:

$$\frac{2x+1}{14} - \frac{3x+4}{7} = \frac{x-1}{2}$$

$$14\left(\frac{2x+1}{14} - \frac{3x+4}{7}\right) = 14\left(\frac{x-1}{2}\right)$$

$$2x + 1 - 6x - 8 = 7x - 7$$

$$-4x - 7 = 7x - 7$$

$$11x = 0$$

$$x = 0$$

The solution set is $\{0\}$.

39. Solving the equation:

$$(x-3)(x-1) - x(x+2) = 7$$

$$x^2 - 4x + 3 - x^2 - 2x = 7$$

$$-6x + 3 = 7$$

$$-6x = 4$$

$$x = -\frac{2}{3}$$

The solution set is $\left\{-\frac{2}{3}\right\}$.

41. Solving the equation:

$$(2y+1)(3y-2) - (6y-1)(y+4) = -20y$$

$$(6y^2 - y - 2) - (6y^2 + 23y - 4) = -20y$$

$$6y^2 - y - 2 - 6y^2 - 23y + 4 = -20y$$

$$-24y + 2 = -20y$$

$$-4y = -2$$

$$y = \frac{1}{2}$$

The solution set is $\left\{\frac{1}{2}\right\}$.

43. Let c represent the grams of carbohydrates in the chicken sandwich, and $2c + 10$ represent the grams of carbohydrates in the pasta salad. The equation is:

$$c + 2c + 10 = 100$$

$$3c + 10 = 100$$

$$3c = 90$$

$$c = 30$$

$$2c + 10 = 70$$

The chicken sandwich had 30 grams of carbohydrates, and the pasta salad had 70 grams of carbohydrates.

45. Let x, $x + 2$, and $x + 4$ represent the three integers. The equation is:

$$4x - (x+4) = 6 + 2(x+2)$$

$$4x - x - 4 = 6 + 2x + 4$$

$$3x - 4 = 2x + 10$$

$$x = 14$$

The integers are 14, 16, and 18.

47. Let x and $x + 1$ represent the two integers. There are two equations:

$$(x+1)^2 - x^2 = 37 \qquad\qquad x^2 - (x+1)^2 = 37$$
$$x^2 + 2x + 1 - x^2 = 37 \qquad\qquad x^2 - x^2 - 2x - 1 = 37$$
$$2x + 1 = 37 \qquad\qquad -2x - 1 = 37$$
$$2x = 36 \qquad\qquad -2x = 38$$
$$x = 18 \qquad\qquad x = -19$$

Since the integers are positive, the integers are 18 and 19.

49. Let x, $x + 1$, $x + 2$, and $x + 3$ represent the four integers. The equation is:

$$(x+2)(x+3) = x(x+1) + 46$$
$$x^2 + 5x + 6 = x^2 + x + 46$$
$$5x + 6 = x + 46$$
$$4x = 40$$
$$x = 10$$

The integers are 10, 11, 12, and 13.

51. Let x represent the measure of angle B, so $3x - 10$ represents the measure of angle A and
$\dfrac{1}{5}(x + 3x - 10) = \dfrac{1}{5}(4x - 10)$ represents the measure of angle C. The equation is:

$$x + (3x - 10) + \frac{1}{5}(4x - 10) = 180$$
$$4x - 10 + \frac{4}{5}x - 2 = 180$$
$$\frac{24}{5}x - 12 = 180$$
$$\frac{24}{5}x = 192$$
$$24x = 960$$
$$x = 40$$

The angles are $30°$, $40°$, and $110°$.

53. Let x represent Barry's normal hourly rate, so $2x$ represents his overtime rate. Since Barry worked 40 hours at his normal rate and 7 hours at his overtime rate, the equation is:

$$40(x) + 7(2x) = 648$$
$$40x + 14x = 648$$
$$54x = 648$$
$$x = 12$$

His normal hourly rate is $12 per hour.

55. Let x represent the number of pennies, so $\dfrac{1}{3}x + 5$ represents the number of nickels and $\dfrac{1}{4}x - 1$ represents the number of dimes. Since there are 80 coins, the equation is:

$$x + \frac{1}{3}x + 5 + \frac{1}{4}x - 1 = 80$$
$$\frac{19}{12}x + 4 = 80$$
$$\frac{19}{12}x = 76$$
$$19x = 912$$
$$x = 48$$

Greg had 48 pennies, 21 nickels, and 11 dimes.

57. Let x represent the number of females and $2x - 8$ represent the number of males. The equation is:

$$x + 2x - 8 = 43$$
$$3x - 8 = 43$$
$$3x = 51$$
$$x = 17$$

There are 17 females and 26 males.

59. Let x represent Janie's age now. The equation is:

$$x - 2 = \frac{1}{2}(x + 9)$$

$$x - 2 = \frac{1}{2}x + \frac{9}{2}$$

$$\frac{1}{2}x = \frac{13}{2}$$

$$x = 13$$

Janie is 13 years old now.

61. Let x represent Pedro's age and $x + 6$ represent Brad's age. The equation relating their ages 5 years ago is:

$$x - 5 = \frac{3}{4}(x + 6 - 5)$$

$$x - 5 = \frac{3}{4}(x + 1)$$

$$x - 5 = \frac{3}{4}x + \frac{3}{4}$$

$$\frac{1}{4}x = \frac{23}{4}$$

$$x = 23$$

Pedro is 23 years old and Brad is 29 years old.

67. Let x, $x + 1$, and $x + 2$ represent the three integers. Then: $x + x + 2 = 2x + 2 = 2(x + 1)$

This verifies the result.

69. **a.** Computing each value:

$$(21)(19) = (20 + 1)(20 - 1) = 20^2 - 1^2 = 400 - 1 = 399$$

$$(39)(41) = (40 - 1)(40 + 1) = 40^2 - 1^2 = 1600 - 1 = 1599$$

$$(22)(18) = (20 + 2)(20 - 2) = 20^2 - 2^2 = 400 - 4 = 396$$

$$(42)(38) = (40 + 2)(40 - 2) = 40^2 - 2^2 = 1600 - 4 = 1596$$

$$(47)(53) = (50 - 3)(50 + 3) = 50^2 - 3^2 = 2500 - 9 = 2491$$

b. Computing:

$$(21)^2 = (20 + 1)^2 = 20^2 + 2(20)(1) + 1^2 = 400 + 40 + 1 = 441$$

$$(32)^2 = (30 + 2)^2 = 30^2 + 2(30)(2) + 2^2 = 900 + 120 + 4 = 1024$$

$$(51)^2 = (50 + 1)^2 = 50^2 + 2(50)(1) + 1^2 = 2500 + 100 + 1 = 2601$$

$$(62)^2 = (60 + 2)^2 = 60^2 + 2(60)(2) + 2^2 = 3600 + 240 + 4 = 3844$$

$$(43)^2 = (40 + 3)^2 = 40^2 + 2(40)(3) + 3^2 = 1600 + 240 + 9 = 1849$$

c. Computing:

$$(29)^2 = (30 - 1)^2 = 30^2 - 2(30)(1) + 1^2 = 900 - 60 + 1 = 841$$

$$(49)^2 = (50 - 1)^2 = 50^2 - 2(50)(1) + 1^2 = 2500 - 100 + 1 = 2401$$

$$(18)^2 = (20 - 2)^2 = 20^2 - 2(20)(2) + 2^2 = 400 - 80 + 4 = 324$$

$$(38)^2 = (40 - 2)^2 = 40^2 - 2(40)(2) + 2^2 = 1600 - 160 + 4 = 1444$$

$$(67)^2 = (70 - 3)^2 = 70^2 - 2(70)(3) + 3^2 = 4900 - 420 + 9 = 4489$$

d. Computing:

$$(15)^2 = (10+5)^2 = 100(1)(2)+25 = 200+25 = 225$$

$$(35)^2 = (30+5)^2 = 100(3)(4)+25 = 1200+25 = 1225$$

$$(45)^2 = (40+5)^2 = 100(4)(5)+25 = 2000+25 = 2025$$

$$(65)^2 = (60+5)^2 = 100(6)(7)+25 = 4200+25 = 4225$$

$$(85)^2 = (80+5)^2 = 100(8)(9)+25 = 7200+25 = 7225$$

71. **a.** Solving the equation:

$$5(x-2)-(2x+3) = 4$$
$$5x-10-2x-3 = 4$$
$$3x-13 = 4$$
$$3x = 17$$
$$x = \frac{17}{3}$$

The solution set is $\left\{\frac{17}{3}\right\}$.

b. Solving the equation:

$$\frac{2x-1}{4} - \frac{3x+1}{5} = 0$$
$$20\left(\frac{2x-1}{4} - \frac{3x+1}{5}\right) = 20 \cdot 0$$
$$10x-5-12x-4 = 0$$
$$-2x = 9$$
$$x = -\frac{9}{2}$$

The solution set is $\left\{-\frac{9}{2}\right\}$.

c. Solving the equation:

$$\frac{4x+1}{3} = \frac{x-3}{2}$$
$$6\left(\frac{4x+1}{3}\right) = 6\left(\frac{x-3}{2}\right)$$
$$8x+2 = 3x-9$$
$$5x = -11$$
$$x = -\frac{11}{5}$$

The solution set is $\left\{-\frac{11}{5}\right\}$.

d. Solving the equation:

$$(x-2)(x+3)-(x-2)(x+1) = 0$$
$$(x^2+x-6)-(x^2-x-2) = 0$$
$$x^2+x-6-x^2+x+2 = 0$$
$$2x-4 = 0$$
$$2x = 4$$
$$x = 2$$

The solution set is $\{2\}$.

e. Solving the equation:

$$(x+1)(x-1)-(x+6)(x-2) = 0$$
$$(x^2-1)-(x^2+4x-12) = 0$$
$$x^2-1-x^2-4x+12 = 0$$
$$-4x+11 = 0$$
$$-4x = -11$$
$$x = \frac{11}{4}$$

The solution set is $\left\{\frac{11}{4}\right\}$.

f. Solving the equation:

$$(x+3)(x+2)-(x+4)(x+1) = 0$$
$$(x^2+5x+6)-(x^2+5x+4) = 0$$
$$x^2+5x+6-x^2-5x-4 = 0$$
$$2 = 0$$

Since this statement is false, there is no solution.

1.2 More Equations and Applications

1. Solving the equation:
$$\frac{x-2}{3} + \frac{x+1}{4} = \frac{1}{6}$$
$$12\left(\frac{x-2}{3} + \frac{x+1}{4}\right) = 12 \cdot \frac{1}{6}$$
$$4x - 8 + 3x + 3 = 2$$
$$7x - 5 = 2$$
$$7x = 7$$
$$x = 1$$
The solution set is $\{1\}$.

3. Solving the equation:
$$\frac{5}{x} + \frac{1}{3} = \frac{8}{x}$$
$$3x\left(\frac{5}{x} + \frac{1}{3}\right) = 3x \cdot \frac{8}{x}$$
$$15 + x = 24$$
$$x = 9$$
The solution set is $\{9\}$.

5. Solving the equation:
$$\frac{1}{3n} + \frac{1}{2n} = \frac{1}{4}$$
$$12n\left(\frac{1}{3n} + \frac{1}{2n}\right) = 12n \cdot \frac{1}{4}$$
$$4 + 6 = 3n$$
$$3n = 10$$
$$n = \frac{10}{3}$$
The solution set is $\left\{\frac{10}{3}\right\}$.

7. Solving the equation:
$$\frac{35-x}{x} = 7 + \frac{3}{x}$$
$$x\left(\frac{35-x}{x}\right) = x\left(7 + \frac{3}{x}\right)$$
$$35 - x = 7x + 3$$
$$-8x = -32$$
$$x = 4$$
The solution set is $\{4\}$.

9. Solving the equation:
$$\frac{n+67}{n} = 5 + \frac{11}{n}$$
$$n\left(\frac{n+67}{n}\right) = n\left(5 + \frac{11}{n}\right)$$
$$n + 67 = 5n + 11$$
$$-4n = -56$$
$$n = 14$$
The solution set is $\{14\}$.

11. Solving the equation:
$$\frac{5}{3x-2} = \frac{1}{x-4}$$
$$(3x-2)(x-4)\left(\frac{5}{3x-2}\right) = (3x-2)(x-4)\left(\frac{1}{x-4}\right)$$
$$5x - 20 = 3x - 2$$
$$2x = 18$$
$$x = 9$$
The solution set is $\{9\}$.

13. Solving the equation:

$$\frac{4}{2y-3} - \frac{7}{3y-5} = 0$$

$$(2y-3)(3y-5)\left(\frac{4}{2y-3} - \frac{7}{3y-5}\right) = (2y-3)(3y-5) \cdot 0$$

$$12y - 20 - 14y + 21 = 0$$

$$-2y + 1 = 0$$

$$-2y = -1$$

$$y = \frac{1}{2}$$

The solution set is $\left\{\frac{1}{2}\right\}$.

15. Solving the equation:

$$\frac{n}{n+1} + 3 = \frac{4}{n+1}$$

$$(n+1)\left(\frac{n}{n+1} + 3\right) = (n+1)\left(\frac{4}{n+1}\right)$$

$$n + 3(n+1) = 4$$

$$n + 3n + 3 = 4$$

$$4n = 1$$

$$n = \frac{1}{4}$$

The solution set is $\left\{\frac{1}{4}\right\}$.

17. Solving the equation:

$$\frac{3x}{2x-1} - 4 = \frac{x}{2x-1}$$

$$(2x-1)\left(\frac{3x}{2x-1} - 4\right) = (2x-1)\left(\frac{x}{2x-1}\right)$$

$$3x - 8x + 4 = x$$

$$-5x + 4 = x$$

$$-6x = -4$$

$$x = \frac{2}{3}$$

The solution set is $\left\{\frac{2}{3}\right\}$.

19. Solving the equation:

$$\frac{3}{x+3} - \frac{1}{x-2} = \frac{5}{2x+6}$$

$$2(x+3)(x-2)\left(\frac{3}{x+3} - \frac{1}{x-2}\right) = 2(x+3)(x-2)\left(\frac{5}{2(x+3)}\right)$$

$$6x - 12 - 2x - 6 = 5x - 10$$

$$4x - 18 = 5x - 10$$

$$-x = 8$$

$$x = -8$$

The solution set is $\{-8\}$.

21. Solving the equation:

$$\frac{n}{n-3} - \frac{3}{2} = \frac{3}{n-3}$$

$$2(n-3)\left(\frac{n}{n-3} - \frac{3}{2}\right) = 2(n-3)\left(\frac{3}{n-3}\right)$$

$$2n - 3n + 9 = 6$$

$$-n + 9 = 6$$

$$-n = -3$$

$$n = 3 \quad (\text{does not check})$$

There is no solution, or $\varnothing$.

23. Solving the equation:

$$s = 9 + 0.25s$$
$$0.75s = 9$$
$$s = 12$$

The solution set is $\{12\}$.

25. Solving the equation:

$$0.09x + 0.1(700 - x) = 67$$
$$100[0.09x + 0.1(700 - x)] = 6700$$
$$9x + 7000 - 10x = 6700$$
$$-x + 7000 = 6700$$
$$-x = -300$$
$$x = 300$$

The solution set is $\{300\}$.

27. Solving the equation:

$$0.09x + 0.11(x + 125) = 68.75$$
$$100[0.09x + 0.11(x + 125)] = 6875$$
$$9x + 11x + 1375 = 6875$$
$$20x + 1375 = 6875$$
$$20x = 5500$$
$$x = 275$$

The solution set is $\{275\}$.

29. Solving the equation:

$$0.8(t - 2) = 0.5(9t + 10)$$
$$10[0.8(t - 2)] = 10[0.5(9t + 10)]$$
$$8t - 16 = 45t + 50$$
$$-37t = 66$$
$$t = -\frac{66}{37}$$

The solution set is $\left\{-\frac{66}{37}\right\}$.

31. Solving the equation:

$$0.92 + 0.9(x - 0.3) = 2x - 5.95$$
$$100[0.92 + 0.9(x - 0.3)] = 100[2x - 5.95]$$
$$92 + 90x - 27 = 200x - 595$$
$$90x + 65 = 200x - 595$$
$$-110x = -660$$
$$x = 6$$

The solution set is $\{6\}$.

33. Solving for w:

$$P = 2l + 2w$$
$$P - 2l = 2w$$
$$w = \frac{P - 2l}{2}$$

35. Solving for h:

$$A = 2lw + 2lh + 2wh$$
$$A - 2lw = 2lh + 2wh$$
$$A - 2lw = h(2l + 2w)$$
$$h = \frac{A - 2lw}{2l + 2w}$$

37. Solving for h:

$$A = 2\pi r^2 + 2\pi rh$$
$$A - 2\pi r^2 = 2\pi rh$$
$$h = \frac{A - 2\pi r^2}{2\pi r}$$

39. Solving for F:

$$C = \frac{5}{9}(F - 32)$$
$$9C = 5F - 160$$
$$5F = 9C + 160$$
$$F = \frac{9C + 160}{5} = \frac{9}{5}C + 32$$

41. Solving for T:

$$V = C\left(1 - \frac{T}{N}\right)$$
$$NV = C(N - T)$$
$$NV = CN - CT$$
$$CT = NC - NV$$
$$T = \frac{NC - NV}{C}$$

43. Solving for T:

$$I = kl(T - t)$$
$$I = klT - klt$$
$$klT = I + klt$$
$$T = \frac{I + klt}{kl}$$

45. Solving for R_n:

$$\frac{1}{R_n} = \frac{1}{R_1} + \frac{1}{R_2}$$
$$\frac{1}{R_n} = \frac{R_1 + R_2}{R_1 R_2}$$
$$R_n = \frac{R_1 R_2}{R_1 + R_2}$$

47.   Let C represent the cost of all the meals. The equation is:
$$0.15C = 157.50$$
$$C = 1,050$$
The cost of all the meals was $1,050.

49.   Let $2x$ and $3x$ represent the amount each painter received. The equation is:
$$2x + 3x = 2250$$
$$5x = 2250$$
$$x = 450$$
The painters received $900 and $1,350.

51.   Let $8x$ and x represent the number of two-wheel-drive and four-wheel-drive trucks, respectively. The equation is:
$$8x + x = 189$$
$$9x = 189$$
$$x = 21$$
$$8x = 168$$
There are 168 two-wheel-drive trucks and 21 four-wheel-drive trucks at the dealership.

53.   Let x represent the original price of the mp3 player. The equation is:
$$x - 0.20x = 52$$
$$0.8x = 52$$
$$x = 65$$
The original price of the mp3 player was $65.

55.   Let x represent Laurie's original salary. The equation is:
$$x + 0.07x = 1,647.80$$
$$1.07x = 1,647.80$$
$$x = 1,540$$
Her original salary was $1,540 per month.

57.   The price is: $28 + 0.75(28) = \$49.00$

59.   Let x represent the original price Karla paid. The equation is:
$$x + 0.30x = 97.50$$
$$1.3x = 97.5$$
$$x = 75$$
The original price of the bike was $75.

61.   Let x represent the selling price. The equation is:
$$18 + 0.40x = x$$
$$18 = 0.6x$$
$$x = 30$$
She should charge $30 for the skirts.

63.   Let n represent the number of nickels and $2n + 1$ represent the number of dimes. The equation is:
$$0.05n + 0.10(2n + 1) = 3.60$$
$$100\left[0.05n + 0.10(2n + 1)\right] = 100\left[3.60\right]$$
$$5n + 20n + 10 = 360$$
$$25n = 350$$
$$n = 14$$
Derek has 14 nickels and 29 dimes.

65.   Let d represent the number of dimes, $3d$ the number of quarters, and $70 - 4d$ the number of half-dollars. The equation is:
$$0.10d + 0.25(3d) + 0.50(70 - 4d) = 17.75$$
$$100\left[0.10d + 0.25(3d) + 0.50(70 - 4d)\right] = 100\left[17.75\right]$$
$$10d + 75d + 3500 - 200d = 1775$$
$$-115d + 3500 = 1775$$
$$-115d = -1725$$
$$d = 15$$
There are 15 dimes, 45 quarters, and 10 half-dollars in the collection.

67. Let x represent the amount invested at 4% and $5500 - x$ invested at 6%. The equation is:

$$0.04x + 0.06(5,500 - x) = 290$$
$$100[0.04x + 0.06(5,500 - x)] = 100[290]$$
$$4x + 33,000 - 6x = 29,000$$
$$-2x = -4,000$$
$$x = 2,000$$
$$5,500 - x = 3,500$$

There is $2,000 invested at 4% and $3,500 invested at 6%.

69. Let x represent the amount invested at 5%. The equation is:

$$0.02(2,500) + 0.05x = 250$$
$$50 + 0.05x = 250$$
$$0.05x = 200$$
$$x = 4,000$$

Celia must invest $4,000 at 5% interest.

71. Let w represent the width and $w + 4$ represent the length of the original rectangle, so $w + 2$ and $w + 7$ represent the width and length of the larger rectangle. Since the difference of their areas is 44 square centimeters, the equation is:

$$(w + 2)(w + 7) - w(w + 4) = 44$$
$$w^2 + 9w + 14 - w^2 - 4w = 44$$
$$5w + 14 = 44$$
$$5w = 30$$
$$w = 6$$

The width is 6 cm and the length is 10 cm for the original rectangle.

73. Let x represent the lengths of sides of the original square. Then the new rectangle has length $x + 3$ and width $x - 2$. Finding the area:

$$(x + 3)(x - 2) - x^2 = 8$$
$$x^2 + x - 6 - x^2 = 8$$
$$x - 6 = 8$$
$$x = 14$$

The sides of the square have a length of 14 centimeters.

79. No. Suppose the item is originally priced at $100. A 10% discount results in a price of $90, then a 20% discount results in a price of $72. Since a 30% discount results in a price of $70, the two discounts are not the same.

81. Partially his claim is correct. Based on the selling price, his profit is 10%. However, based on the cost (which is usually used), his profit is 11.1%.

83. **a.** Finding the selling price: $S = \dfrac{0.80}{1 - 0.20} = \$1.00$ **b.** Finding the selling price: $S = \dfrac{8.50}{1 - 0.25} \approx \$11.33$

c. Finding the selling price: $S = \dfrac{50}{1 - 0.40} \approx \$83.33$ **d.** Finding the selling price: $S = \dfrac{200}{1 - 0.50} = \$400$

e. Finding the selling price: $S = \dfrac{18000}{1 - 0.15} \approx \$21,176$

1.3 Quadratic Equations

1. Solving the equation:
$$x^2 - 3x - 28 = 0$$
$$(x - 7)(x + 4) = 0$$
$$x = -4, 7$$

The solution set is $\{-4, 7\}$.

3. Solving the equation:
$$3x^2 + 5x - 12 = 0$$
$$(3x - 4)(x + 3) = 0$$
$$x = -3, \frac{4}{3}$$

The solution set is $\left\{-3, \frac{4}{3}\right\}$.

5. Solving the equation:

$$2x^2 - 3x = 0$$
$$x(2x - 3) = 0$$
$$x = 0, \frac{3}{2}$$

The solution set is $\left\{0, \frac{3}{2}\right\}$.

7. Solving the equation:
$$9y^2 = 12$$
$$y^2 = \frac{12}{9}$$
$$y = \pm\sqrt{\frac{12}{9}}$$
$$y = \pm\frac{2\sqrt{3}}{3}$$

The solution set is $\left\{\pm\frac{2\sqrt{3}}{3}\right\}$.

9. Solving the equation:
$$(2n + 1)^2 = 20$$
$$2n + 1 = \pm\sqrt{20}$$
$$2n + 1 = \pm2\sqrt{5}$$
$$2n = -1 \pm 2\sqrt{5}$$
$$n = \frac{-1 \pm 2\sqrt{5}}{2}$$

The solution set is $\left\{\frac{-1 \pm 2\sqrt{5}}{2}\right\}$.

11. Solving the equation:

$$15n^2 + 19n - 10 = 0$$
$$(5n - 2)(3n + 5) = 0$$
$$n = -\frac{5}{3}, \frac{2}{5}$$

The solution set is $\left\{-\frac{5}{3}, \frac{2}{5}\right\}$.

13. Solving the equation:
$$(x - 2)^2 = -4$$
$$x - 2 = \pm\sqrt{-4}$$
$$x - 2 = \pm2i$$
$$x = 2 \pm 2i$$

The solution set is $\{2 \pm 2i\}$.

15. Solving the equation:
$$10y^2 + 33y - 7 = 0$$
$$(5y - 1)(2y + 7) = 0$$
$$y = -\frac{7}{2}, \frac{1}{5}$$

The solution set is $\left\{-\frac{7}{2}, \frac{1}{5}\right\}$.

17. Solving the equation:
$$2(x + 3)^2 + 8 = -2$$
$$2(x + 3)^2 = -10$$
$$(x + 3)^2 = -5$$
$$x + 3 = \pm\sqrt{-5}$$
$$x + 3 = \pm i\sqrt{5}$$
$$x = -3 \pm i\sqrt{5}$$

19. Solving by completing the square:
$$x^2 - 10x + 24 = 0$$
$$x^2 - 10x = -24$$
$$x^2 - 10x + 25 = -24 + 25$$
$$(x - 5)^2 = 1$$
$$x - 5 = \pm1$$
$$x = 4, 6$$

The solution set is $\{4, 6\}$.

This checks using the sum-and-product of roots.

21. Solving by completing the square:

$$n^2 + 10n - 2 = 0$$
$$n^2 + 10n = 2$$
$$n^2 + 10n + 25 = 2 + 25$$
$$(n+5)^2 = 27$$
$$n + 5 = \pm\sqrt{27} = \pm 3\sqrt{3}$$
$$n = -5 \pm 3\sqrt{3}$$

The solution set is $\left\{-5 \pm 3\sqrt{3}\right\}$.

This checks using the sum-and-product of roots.

25. Solving by completing the square:

$$x^2 + 4x + 6 = 0$$
$$x^2 + 4x = -6$$
$$x^2 + 4x + 4 = -6 + 4$$
$$(x+2)^2 = -2$$
$$x + 2 = \pm\sqrt{-2} = \pm i\sqrt{2}$$
$$x = -2 \pm i\sqrt{2}$$

The solution set is $\left\{-2 \pm i\sqrt{2}\right\}$.

This checks using the sum-and-product of roots.

29. Solving by completing the square:

$$x(x-2) = 288$$
$$x^2 - 2x - 288 = 0$$
$$x^2 - 2x = 288$$
$$x^2 - 2x + 1 = 288 + 1$$
$$(x-1)^2 = 289$$
$$x - 1 = \pm\sqrt{289} = \pm 17$$
$$x = -16, 18$$

The solution set is $\left\{-16, 18\right\}$.

This checks using the sum-and-product of roots.

23. Solving by completing the square:

$$y^2 - 3y = -1$$
$$y^2 - 3y + \frac{9}{4} = -1 + \frac{9}{4}$$
$$\left(y - \frac{3}{2}\right)^2 = \frac{5}{4}$$
$$y - \frac{3}{2} = \pm\sqrt{\frac{5}{4}} = \pm\frac{\sqrt{5}}{2}$$
$$y = \frac{3 \pm \sqrt{5}}{2}$$

The solution set is $\left\{\dfrac{3 \pm \sqrt{5}}{2}\right\}$.

This checks using the sum-and-product of roots.

27. Solving by completing the square:

$$2t^2 + 12t - 5 = 0$$
$$t^2 + 6t = \frac{5}{2}$$
$$t^2 + 6t + 9 = \frac{5}{2} + 9$$
$$(t+3)^2 = \frac{23}{2}$$
$$t + 3 = \pm\sqrt{\frac{23}{2}} \cdot \frac{\sqrt{2}}{\sqrt{2}} = \pm\frac{\sqrt{46}}{2}$$
$$t = -3 \pm \frac{\sqrt{46}}{2} = \frac{-6 \pm \sqrt{46}}{2}$$

The solution set is $\left\{\dfrac{-6 \pm \sqrt{46}}{2}\right\}$.

This checks using the sum-and-product of roots.

31. Solving by completing the square:

$$3n^2 + 5n - 1 = 0$$
$$n^2 + \frac{5}{3}n = \frac{1}{3}$$
$$n^2 + \frac{5}{3}n + \frac{25}{36} = \frac{1}{3} + \frac{25}{36}$$
$$\left(n + \frac{5}{6}\right)^2 = \frac{37}{36}$$
$$n + \frac{5}{6} = \pm\sqrt{\frac{37}{36}} = \pm\frac{\sqrt{37}}{6}$$
$$n = \frac{-5 \pm \sqrt{37}}{6}$$

The solution set is $\left\{\dfrac{-5 \pm \sqrt{37}}{6}\right\}$.

This checks using the sum-and-product of roots.

33. Let $a = 1$, $b = -3$, and $c = -54$ in the quadratic formula:

$$n = \frac{-(-3) \pm \sqrt{(-3)^2 - 4(1)(-54)}}{2(1)} = \frac{3 \pm \sqrt{225}}{2} = \frac{3 \pm 15}{2} = -6, 9$$

The solution set is $\left\{-6, 9\right\}$.

35. Write the equation as $3x^2 + 16x + 5 = 0$. Let $a = 3$, $b = 16$, and $c = 5$ in the quadratic formula:

$$x = \frac{-16 \pm \sqrt{(16)^2 - 4(3)(5)}}{2(3)} = \frac{-16 \pm \sqrt{196}}{6} = \frac{-16 \pm 14}{6} = -5, -\frac{1}{3}$$

The solution set is $\left\{-5, -\frac{1}{3}\right\}$.

37. Let $a = 1$, $b = -2$, and $c = -4$ in the quadratic formula: $x = \frac{2 \pm \sqrt{(-2)^2 - 4(1)(-4)}}{2(1)} = \frac{2 \pm \sqrt{20}}{2} = \frac{2 \pm 2\sqrt{5}}{2} = 1 \pm \sqrt{5}$

The solution set is $\left\{1 \pm \sqrt{5}\right\}$.

39. Write the equation as $2a^2 - 6a + 1 = 0$. Let $a = 2$, $b = -6$, and $c = 1$ in the quadratic formula:

$$x = \frac{6 \pm \sqrt{(-6)^2 - 4(2)(1)}}{2(2)} = \frac{6 \pm \sqrt{28}}{4} = \frac{6 \pm 2\sqrt{7}}{4} = \frac{3 \pm \sqrt{7}}{2}$$

The solution set is $\left\{\frac{3 \pm \sqrt{7}}{2}\right\}$.

41. Write the equation as $n^2 - 3n + 7 = 0$. Let $a = 1$, $b = -3$, and $c = 7$ in the quadratic formula:

$$n = \frac{3 \pm \sqrt{(-3)^2 - 4(1)(7)}}{2(1)} = \frac{3 \pm \sqrt{-19}}{2} = \frac{3 \pm i\sqrt{19}}{2}$$

The solution set is $\left\{\frac{3 \pm i\sqrt{19}}{2}\right\}$.

43. Write the equation as $x^2 - 8x + 4 = 0$. Let $a = 1$, $b = -8$, and $c = 4$ in the quadratic formula:

$$x = \frac{8 \pm \sqrt{(-8)^2 - 4(1)(4)}}{2(1)} = \frac{8 \pm \sqrt{48}}{2} = \frac{8 \pm 4\sqrt{3}}{2} = 4 \pm 2\sqrt{3}$$

The solution set is $\left\{4 \pm 2\sqrt{3}\right\}$.

45. Let $a = 4$, $b = -4$, and $c = 1$ in the quadratic formula: $x = \frac{4 \pm \sqrt{(-4)^2 - 4(4)(1)}}{2(4)} = \frac{4 \pm \sqrt{0}}{8} = \frac{1}{2}$

The solution set is $\left\{\frac{1}{2}\right\}$.

47. Solving by factoring:

$$8x^2 + 10x - 3 = 0$$
$$(4x - 1)(2x + 3) = 0$$
$$x = -\frac{3}{2}, \frac{1}{4}$$

The solution set is $\left\{-\frac{3}{2}, \frac{1}{4}\right\}$.

49. Solving by factoring:

$$x^2 + 2x = 168$$
$$x^2 + 2x - 168 = 0$$
$$(x + 14)(x - 12) = 0$$
$$x = -14, 12$$

The solution set is $\left\{-14, 12\right\}$.

51. Let $a = 2$, $b = -3$, and $c = 7$ in the quadratic formula: $x = \frac{3 \pm \sqrt{(-3)^2 - 4(2)(7)}}{2(2)} = \frac{3 \pm \sqrt{-47}}{4} = \frac{3 \pm i\sqrt{47}}{4}$

The solution set is $\left\{\frac{3 \pm i\sqrt{47}}{4}\right\}$.

53. Solving by taking square roots:
$$(3n - 1)^2 + 2 = 18$$
$$(3n - 1)^2 = 16$$
$$3n - 1 = \pm\sqrt{16}$$
$$3n - 1 = -4, 4$$
$$3n = -3, 5$$
$$n = -1, \frac{5}{3}$$

The solution set is $\left\{-1, \frac{5}{3}\right\}$.

55. Write the equation as $4y^2 + 4y - 1 = 0$. Let $a = 4$, $b = 4$, and $c = -1$ in the quadratic formula:

$$x = \frac{-4 \pm \sqrt{4^2 - 4(4)(-1)}}{2(4)} = \frac{-4 \pm \sqrt{32}}{8} = \frac{-4 \pm 4\sqrt{2}}{8} = \frac{-1 \pm \sqrt{2}}{2}$$

The solution set is $\left\{\dfrac{-1 \pm \sqrt{2}}{2}\right\}$.

57. Solving by completing the square:
$$x^2 - 16x + 14 = 0$$
$$x^2 - 16x = -14$$
$$x^2 - 16x + 64 = -14 + 64$$
$$(x - 8)^2 = 50$$
$$x - 8 = \pm\sqrt{50} = \pm 5\sqrt{2}$$
$$x = 8 \pm 5\sqrt{2}$$

The solution set is $\left\{8 \pm 5\sqrt{2}\right\}$.

59. Solving by completing the square:
$$t^2 + 20t = 25$$
$$t^2 + 20t + 100 = 25 + 100$$
$$(t + 10)^2 = 125$$
$$t + 10 = \pm\sqrt{125}$$
$$t + 10 = \pm 5\sqrt{5}$$
$$t = -10 \pm 5\sqrt{5}$$

The solution set is $\left\{-10 \pm 5\sqrt{5}\right\}$.

61. Let $a = 5$, $b = -2$, and $c = -1$ in the quadratic formula: $x = \dfrac{2 \pm \sqrt{(-2)^2 - 4(5)(-1)}}{2(5)} = \dfrac{2 \pm \sqrt{24}}{10} = \dfrac{2 \pm 2\sqrt{6}}{10} = \dfrac{1 \pm \sqrt{6}}{5}$

The solution set is $\left\{\dfrac{1 \pm \sqrt{6}}{5}\right\}$.

63.
  a. Finding the discriminant: $20^2 - 4(4)(25) = 400 - 400 = 0$
  Since $D = 0$, the equation has two equal real solutions.
  b. Finding the discriminant: $4^2 - 4(1)(7) = 16 - 28 = -12$
  Since $D < 0$, the equation has two complex but nonreal solutions.
  c. Finding the discriminant: $(-18)^2 - 4(1)(81) = 324 - 324 = 0$
  Since $D = 0$, the equation has two equal real solutions.
  d. Finding the discriminant: $(-31)^2 - 4(36)(3) = 961 - 432 = 529$
  Since $D > 0$, the equation has two unequal real solutions.
  e. Finding the discriminant: $5^2 - 4(2)(7) = 25 - 56 = -31$
  Since $D < 0$, the equation has two complex but nonreal solutions.
  f. First write the equation as $16x^2 - 40x + 25 = 0$.
  Finding the discriminant: $(-40)^2 - 4(16)(25) = 1600 - 1600 = 0$
  Since $D = 0$, the equation has two equal real solutions.
  g. Finding the discriminant: $(-4)^2 - 4(6)(-7) = 16 + 168 = 184$
  Since $D > 0$, the equation has two unequal real solutions.

h.   Finding the discriminant: $(-2)^2 - 4(5)(-4) = 4 + 80 = 84$

Since $D > 0$, the equation has two unequal real solutions.

65.   Let x and $x + 1$ represent the whole numbers. The equation is:

$$x^2 + (x+1)^2 = 265$$
$$x^2 + x^2 + 2x + 1 = 265$$
$$2x^2 + 2x - 264 = 0$$
$$x^2 + x - 132 = 0$$
$$(x+12)(x-11) = 0$$
$$x = 11 \quad (x = -12 \text{ is not a whole number})$$

The numbers are 11 and 12.

67.   Let b represent the base of the sail, and $b + 7$ represent the height. Using the area formula:

$$\frac{1}{2}b(b+7) = 30$$
$$b^2 + 7b = 60$$
$$b^2 + 7b - 60 = 0$$
$$(b+12)(b-5) = 0$$
$$b = 5 \quad (b = -12 \text{ is impossible})$$

The base of the sail is 5 feet, so the height is 12 feet.

69.   Let x and $34 - x$ represent the legs of the triangle. Using the Pythagorean theorem:

$$x^2 + (34-x)^2 = 26^2$$
$$x^2 + 1156 - 68x + x^2 = 676$$
$$2x^2 - 68x + 480 = 0$$
$$x^2 - 34x + 240 = 0$$
$$(x-24)(x-10) = 0$$
$$x = 10, 24$$

The legs are 10 meters and 24 meters.

71.   Let w and l represent the width and length, respectively. The perimeter is:

$$2w + 2l = 44$$
$$w + l = 22$$
$$l = 22 - w$$

Using the area formula:

$$w(22-w) = 112$$
$$22w - w^2 = 112$$
$$0 = w^2 - 22w + 112$$
$$0 = (w-8)(w-14)$$
$$w = 8, 14$$

The width is 8 inches and the length is 14 inches.

73.   Let w represent the width and $2w + 4$ represent the length. Using the area formula:

$$w(2w+4) = 126$$
$$2w^2 + 4w = 126$$
$$2w^2 + 4w - 126 = 0$$
$$w^2 + 2w - 63 = 0$$
$$(w+9)(w-7) = 0$$
$$w = 7 \quad (w = -9 \text{ is impossible})$$

The width is 7 meters and the length is 18 meters.

75. Let w represent the width of the sidewalk, so $12 + 2w$ and $20 + 2w$ represent the dimensions of the larger area. Since the area of the sidewalk is the difference of the two rectangular areas:

$$(12 + 2w)(20 + 2w) - 12 \cdot 20 = 68$$
$$240 + 64w + 4w^2 - 240 = 68$$
$$4w^2 + 64w - 68 = 0$$
$$w^2 + 16w - 17 = 0$$
$$(w + 17)(w - 1) = 0$$
$$w = 1 \quad (w = -17 \text{ is impossible})$$

The width of the sidewalk is 1 meter.

77. Let w and $w + 4$ represent the width and length, respectively. Since 4 inches is removed from each side, the dimensions of the box are $w - 4$ by w by 2. Using the volume formula:

$$2w(w - 4) = 42$$
$$2w^2 - 8w = 42$$
$$w^2 - 4w - 21 = 0$$
$$(w - 7)(w + 3) = 0$$
$$w = 7 \quad (w = -3 \text{ is impossible})$$

The width is 7 inches and the length is 11 inches.

79. Let r represent the radius of the circle. The equation is:

$$\pi r^2 = 4 \cdot 2\pi r$$
$$\pi r^2 - 8\pi r = 0$$
$$\pi r(r - 8) = 0$$
$$r = 8 \quad (r = 0 \text{ is impossible})$$

The length of the radius is 8 units.

85. **a.** Solving for r:

$$A = \pi r^2$$
$$r^2 = \frac{A}{\pi}$$
$$r = \sqrt{\frac{A}{\pi}}$$
$$r = \frac{\sqrt{A\pi}}{\pi}$$

b. Solving for c:

$$E = c^2 m - c^2 m_0$$
$$E = c^2 (m - m_0)$$
$$c^2 = \frac{E}{m - m_0}$$
$$c = \sqrt{\frac{E}{m - m_0}}$$
$$c = \frac{\sqrt{E(m - m_0)}}{m - m_0}$$

c. Solving for t:

$$s = \frac{1}{2} g t^2$$
$$t^2 = \frac{2s}{g}$$
$$t = \sqrt{\frac{2s}{g}}$$
$$t = \frac{\sqrt{2gs}}{g}$$

d. Solving for x:

$$\frac{x^2}{a^2} + \frac{y^2}{b^2} = 1$$
$$b^2 x^2 + a^2 y^2 = a^2 b^2$$
$$b^2 x^2 = a^2 b^2 - a^2 y^2$$
$$x^2 = \frac{a^2(b^2 - y^2)}{b^2}$$
$$x = \frac{a\sqrt{b^2 - y^2}}{b}$$

e. Solving for y:

$$\frac{x^2}{a^2} - \frac{y^2}{b^2} = 1$$

$$b^2 x^2 - a^2 y^2 = a^2 b^2$$

$$b^2 x^2 - a^2 b^2 = a^2 y^2$$

$$y^2 = \frac{b^2 \left(x^2 - a^2 \right)}{a^2}$$

$$y = \frac{b\sqrt{x^2 - a^2}}{a}$$

f. Solving for t:

$$s = \frac{1}{2} gt^2 + V_0 t$$

$$2s = gt^2 + 2V_0 t$$

$$gt^2 + 2V_0 t - 2s = 0$$

Let $a = g$, $b = 2V_0$, and $c = -2s$ in the quadratic formula:

$$t = \frac{-2V_0 \pm \sqrt{\left(2V_0\right)^2 - 4g(-2s)}}{2g} = \frac{-2V_0 \pm \sqrt{4V_0^2 + 8gs}}{2g} = \frac{-2V_0 \pm 2\sqrt{V_0^2 + 2gs}}{2g} = \frac{-V_0 + \sqrt{V_0^2 + 2gs}}{g}$$

Note that we chose the positive root here since $t > 0$.

87. Solving the equation:

$$(-k)^2 - 4(4)(1) = 0$$

$$k^2 - 16 = 0$$

$$k^2 = 16$$

$$k = \pm 4$$

89. **a.** Let $a = 1$, $b = -6$, and $c = -10$ in the quadratic formula: $x = \dfrac{6 \pm \sqrt{(-6)^2 - 4(1)(-10)}}{2(1)} = \dfrac{6 \pm \sqrt{76}}{2} \approx -1.359, 7.359$

b. Let $a = 1$, $b = -16$, and $c = -24$ in the quadratic formula:

$$x = \frac{16 \pm \sqrt{(-16)^2 - 4(1)(-24)}}{2(1)} = \frac{16 \pm \sqrt{352}}{2} \approx -1.381, 17.381$$

c. Let $a = 1$, $b = 6$, and $c = -44$ in the quadratic formula:

$$x = \frac{-6 \pm \sqrt{(6)^2 - 4(1)(-44)}}{2(1)} = \frac{-6 \pm \sqrt{212}}{2} \approx -10.280, 4.280$$

d. Let $a = 1$, $b = 10$, and $c = -46$ in the quadratic formula:

$$x = \frac{-10 \pm \sqrt{(10)^2 - 4(1)(-46)}}{2(1)} = \frac{-10 \pm \sqrt{284}}{2} \approx -13.426, 3.426$$

e. Let $a = 1$, $b = 8$, and $c = 2$ in the quadratic formula: $x = \dfrac{-8 \pm \sqrt{(8)^2 - 4(1)(2)}}{2(1)} = \dfrac{-8 \pm \sqrt{56}}{2} \approx -7.742, -0.258$

f. Let $a = 1$, $b = 9$, and $c = 3$ in the quadratic formula: $x = \dfrac{-9 \pm \sqrt{(9)^2 - 4(1)(3)}}{2(1)} = \dfrac{-9 \pm \sqrt{69}}{2} \approx -8.653, -0.347$

g.   Let $a = 4$, $b = -6$, and $c = 1$ in the quadratic formula: $x = \dfrac{6 \pm \sqrt{(-6)^2 - 4(4)(1)}}{2(4)} = \dfrac{6 \pm \sqrt{20}}{8} \approx 0.191, 1.309$

h.   Let $a = 5$, $b = -9$, and $c = 1$ in the quadratic formula: $x = \dfrac{9 \pm \sqrt{(-9)^2 - 4(5)(1)}}{2(5)} = \dfrac{9 \pm \sqrt{61}}{10} \approx 0.119, 1.681$

i.   Let $a = 2$, $b = -11$, and $c = -5$ in the quadratic formula:

$$x = \dfrac{11 \pm \sqrt{(-11)^2 - 4(2)(-5)}}{2(2)} = \dfrac{11 \pm \sqrt{161}}{4} \approx -0.422, 5.922$$

j.   Let $a = 3$, $b = -12$, and $c = -10$ in the quadratic formula:

$$x = \dfrac{12 \pm \sqrt{(-12)^2 - 4(3)(-10)}}{2(3)} = \dfrac{12 \pm \sqrt{264}}{6} \approx -0.708, 4.708$$

91. **a.**   Let $a = 4$, $b = -12$, and $c = 9$ in the quadratic formula: $x = \dfrac{12 \pm \sqrt{(-12)^2 - 4(4)(9)}}{2(4)} = \dfrac{12 \pm \sqrt{0}}{8} = 1.5$

b.   Let $a = 2$, $b = -1$, and $c = 2$ in the quadratic formula: $x = \dfrac{1 \pm \sqrt{(-1)^2 - 4(2)(2)}}{2(2)} = \dfrac{1 \pm \sqrt{-15}}{4}$, which are not real

c.   Let $a = -3$, $b = 2$, and $c = -4$ in the quadratic formula:

$$x = \dfrac{-2 \pm \sqrt{(2)^2 - 4(-3)(-4)}}{2(-3)} = \dfrac{-2 \pm \sqrt{-44}}{-6}$$, which are not real

d.   Let $a = 2$, $b = -23$, and $c = -12$ in the quadratic formula:

$$x = \dfrac{23 \pm \sqrt{(-23)^2 - 4(2)(-12)}}{2(2)} = \dfrac{23 \pm \sqrt{625}}{4} = \dfrac{23 \pm 25}{4} = -0.5, 12$$

e.   Let $a = 1$, $b = 2\sqrt{5}$, and $c = 5$ in the quadratic formula:

$$x = \dfrac{-2\sqrt{5} \pm \sqrt{\left(2\sqrt{5}\right)^2 - 4(1)(5)}}{2(1)} = \dfrac{-2\sqrt{5} \pm \sqrt{0}}{2} = -\sqrt{5} \approx -2.236$$

f.   Let $a = -1$, $b = 2\sqrt{3}$, and $c = -3$ in the quadratic formula:

$$x = \dfrac{-2\sqrt{3} \pm \sqrt{\left(2\sqrt{3}\right)^2 - 4(-1)(-3)}}{2(-1)} = \dfrac{-2\sqrt{3} \pm \sqrt{0}}{-2} = -\sqrt{3} \approx -1.732$$

1.4 Applications of Linear and Quadratic Equations

1.   The LCM is $2(x+4)(x-4)$. Multiplying by the LCM:

$$2(x+4)(x-4) \cdot \dfrac{x}{2(x-4)} + 2(x+4)(x-4) \cdot \dfrac{16}{(x+4)(x-4)} = 2(x+4)(x-4) \cdot \dfrac{1}{2}$$
$$x(x+4) + 32 = (x+4)(x-4)$$
$$x^2 + 4x + 32 = x^2 - 16$$
$$4x = -48$$
$$x = -12$$

The solution set is $\{-12\}$.

3.   The LCM is $2(t+3)(t-3)$. Multiplying by the LCM:

$$2(t+3)(t-3)\cdot\frac{5t}{2(t+3)} - 2(t+3)(t-3)\cdot\frac{4}{(t+3)(t-3)} = 2(t+3)(t-3)\cdot\frac{5}{2}$$
$$5t(t-3)-8 = 5(t+3)(t-3)$$
$$5t^2-15t-8 = 5t^2-45$$
$$-15t = -37$$
$$t = \frac{37}{15}$$

The solution set is $\left\{\dfrac{37}{15}\right\}$.

5.   The LCM is $n(n-2)$. Multiplying by the LCM:

$$n(n-2)\cdot 2 + n(n-2)\cdot\frac{4}{n-2} = n(n-2)\cdot\frac{8}{n(n-2)}$$
$$2n^2-4n+4n = 8$$
$$2n^2 = 8$$
$$n^2 = 4$$
$$n = -2 \quad (n=2 \text{ does not check})$$

The solution set is $\{-2\}$.

7.   The LCM is $(a+2)(a+4)$. Multiplying by the LCM:

$$(a+2)(a+4)\cdot\frac{a}{a+2} + (a+2)(a+4)\cdot\frac{3}{a+4} = (a+2)(a+4)\cdot\frac{14}{(a+2)(a+4)}$$
$$a(a+4)+3(a+2) = 14$$
$$a^2+4a+3a+6 = 14$$
$$a^2+7a-8 = 0$$
$$(a+8)(a-1) = 0$$
$$a = -8,1$$

The solution set is $\{-8,1\}$.

9.   The LCM is $4(3x+2)(3x-2)$. Multiplying by the LCM:

$$4(3x+2)(3x-2)\cdot\frac{-2}{3x+2} + 4(3x+2)(3x-2)\cdot\frac{x-1}{(3x+2)(3x-2)} = 4(3x+2)(3x-2)\cdot\frac{3}{4(3x-2)}$$
$$-8(3x-2)+4(x-1) = 3(3x+2)$$
$$-24x+16+4x-4 = 9x+6$$
$$-20x+12 = 9x+6$$
$$-29x = -6$$
$$x = \frac{6}{29}$$

The solution set is $\left\{\dfrac{6}{29}\right\}$.

11.   The LCM is $(n-3)(2n-3)$. Multiplying by the LCM:

$$(n-3)(2n-3)\cdot\frac{n}{2n-3} + (n-3)(2n-3)\cdot\frac{1}{n-3} = (n-3)(2n-3)\cdot\frac{n^2-n-3}{(n-3)(2n-3)}$$
$$n(n-3)+1(2n-3) = n^2-n-3$$
$$n^2-3n+2n-3 = n^2-n-3$$
$$n^2-n-3 = n^2-n-3$$

Since this last statement is an identity, any real number except $n = \dfrac{3}{2},3$ is a solution to the equation.

13. The LCM is $(3y+2)(3y-2)(y-2)$. Multiplying by the LCM:

$$\frac{(3y+2)(3y-2)(y-2)(3y+1)}{(3y+2)(y-2)} + \frac{9(3y+2)(3y-2)(y-2)}{(3y+2)(3y-2)} = \frac{(3y+2)(3y-2)(y-2)(2y-2)}{(3y-2)(y-2)}$$

$$(3y-2)(3y+1) + 9(y-2) = (3y+2)(2y-2)$$

$$9y^2 - 3y - 2 + 9y - 18 = 6y^2 - 2y - 4$$

$$3y^2 + 8y - 16 = 0$$

$$(3y-4)(y+4) = 0$$

$$y = -4, \frac{4}{3}$$

The solution set is $\left\{-4, \frac{4}{3}\right\}$.

15. The LCM is $(2x-1)(x+4)(x-3)$. Multiplying by the LCM:

$$\frac{(2x-1)(x+4)(x-3)(x+1)}{(2x-1)(x+4)} - \frac{x(2x-1)(x+4)(x-3)}{(2x-1)(x-3)} = \frac{(2x-1)(x+4)(x-3)}{(x+4)(x-3)}$$

$$(x-3)(x+1) - x(x+4) = 2x-1$$

$$x^2 - 2x - 3 - x^2 - 4x = 2x - 1$$

$$-6x - 3 = 2x - 1$$

$$-8x = 2$$

$$x = -\frac{1}{4}$$

The solution set is $\left\{-\frac{1}{4}\right\}$.

17. The LCM is $(4x-3)(3x+5)$. Multiplying by the LCM:

$$(4x-3)(3x+5) \cdot \frac{7x+2}{(4x-3)(3x+5)} - (4x-3)(3x+5) \cdot \frac{1}{3x+5} = (4x-3)(3x+5) \cdot \frac{2}{4x-3}$$

$$(7x+2) - 1(4x-3) = 2(3x+5)$$

$$7x + 2 - 4x + 3 = 6x + 10$$

$$-3x = 5$$

$$x = -\frac{5}{3} \quad \text{(which does not check)}$$

The solution set is empty, or $\varnothing$.

19. The LCM is $(x+2)(x+4)$. Multiplying by the LCM:

$$\frac{x(x+2)(x+4)}{x+2} - \frac{3(x+2)(x+4)}{x+4} = \frac{6(x+2)(x+4)}{(x+2)(x+4)}$$

$$x(x+4) - 3(x+2) = 6$$

$$x^2 + 4x - 3x - 6 = 6$$

$$x^2 + x - 12 = 0$$

$$(x+4)(x-3) = 0$$

$$x = 3 \quad (x = -4 \text{ results in a 0 denominator})$$

The solution set is $\{3\}$.

21.   Let x represent the number of rows and $2x - 4$ represent the number of trees in each row. The equation is:

$$x(2x - 4) = 126$$
$$2x^2 - 4x = 126$$
$$x^2 - 2x - 63 = 0$$
$$(x - 9)(x + 7) = 0$$
$$x = 9 \quad (x = -7 \text{ is impossible})$$

There are 9 rows and 14 trees in each row.

23.   Let t represent the time. Jill's distance is $50t$ while Russ' distance is $52t$, so the equation is:

$$50t + 52t = 459$$
$$102t = 459$$
$$t = \frac{9}{2} = 4\frac{1}{2}$$

It will take them $4\frac{1}{2}$ hours to meet.

25.   Let t represent her time riding out into the country, and $5\frac{5}{6} - t$ represent the time riding back. Since

the distance riding out is the same as the distance riding back, the equation is:

$$20t = 15\left(5\frac{5}{6} - t\right)$$
$$20t = \frac{175}{2} - 15t$$
$$35t = \frac{175}{2}$$
$$70t = 175$$
$$t = \frac{5}{2}$$

Thus the distance she rode out is $20\left(\dfrac{5}{2}\right) = 50$ miles.

27.   Let r represent the rate of the freight train, and $r + 20$ represent the rate of the express train.
Setting up the equation using time:

$$\frac{300}{r} = \frac{280}{r + 20} + 2$$
$$\frac{300r(r + 20)}{r} = \frac{280r(r + 20)}{r + 20} + 2r(r + 20)$$
$$300r + 6000 = 280r + 2r^2 + 40r$$
$$0 = 2r^2 + 20r - 6000$$
$$0 = r^2 + 10r - 3000$$
$$0 = (r + 60)(r - 50)$$
$$r = 50 \quad (r = -60 \text{ is impossible})$$

The freight train is traveling 50 mph and the express train is traveling 70 mph.

29.   Let x represent the amount of pure alcohol to add. The equation is:

$$0.40(6) + x = 0.60(6 + x)$$
$$2.4 + x = 3.6 + 0.6x$$
$$0.4x = 1.2$$
$$x = 3$$

Thus 3 liters of pure alcohol should be added.

31. Let x represent the amount of 50% alcohol and $10.5 - x$ represent the amount of 80% alcohol to use. The equation is:

$$0.50(x) + 0.80(10.5 - x) = 0.70(10.5)$$
$$0.5x + 8.4 - 0.8x = 7.35$$
$$-0.3x = -1.05$$
$$x = 3.5$$

The mixture contains 3.5 liters of 50% alcohol and 7 liters of 80% alcohol solutions.

33. Let x represent the amount that needs to be drained. The equation is:

$$0.40(10 - x) + x = 0.70(10)$$
$$4 - 0.4x + x = 7$$
$$0.6x = 3$$
$$x = 5$$

Thus 5 quarts should be drained and replaced with pure antifreeze.

35. Let t represent the time to fill the tank with both pipes open. The equation is:

$$\frac{1}{4}t + \frac{1}{6}t = 1$$
$$12 \cdot \frac{1}{4}t + 12 \cdot \frac{1}{6}t = 12 \cdot 1$$
$$3t + 2t = 12$$
$$5t = 12$$
$$t = 2\frac{2}{5}$$

It will take $2\frac{2}{5}$ hours to fill the tank with both pipes open.

37. Let t represent the time for the tank to fill. The equation is:

$$\frac{1}{10}t - \frac{1}{12}t = 1$$
$$60 \cdot \frac{1}{10}t - 60 \cdot \frac{1}{12}t = 60 \cdot 1$$
$$6t - 5t = 60$$
$$t = 60$$

The tank will overflow in 60 minutes.

39. Let t represent the time they work together. The equation is:

$$\frac{1}{20}(t + 5) + \frac{1}{30}t = 1$$
$$60 \cdot \frac{1}{20}(t + 5) + 60 \cdot \frac{1}{30}t = 60 \cdot 1$$
$$3t + 15 + 2t = 60$$
$$5t + 15 = 60$$
$$5t = 45$$
$$t = 9$$

They worked together for 9 hours.

41. Let t represent the time for the professor and $3t$ represent the time for the student assistant. The equation is:

$$\frac{1}{t}(2) + \frac{1}{3t}(2) = 1$$
$$3t \cdot \frac{2}{t} + 3t \cdot \frac{2}{3t} = 3t \cdot 1$$
$$6 + 2 = 3t$$
$$3t = 8$$
$$t = \frac{8}{3}$$

It would take the student assistant $3 \cdot \frac{8}{3} = 8$ hours to grade the tests alone.

43. Let x represent the amount of candy bars Angie bought, so $\dfrac{14}{x}$ represents their price. Setting up the equation with the reduced price:

$$\left(\frac{14}{x}-0.25\right)(x+1)=14$$
$$14-0.25x+\frac{14}{x}-0.25=14$$
$$13.75x-0.25x^2+14=14x$$
$$-0.25x^2-0.25x+14=0$$
$$x^2+x-56=0$$
$$(x+8)(x-7)=0$$
$$x=7 \quad (x=-8 \text{ is impossible})$$

Angie bought 7 candy bars.

45. Let x represent how long he anticipated to paint the house, so $\dfrac{480}{x}$ represents his anticipated hourly wage.

Setting up the equation:

$$\left(\frac{480}{x}-0.50\right)(x+4)=480$$
$$480-0.50x+\frac{1920}{x}-2=480$$
$$478x-0.5x^2+1920=480x$$
$$-0.5x^2-2x+1920=0$$
$$x^2+4x-3840=0$$
$$(x+64)(x-60)=0$$
$$x=60 \quad (x=-64 \text{ is impossible})$$

Todd anticipated it would take him 60 hours to paint the house.

1.5 Miscellaneous Equations

1. Solving the equation:
$$x^3+8=0$$
$$(x+2)(x^2-2x+4)=0$$

One solution is $x=-2$. Using the quadratic formula: $x=\dfrac{2\pm\sqrt{4-4(4)}}{2}=\dfrac{2\pm\sqrt{-12}}{2}=\dfrac{2\pm2i\sqrt{3}}{2}=1\pm i\sqrt{3}$

The solution set is $\left\{-2, 1\pm i\sqrt{3}\right\}$.

3. Solving the equation:
$$x^3=1$$
$$x^3-1=0$$
$$(x-1)(x^2+x+1)=0$$

One solution is $x=1$. Using the quadratic formula: $x=\dfrac{-1\pm\sqrt{1-4(1)}}{2}=\dfrac{-1\pm\sqrt{-3}}{2}=\dfrac{-1\pm i\sqrt{3}}{2}$

The solution set is $\left\{1, \dfrac{-1\pm i\sqrt{3}}{2}\right\}$.

5. Solving the equation:

$$x^3 + x^2 - 4x - 4 = 0$$
$$x^2(x+1) - 4(x+1) = 0$$
$$(x+1)(x^2 - 4) = 0$$
$$(x+1)(x+2)(x-2) = 0$$
$$x = -2, -1, 2$$

The solution set is $\{-2, -1, 2\}$.

7. Solving the equation:

$$2x^3 - 3x^2 + 2x - 3 = 0$$
$$x^2(2x-3) + 1(2x-3) = 0$$
$$(2x-3)(x^2+1) = 0$$
$$x = \frac{3}{2}, \pm\sqrt{-1} = \pm i$$

The solution set is $\left\{ \frac{3}{2}, \pm i \right\}$.

9. Solving the equation:

$$8x^5 + 10x^4 = 4x^3 + 5x^2$$
$$8x^5 + 10x^4 - 4x^3 - 5x^2 = 0$$
$$2x^4(4x+5) - x^2(4x+5) = 0$$
$$x^2(4x+5)(2x^2 - 1) = 0$$
$$x = -\frac{5}{4}, 0, \pm\sqrt{\frac{1}{2}} = \pm\frac{\sqrt{2}}{2}$$

The solution set is $\left\{ -\frac{5}{4}, 0, \pm\frac{\sqrt{2}}{2} \right\}$.

11. Solving the equation:

$$x^{3/2} = 4x$$
$$\left(x^{3/2}\right)^2 = (4x)^2$$
$$x^3 = 16x^2$$
$$x^3 - 16x^2 = 0$$
$$x^2(x-16) = 0$$
$$x = 0, 16$$

The solution set is $\{0, 16\}$.

13. Solving the equation:

$$n^{-2} = n^{-3}$$
$$\frac{1}{n^2} = \frac{1}{n^3}$$
$$n^3 \cdot \frac{1}{n^2} = n^3 \cdot \frac{1}{n^3}$$
$$n = 1$$

The solution set is $\{1\}$.

15. Solving the equation:

$$\sqrt{3x-2} = 4$$
$$\left(\sqrt{3x-2}\right)^2 = (4)^2$$
$$3x - 2 = 16$$
$$3x = 18$$
$$x = 6$$

The solution set is $\{6\}$.

17. Solving the equation:

$$\sqrt{3x-8} - \sqrt{x-2} = 0$$
$$\sqrt{3x-8} = \sqrt{x-2}$$
$$\left(\sqrt{3x-8}\right)^2 = \left(\sqrt{x-2}\right)^2$$
$$3x - 8 = x - 2$$
$$2x = 6$$
$$x = 3$$

The solution set is $\{3\}$.

19. Solving the equation:

$$\sqrt{4x-3} = -2$$
$$\left(\sqrt{4x-3}\right)^2 = (-2)^2$$
$$4x - 3 = 4$$
$$4x = 7$$
$$x = \frac{7}{4} \quad \text{(does not check)}$$

The solution set is empty, or $\varnothing$.

21. Solving the equation:

$$\sqrt{2n+3} - 2 = -1$$
$$\sqrt{2n+3} = 1$$
$$\left(\sqrt{2n+3}\right)^2 = (1)^2$$
$$2n + 3 = 1$$
$$2n = -2$$
$$n = -1$$

The solution set is $\{-1\}$.

23. Solving the equation:

$$\sqrt{4x-1}-3=2$$
$$\sqrt{4x-1}=5$$
$$\left(\sqrt{4x-1}\right)^2=(5)^2$$
$$4x-1=25$$
$$4x=26$$
$$x=\frac{13}{2}$$

The solution set is $\left\{\frac{13}{2}\right\}$.

25. Solving the equation:

$$\sqrt[3]{2x+3}+5=2$$
$$\sqrt[3]{2x+3}=-3$$
$$\left(\sqrt[3]{2x+3}\right)^3=(-3)^3$$
$$2x+3=-27$$
$$2x=-30$$
$$x=-15$$

The solution set is $\{-15\}$.

27. Solving the equation:

$$2\sqrt{n}+3=n$$
$$2\sqrt{n}=n-3$$
$$\left(2\sqrt{n}\right)^2=(n-3)^2$$
$$4n=n^2-6n+9$$
$$n^2-10n+9=0$$
$$(n-1)(n-9)=0$$
$$n=9 \quad (n=1 \text{ does not check})$$

The solution set is $\{9\}$.

29. Solving the equation:

$$\sqrt{3x-2}=3x-2$$
$$\left(\sqrt{3x-2}\right)^2=(3x-2)^2$$
$$3x-2=9x^2-12x+4$$
$$9x^2-15x+6=0$$
$$3x^2-5x+2=0$$
$$(3x-2)(x-1)=0$$
$$x=\frac{2}{3},1$$

The solution set is $\left\{\frac{2}{3},1\right\}$.

31. Solving the equation:

$$\sqrt{2t-1}+2=t$$
$$\sqrt{2t-1}=t-2$$
$$\left(\sqrt{2t-1}\right)^2=(t-2)^2$$
$$2t-1=t^2-4t+4$$
$$0=t^2-6t+5$$
$$0=(t-5)(t-1)$$
$$t=5 \quad (t=1 \text{ does not check})$$

The solution set is $\{5\}$.

33. Solving the equation:

$$\sqrt{x+2}-1=\sqrt{x-3}$$
$$\left(\sqrt{x+2}-1\right)^2=\left(\sqrt{x-3}\right)^2$$
$$x+2-2\sqrt{x+2}+1=x-3$$
$$x+3-2\sqrt{x+2}=x-3$$
$$-2\sqrt{x+2}=-6$$
$$\sqrt{x+2}=3$$
$$x+2=9$$
$$x=7$$

The solution set is $\{7\}$.

35. Solving the equation:

$$\sqrt{7n+23}-\sqrt{3n+7}=2$$
$$\sqrt{7n+23}=\sqrt{3n+7}+2$$
$$\left(\sqrt{7n+23}\right)^2=\left(\sqrt{3n+7}+2\right)^2$$
$$7n+23=3n+7+4\sqrt{3n+7}+4$$
$$7n+23=3n+11+4\sqrt{3n+7}$$
$$4n+12=4\sqrt{3n+7}$$
$$n+3=\sqrt{3n+7}$$
$$(n+3)^2=3n+7$$
$$n^2+6n+9=3n+7$$
$$n^2+3n+2=0$$
$$(n+2)(n+1)=0$$
$$n=-2,-1$$

The solution set is $\{-2,-1\}$.

37. Solving the equation:

$$\sqrt{3x+1}+\sqrt{2x+4}=3$$
$$\sqrt{3x+1}=3-\sqrt{2x+4}$$
$$\left(\sqrt{3x+1}\right)^2=\left(3-\sqrt{2x+4}\right)^2$$
$$3x+1=9-6\sqrt{2x+4}+2x+4$$
$$3x+1=2x+13-6\sqrt{2x+4}$$
$$x-12=-6\sqrt{2x+4}$$
$$(x-12)^2=36(2x+4)$$
$$x^2-24x+144=72x+144$$
$$x^2-96x=0$$
$$x(x-96)=0$$
$$x=0 \quad (x=96 \text{ does not check})$$

The solution set is $\{0\}$.

39. Solving the equation:

$$\sqrt{x-2} - \sqrt{2x-11} = \sqrt{x-5}$$

$$\left(\sqrt{x-2} - \sqrt{2x-11}\right)^2 = \left(\sqrt{x-5}\right)^2$$

$$x - 2 - 2\sqrt{x-2}\sqrt{2x-11} + 2x - 11 = x - 5$$

$$3x - 13 - 2\sqrt{x-2}\sqrt{2x-11} = x - 5$$

$$-2\sqrt{x-2}\sqrt{2x-11} = -2x + 8$$

$$\sqrt{x-2}\sqrt{2x-11} = x - 4$$

$$(x-2)(2x-11) = (x-4)^2$$

$$2x^2 - 15x + 22 = x^2 - 8x + 16$$

$$x^2 - 7x + 6 = 0$$

$$(x-6)(x-1) = 0$$

$$x = 6 \quad (x = 1 \text{ does not check})$$

The solution set is $\{6\}$.

41. Solving the equation:

$$\sqrt{1 + 2\sqrt{x}} = \sqrt{x+1}$$

$$\left(\sqrt{1 + 2\sqrt{x}}\right)^2 = \left(\sqrt{x+1}\right)^2$$

$$1 + 2\sqrt{x} = x + 1$$

$$2\sqrt{x} = x$$

$$\left(2\sqrt{x}\right)^2 = x^2$$

$$4x = x^2$$

$$0 = x^2 - 4x$$

$$0 = x(x-4)$$

$$x = 0, 4$$

The solution set is $\{0, 4\}$.

43. Solving the equation:

$$x^4 - 5x^2 + 4 = 0$$

$$\left(x^2 - 1\right)\left(x^2 - 4\right) = 0$$

$$x^2 = 1, 4$$

$$x = \pm 1, \pm 2$$

The solution set is $\{\pm 1, \pm 2\}$.

45. Solving the equation:

$$2n^4 - 9n^2 + 4 = 0$$

$$\left(2n^2 - 1\right)\left(n^2 - 4\right) = 0$$

$$n^2 = \frac{1}{2}, 4$$

$$n = \pm\sqrt{\frac{1}{2}} = \pm\frac{\sqrt{2}}{2}, \pm 2$$

The solution set is $\left\{\pm\frac{\sqrt{2}}{2}, \pm 2\right\}$.

47. Solving the equation:

$$x^4 - 2x^2 - 35 = 0$$

$$\left(x^2 + 5\right)\left(x^2 - 7\right) = 0$$

$$x^2 = -5, 7$$

$$x = \pm i\sqrt{5}, \pm\sqrt{7}$$

The solution set is $\left\{\pm i\sqrt{5}, \pm\sqrt{7}\right\}$.

49. Solving the equation:

$$x^4 - 4x^2 + 1 = 0$$

$$x^4 - 4x^2 = -1$$

$$x^4 - 4x^2 + 4 = -1 + 4$$

$$\left(x^2 - 2\right)^2 = 3$$

$$x^2 - 2 = \pm\sqrt{3}$$

$$x^2 = 2 \pm \sqrt{3}$$

$$x = \pm\sqrt{2 \pm \sqrt{3}}$$

The solution set is $\left\{\pm\sqrt{2 + \sqrt{3}}, \pm\sqrt{2 - \sqrt{3}}\right\}$.

51. Solving the equation:

$$x^{2/3} + 3x^{1/3} - 10 = 0$$

$$\left(x^{1/3} + 5\right)\left(x^{1/3} - 2\right) = 0$$

$$x^{1/3} = -5, 2$$

$$x = -125, 8$$

The solution set is $\{-125, 8\}$.

53. Solving the equation:

$$6x^{2/3} - 5x^{1/3} - 6 = 0$$
$$\left(3x^{1/3} + 2\right)\left(2x^{1/3} - 3\right) = 0$$
$$x^{1/3} = -\frac{2}{3}, \frac{3}{2}$$
$$x = -\frac{8}{27}, \frac{27}{8}$$

The solution set is $\left\{-\dfrac{8}{27}, \dfrac{27}{8}\right\}$.

57. Solving the equation:

$$x - 11\sqrt{x} + 30 = 0$$
$$\left(\sqrt{x} - 5\right)\left(\sqrt{x} - 6\right) = 0$$
$$\sqrt{x} = 5, 6$$
$$x = 25, 36$$

The solution set is $\{25, 36\}$.

61. Solving the equation:

$$4x^{-4} - 17x^{-2} + 4 = 0$$
$$\left(4x^{-2} - 1\right)\left(x^{-2} - 4\right) = 0$$
$$x^{-2} = \frac{1}{4}, 4$$
$$x^{2} = 4, \frac{1}{4}$$
$$x = \pm\frac{1}{2}, \pm 2$$

The solution set is $\left\{\pm\dfrac{1}{2}, \pm 2\right\}$.

65. Solving the equation:

$$10x^{-2} + 13x^{-1} - 3 = 0$$
$$\left(5x^{-1} - 1\right)\left(2x^{-1} + 3\right) = 0$$
$$x^{-1} = -\frac{3}{2}, \frac{1}{5}$$
$$x = -\frac{2}{3}, 5$$

The solution set is $\left\{-\dfrac{2}{3}, 5\right\}$.

67. Substituting $s = 13$ and $r = 5$:

$$13 = \sqrt{5^2 + h^2}$$
$$25 + h^2 = 169$$
$$h^2 = 144$$
$$h = 12$$

The altitude is 12 inches.

55. Solving the equation:

$$x^{-2} + 4x^{-1} - 12 = 0$$
$$\left(x^{-1} + 6\right)\left(x^{-1} - 2\right) = 0$$
$$x^{-1} = -6, 2$$
$$x = -\frac{1}{6}, \frac{1}{2}$$

The solution set is $\left\{-\dfrac{1}{6}, \dfrac{1}{2}\right\}$.

59. Solving the equation:

$$x + 3\sqrt{x} - 10 = 0$$
$$\left(\sqrt{x} + 5\right)\left(\sqrt{x} - 2\right) = 0$$
$$\sqrt{x} = 2 \quad \left(\sqrt{x} = -5 \text{ is impossible}\right)$$
$$x = 4$$

The solution set is $\{4\}$.

63. Solving the equation:

$$x^{-4/3} - 5x^{-2/3} + 4 = 0$$
$$\left(x^{-2/3} - 4\right)\left(x^{-2/3} - 1\right) = 0$$
$$x^{-2/3} = 1, 4$$
$$x^{2/3} = 1, \frac{1}{4}$$
$$x^{2} = 1, \frac{1}{64}$$
$$x = \pm 1, \pm\frac{1}{8}$$

The solution set is $\left\{\pm 1, \pm\dfrac{1}{8}\right\}$.

69. Substituting $f = 0.35$ and $S = 58$:

$$58 = \sqrt{30(0.35)D}$$
$$10.5D = 58^2$$
$$D = \frac{58^2}{10.5} \approx 320$$

The car will skid 320 feet.

75. **a.** Solving the equation:
$$x^4 - 3x^2 + 1 = 0$$
$$x^2 = \frac{3 \pm \sqrt{9-4}}{2} = \frac{3 \pm \sqrt{5}}{2}$$
$$x = \pm\sqrt{\frac{3-\sqrt{5}}{2}} \approx \pm 0.62, \pm\sqrt{\frac{3+\sqrt{5}}{2}} \approx \pm 1.62$$

The solution set is $\{\pm 0.62, \pm 1.62\}$.

b. Solving the equation:
$$x^4 - 5x^2 + 2 = 0$$
$$x^2 = \frac{5 \pm \sqrt{25-8}}{2} = \frac{5 \pm \sqrt{17}}{2}$$
$$x = \pm\sqrt{\frac{5-\sqrt{17}}{2}} \approx \pm 0.66, \pm\sqrt{\frac{5+\sqrt{17}}{2}} \approx \pm 2.14$$

The solution set is $\{\pm 0.66, \pm 2.14\}$.

c. Solving the equation:
$$2x^4 - 7x^2 + 2 = 0$$
$$x^2 = \frac{7 \pm \sqrt{49-16}}{4} = \frac{7 \pm \sqrt{33}}{4}$$
$$x = \pm\sqrt{\frac{7-\sqrt{33}}{4}} \approx \pm 0.56, \pm\sqrt{\frac{7+\sqrt{33}}{4}} \approx \pm 1.78$$

The solution set is $\{\pm 0.56, \pm 1.78\}$.

d. Solving the equation:
$$3x^4 - 9x^2 + 1 = 0$$
$$x^2 = \frac{9 \pm \sqrt{81-12}}{6} = \frac{9 \pm \sqrt{69}}{6}$$
$$x = \pm\sqrt{\frac{9-\sqrt{69}}{6}} \approx \pm 0.34, \pm\sqrt{\frac{9+\sqrt{69}}{6}} \approx \pm 1.70$$

The solution set is $\{\pm 0.34, \pm 1.70\}$.

e. Solving the equation:
$$x^4 - 100x^2 + 2304 = 0$$
$$\left(x^2 - 64\right)\left(x^2 - 36\right) = 0$$
$$x^2 = 36, 64$$
$$x = \pm 6, \pm 8$$

The solution set is $\{\pm 6, \pm 8\}$.

f. Solving the equation:
$$4x^4 - 373x^2 + 3969 = 0$$
$$\left(4x^2 - 49\right)\left(x^2 - 81\right) = 0$$
$$x^2 = \tfrac{49}{4}, 81$$
$$x = \pm\tfrac{7}{2}, \pm 9$$

The solution set is $\{\pm 3.5, \pm 9\}$.

1.6 Inequalities

1. The solution set is $(-\infty, -2]$. Sketching the graph:

3. The solution set is $(1, 4)$. Sketching the graph:

5. The solution set is $(0, 2)$. Sketching the graph:

7. The solution set is $[-2, -1]$. Sketching the graph:

9. The solution set is $(-\infty, 1) \cup (3, \infty)$. Sketching the graph:

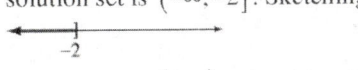
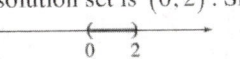
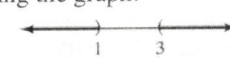

11. The solution set is $(-2, \infty)$. Sketching the graph:

13. The solution set is $(-\infty, \infty)$. Sketching the graph:

15. The solution set is $(4, \infty)$. Sketching the graph:

17. The solution set is $[-3, 2)$. Sketching the graph:

19. Solving the inequality:

$$-17 \le 3x - 2 \le 10$$
$$-15 \le 3x \le 12$$
$$-5 \le x \le 4$$

The solution set is $[-5, 4]$.

21. Solving the inequality:

$$2 > 2x - 1 > -3$$
$$3 > 2x > -2$$
$$\frac{3}{2} > x > -1$$

The solution set is $\left(-1, \frac{3}{2}\right)$.

23. Solving the inequality:

$$-4 < \frac{x-1}{3} < 4$$
$$-12 < x - 1 < 12$$
$$-11 < x < 13$$

The solution set is $(-11, 13)$.

25. Solving the inequality:

$$-3 < 2 - x < 3$$
$$-5 < -x < 1$$
$$5 > x > -1$$

The solution set is $(-1, 5)$.

27. Solving the inequality:

$$-2x + 1 > 5$$
$$-2x > 4$$
$$x < -2$$

The solution set is $(-\infty, -2)$.

29. Solving the inequality:

$$-3n + 5n - 2 \ge 8n - 7 - 9n$$
$$2n - 2 \ge -n - 7$$
$$3n \ge -5$$
$$n \ge -\frac{5}{3}$$

The solution set is $\left[-\frac{5}{3}, \infty\right)$.

31. Solving the inequality:

$$6(2t - 5) - 2(4t - 1) \ge 0$$
$$12t - 30 - 8t + 2 \ge 0$$
$$4t - 28 \ge 0$$
$$4t \ge 28$$
$$t \ge 7$$

The solution set is $[7, \infty)$.

33. Solving the inequality:

$$\frac{2}{3}x - \frac{3}{4} \le \frac{1}{4}x + \frac{2}{3}$$
$$12\left(\frac{2}{3}x - \frac{3}{4}\right) \le 12\left(\frac{1}{4}x + \frac{2}{3}\right)$$
$$8x - 9 \le 3x + 8$$
$$5x \le 17$$
$$x \le \frac{17}{5}$$

The solution set is $\left(-\infty, \frac{17}{5}\right]$.

35. Solving the inequality:

$$\frac{n+2}{4} + \frac{n-3}{8} < 1$$
$$8\left(\frac{n+2}{4} + \frac{n-3}{8}\right) < 8(1)$$
$$2(n+2) + (n-3) < 8$$
$$2n + 4 + n - 3 < 8$$
$$3n + 1 < 8$$
$$3n < 7$$
$$n < \frac{7}{3}$$

The solution set is $\left(-\infty, \frac{7}{3}\right)$.

37. Solving the inequality:

$$\frac{x}{2} - \frac{x-1}{5} \ge \frac{x+2}{10} - 4$$
$$10\left(\frac{x}{2} - \frac{x-1}{5}\right) \ge 10\left(\frac{x+2}{10} - 4\right)$$
$$5x - 2(x-1) \ge x + 2 - 40$$
$$5x - 2x + 2 \ge x - 38$$
$$3x + 2 \ge x - 38$$
$$2x \ge -40$$
$$x \ge -20$$

The solution set is $[-20, \infty)$.

39. Solving the inequality:

$$0.09x + 0.1(x + 200) > 77$$
$$0.09x + 0.1 + 20 > 77$$
$$0.19x > 57$$
$$x > 300$$

The solution set is $(300, \infty)$.

41. Solving the inequality:

$$0 < \frac{5x - 1}{3} < 2$$
$$0 < 5x - 1 < 6$$
$$1 < 5x < 7$$
$$\frac{1}{5} < x < \frac{7}{5}$$

The solution set is $\left(\frac{1}{5}, \frac{7}{5} \right)$.

43. Solving the inequality:

$$3 \geq \frac{7 - x}{2} \geq 1$$
$$6 \geq 7 - x \geq 2$$
$$-1 \geq -x \geq -5$$
$$1 \leq x \leq 5$$

The solution set is $[1, 5]$.

45. Let r represent the rate to invest his money. The inequality is:

$$0.04(5,000) + r(5,000) > 500$$
$$200 + 5000r > 500$$
$$5000r > 300$$
$$r > 0.06$$

The rate must be greater than 6%.

47. Let x represent her score on the fifth exam. The inequality is:

$$\frac{94 + 84 + 86 + 88 + x}{5} \geq 90$$
$$\frac{352 + x}{5} \geq 90$$
$$352 + x \geq 450$$
$$x \geq 98$$

Rhonda must score 98 or higher on the exam.

49. Solving the inequality:

$$41 \leq \frac{9}{5}C + 32 \leq 59$$
$$9 \leq \frac{9}{5}C \leq 27$$
$$45 \leq 9C \leq 135$$
$$5 \leq C \leq 15$$

The temperature is between 5°C and 15°C, inclusive.

51. Solving the inequality:

$$80 \leq \frac{100M}{11} \leq 140$$
$$880 \leq 100M \leq 1540$$
$$8.8 \leq M \leq 15.4$$

The mental-age range is between 8.8 and 15.4, inclusive.

53. Solving the inequality:

$$\frac{x - 8.7}{1.2} > 2.5$$
$$x - 8.7 > 3$$
$$x > 11.7$$

The score must be greater than 11.7.

1.7 Quadratic Inequalities and Inequalities Involving Quotients

1. Factoring the inequality:
$$x^2 + 3x - 4 < 0$$
$$(x+4)(x-1) < 0$$
Forming a sign chart:

The solution set is $(-4, 1)$.

3. Factoring the inequality:
$$x^2 - 2x - 15 > 0$$
$$(x-5)(x+3) > 0$$
Forming a sign chart:

The solution set is $(-\infty, -3) \cup (5, \infty)$.

5. Factoring the inequality:
$$n^2 - n \le 2$$
$$n^2 - n - 2 \le 0$$
$$(n-2)(n+1) \le 0$$
Forming a sign chart:

The solution set is $[-1, 2]$.

7. Factoring the inequality:

$$3t^2 + 11t - 4 > 0$$
$$(3t-1)(t+4) > 0$$

Forming a sign chart:

The solution set is $(-\infty, -4) \cup \left(\dfrac{1}{3}, \infty\right)$.

9. Factoring the inequality:
$$15x^2 - 26x + 8 \le 0$$
$$(5x-2)(3x-4) \le 0$$

Forming a sign chart:

The solution set is $\left[\dfrac{2}{5}, \dfrac{4}{3}\right]$.

11. Factoring the inequality:
$$4x^2 - 4x + 1 > 0$$
$$(2x-1)^2 > 0$$

The solution set is $\left(-\infty, \dfrac{1}{2}\right) \cup \left(\dfrac{1}{2}, \infty\right)$.

13. Factoring the inequality:
$$4 - x^2 < 0$$
$$(2+x)(2-x) < 0$$

Forming a sign chart:

The solution set is $(-\infty, -2) \cup (2, \infty)$.

15. Factoring the inequality:
$$4\left(x^2 - 36\right) < 0$$
$$4(x+6)(x-6) < 0$$

Forming a sign chart:

The solution set is $(-6, 6)$.

17. Factoring the inequality:
$$5x^2 + 20 > 0$$
$$5\left(x^2 + 4\right) > 0$$

Since this inequality is always true, the solution set is $(-\infty, \infty)$.

19. Factoring the inequality:
$$x^2 - 2x \ge 0$$
$$x(x-2) \ge 0$$
Forming a sign chart:

The solution set is $(-\infty, 0] \cup [2, \infty)$.

21. Factoring the inequality:
$$3x^3 + 12x^2 > 0$$
$$3x^2(x+4) > 0$$
Forming a sign chart:

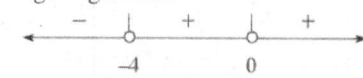

The solution set is $(-4, 0) \cup (0, \infty)$.

23. Factoring the inequality:

$$(x+1)(x-3) > (x+1)(2x-1)$$
$$x^2 - 2x - 3 > 2x^2 + x - 1$$
$$0 > x^2 + 3x + 2$$
$$(x+2)(x+1) < 0$$

Forming a sign chart:

The solution set is $(-2,-1)$.

25. Solving the inequality:

$$(x+1)(x-2) \geq (x-4)(x+6)$$
$$x^2 - x - 2 \geq x^2 + 2x - 24$$
$$-3x \geq -22$$
$$x \leq \frac{22}{3}$$

The solution set is $\left(-\infty, \dfrac{22}{3}\right]$.

27. Forming a sign chart:

The solution set is $(-4,1) \cup (2,\infty)$.

29. Forming a sign chart:

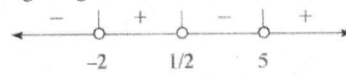

The solution set is $(-\infty,-2] \cup \left[\dfrac{1}{2}, 5\right]$.

31. Factoring the inequality:

$$x^3 - 2x^2 - 24x \geq 0$$
$$x\left(x^2 - 2x - 24\right) \geq 0$$
$$x(x-6)(x+4) \geq 0$$

Forming a sign chart:

The solution set is $[-4,0] \cup [6,\infty)$.

33. Forming a sign chart:

The solution set is $(-3,2) \cup (2,\infty)$.

35. Forming a sign chart:

The solution set is $(-\infty,-1) \cup (5,\infty)$.

37. Forming a sign chart:

The solution set is $\left(-2, \dfrac{1}{2}\right)$.

39. Forming a sign chart:

The solution set is $\left(\dfrac{1}{3}, 3\right]$.

41. Solving the inequality:

$$\frac{n}{n+2} \geq 3$$
$$\frac{n}{n+2} - 3 \geq 0$$
$$\frac{n - 3n - 6}{n+2} \geq 0$$
$$\frac{-2n-6}{n+2} \geq 0$$

Forming a sign chart:

The solution set is $[-3,-2)$.

43. Solving the inequality:

$$\frac{x-1}{x+2} < 2$$
$$\frac{x-1}{x+2} - 2 < 0$$
$$\frac{x-1-2x-4}{x+2} < 0$$
$$\frac{-x-5}{x+2} < 0$$

Forming a sign chart:

The solution set is $(-\infty,-5) \cup (-2,\infty)$.

45. Solving the inequality:

$$\frac{t-3}{t+5} > 1$$
$$\frac{t-3}{t+5} - 1 > 0$$
$$\frac{t-3-t-5}{t+5} > 0$$
$$\frac{-8}{t+5} > 0$$
$$t+5 < 0$$
$$t < -5$$

The solution set is $(-\infty, -5)$.

47. Solving the inequality:

$$\frac{1}{x-2} < \frac{1}{x+3}$$
$$\frac{1}{x-2} - \frac{1}{x+3} < 0$$
$$\frac{x+3-x+2}{(x-2)(x+3)} < 0$$
$$\frac{5}{(x-2)(x+3)} < 0$$

Forming a sign chart:

The solution set is $(-3, 2)$.

1.8 Absolute Value Equations and Inequalities

1. Solving the equation:

$$|x-2| = 6$$
$$x-2 = -6, 6$$
$$x = -4, 8$$

The solution set is $\{-4, 8\}$.

3. Solving the equation:

$$\left|x + \frac{1}{4}\right| = \frac{2}{5}$$
$$x + \frac{1}{4} = -\frac{2}{5}, \frac{2}{5}$$
$$x = -\frac{13}{20}, \frac{3}{20}$$

The solution set is $\left\{-\frac{13}{20}, \frac{3}{20}\right\}$.

5. Solving the equation:

$$|2n-1| = 7$$
$$2n-1 = -7, 7$$
$$2n = -6, 8$$
$$n = -3, 4$$

The solution set is $\{-3, 4\}$.

7. Solving the equation:

$$|3x+4| = 5$$
$$3x+4 = -5, 5$$
$$3x = -9, 1$$
$$x = -3, \frac{1}{3}$$

The solution set is $\left\{-3, \frac{1}{3}\right\}$.

9. There is no solution, or $\varnothing$.

11. Solving the equation:

$$|-3x-2| = 8$$
$$-3x-2 = -8, 8$$
$$-3x = -6, 10$$
$$x = -\frac{10}{3}, 2$$

The solution set is $\left\{-\frac{10}{3}, 2\right\}$.

13. Solving the equation:

$$|x-3| - 2 = 4$$
$$|x-3| = 6$$
$$x-3 = -6, 6$$
$$x = -3, 9$$

The solution set is $\{-3, 9\}$.

15. Solving the equation:

$$|5x+1| = 0$$
$$5x+1 = 0$$
$$5x = -1$$
$$x = -\frac{1}{5}$$

The solution set is $\left\{-\frac{1}{5}\right\}$.

17. Solving the equation:

$$|2n+3| - 7 = -2$$
$$|2n+3| = 5$$
$$2n+3 = -5, 5$$
$$2n = -8, 2$$
$$n = -4, 1$$

The solution set is $\{-4, 1\}$.

19. Solving the equation:

$$\left|\frac{3}{k-1}\right| = 4$$

$$\frac{3}{k-1} = -4, 4$$

$$\frac{k-1}{3} = -\frac{1}{4}, \frac{1}{4}$$

$$k-1 = -\frac{3}{4}, \frac{3}{4}$$

$$k = \frac{1}{4}, \frac{7}{4}$$

The solution set is $\left\{\frac{1}{4}, \frac{7}{4}\right\}$.

21. Solving the equation:

$$|3 - 2x| = 7$$

$$3 - 2x = -7, 7$$

$$-2x = -10, 4$$

$$x = -2, 5$$

The solution set is $\{-2, 5\}$.

23. Solving the equation:

$$3|x - 4| - 2 = 7$$

$$3|x - 4| = 9$$

$$|x - 4| = 3$$

$$x - 4 = -3, 3$$

$$x = 1, 7$$

The solution set is $\{1, 7\}$.

25. Solving the equation:

$$2|3x + 1| - 12 = -4$$

$$2|3x + 1| = 8$$

$$|3x + 1| = 4$$

$$3x + 1 = -4, 4$$

$$3x = -5, 3$$

$$x = -\frac{5}{3}, 1$$

The solution set is $\left\{-\frac{5}{3}, 1\right\}$.

27. Solving the equation:

$$|3x - 1| = |2x + 3|$$

$$3x - 1 = -2x - 3 \qquad \text{or} \qquad 3x - 1 = 2x + 3$$

$$5x = -2 \qquad\qquad\qquad\qquad x = 4$$

$$x = -\frac{2}{5}$$

The solution set is $\left\{-\frac{2}{5}, 4\right\}$.

29. Solving the equation:

$$|-2n + 1| = |-3n - 1|$$

$$-2n + 1 = 3n + 1 \qquad \text{or} \qquad -2n + 1 = -3n - 1$$

$$-5n = 0 \qquad\qquad\qquad\qquad n = -2$$

$$n = 0$$

The solution set is $\{-2, 0\}$.

31. Solving the equation:

$$|x - 2| = |x + 4|$$

$$x - 2 = -x - 4 \qquad \text{or} \qquad x - 2 = x + 4$$

$$2x = -2 \qquad\qquad\qquad\qquad -2 = 4$$

$$x = -1$$

The solution set is $\{-1\}$.

33. Solving the inequality:

$$|x| < 6$$

$$-6 < x < 6$$

The solution set is $(-6, 6)$.

35. Solving the inequality:

$$|x| > 8$$

$$x < -8 \ \text{ or } \ x > 8$$

The solution set is $(-\infty, -8) \cup (8, \infty)$.

37. The solution set is $(-\infty, \infty)$.

39. Solving the inequality:

$$|t - 3| > 5$$

$$t - 3 < -5 \qquad \text{or} \qquad t - 3 > 5$$

$$t < -2 \qquad\qquad\qquad t > 8$$

The solution set is $(-\infty, -2) \cup (8, \infty)$.

41. Solving the inequality:

$$|2x - 1| \le 7$$

$$-7 \le 2x - 1 \le 7$$

$$-6 \le 2x \le 8$$

$$-3 \le x \le 4$$

The solution set is $[-3, 4]$.

43. Solving the inequality:

$$|3n + 2| > 9$$

$$3n + 2 < -9 \qquad \text{or} \qquad 3n + 2 > 9$$

$$3n < -11 \qquad\qquad\qquad 3n > 7$$

$$n < -\frac{11}{3} \qquad\qquad\qquad n > \frac{7}{3}$$

The solution set is $\left(-\infty, -\frac{11}{3}\right) \cup \left(\frac{7}{3}, \infty\right)$.

45. The solution set is $\varnothing$, since $|4x - 3| \ge 0$.

47. Solving the inequality:

$$|3 - 2x| < 4$$

$$-4 < 3 - 2x < 4$$

$$-7 < -2x < 1$$

$$\frac{7}{2} > x > -\frac{1}{2}$$

The solution set is $\left(-\frac{1}{2}, \frac{7}{2}\right)$.

49. The solution set is $(-\infty, \infty)$, since $|7x + 2| \ge 0$.

51. Solving the inequality:

$$|-1 - x| \ge 8$$

$$-1 - x \le -8 \qquad \text{or} \qquad -1 - x \ge 8$$

$$-x \le -7 \qquad\qquad\qquad -x \ge 9$$

$$x \ge 7 \qquad\qquad\qquad x \le -9$$

The solution set is $(-\infty, -9] \cup [7, \infty)$.

53. Solving the inequality:

$$|x + 3| - 2 < 1$$

$$|x + 3| < 3$$

$$-3 < x + 3 < 3$$

$$-6 < x < 0$$

The solution set is $(-6, 0)$.

55. Solving the inequality:

$$|x + 4| - 1 > 1$$

$$|x + 4| > 2$$

$$x + 4 < -2 \qquad \text{or} \qquad x + 4 > 2$$

$$x < -6 \qquad\qquad\qquad x > -2$$

The solution set is $(-\infty, -6) \cup (-2, \infty)$.

57. Solving the inequality:

$$3|x - 2| \ge 6$$

$$|x - 2| \ge 2$$

$$x - 2 \le -2 \qquad \text{or} \qquad x - 2 \ge 2$$

$$x \le 0 \qquad\qquad\qquad x \ge 4$$

The solution set is $(-\infty, 0] \cup [4, \infty)$.

59. Solving the inequality:

$$-2|x + 1| > -10$$

$$|x + 1| < 5$$

$$-5 < x + 1 < 5$$

$$-6 < x < 4$$

The solution set is $(-6, 4)$.

61. Solving the inequality:

$$2|3x - 1| - 3 \ge 5$$

$$2|3x - 1| \ge 8$$

$$|3x - 1| \ge 4$$

$$3x - 1 \le -4 \qquad \text{or} \qquad 3x - 1 \ge 4$$

$$3x \le -3 \qquad\qquad\qquad 3x \ge 5$$

$$x \le -1 \qquad\qquad\qquad x \ge \frac{5}{3}$$

The solution set is $\left(-\infty, -1\right] \cup \left[\frac{5}{3}, \infty\right)$.

63.   Solving the inequality:
$$-2|x+3|-1>4$$
$$-2|x+3|>5$$
$$|x+3|<-\frac{5}{2}$$

Since $|x+3|\geq 0$ this last inequality is impossible, so the solution set is $\varnothing$.

65.   The inequality is equivalent to $-3<\dfrac{x+1}{x-2}<3$. Solving each inequality:

$$\frac{x+1}{x-2}>-3$$
$$\frac{x+1}{x-2}+3>0$$
$$\frac{x+1+3x-6}{x-2}>0$$
$$\frac{4x-5}{x-2}>0$$

$$\frac{x+1}{x-2}<3$$
$$\frac{x+1}{x-2}-3<0$$
$$\frac{x+1-3x+6}{x-2}<0$$
$$\frac{-2x+7}{x-2}<0$$

The solution sets for each inequality are $\left(-\infty,\dfrac{5}{4}\right)\cup(2,\infty)$ and $(-\infty,2)\cup\left(\dfrac{7}{2},\infty\right)$. The intersection of these solution

sets is $\left(-\infty,\dfrac{5}{4}\right)\cup\left(\dfrac{7}{2},\infty\right)$.

67.   The inequality is equivalent to $\dfrac{x-1}{x+3}<-1$ or $\dfrac{x-1}{x+3}>1$. Solving each inequality:

$$\frac{x-1}{x+3}<-1$$
$$\frac{x-1}{x+3}+1<0$$
$$\frac{x-1+x+3}{x+3}<0$$
$$\frac{2x+2}{x+3}<0$$

$$\frac{x-1}{x+3}>1$$
$$\frac{x-1}{x+3}-1>0$$
$$\frac{x-1-x-3}{x+3}>0$$
$$\frac{-4}{x+3}>0$$

The solution sets for each inequality are $(-3,-1)$ or $(-\infty,-3)$. The solution set is $(-\infty,-3)\cup(-3,-1)$.

69.   The inequality is equivalent to $\dfrac{n+2}{n}\leq -4$ or $\dfrac{n+2}{n}\geq 4$. Solving each inequality:

$$\frac{n+2}{n}\leq -4$$
$$\frac{n+2}{n}+4\leq 0$$
$$\frac{n+2+4n}{n}\leq 0$$
$$\frac{5n+2}{n}\leq 0$$

$$\frac{n+2}{n}\geq 4$$
$$\frac{n+2}{n}-4\geq 0$$
$$\frac{n+2-4n}{n}\geq 0$$
$$\frac{-3n+2}{n}\geq 0$$

The solution sets for each inequality are $\left[-\dfrac{2}{5},0\right)$ or $\left(0,\dfrac{2}{3}\right]$. The solution set is $\left[-\dfrac{2}{5},0\right)\cup\left(0,\dfrac{2}{3}\right]$.

71. The inequality is equivalent to $-2 \leq \dfrac{k}{2k-1} \leq 2$. Solving each inequality:

$$\dfrac{k}{2k-1} \geq -2$$

$$\dfrac{k}{2k-1} + 2 \geq 0$$

$$\dfrac{k + 4k - 2}{2k-1} \geq 0$$

$$\dfrac{5k - 2}{2k-1} \geq 0$$

$$\dfrac{k}{2k-1} \leq 2$$

$$\dfrac{k}{2k-1} - 2 \leq 0$$

$$\dfrac{k - 4k + 2}{2k-1} \leq 0$$

$$\dfrac{-3k + 2}{2k-1} \leq 0$$

The solution sets for each inequality are $\left(-\infty, \dfrac{2}{5}\right] \cup \left(\dfrac{1}{2}, \infty\right)$ and $\left(-\infty, \dfrac{1}{2}\right) \cup \left[\dfrac{2}{3}, \infty\right)$. The intersection of these solution

sets is $\left(-\infty, \dfrac{2}{5}\right] \cup \left[\dfrac{2}{3}, \infty\right)$.

Chapter 1 Review Problem Set

1. Solving the equation:

$$2(3x - 1) - 3(x - 2) = 2(x - 5)$$
$$6x - 2 - 3x + 6 = 2x - 10$$
$$3x + 4 = 2x - 10$$
$$x = -14$$

The solution set is $\{-14\}$.

2. Solving the equation:

$$\dfrac{n-1}{4} - \dfrac{2n+3}{5} = 2$$
$$5(n-1) - 4(2n+3) = 20 \cdot 2$$
$$5n - 5 - 8n - 12 = 40$$
$$-3n - 17 = 40$$
$$-3n = 57$$
$$n = -19$$

The solution set is $\{-19\}$.

3. Solving the equation:

$$\dfrac{2}{x+2} + \dfrac{5}{x-4} = \dfrac{7}{2(x-4)}$$

$$\dfrac{2 \cdot 2(x+2)(x-4)}{x+2} + \dfrac{5 \cdot 2(x+2)(x-4)}{x-4} = \dfrac{7 \cdot 2(x+2)(x-4)}{2(x-4)}$$

$$4(x-4) + 10(x+2) = 7(x+2)$$
$$4x - 16 + 10x + 20 = 7x + 14$$
$$14x + 4 = 7x + 14$$
$$7x = 10$$
$$x = \dfrac{10}{7}$$

The solution set is $\left\{\dfrac{10}{7}\right\}$.

4. Solving the equation:

$$0.07x + 0.12(550 - x) = 56$$
$$0.07x + 66 - 0.12x = 56$$
$$-0.05x = -10$$
$$x = 200$$

The solution set is $\{200\}$.

5. Solving the equation:
$$(3x-1)^2 = 16$$
$$3x-1 = \pm\sqrt{16}$$
$$3x-1 = -4, 4$$
$$3x = -3, 5$$
$$x = -1, \frac{5}{3}$$
The solution set is $\left\{-1, \frac{5}{3}\right\}$.

6. Solving the equation:
$$4x^2 - 29x + 30 = 0$$
$$(4x-5)(x-6) = 0$$
$$x = \frac{5}{4}, 6$$
The solution set is $\left\{\frac{5}{4}, 6\right\}$.

7. Solving the equation:
$$x^2 - 6x + 10 = 0$$
$$x^2 - 6x = -10$$
$$x^2 - 6x + 9 = -10 + 9$$
$$(x-3)^2 = -1$$
$$x-3 = \pm i$$
$$x = 3 \pm i$$
The solution set is $\{3 \pm i\}$.

8. Solving the equation:
$$n^2 + 4n = 396$$
$$n^2 + 4n + 4 = 396 + 4$$
$$(n+2)^2 = 400$$
$$n+2 = \pm\sqrt{400}$$
$$n+2 = -20, 20$$
$$n = -22, 18$$
The solution set is $\{-22, 18\}$.

9. Solving the equation:
$$15x^3 + x^2 - 2x = 0$$
$$x\left(15x^2 + x - 2\right) = 0$$
$$x(5x+2)(3x-1) = 0$$
$$x = -\frac{2}{5}, 0, \frac{1}{3}$$
The solution set is $\left\{-\frac{2}{5}, 0, \frac{1}{3}\right\}$.

10. Solving the equation:
$$\frac{t+3}{t-1} - \frac{2t+3}{t-5} = \frac{3-t^2}{(t-5)(t-1)}$$
$$(t+3)(t-5) - (2t+3)(t-1) = 3-t^2$$
$$t^2 - 2t - 15 - 2t^2 - t + 3 = 3 - t^2$$
$$-t^2 - 3t - 12 = 3 - t^2$$
$$-3t = 15$$
$$t = -5$$
The solution set is $\{-5\}$.

11. Solving the equation:
$$\frac{5-x}{2-x} - \frac{3-2x}{2x} = 1$$
$$(5-x)(2x) - (3-2x)(2-x) = 2x(2-x)$$
$$10x - 2x^2 - 6 + 7x - 2x^2 = 4x - 2x^2$$
$$-4x^2 + 17x - 6 = 4x - 2x^2$$
$$-2x^2 + 13x - 6 = 0$$
$$2x^2 - 13x + 6 = 0$$
$$(2x-1)(x-6) = 0$$
$$x = \frac{1}{2}, 6$$
The solution set is $\left\{\frac{1}{2}, 6\right\}$.

12. Solving the equation:
$$x^4 + 4x^2 - 45 = 0$$
$$\left(x^2 + 9\right)\left(x^2 - 5\right) = 0$$
$$x^2 = -9, 5$$
$$x = \pm 3i, \pm\sqrt{5}$$
The solution set is $\left\{\pm 3i, \pm\sqrt{5}\right\}$.

13.   Solving the equation:
$$2n^{-4} - 11n^{-2} + 5 = 0$$
$$\left(2n^{-2} - 1\right)\left(n^{-2} - 5\right) = 0$$
$$n^{-2} = \frac{1}{2}, 5$$
$$n^2 = 2, \frac{1}{5}$$
$$n = \pm\sqrt{2}, \pm\sqrt{\frac{1}{5}} = \pm\frac{\sqrt{5}}{5}$$

The solution set is $\left\{ \pm\sqrt{2}, \pm\frac{\sqrt{5}}{5} \right\}$.

14.   Let $u = x - \dfrac{2}{x}$, so the equation becomes:
$$u^2 + 4u = 5$$
$$u^2 + 4u - 5 = 0$$
$$(u + 5)(u - 1) = 0$$
$$u = 1, -5$$

Solving each equation:

$$x - \frac{2}{x} = 1$$
$$x^2 - 2 = x$$
$$x^2 - x - 2 = 0$$
$$(x - 2)(x + 1) = 0$$
$$x = -1, 2$$

$$x - \frac{2}{x} = -5$$
$$x^2 - 2 = -5x$$
$$x^2 + 5x - 2 = 0$$
$$x = \frac{-5 \pm \sqrt{25 + 8}}{2} = \frac{-5 \pm \sqrt{33}}{2}$$

The solution set is $\left\{ -1, 2, \dfrac{-5 \pm \sqrt{33}}{2} \right\}$.

15.   Solving the equation:

$$\sqrt{5 + 2x} = 1 + \sqrt{2x}$$
$$\left(\sqrt{5 + 2x}\right)^2 = \left(1 + \sqrt{2x}\right)^2$$
$$5 + 2x = 1 + 2\sqrt{2x} + 2x$$
$$4 = 2\sqrt{2x}$$
$$2 = \sqrt{2x}$$
$$4 = 2x$$
$$x = 2$$

The solution set is $\{2\}$.

16.   Solving the equation:
$$\sqrt{3 + 2n} + \sqrt{2 - 2n} = 3$$
$$\sqrt{3 + 2n} = 3 - \sqrt{2 - 2n}$$
$$\left(\sqrt{3 + 2n}\right)^2 = \left(3 - \sqrt{2 - 2n}\right)^2$$
$$3 + 2n = 9 - 6\sqrt{2 - 2n} + 2 - 2n$$
$$4n - 8 = -6\sqrt{2 - 2n}$$
$$(2n - 4)^2 = \left(-3\sqrt{2 - 2n}\right)^2$$
$$4n^2 - 16n + 16 = 9(2 - 2n)$$
$$4n^2 - 16n + 16 = 18 - 18n$$
$$4n^2 + 2n - 2 = 0$$
$$2n^2 + n - 1 = 0$$
$$(2n - 1)(n + 1) = 0$$
$$n = -1, \frac{1}{2}$$

The solution set is $\left\{ -1, \dfrac{1}{2} \right\}$.

17. Solving the equation:
$$\sqrt{3-t} - \sqrt{3+t} = \sqrt{t}$$
$$\sqrt{3-t} = \sqrt{t} + \sqrt{3+t}$$
$$\left(\sqrt{3-t}\right)^2 = \left(\sqrt{t} + \sqrt{3+t}\right)^2$$
$$3-t = t + 2\sqrt{3t + t^2} + 3 + t$$
$$-3t = 2\sqrt{3t + t^2}$$
$$(-3t)^2 = \left(2\sqrt{3t + t^2}\right)^2$$
$$9t^2 = 12t + 4t^2$$
$$5t^2 - 12t = 0$$
$$t(5t - 12) = 0$$
$$t = 0 \qquad \left(t = \frac{12}{5} \text{ does not check}\right)$$

The solution set is $\{0\}$.

18. Solving the equation:
$$|5x - 1| = 7$$
$$5x - 1 = -7, 7$$
$$5x = -6, 8$$
$$x = -\frac{6}{5}, \frac{8}{5}$$

The solution set is $\left\{-\frac{6}{5}, \frac{8}{5}\right\}$.

19. Solving the equation:
$$|2x + 5| = |3x - 7|$$

$$2x + 5 = 3x - 7 \qquad \text{or} \qquad 2x + 5 = -3x + 7$$
$$-x = -12 \qquad\qquad\qquad 5x = 2$$
$$x = 12 \qquad\qquad\qquad\qquad x = \frac{2}{5}$$

The solution set is $\left\{\frac{2}{5}, 12\right\}$.

20. Solving the equation:
$$\left|\frac{-3}{n-1}\right| = 4$$
$$\frac{-3}{n-1} = -4, 4$$
$$\frac{n-1}{-3} = -\frac{1}{4}, \frac{1}{4}$$
$$n - 1 = -\frac{3}{4}, \frac{3}{4}$$
$$n = \frac{1}{4}, \frac{7}{4}$$

The solution set is $\left\{\frac{1}{4}, \frac{7}{4}\right\}$.

21. Solving the equation:

$$x^3 + x^2 - 2x - 2 = 0$$
$$x^2(x+1) - 2(x+1) = 0$$
$$(x+1)(x^2 - 2) = 0$$
$$x = -1, \pm\sqrt{2}$$

The solution set is $\left\{-1, \pm\sqrt{2}\right\}$.

22. Solving the equation:

$$2x^{2/3} + 5x^{1/3} - 12 = 0$$
$$\left(2x^{1/3} - 3\right)\left(x^{1/3} + 4\right) = 0$$
$$x^{1/3} = -4, \frac{3}{2}$$
$$x = -64, \frac{27}{8}$$

The solution set is $\left\{-64, \frac{27}{8}\right\}$.

23. Solving the inequality:

$$3(2-x) + 2(x-4) > -2(x+5)$$
$$6 - 3x + 2x - 8 > -2x - 10$$
$$-x - 2 > -2x - 10$$
$$x > -8$$

The solution set is $(-8, \infty)$.

24. Solving the inequality:

$$\frac{3}{5}x - \frac{1}{3} \le \frac{2}{3}x + \frac{3}{4}$$
$$60\left(\frac{3}{5}x - \frac{1}{3}\right) \le 60\left(\frac{2}{3}x + \frac{3}{4}\right)$$
$$36x - 20 \le 40x + 45$$
$$-4x \le 65$$
$$x \ge -\frac{65}{4}$$

The solution set is $\left[-\frac{65}{4}, \infty\right)$.

25. Solving the inequality:

$$\frac{n-1}{3} - \frac{2n+1}{4} > \frac{1}{6}$$
$$12\left(\frac{n-1}{3} - \frac{2n+1}{4}\right) > 12\left(\frac{1}{6}\right)$$
$$4(n-1) - 3(2n+1) > 2$$
$$4n - 4 - 6n - 3 > 2$$
$$-2n - 7 > 2$$
$$-2n > 9$$
$$n < -\frac{9}{2}$$

The solution set is $\left(-\infty, -\frac{9}{2}\right)$.

26. Solving the inequality:

$$0.08x + 0.09(700 - x) \ge 59$$
$$0.08x + 63 - 0.09x \ge 59$$
$$-0.01x \ge -4$$
$$x \le 400$$

The solution set is $(-\infty, 400]$.

27. Solving the inequality:

$$-16 \le 7x - 2 \le 5$$
$$-14 \le 7x \le 7$$
$$-2 \le x \le 1$$

The solution set is $[-2, 1]$.

28. Solving the equation:

$$5 > \frac{3y+4}{2} > 1$$
$$10 > 3y + 4 > 2$$
$$6 > 3y > -2$$
$$2 > y > -\frac{2}{3}$$

The solution set is $\left(-\frac{2}{3}, 2\right)$.

29. Factoring the inequality:

$$x^2 - 3x - 18 < 0$$
$$(x-6)(x+3) < 0$$

Forming a sign chart:

$$\xleftarrow{\qquad + \qquad \underset{-3}{\circ} \qquad - \qquad \underset{6}{\circ} \qquad + \qquad}\rightarrow$$

The solution set is $(-3, 6)$.

30. Factoring the inequality:

$$n^2 - 5n \ge 14$$
$$n^2 - 5n - 14 \ge 0$$
$$(n-7)(n+2) \ge 0$$

Forming a sign chart:

$$\xleftarrow{\qquad + \qquad \underset{-2}{\circ} \qquad - \qquad \underset{7}{\circ} \qquad + \qquad}\rightarrow$$

The solution set is $(-\infty, -2] \cup [7, \infty)$.

31. Forming a sign chart:

The solution set is $(-\infty, -2) \cup (1, 4)$.

32. Forming a sign chart:

The solution set is $\left[-4, \dfrac{3}{2}\right)$.

33. Forming a sign chart:

The solution set is $\left(-\infty, -\dfrac{1}{5}\right) \cup (2, \infty)$.

34. Solving the inequality:
$$\frac{x-1}{x+3} \geq 2$$
$$\frac{x-1}{x+3} - 2 \geq 0$$
$$\frac{x-1-2x-6}{x+3} \geq 0$$
$$\frac{-x-7}{x+3} \geq 0$$

Forming a sign chart:

The solution set is $[-7, -3)$.

35. Solving the inequality:
$$\frac{t+5}{t-4} < 1$$
$$\frac{t+5}{t-4} - 1 < 0$$
$$\frac{t+5-t+4}{t-4} < 0$$
$$\frac{9}{t-4} < 0$$

The solution set is $(-\infty, 4)$.

36. Solving the inequality:
$$|4x-3| > 5$$
$$4x-3 < -5 \quad \text{or} \quad 4x-3 > 5$$
$$4x < -2 \qquad\qquad 4x > 8$$
$$x < -\frac{1}{2} \qquad\qquad x > 2$$

The solution set is $\left(-\infty, -\dfrac{1}{2}\right) \cup (2, \infty)$.

37. Solving the inequality:
$$|3x+5| \leq 14$$
$$-14 \leq 3x+5 \leq 14$$
$$-19 \leq 3x \leq 9$$
$$-\frac{19}{3} \leq x \leq 3$$

The solution set is $\left[-\dfrac{19}{3}, 3\right]$.

38. Solving the inequality:
$$|-3-2x| < 6$$
$$-6 < -3-2x < 6$$
$$-3 < -2x < 9$$
$$\frac{3}{2} > x > -\frac{9}{2}$$

The solution set is $\left(-\dfrac{9}{2}, \dfrac{3}{2}\right)$.

39. The inequality is equivalent to $\dfrac{x-1}{x} > 2$ or $\dfrac{x-1}{x} < -2$. Solving each inequality:

$$\frac{x-1}{x} < -2$$
$$\frac{x-1}{x} + 2 < 0$$
$$\frac{x-1+2x}{x} < 0$$
$$\frac{3x-1}{x} < 0$$

$$\frac{x-1}{x} > 2$$
$$\frac{x-1}{x} - 2 > 0$$
$$\frac{x-1-2x}{x} > 0$$
$$\frac{-x-1}{x} > 0$$

The solution sets for each inequality are $\left(0, \dfrac{1}{3}\right)$ and $(-1, 0)$. The solution set is $(-1, 0) \cup \left(0, \dfrac{1}{3}\right)$.

40. The inequality is equivalent to $-1 < \dfrac{n+1}{n+2} < 1$. Solving each inequality:

$$\frac{n+1}{n+2} > -1$$
$$\frac{n+1}{n+2} + 1 > 0$$
$$\frac{n+1+n+2}{n+2} > 0$$
$$\frac{2n+3}{n+2} > 0$$

$$\frac{n+1}{n+2} < 1$$
$$\frac{n+1}{n+2} - 1 < 0$$
$$\frac{n+1-n-2}{n+2} < 0$$
$$\frac{-1}{n+2} < 0$$

The solution sets for each inequality are $\left(-\infty, -2\right) \cup \left(-\dfrac{3}{2}, \infty\right)$ and $\left(-2, \infty\right)$. The intersection of these

solution sets is $\left(-\dfrac{3}{2}, \infty\right)$.

41. Let x, $x+2$, and $x+4$ represent the three integers. The equation is:

$$x + (x+2) + (x+4) = 4(x+4) - 31$$
$$3x + 6 = 4x + 16 - 31$$
$$-x = -21$$
$$x = 21$$

The integers are 21, 23, and 25.

42. Let $7x$ represent the number of men and $2x$ represent the number of women on the crew team, respectively. The equation is:

$$7x + 2x = 63$$
$$9x = 63$$
$$x = 7$$
$$7x = 49$$

There are 49 men on the crew team.

43. Let w and l represent the width and length. Since the perimeter is 38:

$$2w + 2l = 38$$
$$w + l = 19$$
$$l = 19 - w$$

Using the area equation:

$$w(19 - w) = 84$$
$$19w - w^2 = 84$$
$$w^2 - 19w + 84 = 0$$
$$(w - 7)(w - 12) = 0$$
$$w = 7, 12$$

The dimensions are 7 cm by 12 cm.

44. Let n represent the number of nickels, $3n$ represent the number of dimes, and $3n - 3$ represent the number of quarters. The equation is:

$$0.05(n) + 0.10(3n) + 0.25(3n - 3) = 13.55$$
$$0.05n + 0.30n + 0.75n - 0.75 = 13.55$$
$$1.10n = 14.3$$
$$n = 13$$

There are 13 nickels, 39 dimes, and 36 quarters in the collection.

45. Let p represent the selling price of the computer flash drives. The equation is:

$$14 + 0.30p = p$$
$$14 = 0.70p$$
$$p = 20$$

The selling price should be $20 for the computer flash drives.

46. Let x represent the quantity to add. The equation is:
$$0.55x + 0.20(15) = 0.40(x + 15)$$
$$0.55x + 3 = 0.4x + 6$$
$$0.15x = 3$$
$$x = 20$$
Thus 20 gallons of 55% glycerine should be added.

47. Let x represent Rosie's age and $47 - x$ represent her mother's age. The equation is:
$$x + 5 = \frac{1}{2}(47 - x + 5)$$
$$2x + 10 = 52 - x$$
$$3x = 42$$
$$x = 14$$
Rosie is 14 years old and her mother is 33 years old.

48. Let x and $8000 - x$ represent the amounts invested in each. The equation is:
$$0.04x + 0.065(8,000 - x) = 432.50$$
$$0.04x + 520 - 0.065x = 432.50$$
$$-0.025x = -87.5$$
$$x = 3,500$$
Kelly invested $3,500 at 4% and $4,500 at 6.5%.

49. Let x represent the score on her fifth exam. The inequality is:
$$\frac{93 + 88 + 89 + 95 + x}{5} \geq 92$$
$$365 + x \geq 460$$
$$x \geq 95$$
She must score at least 95 on her fifth exam.

50. Let t represent the time it takes Amy to clean the house alone, and let $2t$ represent the time it takes Angie to clean the house alone. The equation is:
$$\frac{1}{t}(3) + \frac{1}{2t}(3) = 1$$
$$\frac{3}{t} + \frac{3}{2t} = 1$$
$$2t\left(\frac{3}{t} + \frac{3}{2t}\right) = 2t \cdot 1$$
$$6 + 3 = 2t$$
$$2t = 9$$
$$t = 4.5$$
It takes 4.5 hours for Amy and 9 hours for Angie to clean the house alone.

51. Let t represent the time for Jay to mow the lawn by himself. The equation is:
$$\frac{1}{40}(25) + \frac{1}{t}(10) = 1$$
$$25t + 400 = 40t$$
$$400 = 15t$$
$$t = \frac{400}{15} = 26\frac{2}{3}$$
It would take Jay $26\frac{2}{3}$ minutes to mow the lawn by himself.

52. Let x and $x - 4000$ represent the amounts invested in each. The equation is:
$$0.05x + 0.03(x - 4,000) = 488$$
$$0.05x + 0.03x - 120 = 488$$
$$0.08x = 608$$
$$x = 7,600$$
Melinda invested $7,600 at 5% and $3,600 at 3%.

53. Let r represent Larry's rate and $r + 2$ represent Mike's rate. The equation is:

$$\frac{156}{r} = \frac{108}{r+2} + 1$$
$$156(r+2) = 108r + r(r+2)$$
$$156r + 312 = 108r + r^2 + 2r$$
$$0 = r^2 - 46r - 312$$
$$0 = (r-52)(r+6)$$
$$r = 52 \quad (r = -6 \text{ is impossible})$$

Larry's rate is 52 mph and Mike's rate is 54 mph.

54. Let t represent Cindy's time and $t + 2$ represent Bill's time. The equation is:

$$\frac{1}{t}(2) + \frac{1}{t+2}(3) = 1$$
$$2(t+2) + 3t = t(t+2)$$
$$2t + 4 + 3t = t^2 + 2t$$
$$0 = t^2 - 3t - 4$$
$$0 = (t-4)(t+1)$$
$$t = 4 \quad (t = -1 \text{ is impossible})$$

It would take Cindy 4 hours and Bill 6 hours to do the job alone.

55. Let x and $x + 5$ represent the length of the legs. The equation is:

$$x^2 + (x+5)^2 = 25^2$$
$$x^2 + x^2 + 10x + 25 = 625$$
$$2x^2 + 10x - 600 = 0$$
$$x^2 + 5x - 300 = 0$$
$$(x+20)(x-15) = 0$$
$$x = 15 \quad (x = -20 \text{ is impossible})$$

The legs are 15 cm and 20 cm long.

56. Let w represent the width and $\dfrac{35}{w}$ represent the length. The equation is:

$$(w+3)\left(\frac{35}{w} + 3\right) - 35 = 45$$
$$35(w+3) + 3w(w+3) - 35w = 45w$$
$$35w + 105 + 3w^2 + 9w - 35w = 45w$$
$$3w^2 - 36w + 105 = 0$$
$$w^2 - 12w + 35 = 0$$
$$(w-5)(w-7) = 0$$
$$w = 5, 7$$

The width is 5 inches and the length is 7 inches.

Chapter 1 Test

1. Solving the equation:
$$3(2x-1)-4(x+2)=-7$$
$$6x-3-4x-8=-7$$
$$2x-11=-7$$
$$2x=4$$
$$x=2$$

The solution set is $\{2\}$.

2. Solving the equation:
$$10x^2+13x-3=0$$
$$(5x-1)(2x+3)=0$$
$$x=-\frac{3}{2},\frac{1}{5}$$

The solution set is $\left\{-\frac{3}{2},\frac{1}{5}\right\}$.

3. Solving the equation:
$$(5x+2)^2=25$$
$$5x+2=\pm\sqrt{25}$$
$$5x+2=-5,5$$
$$5x=-7,3$$
$$x=-\frac{7}{5},\frac{3}{5}$$

The solution set is $\left\{-\frac{7}{5},\frac{3}{5}\right\}$.

4. Solving the equation:
$$\frac{3n+4}{4}-\frac{2n-1}{10}=\frac{11}{20}$$
$$5(3n+4)-2(2n-1)=11$$
$$15n+20-4n+2=11$$
$$11n=-11$$
$$n=-1$$

The solution set is $\{-1\}$.

5. Using $a=2$, $b=-1$, and $c=4$ in the quadratic formula: $x=\dfrac{1\pm\sqrt{1-4(2)(4)}}{2(2)}=\dfrac{1\pm\sqrt{-31}}{4}=\dfrac{1\pm i\sqrt{31}}{4}$

The solution set is $\left\{\dfrac{1\pm i\sqrt{31}}{4}\right\}$.

6. Solving the equation:
$$(n-2)(n+7)=-18$$
$$n^2+5n-14=-18$$
$$n^2+5n+4=0$$
$$(n+4)(n+1)=0$$
$$n=-4,-1$$

The solution set is $\{-4,-1\}$.

7. Solving the equation:
$$0.06x+0.08(1400-x)=100$$
$$0.06x+112-0.08x=100$$
$$-0.02x=-12$$
$$x=600$$

The solution set is $\{600\}$.

8. Solving the equation:
$$|3x-4|=7$$
$$3x-4=-7,7$$
$$3x=-3,11$$
$$x=-1,\frac{11}{3}$$

The solution set is $\left\{-1,\frac{11}{3}\right\}$.

9. Using $a=3$, $b=-2$, and $c=-2$ in the quadratic formula: $x=\dfrac{2\pm\sqrt{4-4(3)(-2)}}{2(3)}=\dfrac{2\pm\sqrt{28}}{6}=\dfrac{2\pm2\sqrt{7}}{6}=\dfrac{1\pm\sqrt{7}}{3}$

The solution set is $\left\{\dfrac{1\pm\sqrt{7}}{3}\right\}$.

10. Solving the equation:

$$3x^3 + 21x^2 - 54x = 0$$
$$3x(x^2 + 7x - 18) = 0$$
$$3x(x+9)(x-2) = 0$$
$$x = -9, 0, 2$$

The solution set is $\{-9, 0, 2\}$.

11. Solving the equation:

$$\frac{x}{2x+1} - 1 = \frac{-4}{7(x-2)}$$

$$\frac{7x(2x+1)(x-2)}{2x+1} - 7(2x+1)(x-2) = \frac{-4 \cdot 7(2x+1)(x-2)}{7(x-2)}$$

$$7x(x-2) - 7(2x+1)(x-2) = -4(2x+1)$$
$$7x^2 - 14x - 14x^2 + 21x + 14 = -8x - 4$$
$$-7x^2 + 15x + 18 = 0$$
$$7x^2 - 15x - 18 = 0$$
$$(7x+6)(x-3) = 0$$
$$x = -\frac{6}{7}, 3$$

The solution set is $\left\{ -\frac{6}{7}, 3 \right\}$.

12. Solving the equation:

$$\sqrt{2x} = x - 4$$
$$\left(\sqrt{2x}\right)^2 = (x-4)^2$$
$$2x = x^2 - 8x + 16$$
$$0 = x^2 - 10x + 16$$
$$0 = (x-8)(x-2)$$
$$x = 8 \quad (x = 2 \text{ does not check})$$

The solution set is $\{8\}$.

13. Solving the equation:

$$\sqrt{x+1} + 2 = \sqrt{x}$$
$$\sqrt{x+1} = \sqrt{x} - 2$$
$$\left(\sqrt{x+1}\right)^2 = \left(\sqrt{x} - 2\right)^2$$
$$x + 1 = x - 4\sqrt{x} + 4$$
$$-3 = -4\sqrt{x}$$
$$9 = 16x$$
$$x = \frac{9}{16} \quad (\text{this does not check})$$

The solution set is $\varnothing$.

14. Solving the equation:

$$2n^{-2} + 5n^{-1} - 12 = 0$$
$$(2n^{-1} - 3)(n^{-1} + 4) = 0$$
$$n^{-1} = -4, \frac{3}{2}$$
$$n = -\frac{1}{4}, \frac{2}{3}$$

The solution set is $\left\{ -\frac{1}{4}, \frac{2}{3} \right\}$.

15. Solving the inequality:

$$2(x-1) - 3(3x+1) \geq -6(x-5)$$
$$2x - 2 - 9x - 3 \geq -6x + 30$$
$$-7x - 5 \geq -6x + 30$$
$$-x \geq 35$$
$$x \leq -35$$

The solution set is $(-\infty, -35]$.

16. Solving the inequality:

$$\frac{x-2}{6} - \frac{x+3}{9} > -\frac{1}{2}$$
$$3(x-2) - 2(x+3) > -9$$
$$3x - 6 - 2x - 6 > -9$$
$$x - 12 > -9$$
$$x > 3$$

The solution set is $(3, \infty)$.

17. Solving the inequality:

$$|6x - 4| < 10$$
$$-10 < 6x - 4 < 10$$
$$-6 < 6x < 14$$
$$-1 < x < \frac{7}{3}$$

The solution set is $\left(-1, \frac{7}{3} \right)$.

18. Solving the inequality:
$$|4x+5| \geq 6$$
$$4x+5 \leq -6 \qquad \text{or} \qquad 4x+5 \geq 6$$
$$4x \leq -11 \qquad\qquad 4x \geq 1$$
$$x \leq -\frac{11}{4} \qquad\qquad x \geq \frac{1}{4}$$
The solution set is $\left(-\infty, -\frac{11}{4}\right] \cup \left[\frac{1}{4}, \infty\right)$.

19. Factoring the inequality:
$$2x^2 - 9x - 5 \leq 0$$
$$(2x+1)(x-5) \leq 0$$
Forming a sign chart:

The solution set is $\left[-\frac{1}{2}, 5\right]$.

20. Forming a sign chart:

The solution set is $\left(-\infty, -2\right) \cup \left(\frac{1}{3}, \infty\right)$.

21. Solving the inequality:
$$\frac{x-2}{x+6} \geq 3$$
$$\frac{x-2}{x+6} - 3 \geq 0$$
$$\frac{x-2-3x-18}{x+6} \geq 0$$
$$\frac{-2x-20}{x+6} \geq 0$$
Forming a sign chart:

The solution set $\left[-10, -6\right)$.

22. Let x represent the amount of grapefruit juice to add. The equation is:
$$0.08(30) + x = 0.10(30 + x)$$
$$2.4 + x = 3 + 0.1x$$
$$0.9x = 0.6$$
$$x = \frac{2}{3}$$
Thus $\frac{2}{3}$ cup must be added.

23. Let r represent Tasya's rate and $r + 3$ represent Lian's rate. The equation is

$$\frac{60}{r+3} = \frac{60}{r} - 1$$

$$\frac{60r(r+3)}{r+3} = \frac{60r(r+3)}{r} - r(r+3)$$

$$60r = 60r + 180 - r^2 - 3r$$

$$r^2 + 3r - 180 = 0$$

$$(r+15)(r-12) = 0$$

$$r = 12 \quad (r = -15 \text{ is not possible})$$

Lian's rate is 15 mph.

24. Let x and $x + 2000$ represent the amounts invested in each. The equation is:

$$0.045x + 0.05(x + 2,000) = 860$$

$$0.045x + 0.05x + 100 = 860$$

$$0.095x = 760$$

$$x = 8,000$$

Alexander invested $8,000 at 4.5% and $10,000 at 5%.

25. Let w and l represent the width and length of the rectangle. Therefore:

$$2w + 2l = 46$$

$$w + l = 23$$

$$l = 23 - w$$

The area equation is:

$$w(23 - w) = 126$$

$$23w - w^2 = 126$$

$$0 = w^2 - 23w + 126$$

$$0 = (w - 14)(w - 9)$$

$$w = 9, 14$$

The dimensions are 9 cm by 14 cm.

Chapter 2
Coordinate Geometry and Graphing Techniques

2.1 Coordinate Geometry

1. The distance is $6-(-4)=10$ units.

3. The distance is $-11-(-6)=-5$ units.

5. The distance is $-2-(-7)=5$ units.

7. The distance is $5-(-4)=9$ units.

9. The point is: $1+\frac{2}{3}(9)=1+6=7$

11. The point is: $-3+\frac{1}{3}(10)=-3+\frac{10}{3}=\frac{1}{3}$

13. The point is: $-1+\frac{3}{5}(-10)=-1-6=-7$

15. Using the distance formula: $AB=\sqrt{(7-4)^2+(5-1)^2}=\sqrt{3^2+4^2}=\sqrt{9+16}=\sqrt{25}=5$ units

17. Using the distance formula: $AB=\sqrt{(3+1)^2+(-2-4)^2}=\sqrt{4^2+(-6)^2}=\sqrt{16+36}=\sqrt{52}=2\sqrt{13}$ units

19. Using the distance formula: $AB=\sqrt{(10-2)^2+(7-1)^2}=\sqrt{8^2+6^2}=\sqrt{64+36}=\sqrt{100}=10$ units

21. The midpoint is given by: $M=\left(\frac{1+3}{2},\frac{-1-4}{2}\right)=\left(\frac{4}{2},-\frac{5}{2}\right)=\left(2,-\frac{5}{2}\right)$

23. The midpoint is given by: $M=\left(\frac{6+9}{2},\frac{-4-7}{2}\right)=\left(\frac{15}{2},-\frac{11}{2}\right)$

25. The midpoint is given by: $M=\left(\frac{\frac{1}{2}-\frac{1}{3}}{2},\frac{\frac{1}{3}+\frac{3}{2}}{2}\right)=\left(\frac{\frac{1}{6}}{2},\frac{\frac{11}{6}}{2}\right)=\left(\frac{1}{12},\frac{11}{12}\right)$

27. The coordinates are given by:

    $x=2+\frac{1}{3}(3)=3$ $\qquad y=3+\frac{1}{3}(6)=5$

    The point is (3,5).

29. The coordinates are given by:

    $x=-2+\frac{2}{5}(10)=2$ $\qquad y=1+\frac{2}{5}(10)=5$

    The point is (2,5).

31.   The coordinates are given by:

$$x = -1 + \frac{5}{8}(5) = -1 + \frac{25}{8} = \frac{17}{8} \qquad\qquad y = -2 + \frac{5}{8}(-8) = -2 - 5 = -7$$

The point is $\left(\frac{17}{8}, -7 \right)$.

33.   The coordinates are given by:

$$x = -7 + \frac{5}{6}(6) = -7 + 5 = -2 \qquad\qquad y = 2 + \frac{5}{6}(-6) = 2 - 5 = -3$$

The point is $(-2, -3)$.

35.   The coordinates are given by:

$$x = 6 + \frac{3}{8}(-8) = 6 - 3 = 3 \qquad\qquad y = 8 + \frac{3}{8}(-4) = 8 - \frac{3}{2} = \frac{13}{2}$$

The point is $\left(3, \frac{13}{2} \right)$.

37.   a.   The midpoint is $\left(\dfrac{2+10}{2}, \dfrac{4+13}{2} \right) = \left(6, \dfrac{17}{2} \right)$, so the quarter point is $\left(\dfrac{2+6}{2}, \dfrac{4 + \frac{17}{2}}{2} \right) = \left(4, \dfrac{25}{4} \right)$.

   b.   The coordinates are given by:

$$x = 2 + \frac{1}{4}(8) = 2 + 2 = 4 \qquad\qquad y = 4 + \frac{1}{4}(9) = 4 + \frac{9}{4} = \frac{25}{4}$$

The point is $\left(4, \dfrac{25}{4} \right)$.

39.   Finding the three distances:

$$d_1 = \sqrt{(-2-2)^2 + (7-1)^2} = \sqrt{16+36} = \sqrt{52} = 2\sqrt{13}$$
$$d_2 = \sqrt{(4-2)^2 + (-2-1)^2} = \sqrt{4+9} = \sqrt{13}$$
$$d_3 = \sqrt{(-2-4)^2 + (7+2)^2} = \sqrt{36+81} = \sqrt{117} = 3\sqrt{13}$$

Since $d_1 + d_2 = d_3$, the three points lie on a straight line.

41.   Finding the three distances:

$$d_1 = \sqrt{(2-0)^2 + (-3-3)^2} = \sqrt{4+36} = \sqrt{40} = 2\sqrt{10}$$
$$d_2 = \sqrt{(-4-2)^2 + (-5+3)^2} = \sqrt{36+4} = \sqrt{40} = 2\sqrt{10}$$
$$d_3 = \sqrt{(-4-0)^2 + (-5-3)^2} = \sqrt{16+64} = \sqrt{80} = 4\sqrt{5}$$

Since $d_1 = d_2$, the three points are vertices of an isosceles triangle.

43.   Finding the lengths of each side:

$$d_1 = \sqrt{(-6-0)^2 + (-4-8)^2} = \sqrt{36+144} = \sqrt{180} = 6\sqrt{5}$$
$$d_2 = \sqrt{(6-0)^2 + (5-8)^2} = \sqrt{36+9} = \sqrt{45} = 3\sqrt{5}$$
$$d_3 = \sqrt{(-6-6)^2 + (-4-5)^2} = \sqrt{144+81} = \sqrt{225} = 15$$

The perimeter is given by: $6\sqrt{5} + 3\sqrt{5} + 15 = 15 + 9\sqrt{5}$

45. Finding the distance from each point to the center:

$$d_1 = \sqrt{(-1-4)^2 + (2+5)^2} = \sqrt{25+49} = \sqrt{74}$$

$$d_2 = \sqrt{(-1-6)^2 + (2-7)^2} = \sqrt{49+25} = \sqrt{74}$$

$$d_3 = \sqrt{(-1+8)^2 + (2+3)^2} = \sqrt{49+25} = \sqrt{74}$$

Since $d_1 = d_2 = d_3$, the three points lie on a circle with a center at $(-1,2)$.

47. Using the distance formula:

$$\sqrt{(x+2)^2 + (-2+14)^2} = 13$$
$$(x+2)^2 + 144 = 169$$
$$(x+2)^2 = 25$$
$$x+2 = \pm\sqrt{25}$$
$$x+2 = -5, 5$$
$$x = -7, 3$$

49. Since this point P must lie $\dfrac{2}{3}$ of the distance from A to B, its coordinates are:

$$x = -1 + \frac{2}{3}(6) = -1 + 4 = 3 \qquad\qquad y = 2 + \frac{2}{3}(9) = 2 + 6 = 8$$

The coordinates of P are $(3,8)$.

51. The midpoint of AC is $\left(\dfrac{1+6}{2}, \dfrac{1+4}{2}\right) = \left(\dfrac{7}{2}, \dfrac{5}{2}\right)$, and the midpoint of BD is $\left(\dfrac{5+2}{2}, \dfrac{1+4}{2}\right) = \left(\dfrac{7}{2}, \dfrac{5}{2}\right)$,

so the two diagonals bisect each other.

2.2 Graphing Techniques: Linear Equations and Inequalities

1. The x-intercept is $(4,0)$ and the y-intercept is $(0,-2)$:

3. The x-intercept is $(2,0)$ and the y-intercept is $(0,3)$:

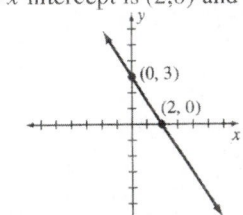

5. The x-intercept is $(5,0)$ and the y-intercept is $(0,-4)$:

7. The x-intercept is $(3,0)$ and the y-intercept is $(0,-3)$:

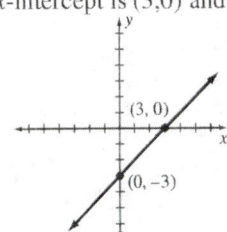

9. The x-intercept is $\left(\dfrac{1}{3},0\right)$ and the y-intercept is $(0,-1)$:

11. The x-intercept is $(0,0)$ and the y-intercept is $(0,0)$:

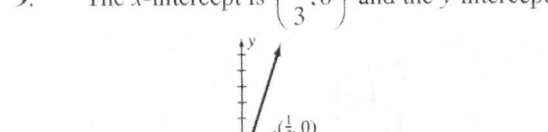

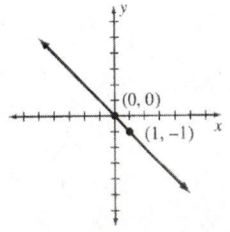

13. The x-intercept is $(0,0)$ and the y-intercept is the y-axis:

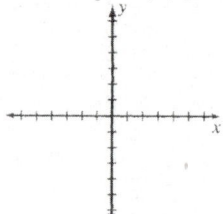

15. The x-intercept is $(0,0)$ and the y-intercept is $(0,0)$:

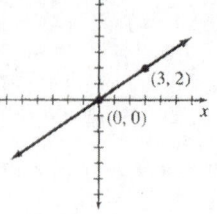

17. There is no x-intercept and the y-intercept is $(0,4)$;

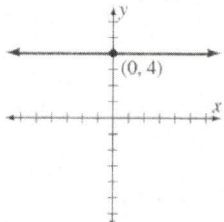

19. The x-intercept is $(-10,0)$ and the y-intercept is $(0,-2)$:

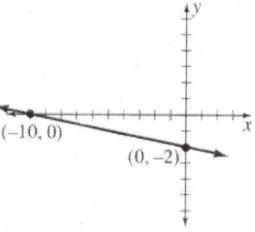

21. The x-intercept is $(0,0)$ and the y-intercept is $(0,0)$:

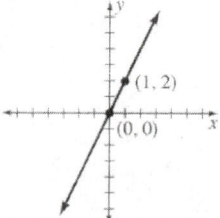

23. The x-intercept is $(-2,0)$ and there is no y-intercept:

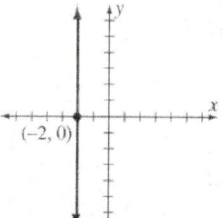

25. Graphing the inequality $x + 2y > 4$:

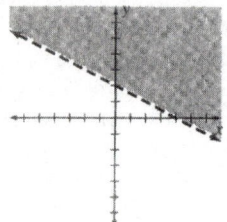

27. Graphing the inequality $3x - 2y < 6$:

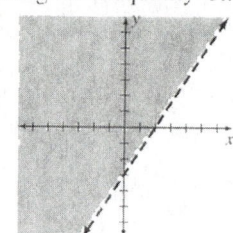

29. Graphing the inequality $2x + 5y \le 10$:

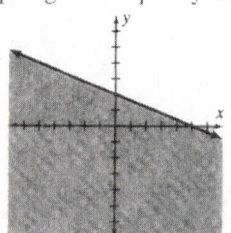

31. Graphing the inequality $y > -x - 1$:

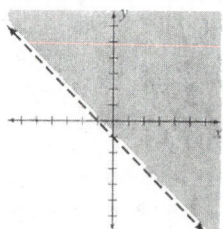

33. Graphing the inequality $y \le -x$:

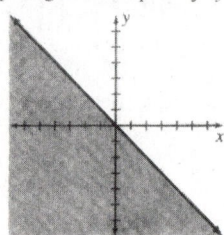

35. Graphing the inequality $x + 2y < 0$:

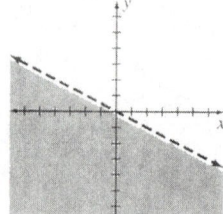

37. Graphing the inequality $x > -1$:

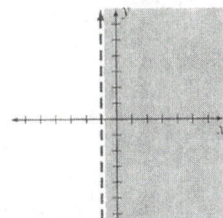

39. Graphing the inequality $y < \frac{2}{3}x - 4$:

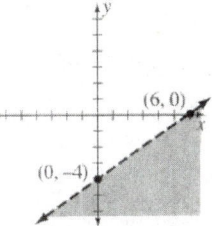

41. Graphing the inequality $y \geq -\frac{1}{2}x + 6$:

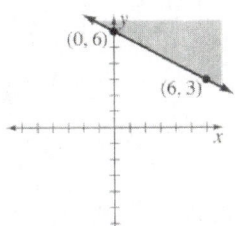

43. Graphing the inequality $x + 4 > 0$:

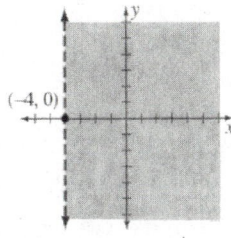

45. Graphing the inequality $3x - y < 0$:

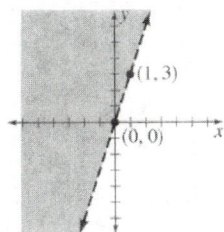

49. Graphing the equation $|x - y| = 2$:

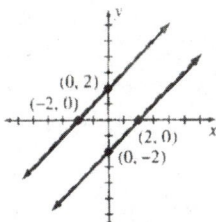

51. Graphing the inequality $|x - 2y| \leq 4$:

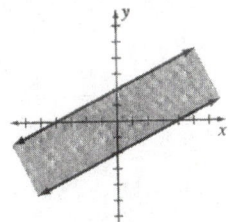

53. Graphing the inequality $|2x + 3y| > 6$:

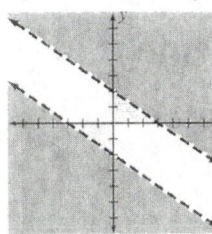

55. Graphing the equation $y = |x| - 1$:

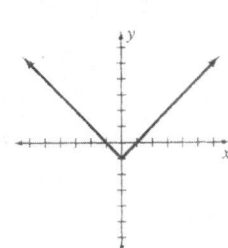

57. Graphing the equation $y = |x + 2|$:

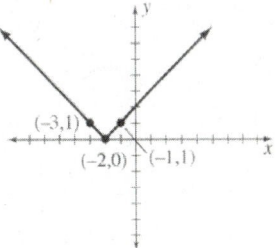

59. Graphing the equation $y = 2|x|$:

2.3 Determining the Equation of a Line

1.   Finding the slope: $m = \dfrac{4-1}{7-3} = \dfrac{3}{4}$

3.   Finding the slope: $m = \dfrac{-6+1}{-1+2} = -5$

5.   Finding the slope: $m = \dfrac{2-2}{-2+4} = 0$

7.   Finding the slope: $m = \dfrac{b-0}{0-a} = -\dfrac{b}{a}$

9.   Solving the equation:
$$\dfrac{7-4}{x+2} = \dfrac{2}{9}$$
$$2x+4 = 27$$
$$2x = 23$$
$$x = \dfrac{23}{2}$$

11.   Solving the equation:
$$\dfrac{-6-4}{2-x} = -\dfrac{9}{4}$$
$$-18+9x = -40$$
$$9x = -22$$
$$x = -\dfrac{22}{9}$$

13.   Solving for y:
$$2x - 3y = 4$$
$$-3y = -2x + 4$$
$$y = \dfrac{2}{3}x - \dfrac{4}{3}$$
Thus $m = \dfrac{2}{3}, b = -\dfrac{4}{3}$.

15.   Solving for y:
$$x - 2y = 7$$
$$-2y = -x + 7$$
$$y = \dfrac{1}{2}x - \dfrac{7}{2}$$
Thus $m = \dfrac{1}{2}, b = -\dfrac{7}{2}$.

17.   The slope and y-intercept are $m = -3, b = 0$.

19.   Solving for y:
$$7x - 5y = 12$$
$$-5y = -7x + 12$$
$$y = \dfrac{7}{5}x - \dfrac{12}{5}$$
Thus $m = \dfrac{7}{5}, b = -\dfrac{12}{5}$.

21.   Graphing the line $y = \dfrac{3}{4}x + 2$:

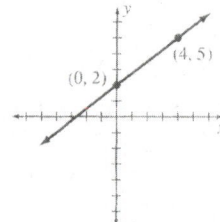

23.   Graphing the line $y = -\dfrac{4}{5}x + 1$:

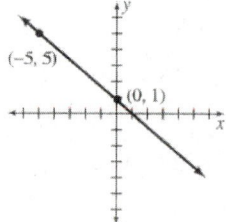

25.   Graphing the line $y = -2x + \dfrac{5}{4}$:

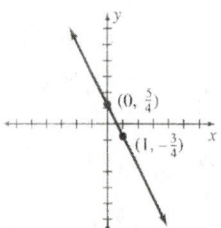

27.   Using the point-slope formula:
$$y - 4 = \dfrac{1}{3}(x - 2)$$
$$3y - 12 = x - 2$$
$$x - 3y = -10$$

29.   Using the point-slope formula:
$$y + 2 = 2(x + 1)$$
$$y + 2 = 2x + 2$$
$$2x - y = 0$$

31. Using the point-slope formula:
$$y + 3 = -\frac{2}{3}(x - 4)$$
$$3y + 9 = -2x + 8$$
$$2x + 3y = -1$$

33. Using the point-slope formula:
$$y + 2 = 0(x - 5)$$
$$y + 2 = 0$$
$$y = -2$$
$$0x + y = -2$$

35. This is a vertical line, so its equation is $x = 3$, or $x + 0y = 3$.

37. The equation is $y = \frac{1}{2}x + 3$.

39. The equation is $y = -\frac{3}{7}x + 2$.

41. The equation is $y = 4x + \frac{3}{2}$.

43. The equation is $y = 0x + \frac{1}{4}$, or $y = \frac{1}{4}$.

45. First find the slope: $m = \frac{8 - 3}{9 - 2} = \frac{5}{7}$

Using the point-slope formula:
$$y - 3 = \frac{5}{7}(x - 2)$$
$$7y - 21 = 5x - 10$$
$$5x - 7y = -11$$

47. First find the slope: $m = \frac{2 - 7}{5 + 1} = -\frac{5}{6}$

Using the point-slope formula:
$$y - 2 = -\frac{5}{6}(x - 5)$$
$$6y - 12 = -5x + 25$$
$$5x + 6y = 37$$

49. First find the slope: $m = \frac{3 - 2}{-1 - 4} = -\frac{1}{5}$. Using the point-slope formula:
$$y - 2 = -\frac{1}{5}(x - 4)$$
$$5y - 10 = -x + 4$$
$$x + 5y = 14$$

51. First find the slope: $m = \frac{-3 - 3}{-7 - 4} = 0$. This is a horizontal line, so its equation is $y = -3$, or $0x + y = -3$.

53. First find the slope: $m = \frac{-7 - 6}{-2 + 2}$, which is undefined. This is a vertical line, so its equation is $x = -2$, or $x + 0y = -2$.

55. Since the slopes are the same, the lines are parallel.

57. Finding the slope of each line:
$$5x - 7y = 14$$
$$-7y = -5x + 14$$
$$y = \frac{5}{7}x - 2$$

$$7x + 5y = 12$$
$$5y = -7x + 12$$
$$y = -\frac{7}{5}x + \frac{12}{5}$$

Since the slopes are opposite reciprocals, the lines are perpendicular.

59. Finding the slope of each line:
$$4x + 9y = 13$$
$$9y = -4x + 13$$
$$y = -\frac{4}{9}x + \frac{13}{9}$$

$$-4x + y = 11$$
$$y = 4x + 11$$

The slopes are not the same, and are not opposite reciprocals, so these are intersecting lines that are not perpendicular.

61. Finding the slope of each line:
$$x + y = 0$$
$$y = -x$$

$$x - y = 0$$
$$y = x$$

Since the slopes are opposite reciprocals, the lines are perpendicular.

63. Two points are (4,0) and (0,–5), so the slope is: $m = \frac{-5 - 0}{0 - 4} = \frac{5}{4}$

The equation is $y = \frac{5}{4}x - 5$, which is $5x - 4y = 20$.

65. The slope is undefined, so the equation is $x = -4$, or $x + 0y = -4$.

67.   The line $5x + 2y = 1$ is equivalent to $y = -\dfrac{5}{2}x + \dfrac{1}{2}$, so the parallel slope is $-\dfrac{5}{2}$. Using the point-slope formula:

$$y + 3 = -\frac{5}{2}(x - 4)$$
$$2y + 6 = -5x + 20$$
$$5x + 2y = 14$$

69.   The line $x - 4y = 7$ is equivalent to $y = \dfrac{1}{4}x - \dfrac{7}{4}$, so the perpendicular slope is -4. Using the point-slope formula:

$$y - 6 = -4(x + 2)$$
$$y - 6 = -4x - 8$$
$$4x + y = -2$$

71.   Since this must be a vertical line, its equation is $x = 1$, or $x + 0y = 1$.

73.   Since $y = 3$ is a horizontal line, the perpendicular line will be a vertical line, which is $x = -3$, or $x + 0y = -3$.

75.   The midpoint is given by: $\left(\dfrac{2 + 10}{2}, \dfrac{6 - 4}{2}\right) = \left(\dfrac{12}{2}, \dfrac{2}{2}\right) = (6, 1)$

The slope between the points is given by: $m = \dfrac{-4 - 6}{10 - 2} = \dfrac{-10}{8} = -\dfrac{5}{4}$

Using the point-slope formula with the perpendicular slope $\dfrac{4}{5}$:

$$y - 1 = \frac{4}{5}(x - 6)$$
$$5y - 5 = 4x - 24$$
$$-4x + 5y = -19$$
$$4x - 5y = 19$$

77.   The midpoint is given by: $\left(\dfrac{-4 + 6}{2}, \dfrac{0 + 2}{2}\right) = \left(\dfrac{2}{2}, \dfrac{2}{2}\right) = (1, 1)$

The slope between the points is given by: $m = \dfrac{2 - 0}{6 + 4} = \dfrac{2}{10} = \dfrac{1}{5}$

Using the point-slope formula with the perpendicular slope -5:

$$y - 1 = -5(x - 1)$$
$$y - 1 = -5x + 5$$
$$5x + y = 6$$

79.   Since it must rise 2%, it rises $0.02(5280) = 105.6$ feet.

81.   Finding the run:

$$\frac{3}{5} = \frac{19}{x}$$
$$3x = 95$$
$$x \approx 32 \text{ centimeters}$$

83.   The vertical drop must be $0.0225(45) \approx 1.0$ feet.

85.   The midpoint of $(1, -6)$ and $(3, 1)$ is $\left(2, -\dfrac{5}{2}\right)$. Finding the slope to the opposite vertex: $\left(2, -\dfrac{5}{2}\right)$

Using the point-slope formula:

$$y - 2 = -\frac{9}{8}(x + 2)$$
$$8y - 16 = -9x - 18$$
$$9x + 8y = -2$$

The midpoint of $(3, 1)$ and $(-2, 2)$ is $\left(\dfrac{1}{2}, \dfrac{3}{2}\right)$. Finding the slope to the opposite vertex: $m = \dfrac{\dfrac{3}{2} + 6}{\dfrac{1}{2} - 1} = \dfrac{\dfrac{15}{2}}{-\dfrac{1}{2}} = -15$

Using the point-slope formula:
$$y + 6 = -15(x - 1)$$
$$y + 6 = -15x + 15$$
$$15x + y = 9$$

The midpoint of $(1,-6)$ and $(-2,2)$ is $\left(-\dfrac{1}{2}, -2\right)$. Finding the slope to the opposite vertex: $m = \dfrac{-2-1}{-\dfrac{1}{2}-3} = \dfrac{-3}{-\dfrac{7}{2}} = \dfrac{6}{7}$

Using the point-slope formula:
$$y - 1 = \frac{6}{7}(x - 3)$$
$$7y - 7 = 6x - 18$$
$$6x - 7y = 11$$

87. Finding the slope between each pair of points:

$$m_1 = \frac{-2-6}{2-6} = \frac{8}{4} = 2 \qquad\qquad m_2 = \frac{-5+2}{-8-2} = \frac{3}{10}$$

$$m_3 = \frac{3+5}{-4+8} = \frac{8}{4} = 2 \qquad\qquad m_4 = \frac{3-6}{-4-6} = \frac{3}{10}$$

Since $m_1 = m_3$ and $m_2 = m_4$, the points are vertices of a parallelogram.

89. Finding the slope between each pair of points:

$$m_1 = \frac{-1-7}{-2-0} = 4 \qquad\qquad m_2 = \frac{-2+1}{2+2} = -\frac{1}{4}$$

$$m_3 = \frac{6+2}{4-2} = 4 \qquad\qquad m_4 = \frac{6-7}{4-0} = -\frac{1}{4}$$

Since $m_1 = m_3$ and $m_2 = m_4$, while m_1 and m_2 are opposite reciprocals, the vertices are of a rectangle.

95. **a.** Finding the equation:
$$\frac{y-3}{x-4} = \frac{6-3}{5-4}$$
$$\frac{y-3}{x-4} = \frac{3}{1}$$
$$y - 3 = 3x - 12$$
$$3x - y = 9$$

b. Finding the equation:
$$\frac{y-5}{x+3} = \frac{-1-5}{2+3}$$
$$\frac{y-5}{x+3} = \frac{-6}{5}$$
$$5y - 25 = -6x - 18$$
$$6x + 5y = 7$$

c. Finding the equation:
$$\frac{y-0}{x-0} = \frac{2-0}{-7-0}$$
$$\frac{y}{x} = \frac{2}{-7}$$
$$-7y = 2x$$
$$2x + 7y = 0$$

d. Finding the equation:
$$\frac{y+4}{x+3} = \frac{-1+4}{5+3}$$
$$\frac{y+4}{x+3} = \frac{3}{8}$$
$$8y + 32 = 3x + 9$$
$$3x - 8y = 23$$

99. **a.** The form is $2x - y = k$. Substituting the point $(5,6)$:
$$2(5) - 6 = k$$
$$k = 4$$
The equation is $2x - y = 4$.

b. The form is $3x + 7y = k$. Substituting the point $(-3,4)$:
$$3(-3) + 7(4) = k$$
$$k = 19$$
The equation is $3x + 7y = 19$.

c. The form is $5x + 2y = k$. Substituting the point $(2,-4)$:
$$5(2) + 2(-4) = k$$
$$k = 2$$
The equation is $5x + 2y = 2$.

d.   The form is $6x - 4y = k$. Substituting the point $(-3,-5)$:

$$6(-3) - 4(-5) = k$$
$$k = 2$$

The equation is $6x - 4y = 2$, or $3x - 2y = 1$.

2.4 More on Graphing

1.   For the x-axis the symmetric point is $(4,-3)$, for the y-axis the symmetric point is $(-4,3)$, and for the origin the symmetric point is $(-4,-3)$.

3.   For the x-axis the symmetric point is $(-6,1)$, for the y-axis the symmetric point is $(6,-1)$, and for the origin the symmetric point is $(6,1)$.

5.   For the x-axis the symmetric point is $(0,-4)$, for the y-axis the symmetric point is $(0,4)$, and for the origin the symmetric point is $(0,-4)$.

7.   Since $(-x, y)$ results in the same equation, the graph has y-axis symmetry.

9.   Since $(x,-y)$ results in the same equation, the graph has x-axis symmetry.

11.   Since $(-x,y)$, $(x,-y)$, and $(-x,-y)$ result in the same equation, the graph has x-axis, y-axis, and origin symmetry.

13.   The graph does not possess any of the three types of symmetry.

15.   Since $(-x,-y)$ results in the same equation, the graph has origin symmetry.

17.   The graph does not possess any of the three types of symmetry.

19.   Since $(-x, y)$ results in the same equation, the graph has y-axis symmetry.

21.   Since $(-x,-y)$ results in the same equation, the graph has origin symmetry.

23.   Since $(x,-y)$ results in the same equation, the graph has x-axis symmetry.

25.   Since $(-x,y)$, $(x,-y)$, and $(-x,-y)$ result in the same equation, the graph has x-axis, y-axis, and origin symmetry.

27.   The graph does not possess any of the three types of symmetry.

29.   Since $(-x,-y)$ results in the same equation, the graph has origin symmetry.

31.   Graphing the equation $y = x^2$:

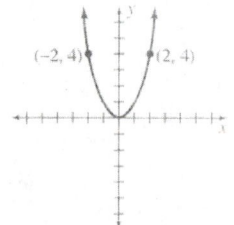

33.   Graphing the equation $y = x^2 + 2$:

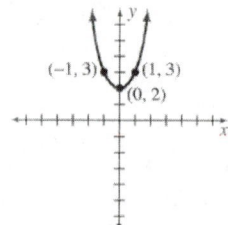

35.   Graphing the equation $xy = 4$:

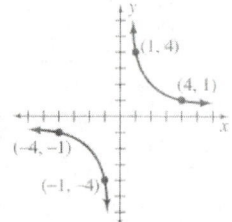

37.   Graphing the equation $y = -x^3$:

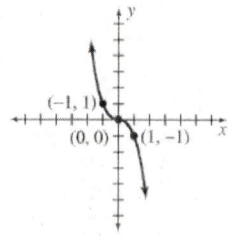

39. Graphing the equation $y^2 = x^3$:

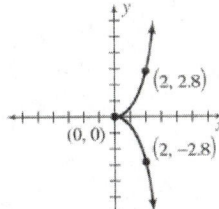

41. Graphing the equation $y^2 - x^2 = 4$:

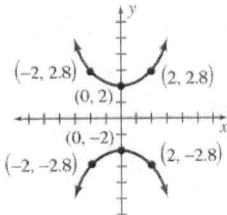

43. Graphing the equation $y = -\sqrt{x}$:

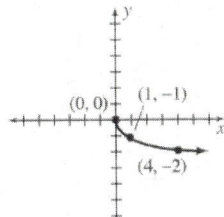

45. Graphing the equation $x^2 y = 4$:

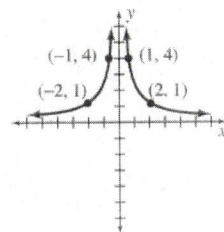

47. Graphing the equation $x^2 + 2y^2 = 8$:

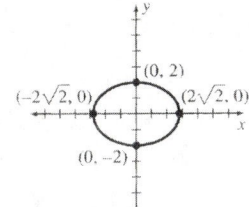

49. Graphing the equation $y = \dfrac{4}{x^2 + 1}$:

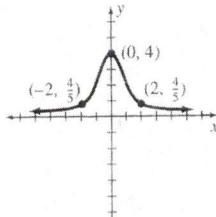

51. Graphing the equation $y = \sqrt{x - 2}$

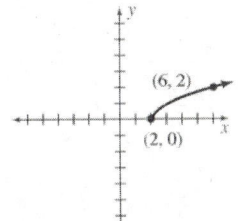

53. Graphing the equation $-xy = 3$:

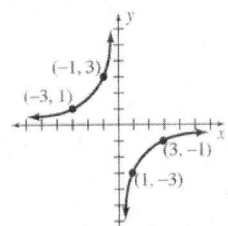

55. Graphing the equation $x = y^2 + 2$:

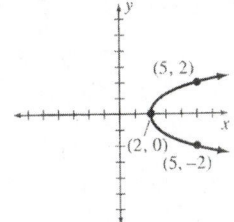

57. Graphing the equation $x = -y^2 - 1$:

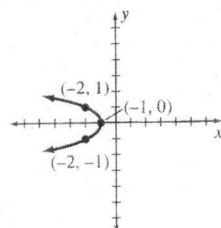

2.5   Circles, Ellipses, and Hyperbolas

1.   The equation is:
$$(x-2)^2 + (y-3)^2 = 5^2$$
$$x^2 - 4x + 4 + y^2 - 6y + 9 = 25$$
$$x^2 + y^2 - 4x - 6y - 12 = 0$$

3.   The equation is:
$$(x+1)^2 + (y+5)^2 = 3^2$$
$$x^2 + 2x + 1 + y^2 + 10y + 25 = 9$$
$$x^2 + y^2 + 2x + 10y + 17 = 0$$

5.   The equation is:
$$(x-3)^2 + (y-0)^2 = 3^2$$
$$x^2 - 6x + 9 + y^2 = 9$$
$$x^2 + y^2 - 6x = 0$$

7.   The equation is:
$$(x-0)^2 + (y-0)^2 = 7^2$$
$$x^2 + y^2 = 49$$
$$x^2 + y^2 - 49 = 0$$

9.   Completing the square:
$$x^2 + y^2 - 6x - 10y + 30 = 0$$
$$\left(x^2 - 6x + 9\right) + \left(y^2 - 10y + 25\right) = -30 + 9 + 25$$
$$(x-3)^2 + (y-5)^2 = 4$$

The center is $(3,5)$ and the radius is 4.

11.   Completing the square:
$$x^2 + y^2 + 10x + 1 = 0$$
$$\left(x^2 + 10x + 25\right) + y^2 = -1 + 25$$
$$(x+5)^2 + y^2 = 24$$

The center is $(-5,0)$ and the radius is $\sqrt{24} = 2\sqrt{6}$.

13.   Completing the square:
$$x^2 + y^2 - 10x = 0$$
$$\left(x^2 - 10x + 25\right) + y^2 = 25$$
$$(x-5)^2 + y^2 = 25$$

The center is $(5,0)$ and the radius is 5.

15.   The center is $(0,0)$ and the radius is $\sqrt{8} = 2\sqrt{2}$.

15.   The center is $(0,0)$ and the radius is $\sqrt{8} = 2\sqrt{2}$.

17.   Completing the square:
$$4x^2 + 4y^2 - 4x - 8y - 11 = 0$$
$$x^2 + y^2 - x - 2y - \frac{11}{4} = 0$$
$$\left(x^2 - x + \frac{1}{4}\right) + \left(y^2 - 2y + 1\right) = \frac{11}{4} + \frac{1}{4} + 1$$
$$\left(x - \frac{1}{2}\right)^2 + (y-1)^2 = 4$$

The center is $\left(\frac{1}{2}, 1\right)$ and the radius is 2.

19.   Completing the square:
$$x^2 + y^2 - 4y - 2 = 0$$
$$x^2 + \left(y^2 - 4y + 4\right) = 2 + 4$$
$$x^2 + (y-2)^2 = 6$$

The center is $(0,2)$ and the radius is $\sqrt{6}$.

21.   Completing the square:
$$3x^2 + 3y^2 - 6x + 12y - 1 = 0$$
$$x^2 + y^2 - 2x + 4y - \frac{1}{3} = 0$$
$$\left(x^2 - 2x + 1\right) + \left(y^2 + 4y + 4\right) = \frac{1}{3} + 1 + 4$$
$$(x-1)^2 + (y+2)^2 = \frac{16}{3}$$

The center is $(1,-2)$ and the radius is $\sqrt{\frac{16}{3}} = \frac{4}{\sqrt{3}} = \frac{4\sqrt{3}}{3}$.

23. Completing the square:
$$2x^2 + 2y^2 + 6x + 14y + 4 = 0$$
$$x^2 + y^2 + 3x + 7y + 2 = 0$$
$$\left(x^2 + 3x + \frac{9}{4}\right) + \left(y^2 + 7y + \frac{49}{4}\right) = -2 + \frac{9}{4} + \frac{49}{4}$$
$$\left(x + \frac{3}{2}\right)^2 + \left(y + \frac{7}{2}\right)^2 = \frac{25}{2}$$

The center is $\left(-\frac{3}{2}, -\frac{7}{2}\right)$ and the radius is $\sqrt{\frac{25}{2}} = \frac{5}{\sqrt{2}} = \frac{5\sqrt{2}}{2}$.

25. The radius of the circle is: $r = \sqrt{(6-0)^2 + (-8-0)^2} = \sqrt{36 + 64} = \sqrt{100} = 10$. Therefore the equation is:
$$(x - 6)^2 + (y + 8)^2 = 100$$
$$x^2 - 12x + 36 + y^2 + 16y + 64 = 100$$
$$x^2 + y^2 - 12x + 16y = 0$$

27. The midpoint of the line segment is $(3,3)$, which is the center of the circle. Using the distance
formula: $r = \sqrt{(-4-3)^2 + (9-3)^2} = \sqrt{49 + 36} = \sqrt{85}$. The equation is:
$$(x - 3)^2 + (y - 3)^2 = 85$$
$$x^2 - 6x + 9 + y^2 - 6y + 9 = 85$$
$$x^2 + y^2 - 6x - 6y - 67 = 0$$

29. The center must be $(7,-7)$ and the radius is 7, so the equation is:
$$(x - 7)^2 + (y + 7)^2 = 7^2$$
$$x^2 - 14x + 49 + y^2 + 14y + 49 = 49$$
$$x^2 + y^2 - 14x + 14y + 49 = 0$$

31. For one circle the center is $(-3,-5)$ with radius = 5, and for the other circle the center is $(-3,5)$ with radius = 5.
Finding the equations:
$$(x + 3)^2 + (y + 5)^2 = 5^2 \qquad\qquad (x + 3)^2 + (y - 5)^2 = 5^2$$
$$x^2 + 6x + 9 + y^2 + 10y + 25 = 25 \qquad x^2 + 6x + 9 + y^2 - 10y + 25 = 25$$
$$x^2 + y^2 + 6x + 10y + 9 = 0 \qquad\qquad x^2 + y^2 + 6x - 10y + 9 = 0$$

33. The equation is equivalent to $\dfrac{x^2}{4} + \dfrac{y^2}{9} = 1$:

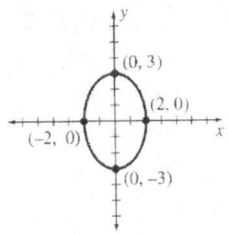

35. The equation is equivalent to $\dfrac{y^2}{9} - \dfrac{x^2}{9} = 1$:

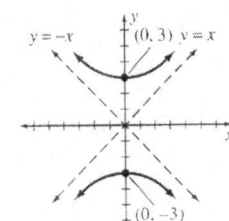

37. First complete the square:

$$x^2 + y^2 - 4x = 0$$

$$\left(x^2 - 4x + 4\right) + y^2 = 4$$

$$(x - 2)^2 + y^2 = 4$$

Graphing the equation:

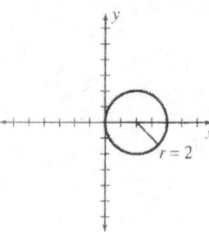

39. The equation is equivalent to $\dfrac{x^2}{36} + \dfrac{y^2}{4} = 1$:

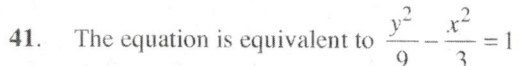

41. The equation is equivalent to $\dfrac{y^2}{9} - \dfrac{x^2}{3} = 1$:

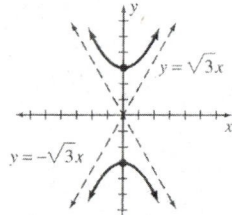

43. First complete the square:

$$x^2 + y^2 + 4x + 6y - 12 = 0$$

$$\left(x^2 + 4x + 4\right) + \left(y^2 + 6y + 9\right) = 12 + 4 + 9$$

$$(x + 2)^2 + (y + 3)^2 = 25$$

Graphing the equation:

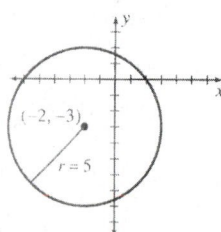

45. The equation is equivalent to $\dfrac{x^2}{3} + \dfrac{y^2}{4} = 1$:

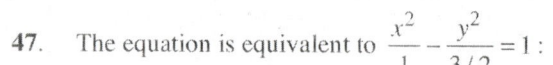

47. The equation is equivalent to $\dfrac{x^2}{1} - \dfrac{y^2}{3/2} = 1$:

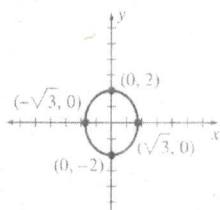

49. Graphing the equation $xy = 2$:

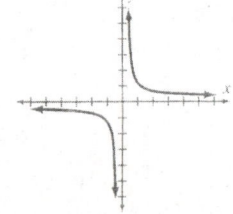

51. Graphing the equation $xy = -3$:

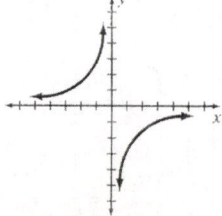

55.   a.   The center is $\left(\dfrac{-2}{-2},\dfrac{-8}{-2}\right)=(1,4)$ and the radius is $r=\sqrt{1^2+4^2-8}=\sqrt{9}=3$.

b.   The center is $\left(\dfrac{4}{-2},\dfrac{-14}{-2}\right)=(-2,7)$ and the radius is $r=\sqrt{(-2)^2+7^2-49}=\sqrt{4}=2$.

c.   The center is $\left(\dfrac{12}{-2},\dfrac{8}{-2}\right)=(-6,-4)$ and the radius is $r=\sqrt{(-6)^2+(-4)^2+12}=\sqrt{64}=8$.

d.   The center is $\left(\dfrac{-16}{-2},\dfrac{20}{-2}\right)=(8,-10)$ and the radius is $r=\sqrt{8^2+(-10)^2-115}=\sqrt{49}=7$.

e.   The center is $\left(\dfrac{0}{-2},\dfrac{-12}{-2}\right)=(0,6)$ and the radius is $r=\sqrt{0^2+6^2+45}=\sqrt{81}=9$.

f.   The center is $\left(\dfrac{14}{-2},\dfrac{0}{-2}\right)=(-7,0)$ and the radius is $r=\sqrt{(-7)^2+0^2-0}=\sqrt{49}=7$.

Chapter 2 Review Problem Set

1.   The point is: $-4+\dfrac{3}{5}(15)=-4+9=5$

2.   The point is: $3+\dfrac{4}{9}(-18)=3-8=-5$

3.   The coordinates are:
$$x=-1+\frac{5}{6}(12)=-1+10=9 \qquad y=-3+\frac{5}{6}(4)=-3+\frac{10}{3}=\frac{1}{3}$$
The point is $\left(9,\dfrac{1}{3}\right)$.

4.   Let (x,y) represent the coordinates of the other endpoint. Using the midpoint formula:
$$\frac{8+x}{2}=3 \qquad\qquad \frac{14+y}{2}=10$$
$$8+x=6 \qquad\qquad 14+y=20$$
$$x=-2 \qquad\qquad\quad y=6$$
The other endpoint is $(-2,6)$.

5.   Finding the distances:
$$d_1=\sqrt{(6-2)^2+(4-2)^2}=\sqrt{16+4}=\sqrt{20}=2\sqrt{5}$$
$$d_2=\sqrt{(5-6)^2+(6-4)^2}=\sqrt{1+4}=\sqrt{5}$$
$$d_3=\sqrt{(5-2)^2+(6-2)^2}=\sqrt{9+16}=\sqrt{25}=5$$
Since $d_1^2+d_2^2=d_3^2$, the points are vertices of a right triangle.

6.   Finding the distances:
$$d_1=\sqrt{(1+3)^2+(3-1)^2}=\sqrt{16+4}=\sqrt{20}=2\sqrt{5}$$
$$d_2=\sqrt{(9-1)^2+(7-3)^2}=\sqrt{64+16}=\sqrt{80}=4\sqrt{5}$$
$$d_3=\sqrt{(9+3)^2+(7-1)^2}=\sqrt{144+36}=\sqrt{180}=6\sqrt{5}$$
Since $d_1+d_2=d_3$, the points lie in a straight line.

7.   Since $(x,-y)$ results in the same equation, the graph has x-axis symmetry.

8.   The graph does not possess any of the three types of symmetry.

9.   Since $(-x,y)$, $(x,-y)$, and $(-x,-y)$ result in the same equation, the graph has x-axis, y-axis, and origin symmetry.

10.   Since $(-x,y)$ results in the same equation, the graph has y-axis symmetry.

11.   Since $(-x,-y)$ results in the same equation, the graph has origin symmetry.

12. Since $(-x, y)$ results in the same equation, the graph has y-axis symmetry.

13. First complete the square:

$$x^2 + y^2 - 6x + 4y - 3 = 0$$
$$\left(x^2 - 6x + 9\right) + \left(y^2 + 4y + 4\right) = 3 + 9 + 4$$
$$(x - 3)^2 + (y + 2)^2 = 16$$

Graphing the equation:

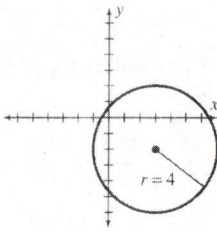

14. The equation is equivalent to $\dfrac{x^2}{16} + \dfrac{y^2}{4} = 1$:

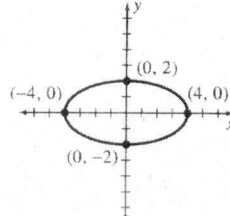

15. The equation is equivalent to $\dfrac{x^2}{16} - \dfrac{y^2}{4} = 1$:

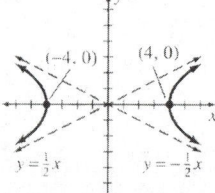

16. Graphing the equation $-2x + 3y = 6$:

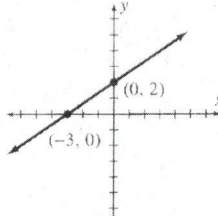

17. Graphing the inequality $2x - y < 4$:

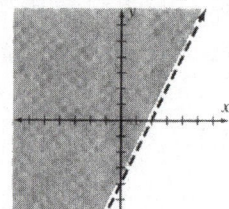

18. Graphing the equation $x^2 y^2 = 4$:

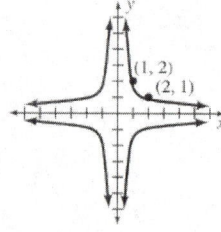

19. The equation is equivalent to $\dfrac{y^2}{2} - \dfrac{x^2}{8/3} = 1$:

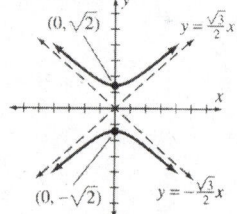

20. First complete the square:

$$x^2 + y^2 + 10y = 0$$
$$x^2 + \left(y^2 + 10y + 25\right) = 25$$
$$x^2 + (y+5)^2 = 5$$

Graphing the equation:

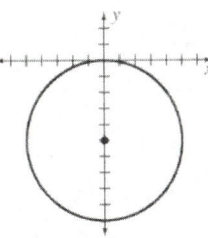

21. The equation is equivalent to $\dfrac{x^2}{4} + \dfrac{y^2}{18} = 1$:

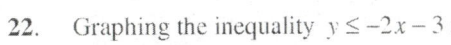

22. Graphing the inequality $y \leq -2x - 3$:

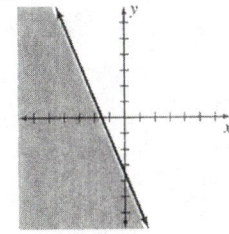

23. Finding the slope: $m = \dfrac{6+4}{-5+3} = \dfrac{10}{-2} = -5$

24. Solving for y:

$$5x - 7y = 12$$
$$-7y = -5x + 12$$
$$y = \frac{5}{7}x - \frac{12}{7}$$

The slope is $\dfrac{5}{7}$.

25. Using the point-slope formula:

$$y - 2 = -\frac{3}{4}(x - 7)$$
$$4y - 8 = -3x + 21$$
$$3x + 4y = 29$$

26. First find the slope: $m = \dfrac{6+2}{1+3} = \dfrac{8}{4} = 2$. Using the point-slope formula:

$$y - 6 = 2(x - 1)$$
$$y - 6 = 2x - 2$$
$$2x - y = -4$$

27. The line $4x + 3y = 17$ is equivalent to $y = -\dfrac{4}{3}x + \dfrac{17}{3}$, so the slope is $-\dfrac{4}{3}$. Using the point-slope formula:

$$y + 4 = -\frac{4}{3}(x - 2)$$
$$3y + 12 = -4x + 8$$
$$4x + 3y = -4$$

28.   The line $2x - y = 7$ is equivalent to $y = 2x - 7$, so the perpendicular slope is $-\dfrac{1}{2}$. Using the point-slope formula:

$$y - 4 = -\frac{1}{2}(x + 5)$$
$$2y - 8 = -x - 5$$
$$x + 2y = 3$$

29.   The equation is:
$$(x - 5)^2 + (y + 6)^2 = 1^2$$
$$x^2 - 10x + 25 + y^2 + 12y + 36 = 1$$
$$x^2 + y^2 - 10x + 12y + 60 = 0$$

30.   The midpoint of the diameter is $(2, 3)$, which is the center of the circle. Using the distance formula:
$$r = \sqrt{(6 - 2)^2 + (2 - 3)^2} = \sqrt{16 + 1} = \sqrt{17}$$

The equation is:
$$(x - 2)^2 + (y - 3)^2 = 17$$
$$x^2 - 4x + 4 + y^2 - 6y + 9 = 17$$
$$x^2 + y^2 - 4x - 6y - 4 = 0$$

31.   Using the distance formula: $r = \sqrt{(-5 - 0)^2 + (12 - 0)^2} = \sqrt{25 + 144} = \sqrt{169} = 13$. The equation is:
$$(x + 5)^2 + (y - 12)^2 = 13^2$$
$$x^2 + 10x + 25 + y^2 - 24y + 144 = 169$$
$$x^2 + y^2 + 10x - 24y = 0$$

32.   The center is $(-4, -4)$, so the equation is:
$$(x + 4)^2 + (y + 4)^2 = 4^2$$
$$x^2 + 8x + 16 + y^2 + 8y + 16 = 16$$
$$x^2 + y^2 + 8x + 8y + 16 = 0$$

Chapter 2 Test

1.   The point is: $-4 + \dfrac{2}{3}(18) = -4 + 12 = 8$

2.   The coordinates of the point are:

$$x = 2 + \frac{3}{4}(-8) = 2 - 6 = -4 \qquad\qquad y = -3 + \frac{3}{4}(12) = -3 + 9 = 6$$

The point is $(-4, 6)$.

3.   Let (x, y) represent the coordinates of the other point. Using the midpoint formula:

$$\frac{x - 2}{2} = 2 \qquad\qquad\qquad \frac{y - 1}{2} = -\frac{5}{2}$$
$$x - 2 = 4 \qquad\qquad\qquad\quad y - 1 = -5$$
$$x = 6 \qquad\qquad\qquad\qquad y = -4$$

The other endpoint is $(6, -4)$.

4.   Finding the slope: $m = \dfrac{-6 + 2}{5 + 4} = -\dfrac{4}{9}$

5. Solving for y:

$$2x - 7y = -9$$
$$-7y = -2x - 9$$
$$y = \frac{2}{7}x + \frac{9}{7}$$

The slope is $\frac{2}{7}$.

6. The equation is:

$$y = -\frac{3}{4}x - 3$$
$$4y = -3x - 12$$
$$3x + 4y = -12$$

7. Finding the slope: $m = \dfrac{7+4}{4-1} = \dfrac{11}{3}$. Using the point-slope formula:

$$y - 7 = \tfrac{11}{3}(x - 4)$$
$$3y - 21 = 11x - 44$$
$$11x - 3y = 23$$

8. The line $x - 5y = 5$ is equivalent to $y = \frac{1}{5}x - 1$, so the parallel slope is $\frac{1}{5}$. Using the point-slope formula:

$$y - 4 = \frac{1}{5}(x + 1)$$
$$5y - 20 = x + 1$$
$$x - 5y = -21$$

9. The line $4x + 7y = 3$ is equivalent to $y = -\frac{4}{7}x + \frac{3}{7}$, so the perpendicular slope is $\frac{7}{4}$. Using the point-slope formula:

$$y - 5 = \frac{7}{4}(x - 3)$$
$$4y - 20 = 7x - 21$$
$$7x - 4y = 1$$

10. For the line to be perpendicular to the x-axis, the line is vertical, so its equation is $x = -2$, or $x + 0y = -2$.

11. The equation is:

$$(x + 3)^2 + (y + 6)^2 = 4^2$$
$$x^2 + 6x + 9 + y^2 + 12y + 36 = 16$$
$$x^2 + y^2 + 6x + 12y + 29 = 0$$

12. The center is the midpoint of the endpoints, which is $(2,4)$. Using the distance formula:

$$r = \sqrt{(5-2)^2 + (5-4)^2} = \sqrt{9+1} = \sqrt{10}$$

The equation is:

$$(x - 2)^2 + (y - 4)^2 = 10$$
$$x^2 - 4x + 4 + y^2 - 8y + 16 = 10$$
$$x^2 + y^2 - 4x - 8y + 10 = 0$$

13. Using the distance formula: $r = \sqrt{(4-0)^2 + (-3-0)^2} = \sqrt{16+9} = \sqrt{25} = 5$. The equation is:

$$(x - 4)^2 + (y + 3)^2 = 5^2$$
$$x^2 - 8x + 16 + y^2 + 6y + 9 = 25$$
$$x^2 + y^2 - 8x + 6y = 0$$

14. Completing the square:

$$x^2 + 16x + y^2 - 10y + 80 = 0$$
$$\left(x^2 + 16x + 64\right) + \left(y^2 - 10y + 25\right) = -80 + 64 + 25$$
$$(x + 8)^2 + (y - 5)^2 = 9$$

The center is $(-8,5)$ and the radius is 3.

15. Using the distance formula:

$$d_1 = \sqrt{(5-3)^2 + (-2-2)^2} = \sqrt{4+16} = \sqrt{20} = 2\sqrt{5}$$

$$d_2 = \sqrt{(-1-5)^2 + (-1+2)^2} = \sqrt{36+1} = \sqrt{37}$$

$$d_3 = \sqrt{(-1-3)^2 + (-1-2)^2} = \sqrt{16+9} = \sqrt{25} = 5$$

16. Setting $y = 0$:

$$x^2 - 6x + 5 = 0$$
$$(x-5)(x-1) = 0$$
$$x = 1, 5$$

17. Setting $x = 0$:

$$12y^2 = 36$$
$$y^2 = 3$$
$$y = \pm\sqrt{3}$$

18. The equation is equivalent to $\dfrac{x^2}{2} + \dfrac{y^2}{9} = 1$, so $a = 3$. Thus the major axis has a length of 6.

19. The equation is equivalent to $\dfrac{x^2}{16/3} - \dfrac{y^2}{3} = 1$, so $a = \sqrt{\dfrac{16}{3}} = \dfrac{4\sqrt{3}}{3}$ and $b = \sqrt{3}$. Thus the asymptotes

have equations: $y = \pm\dfrac{b}{a}x = \pm\dfrac{\sqrt{3}}{\dfrac{4\sqrt{3}}{3}}x = \pm\dfrac{3}{4}x$

20.
 a. Since $(x, -y)$ results in the same equation, the graph has x-axis symmetry.

 b. Since $(-x, -y)$ results in the same equation, the graph has origin symmetry.

 c. Since $(-x, y)$ results in the same equation, the graph has y-axis symmetry.

 d. Since $(-x, y)$, $(x, -y)$, and $(-x, -y)$ result in the same equation, the graph has x-axis, y-axis, and origin symmetry.

21. Graphing the inequality $3x - y \le 6$:

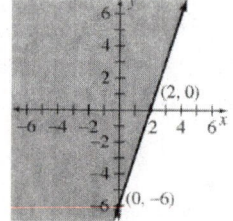

22. The equation is equivalent to $\dfrac{y^2}{9} - \dfrac{x^2}{9/2} = 1$:

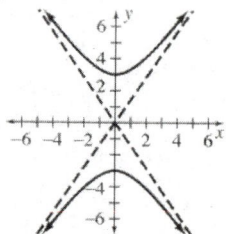

23. Graphing the equation $x = y^2 - 4$:

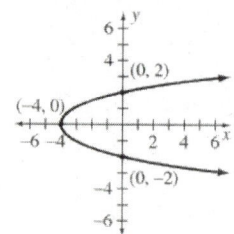

24. The equation is equivalent to $\dfrac{x^2}{15} + \dfrac{y^2}{9} = 1$:

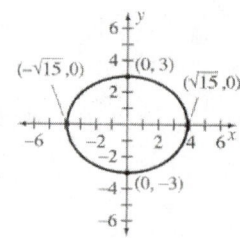

25. First complete the square:

$$x^2 + 4x + y^2 - 12 = 0$$

$$\left(x^2 + 4x + 4\right) + y^2 = 12 + 4$$

$$(x+2)^2 + y^2 = 16$$

Graphing the equation:

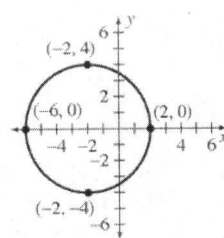

Chapters 0-2 Cumulative Review Problem Set

1. Evaluating: $3^{-3} = \dfrac{1}{3^3} = \dfrac{1}{27}$

2. Evaluating: $-4^{-2} = -\dfrac{1}{4^2} = -\dfrac{1}{16}$

3. Evaluating: $\left(\dfrac{2}{3}\right)^{-2} = \left(\dfrac{3}{2}\right)^2 = \dfrac{9}{4}$

4. Evaluating: $-\sqrt[3]{\dfrac{8}{27}} = -\dfrac{2}{3}$

5. Evaluating: $\left(\dfrac{1}{27}\right)^{-2/3} = (27)^{2/3} = \left(\sqrt[3]{27}\right)^2 = (3)^2 = 9$

6. Evaluating: $\dfrac{1}{\left(\dfrac{3}{4}\right)^{-2}} = \left(\dfrac{3}{4}\right)^2 = \dfrac{9}{16}$

7. Simplifying: $\left(5x^{-3}y^{-2}\right)\left(4xy^{-1}\right) = 20x^{-2}y^{-3} = \dfrac{20}{x^2y^3}$

8. Simplifying: $\left(-7a^{-3}b^2\right)\left(8a^4b^{-3}\right) = -56ab^{-1} = -\dfrac{56a}{b}$

9. Simplifying: $\left(\dfrac{1}{2}x^{-2}y^{-1}\right)^{-2} = \left(\dfrac{1}{2}\right)^{-2}x^4y^2 = 4x^4y^2$

10. Simplifying: $\dfrac{80x^{-3}y^{-4}}{16xy^{-6}} = 5x^{-4}y^2 = \dfrac{5y^2}{x^4}$

11. Simplifying: $\left(\dfrac{102x^{2/3}y^{3/4}}{6xy^{-1}}\right)^{-1} = \left(17x^{-1/3}y^{7/4}\right)^{-1} = 17^{-1}x^{1/3}y^{-7/4} = \dfrac{x^{1/3}}{17y^{7/4}}$

12. Simplifying: $\left(\dfrac{14a^3b^{-4}}{7a^{-1}b^3}\right)^2 = \left(2a^4b^{-7}\right)^2 = 4a^8b^{-14} = \dfrac{4a^8}{b^{14}}$

13. Simplifying: $-5\sqrt{72} = -5\sqrt{36 \cdot 2} = -30\sqrt{2}$

14. Simplifying: $2\sqrt{27x^3y^2} = 2\sqrt{9x^2y^2 \cdot 3x} = 6xy\sqrt{3x}$

15. Simplifying: $\sqrt[3]{56x^4y^7} = \sqrt[3]{8x^3y^6 \cdot 7xy} = 2xy^2\sqrt[3]{7xy}$

16. Simplifying: $\dfrac{3\sqrt{18}}{5\sqrt{12}} = \dfrac{3}{5}\sqrt{\dfrac{18}{12}} = \dfrac{3}{5}\sqrt{\dfrac{3}{2}} = \dfrac{3\sqrt{3}}{5\sqrt{2}} \cdot \dfrac{\sqrt{2}}{\sqrt{2}} = \dfrac{3\sqrt{6}}{10}$

17. Simplifying: $\sqrt{\dfrac{3x}{7y}} = \dfrac{\sqrt{3x}}{\sqrt{7y}} \cdot \dfrac{\sqrt{7y}}{\sqrt{7y}} = \dfrac{\sqrt{21xy}}{7y}$

18. Simplifying: $\dfrac{5}{\sqrt{2}-3} = \dfrac{5}{\sqrt{2}-3} \cdot \dfrac{\sqrt{2}+3}{\sqrt{2}+3} = \dfrac{5\sqrt{2}+15}{2-9} = -\dfrac{5\left(\sqrt{2}+3\right)}{7}$

19. Simplifying: $\dfrac{3\sqrt{7}}{2\sqrt{2}-\sqrt{6}} = \dfrac{3\sqrt{7}}{2\sqrt{2}-\sqrt{6}} \cdot \dfrac{2\sqrt{2}+\sqrt{6}}{2\sqrt{2}+\sqrt{6}} = \dfrac{6\sqrt{14}+3\sqrt{42}}{8-6} = \dfrac{6\sqrt{14}+3\sqrt{42}}{2}$

20. Simplifying: $\dfrac{4\sqrt{x}}{\sqrt{x}+3\sqrt{y}} = \dfrac{4\sqrt{x}}{\sqrt{x}+3\sqrt{y}} \cdot \dfrac{\sqrt{x}-3\sqrt{y}}{\sqrt{x}-3\sqrt{y}} = \dfrac{4x-12\sqrt{xy}}{x-9y}$

21. Simplifying: $\dfrac{12x^2y}{18x} \cdot \dfrac{9x^3y^3}{16xy^2} = \dfrac{3x^5y^4}{8x^2y^2} = \dfrac{3x^3y^2}{8}$

22. Simplifying: $\dfrac{-15ab^2}{14a^3b} \div \dfrac{20a}{7b^2} = \dfrac{-15ab^2}{14a^3b} \cdot \dfrac{7b^2}{20a} = \dfrac{-3ab^4}{8a^4b} = -\dfrac{3b^3}{8a^3}$

23. Simplifying: $\dfrac{3x^2+5x-2}{x^2-4} \cdot \dfrac{5x^2-9x-2}{3x^2-x} = \dfrac{(3x-1)(x+2)}{(x+2)(x-2)} \cdot \dfrac{(5x+1)(x-2)}{x(3x-1)} = \dfrac{5x+1}{x}$

24. Simplifying: $\dfrac{2x-1}{4} + \dfrac{3x+2}{6} - \dfrac{x-1}{8} = \dfrac{12x-6}{24} + \dfrac{12x+8}{24} - \dfrac{3x-3}{24} = \dfrac{12x-6+12x+8-3x+3}{24} = \dfrac{21x+5}{24}$

25. Simplifying: $\dfrac{5}{3n^2} - \dfrac{2}{n} + \dfrac{3}{2n} = \dfrac{10}{6n^2} - \dfrac{12n}{6n^2} + \dfrac{9n}{6n^2} = \dfrac{10-12n+9n}{6n^2} = \dfrac{10-3n}{6n^2}$

26. Simplifying:

$$\dfrac{5x}{x^2+6x-27} + \dfrac{3}{x^2-9} = \dfrac{5x}{(x+9)(x-3)} + \dfrac{3}{(x+3)(x-3)}$$

$$= \dfrac{5x(x+3)}{(x+9)(x+3)(x-3)} + \dfrac{3(x+9)}{(x+9)(x+3)(x-3)}$$

$$= \dfrac{5x^2+15x+3x+27}{(x+9)(x+3)(x-3)}$$

$$= \dfrac{5x^2+18x+27}{(x+9)(x+3)(x-3)}$$

27. Solving the equation:
$$3(-2x-1)-2(3x+4) = -4(2x-3)$$
$$-6x-3-6x-8 = -8x+12$$
$$-12x-11 = -8x+12$$
$$-4x = 23$$
$$x = -\dfrac{23}{4}$$

The solution set is $\left\{-\dfrac{23}{4}\right\}$.

28. Solving the equation:
$$(2x-1)(3x+4) = (x+2)(6x-5)$$
$$6x^2+5x-4 = 6x^2+7x-10$$
$$5x-4 = 7x-10$$
$$-2x = -6$$
$$x = 3$$

The solution set is $\{3\}$.

29. Solving the equation:
$$\dfrac{3x-1}{4} - \dfrac{2x-1}{5} = \dfrac{1}{10}$$
$$100 \cdot \dfrac{3x-1}{4} - 100 \cdot \dfrac{2x-1}{5} = 100 \cdot \dfrac{1}{10}$$
$$75x-25-40x+20 = 10$$
$$35x-5 = 10$$
$$35x = 15$$
$$x = \dfrac{3}{7}$$

The solution set is $\left\{\dfrac{3}{7}\right\}$.

30. Solving the equation:
$$9x^2-4 = 0$$
$$(3x+2)(3x-2) = 0$$
$$x = -\dfrac{2}{3}, \dfrac{2}{3}$$

The solution set is $\left\{-\dfrac{2}{3}, \dfrac{2}{3}\right\}$.

31. Solving the equation:
$$5x^3 + 10x^2 - 40x = 0$$
$$5x(x^2 + 2x - 8) = 0$$
$$5x(x+4)(x-2) = 0$$
$$x = -4, 0, 2$$

The solution set is $\{-4, 0, 2\}$.

32. Solving the equation:
$$7t^2 - 31t + 12 = 0$$
$$(7t - 3)(t - 4) = 0$$
$$t = \frac{3}{7}, 4$$

The solution set is $\left\{ \frac{3}{7}, 4 \right\}$.

33. Solving the equation:
$$x^4 + 15x^2 - 16 = 0$$
$$(x^2 + 16)(x^2 - 1) = 0$$
$$x^2 = -16, 1$$
$$x = \pm 4i, \pm 1$$

The solution set is $\{\pm 4i, \pm 1\}$.

34. Solving the equation:
$$|5x - 2| = 3$$
$$5x - 2 = -3, 3$$
$$5x = -1, 5$$
$$x = -\frac{1}{5}, 1$$

The solution set is $\left\{ -\frac{1}{5}, 1 \right\}$.

35. Using $a = 2$, $b = -3$, and $c = -1$ in the quadratic formula: $x = \dfrac{3 \pm \sqrt{9 - 4(2)(-1)}}{4} = \dfrac{3 \pm \sqrt{17}}{4}$

The solution set is $\left\{ \dfrac{3 \pm \sqrt{17}}{4} \right\}$.

36. Solving the equation:
$$(3x - 2)(x + 4) = (2x - 1)(x - 1)$$
$$3x^2 + 10x - 8 = 2x^2 - 3x + 1$$
$$x^2 + 13x - 9 = 0$$

Using $a = 1$, $b = 13$, and $c = -9$ in the quadratic formula: $x = \dfrac{-13 \pm \sqrt{169 - 4(1)(-9)}}{2} = \dfrac{-13 \pm \sqrt{205}}{2}$

The solution set is $\left\{ \dfrac{-13 \pm \sqrt{205}}{2} \right\}$.

37. Solving the equation:
$$\sqrt{5 - t} + 1 = \sqrt{7 + 2t}$$
$$\left(\sqrt{5 - t} + 1\right)^2 = \left(\sqrt{7 + 2t}\right)^2$$
$$5 - t + 2\sqrt{5 - t} + 1 = 7 + 2t$$
$$2\sqrt{5 - t} = 3t + 1$$
$$\left(2\sqrt{5 - t}\right)^2 = (3t + 1)^2$$
$$20 - 4t = 9t^2 + 6t + 1$$
$$0 = 9t^2 + 10t - 19$$
$$0 = (t - 1)(9t + 19)$$
$$t = 1 \quad \left(t = -\frac{19}{9} \text{ does not check} \right)$$

The solution set is $\{1\}$.

38.   Solving the equation:

$$(2x-1)^2 + 4 = 0$$
$$(2x-1)^2 = -4$$
$$2x-1 = \pm 2i$$
$$2x = 1 \pm 2i$$
$$x = \frac{1}{2} \pm i$$

The solution set is $\left\{\dfrac{1}{2} \pm i\right\}$.

39.   Solving the inequality:

$$-2(x-1) + (3-2x) > 4(x+1)$$
$$-2x + 2 + 3 - 2x > 4x + 4$$
$$-4x + 5 > 4x + 4$$
$$-8x > -1$$
$$x < \frac{1}{8}$$

The solution set is $\left(-\infty, \dfrac{1}{8}\right)$.

40.   Solving the inequality:

$$2n + 1 + \frac{3n-1}{4} \geq \frac{n-1}{2}$$
$$4(2n+1) + 4 \cdot \frac{3n-1}{4} \geq 4 \cdot \frac{n-1}{2}$$
$$8n + 4 + 3n - 1 \geq 2n - 2$$
$$11n + 3 \geq 2n - 2$$
$$9n \geq -5$$
$$n \geq -\frac{5}{9}$$

The solution set is $\left[-\dfrac{5}{9}, \infty\right)$.

41.   Solving the inequality:

$$0.09x + 0.12(450 - x) \geq 46.5$$
$$0.09x + 54 - 0.12x \geq 46.5$$
$$-0.03x \geq -7.5$$
$$x \leq 250$$

The solution set is $\left(-\infty, 250\right]$.

42.   Factoring the inequality:

$$n^2 + 5n \geq 24$$
$$n^2 + 5n - 24 \geq 0$$
$$(n+8)(n-3) \geq 0$$

Forming a sign chart:

$$\begin{array}{c c c} + & - & + \\ \hline & -8 \quad\quad\quad 3 & \end{array}$$

The solution set is $(-\infty, -8) \cup (3, \infty)$.

43.   Factoring the inequality:

$$6x^2 + 7x - 3 < 0$$
$$(3x-1)(2x+3) < 0$$

Forming a sign chart:

$$\begin{array}{c c c} + & - & + \\ \hline & -3/2 \quad\quad\quad 1/3 & \end{array}$$

The solution set is $\left(-\dfrac{3}{2}, \dfrac{1}{3}\right)$.

44.   Forming a sign chart:

$$\begin{array}{c c c c} - & + & - & + \\ \hline & -3 \quad\; 1/2 \quad\; 4 & & \end{array}$$

The solution set is $\left(-3, \dfrac{1}{2}\right) \cup (4, \infty)$.

45.   Forming a sign chart:

$$\begin{array}{c c c} + & - & + \\ \hline & -1 \quad\quad\quad 2/3 & \end{array}$$

The solution set is $\left(-1, \dfrac{2}{3}\right]$.

46.   Solving the inequality:

$$\frac{x+5}{x-1} \geq 2$$
$$\frac{x+5}{x-1} - 2 \geq 0$$
$$\frac{x+5-2x+2}{x-1} \geq 0$$
$$\frac{-x+7}{x-1} \geq 0$$

Forming a sign chart:

$$\begin{array}{c c c} - & + & - \\ \hline & 1 \quad\quad\quad 7 & \end{array}$$

The solution set is $\left(1, 7\right]$.

47. Solving the inequality:

$$|3x - 1| > 5$$

$$3x - 1 < -5 \qquad \text{or} \qquad 3x - 1 > 5$$

$$3x \le -4 \qquad\qquad\qquad 3x > 6$$

$$x \le -\frac{4}{3} \qquad\qquad\qquad x > 2$$

The solution set is $\left(-\infty, -\dfrac{4}{3}\right] \cup (2, \infty)$.

48. Solving the inequality:

$$|5x - 3| < 12$$

$$-12 < 5x - 3 < 12$$

$$-9 < 5x < 15$$

$$-\frac{9}{5} < x < 3$$

The solution set is $\left(-\dfrac{9}{5}, 3\right)$.

49. The equation is equivalent to $\dfrac{x^2}{36} + \dfrac{y^2}{9} = 1$:

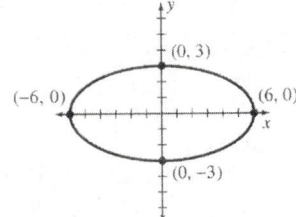

50. The equation is equivalent to $\dfrac{x^2}{1} - \dfrac{y^2}{4} = 1$:

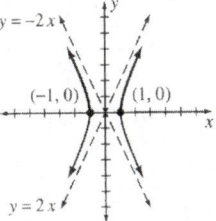

51. Graphing the equation $y = -x^3 - 1$:

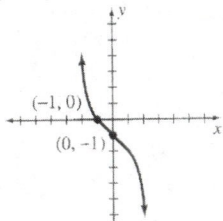

52. Graphing the equation $y = -x + 3$:

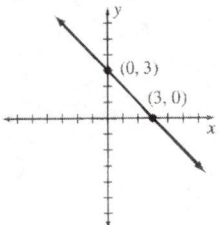

53. The equation is equivalent to $\dfrac{y^2}{9} - \dfrac{x^2}{9/5} = 1$:

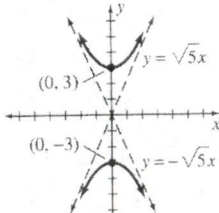

54. Graphing the equation $y = -\dfrac{3}{4}x - 1$:

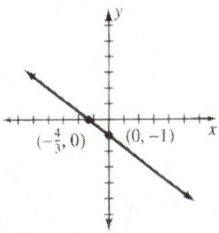

55. Completing the square:

$$x^2 + y^2 + 14x - 8y + 56 = 0$$

$$\left(x^2 + 14x + 49\right) + \left(y^2 - 8y + 16\right) = -56 + 49 + 16$$

$$(x + 7)^2 + (y - 4)^2 = 9$$

The center is $(-7, 4)$ and the radius is 3.

56. The line $3x - 4y = 17$ is equivalent to $y = \dfrac{3}{4}x - \dfrac{17}{4}$, so its slope is $\dfrac{3}{4}$. Using the point-slope formula:

$$y - 8 = \frac{3}{4}(x - 2)$$

$$4y - 32 = 3x - 6$$

$$3x - 4y = -26$$

57. The coordinates are given by:

$$x = -3 + \frac{1}{5}(5) = -3 + 1 = -2 \qquad\qquad y = 4 + \frac{1}{5}(10) = 4 + 2 = 6$$

The point is $(-2, 6)$.

58. The midpoint is $(1,7)$. Finding the slope: $m = \dfrac{10-4}{5+3} = \dfrac{6}{8} = \dfrac{3}{4}$

The perpendicular slope is $-\dfrac{4}{3}$. Using the point-slope formula:

$$y - 7 = -\frac{4}{3}(x - 1)$$
$$3y - 21 = -4x + 4$$
$$4x + 3y = 25$$

59. To obtain a profit which is 30% of the cost, the equation is: $22 + 0.30(22) = \$28.60$

To obtain a profit which is 30% of the selling price, the equation is:
$$22 + 0.30x = x$$
$$22 = 0.70x$$
$$x = \$31.43$$

60. Let x represent the amount invested at 5% and $7500 - x$ the amount invested at 6%. The equation is:
$$0.05(x) + 0.06(7500 - x) = 420$$
$$0.05x + 450 - 0.06x = 420$$
$$-0.01x = -30$$
$$x = 3000$$

Thus $3000 was invested at 5% and $4500 was invested at 6%.

61. Let w represent the width and $2w - 1$ represent the length. The equation is:
$$w(2w - 1) = 36$$
$$2w^2 - w - 36 = 0$$
$$(2w - 9)(w + 4) = 0$$
$$w = \frac{9}{2} \quad (w = -4 \text{ is impossible})$$

The width is 4.5 inches and the length is 8 inches.

62. Let h represent the altitude and $3h - 4$ represent the length of the side. The equation is:
$$\frac{1}{2}h(3h - 4) = 80$$
$$3h^2 - 4h = 160$$
$$3h^2 - 4h - 160 = 0$$
$$(3h + 20)(h - 8) = 0$$
$$h = 8 \quad \left(h = -\frac{20}{3} \text{ is impossible}\right)$$

The altitude is 8 cm and the side is 20 cm.

63. Let x represent the amount of pure acid to use. The equation is:
$$0.30(40) + x = 0.50(40 + x)$$
$$12 + x = 20 + 0.5x$$
$$0.5x = 8$$
$$x = 16$$

Thus 16 ml should be added.

64. Let t represent her time riding out, and $5 - t$ represent her time riding back. The distances are the same, so the equation is:
$$15(t) = 10(5 - t)$$
$$15t = 50 - 10t$$
$$25t = 50$$
$$t = 2$$

She rode a distance of 30 miles out.

65.   Note that 1 hr 12 minutes is $\frac{6}{5}$ hours. Let t represent the time to fill the pool for the other pipe. The equation is:

$$\frac{1}{2}\left(\frac{6}{5}\right)+\frac{1}{t}\left(\frac{6}{5}\right)=1$$

$$\frac{3}{5}+\frac{6}{5t}=1$$

$$3t+6=5t$$

$$6=2t$$

$$t=3$$

It will take the second pipe 3 hours to fill the pool by itself.

Chapter 3
Functions

3.1 Concept of a Function

1. Yes, these ordered pairs represent a function.
3. No, these ordered pairs do not represent a function, since both 0 and 1 are assigned two different values.
5. Yes, these ordered pairs represent a function.
7. This graph passes the vertical line test, so it represents the graph of a function.
9. This graph does not pass the vertical line test, so it does not represent the graph of a function.
11. This graph passes the vertical line test, so it represents the graph of a function.
14. This graph does not pass the vertical line test, so it does not represent the graph of a function.

15. Finding the values:
$$f(3) = -2(3) + 5 = -1$$
$$f(5) = -2(5) + 5 = -5$$
$$f(-2) = -2(-2) + 5 = 9$$

17. Finding the values:
$$g(3) = -2(3)^2 + (3) - 5 = -20$$
$$g(-1) = -2(-1)^2 + (-1) - 5 = -8$$
$$g(2a) = -2(2a)^2 + (2a) - 5 = -8a^2 + 2a - 5$$

19. Finding the values:
$$h(3) = \frac{2}{3}(3) - \frac{3}{4} = 2 - \frac{3}{4} = \frac{5}{4}$$
$$h(4) = \frac{2}{3}(4) - \frac{3}{4} = \frac{8}{3} - \frac{3}{4} = \frac{23}{12}$$
$$h\left(-\frac{1}{2}\right) = \frac{2}{3}\left(-\frac{1}{2}\right) - \frac{3}{4} = -\frac{1}{3} - \frac{3}{4} = -\frac{13}{12}$$

21. Finding the values:
$$f(5) = \sqrt{2(5) - 1} = \sqrt{9} = 3$$
$$f\left(\frac{1}{2}\right) = \sqrt{2\left(\frac{1}{2}\right) - 1} = \sqrt{0} = 0$$
$$f(23) = \sqrt{2(23) - 1} = \sqrt{45} = 3\sqrt{5}$$

23. Finding the values:
$$f(a) = -2(a) + 7 = -2a + 7$$
$$f(a+2) = -2(a+2) + 7 = -2a - 4 + 7 = -2a + 3$$
$$f(a+h) = -2(a+h) + 7 = -2a - 2h + 7$$

25. Finding the values:
$$f(-a) = (-a)^2 - 4(-a) + 10 = a^2 + 4a + 10$$
$$f(a-4) = (a-4)^2 - 4(a-4) + 10 = a^2 - 8a + 16 - 4a + 16 + 10 = a^2 - 12a + 42$$
$$f(a+h) = (a+h)^2 - 4(a+h) + 10 = a^2 + 2ah + h^2 - 4a - 4h + 10$$

27. Finding the values:

$$f(-x) = (-x)^2 + 3(-x) + 5 = x^2 - 3x + 5$$
$$f(3x) = (3x)^2 + 3(3x) + 5 = 9x^2 + 9x + 5$$
$$f(1-x) = (1-x)^2 + 3(1-x) + 5 = 1 - 2x + x^2 + 3 - 3x + 5 = x^2 - 5x + 9$$

29. Finding the values:

$$f(4) = 4$$
$$f(10) = 10$$
$$f(-3) = (-3)^2 = 9$$
$$f(-5) = (-5)^2 = 25$$

31. Finding the values:

$$f(3) = 2(3) = 6$$
$$f(5) = 2(5) = 10$$
$$f(-3) = -2(-3) = 6$$
$$f(-5) = -2(-5) = 10$$

33. Finding the values:

$$f(2) = 1 \qquad f(0) = 0 \qquad f\left(-\frac{1}{2}\right) = 0 \qquad f(-4) = -1$$

35. The domain is $\left\{x \mid x \geq \dfrac{4}{3}\right\}$ and the range is $\{f(x) \mid f(x) \geq 0\}$.

37. The domain is $\{x \mid x \text{ is any real number}\}$ and the range is $\{f(x) \mid f(x) \geq -2\}$.

39. The domain is $\{x \mid x \text{ is any real number}\}$ and the range is $\{f(x) \mid f(x) \geq 0\}$.

41. The domain is $\{x \mid x \geq 0\}$ and the range is $\{f(x) \mid f(x) \leq 0\}$.

43. The domain is $\{x \mid x \text{ is any real number}\}$ and the range is $\{f(x) \mid f(x) = 6\}$.

45. The domain is $\{x \mid x \geq -4\}$ and the range is $\{f(x) \mid f(x) \geq -2\}$.

47. The domain is $\{x \mid x \text{ is any real number}\}$ and the range is $\{f(x) \mid f(x) \leq -6\}$.

49. The domain is $\{x \mid x \neq -2\}$.

51. The domain is $\left\{x \mid x \neq -4, \dfrac{1}{2}\right\}$.

53. Since $x^2 - 4 = (x+2)(x-2)$, the domain is $\{x \mid x \neq -2, 2\}$.

55. Since $x^2 - x - 12 = (x+3)(x-4)$, the domain is $\{x \mid x \neq -3, 4\}$.

57. Since $6x^2 + 13x - 5 = (3x-1)(2x+5)$, the domain is $\left\{x \mid x \neq -\dfrac{5}{2}, \dfrac{1}{3}\right\}$.

59. Since $x^2 + 1 \neq 0$, the domain is $\{x \mid x \text{ is any real number}\}$.

61. The quantity inside the radical must be positive, so:

$$-4x + 3 \geq 0$$
$$-4x \geq -3$$
$$x \leq \frac{3}{4}$$

The domain is $\left\{x \mid x \leq \dfrac{3}{4}\right\}$.

63. Solving the inequality:

$$x^2 - 16 \geq 0$$
$$(x+4)(x-4) \geq 0$$

Forming the sign chart:

The domain is $(-\infty, -4] \cup [4, \infty)$.

65. The quantity inside the radical is always positive, so the domain is $(-\infty, \infty)$.

67. Solving the inequality:
$$x^2 - 3x - 40 \geq 0$$
$$(x - 8)(x + 5) \geq 0$$
Forming the sign chart:

The domain is $(-\infty, -5] \cup [8, \infty)$.

69. Solving the inequality:
$$8x^2 + 6x - 35 \geq 0$$
$$(4x - 7)(2x + 5) \geq 0$$
Forming the sign chart:

The domain is $\left(-\infty, -\dfrac{5}{2}\right] \cup \left[\dfrac{7}{4}, \infty\right)$.

71. Solving the inequality:
$$1 - x^2 \geq 0$$
$$(1 + x)(1 - x) \geq 0$$
Forming a sign chart:

The domain is $[-1, 1]$.

73. Since $f(-x) = (-x)^2 = x^2 = f(x)$, the function is even.

75. Since $f(-x) = (-x)^2 + 1 = x^2 + 1 = f(x)$, the function is even.

77. Since $f(-x) = (-x)^2 + (-x) = x^2 - x$, the function is neither.

79. Since $f(-x) = (-x)^5 = -x^5 = -f(x)$, the function is odd.

81. Since $f(-x) = -(-x)^3 = x^3 = -f(x)$, the function is odd.

83. Evaluating the charges separately:

color copies: $c(20) = 0.89(20) = \$17.80$

black and white copies: $b(210) = 0.06(210) = \$12.60$

The total cost is $\$17.80 + \$12.60 = \$30.40$.

85. Evaluating the areas:
$$A(2) = \pi(2)^2 = 4\pi \approx 12.57$$
$$A(3) = \pi(3)^2 = 9\pi \approx 28.27$$
$$A(12) = \pi(12)^2 = 144\pi \approx 452.39$$
$$A(17) = \pi(17)^2 = 289\pi \approx 907.92$$

87. Finding the heights:
$$h(1) = 64(1) - 16(1)^2 = 48$$
$$h(2) = 64(2) - 16(2)^2 = 64$$
$$h(3) = 64(3) - 16(3)^2 = 48$$
$$h(4) = 64(4) - 16(4)^2 = 0$$

89. Computing the interest earned:
$$I(0.11) = 500(0.11) = \$55$$
$$I(0.12) = 500(0.12) = \$60$$
$$I(0.135) = 500(0.135) = \$67.50$$
$$I(0.15) = 500(0.15) = \$75$$

91. Computing the surface areas:
$$A(2) = 2\pi(2)^2 + 16\pi(2) = 40\pi \approx 125.66$$
$$A(4) = 2\pi(4)^2 + 16\pi(4) = 96\pi \approx 301.59$$
$$A(8) = 2\pi(8)^2 + 16\pi(8) = 256\pi \approx 804.25$$

95. Any function of the form $f(x) = mx$ satisfies this property.

3.2 Linear Functions and Applications

1.   Graphing the function $f(x) = 2x - 4$:

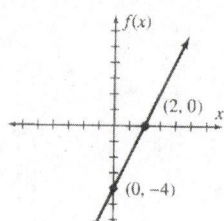

3.   Graphing the function $f(x) = -x + 3$:

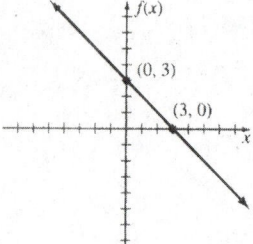

5.   Graphing the function $f(x) = 3x + 9$:

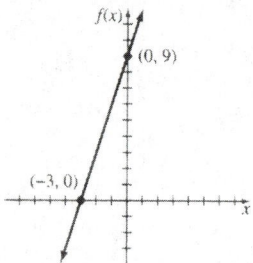

7.   Graphing the function $f(x) = -4x - 4$:

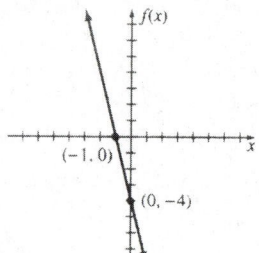

9.   Graphing the function $f(x) = -3x$:

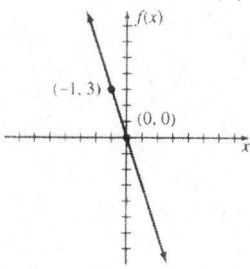

11.   Graphing the function $f(x) = -3$:

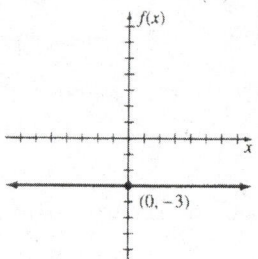

13.   Graphing the function $f(x) = \dfrac{1}{2}x + 3$:

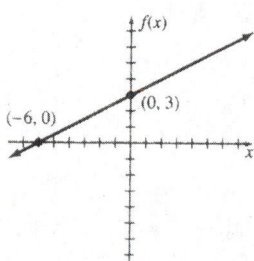

15.   Graphing the function $f(x) = -\dfrac{3}{4}x$:

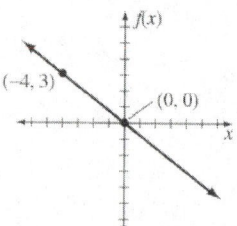

17.   Graphing the function $f(x) = -\dfrac{1}{3}x$:

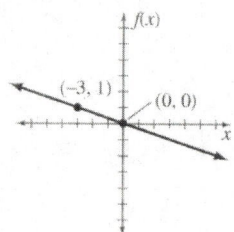

19.   Graphing the function $f(x) = \dfrac{1}{4}x + 2$:

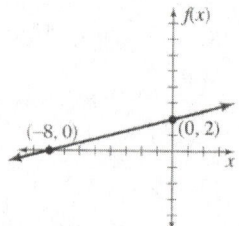

21. Graphing the piecewise function:

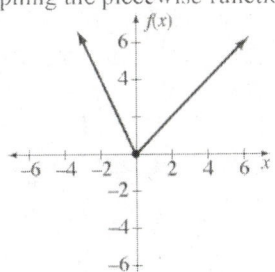

23. Graphing the piecewise function:

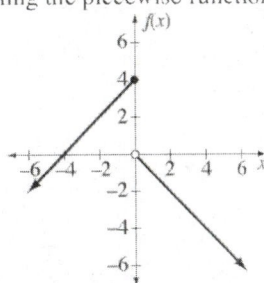

25. Graphing the piecewise function:

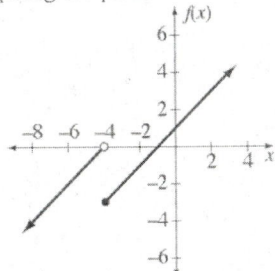

27. Graphing the piecewise function:

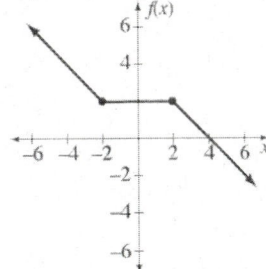

29. Graphing the piecewise function:

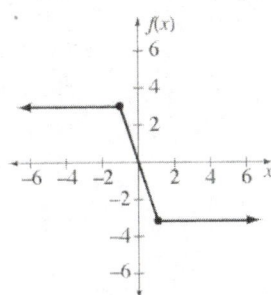

31. Using the point-slope formula:

$$f(x) - 3 = \frac{2}{3}(x + 1)$$

$$f(x) - 3 = \frac{2}{3}x + \frac{2}{3}$$

$$f(x) = \frac{2}{3}x + \frac{11}{3}$$

33. First find the slope: $m = \dfrac{-6 + 1}{2 + 3} = -1$

Using the point-slope formula:

$$f(x) + 1 = -1(x + 3)$$
$$f(x) + 1 = -x - 3$$
$$f(x) = -x - 4$$

35. The perpendicular slope is $-\dfrac{1}{5}$.

Using the point-slope formula:

$$f(x) - 3 = -\frac{1}{5}(x - 6)$$

$$f(x) - 3 = -\frac{1}{5}x + \frac{6}{5}$$

$$f(x) = -\frac{1}{5}x + \frac{21}{5}$$

37.   **a.**   Substituting $h = 93$: $c(93) = 0.0045(93) \approx \$0.42$

      b.   Graphing the function $c(h) = 0.0045h$:

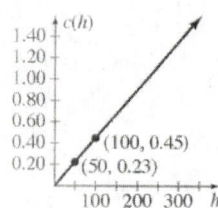

      c.   The cost is approximately $1.00.

      d.   Substituting $h = 225$: $c(225) = 0.0045(225) \approx \$1.01$

39.   First find the slope: $m = \dfrac{117.50 - 80}{350 - 200} = \dfrac{37.5}{150} = 0.25$. Using the point-slope formula:

$$f(x) - 80 = 0.25(x - 200)$$
$$f(x) - 80 = 0.25x - 50$$
$$f(x) = 0.25x + 30$$

41.   Evaluating the function:

$$g(150) = \$26.00 \qquad\qquad g(230) = 26 + 0.15(230 - 200) = \$30.50$$
$$g(360) = 26 + 0.15(360 - 200) = \$50.00 \qquad g(430) = 26 + 0.15(430 - 200) = \$60.50$$

43.   The function is: $f(p) = p - 0.20p = 0.8p$. Finding the sale prices:

$$f(9.50) = 0.8(9.5) = \$7.60 \qquad\qquad f(15) = 0.8(15) = \$12$$
$$f(75) = 0.8(75) = \$60 \qquad\qquad f(12.50) = 0.8(12.5) = \$10$$
$$f(750) = 0.8(750) = \$600$$

45.   **a.**   Finding the slope: $m = \dfrac{3015 - 2015}{15 - 10} = \dfrac{1000}{5} = 200$

        Using the point-slope formula:

$$y - 2015 = 200(x - 10)$$
$$y - 2015 = 200x - 2000$$
$$y = 200x + 15$$
$$f(x) = 200x + 15$$

      b.   Finding the slope: $m = \dfrac{13 - 8}{10 - 6} = \dfrac{5}{4}$

        Using the point-slope formula:

$$y - 8 = \frac{5}{4}(x - 6)$$
$$y - 8 = \frac{5}{4}x - \frac{15}{2}$$
$$y = \frac{5}{4}x + \frac{1}{2}$$
$$f(x) = \frac{5}{4}x + \frac{1}{2}$$

c. Finding the slope: $m = \dfrac{23 - 50}{-5 - 10} = \dfrac{-27}{-15} = \dfrac{9}{5}$

Using the point-slope formula:

$$y - 50 = \frac{9}{5}(x - 10)$$

$$y - 50 = \frac{9}{5}x - 18$$

$$y = \frac{9}{5}x + 32$$

$$f(x) = \frac{9}{5}x + 32$$

3.3 Quadratic Functions and Applications

1. Graphing the function $f(x) = x^2 + 1$:

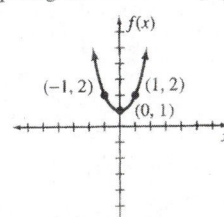

3. Graphing the function $f(x) = 3x^2$:

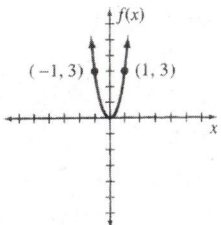

5. Graphing the function $f(x) = -x^2 + 2$:

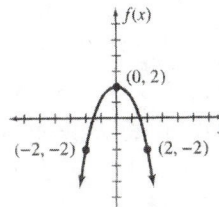

7. Graphing the function $f(x) = (x + 2)^2$:

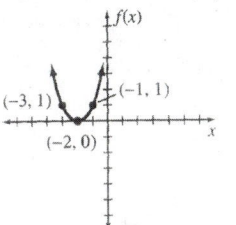

9. Graphing the function $f(x) = -2(x + 1)^2$:

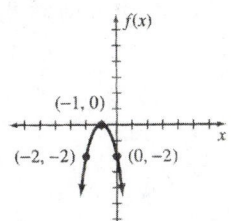

11. Graphing the function $f(x) = (x - 1)^2 + 2$:

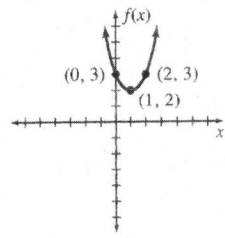

13. Graphing the function $f(x) = \dfrac{1}{2}(x - 2)^2 - 3$:

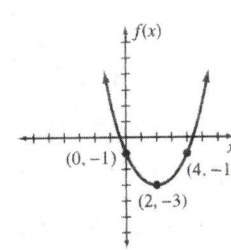

15. Completing the square: $f(x) = x^2 + 2x + 4 = (x^2 + 2x + 1) + 4 - 1 = (x + 1)^2 + 3$. Graphing the function:

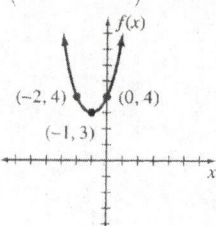

17. Completing the square: $f(x) = x^2 - 3x + 1 = \left(x^2 - 3x + \dfrac{9}{4}\right) + 1 - \dfrac{9}{4} = \left(x - \dfrac{3}{2}\right)^2 - \dfrac{5}{4}$. Graphing the function:

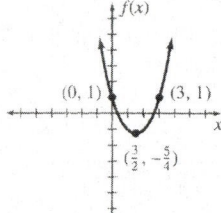

19. Completing the square: $f(x) = 2x^2 + 4x = 2(x^2 + 2x + 1) - 2 = 2(x + 1)^2 - 2$. Graphing the function:

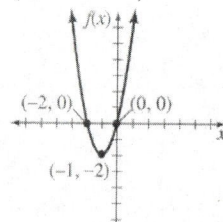

21. Completing the square: $f(x) = -x^2 - 2x + 1 = -(x^2 + 2x + 1) + 1 + 1 = -(x + 1)^2 + 2$. Graphing the function:

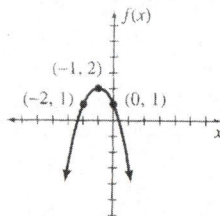

23. Completing the square: $f(x) = 2x^2 - 2x + 3 = 2\left(x^2 - x + \dfrac{1}{4}\right) - \dfrac{1}{2} + 3 = 2\left(x - \dfrac{1}{2}\right)^2 + \dfrac{5}{2}$. Graphing the function:

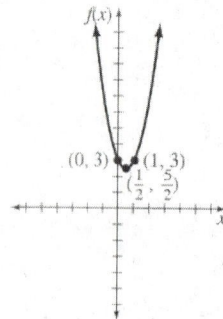

25.   Completing the square: $f(x) = -2x^2 - 5x + 1 = -2\left(x^2 + \dfrac{5}{2}x + \dfrac{25}{16}\right) + 1 + \dfrac{25}{8} = -2\left(x + \dfrac{5}{4}\right)^2 + \dfrac{33}{8}$.

Graphing the function:

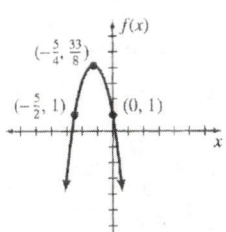

27.   The vertex is $(4,-1)$ and the parabola is pointed up: 29.   The vertex is $(-5,2)$ and the parabola is pointed up:

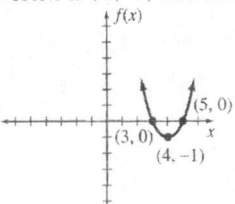   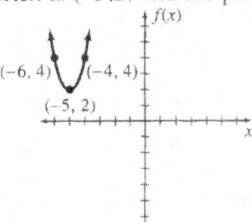

31.   The vertex is $(2,-3)$ and the parabola is pointed down: 33.   The vertex is $(1,-2)$ and the parabola is pointed down:

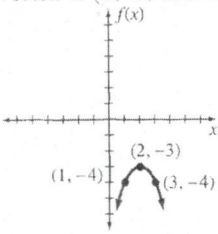   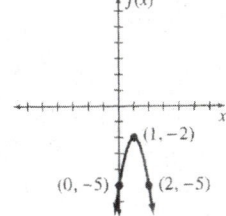

35.   The vertex is $\left(-\dfrac{3}{2}, -\dfrac{13}{4}\right)$ and the parabola is pointed up:

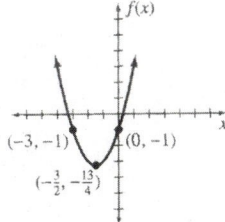

37.   The vertex is $\left(\dfrac{5}{4}, \dfrac{33}{8}\right)$ and the parabola is pointed down:

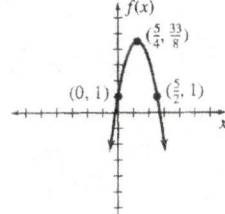

39.   The vertex is $(0,3)$ and the parabola is pointed down:

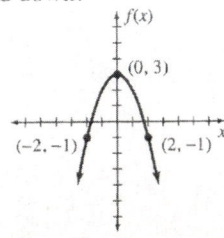

41.   The vertex is $\left(-\dfrac{1}{2},-\dfrac{5}{4}\right)$ and the parabola is pointed up:

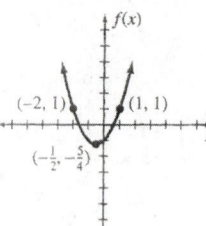

43.   The vertex is $(1,3)$ and the parabola is pointed down:

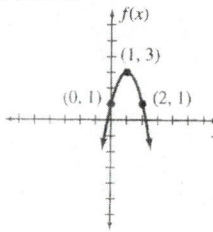

45.   The vertex is $\left(-\dfrac{5}{2},\dfrac{3}{2}\right)$ and the parabola is pointed down:

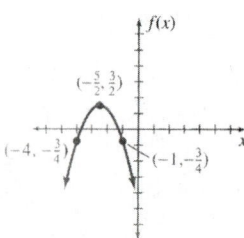

47.   Finding the x-intercepts:
$$3x^2 - 12 = 0$$
$$3x^2 = 12$$
$$x^2 = 4$$
$$x = \pm\sqrt{4} = \pm 2$$
The vertex is $(0,-12)$.

51.   Finding the x-intercepts:
$$x^2 - 8x + 15 = 0$$
$$(x-3)(x-5) = 0$$
$$x = 3, 5$$

The vertex is $(4,-1)$.

55.   Finding the x-intercepts:

$$-x^2 + 9x - 21 = 0$$
$$x^2 - 9x + 21 = 0$$
$$\left(x - \dfrac{9}{2}\right)^2 = -21 + \dfrac{81}{4}$$
$$\left(x - \dfrac{9}{2}\right)^2 = -\dfrac{3}{4}$$

There are no x-intercepts. The vertex is $\left(\dfrac{9}{2},-\dfrac{3}{4}\right)$.

49.   Finding the x-intercepts:

$$5x^2 - 10x = 0$$
$$5x(x-2) = 0$$
$$x = 0, 2$$

The vertex is $(1,-5)$.

53.   Finding the x-intercepts:
$$-x^2 + 10x - 24 = 0$$
$$x^2 - 10x + 24 = 0$$
$$(x-4)(x-6) = 0$$
$$x = 4, 6$$
The vertex is $(5,1)$.

57.   Finding the x-intercepts:
$$-4x^2 + 4x + 4 = 0$$
$$x^2 - x - 1 = 0$$
$$x^2 - x + \dfrac{1}{4} = 1 + \dfrac{1}{4}$$
$$\left(x - \dfrac{1}{2}\right)^2 = \dfrac{5}{4}$$
$$x - \dfrac{1}{2} = \pm\dfrac{\sqrt{5}}{2}$$
$$x = \dfrac{1 \pm \sqrt{5}}{2}$$

The vertex is $\left(\dfrac{1}{2}, 5\right)$.

59. Finding the zeros:

$$x^2 + 3x - 88 = 0$$
$$(x+11)(x-8) = 0$$
$$x = -11, 8$$

61. Finding the zeros:

$$4x^2 - 48x + 108 = 0$$
$$4(x^2 - 12x + 27) = 0$$
$$4(x-3)(x-9) = 0$$
$$x = 3, 9$$

63. Finding the zeros:

$$x^2 - 4x + 11 = 0$$
$$x^2 - 4x + 4 = -11 + 4$$
$$(x-2)^2 = -7$$
$$x - 2 = \pm i\sqrt{7}$$
$$x = 2 \pm i\sqrt{7}$$

65. Finding the vertex: $x = \dfrac{-280}{-4} = 70$. Thus 70 items should be sold to maximize the profit.

67. Finding the vertex: $x = \dfrac{-96}{-32} = 3$. Since $f(3) = 96(3) - 16(3)^2 = 144$, the highest point reached is 144 feet.

69. Let x and $50 - x$ represent the two numbers. The function is: $f(x) = x(50 - x) = -x^2 + 50x$

Finding the vertex: $x = \dfrac{-50}{-2} = 25$. The two numbers are 25 and 25.

71. Let x and y represent the two dimensions. Since the perimeter is 240 meters:

$$2x + 2y = 240$$
$$x + y = 120$$
$$y = 120 - x$$

The area function is: $A(x) = x(120 - x) = -x^2 + 120x$. Finding the vertex: $x = \dfrac{-120}{-2} = 60$

The dimensions are 60 m by 60 m.

73. Let x represent the number of \$0.25 decreases in the monthly rate. The revenue function is:

$$f(x) = (1000 + 20x)(15 - 0.25x) = -5x^2 + 50x + 15000$$

Finding the vertex: $x = \dfrac{-50}{-10} = 5$. At a rate of \$13.75 they will have 1100 subscribers to achieve a maximum revenue.

3.4 Transformations of Some Basic Curves

1. Graphing the function $f(x) = x^4 + 2$:

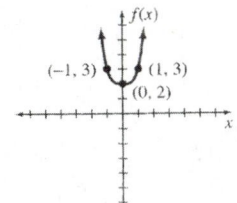

3. Graphing the function $f(x) = (x-2)^4$:

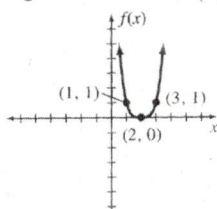

5. Graphing the function $f(x) = -x^3$:

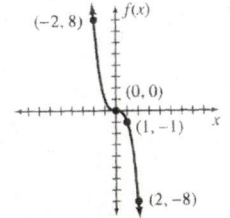

7. Graphing the function $f(x) = (x+2)^3$:

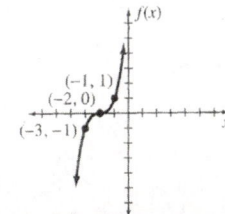

9.   Graphing the function $f(x) = |x-1| + 2$:

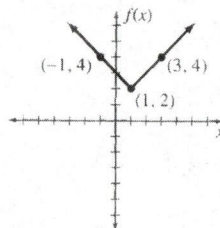

11.   Graphing the function $f(x) = |x+1| - 3$:

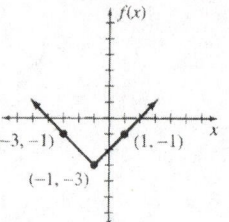

13.   Graphing the function $f(x) = -(x+3)^2 + 4$:

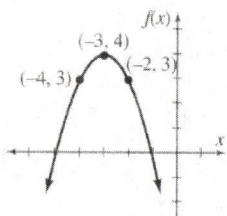

15.   Graphing the function $f(x) = -|x-2| - 1$:

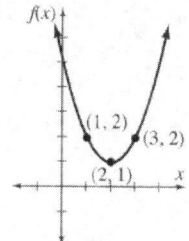

17.   Graphing the function $f(x) = 2x^2 + 4$:

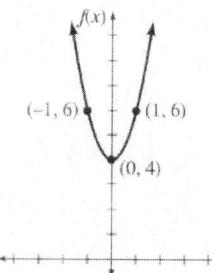

19.   Graphing the function $f(x) = -2\sqrt{x}$:

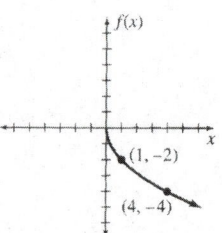

21.   Graphing the function $f(x) = \sqrt{x+2} - 3$:

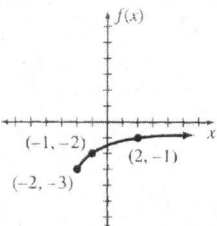

23.   Graphing the function $f(x) = \sqrt{2-x}$:

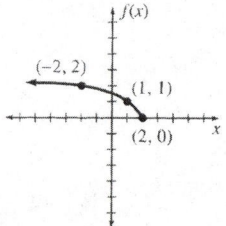

25.   Graphing the function $f(x) = -2x^4 + 1$:

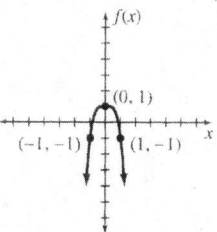

27.   Graphing the function $f(x) = -2x^3$:

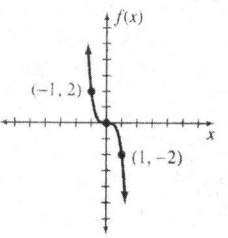

29.   Graphing the function $f(x) = 3(x-2)^3 - 1$:

31.   **a.**   Graphing the function $y = f(x) + 3$:

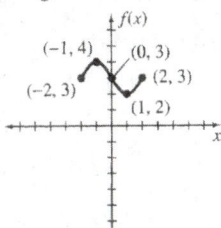

b.   Graphing the function $y = f(x - 2)$:

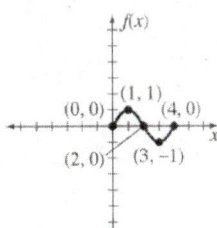

c.   Graphing the function $y = -f(x)$:

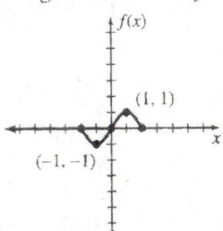

d.   Graphing the function $y = f(x + 3) - 4$:

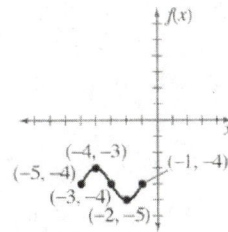

33.   Graphing the function:

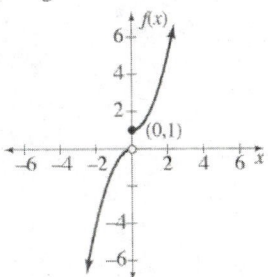

35.   Graphing the function:

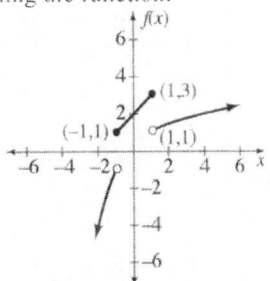

37.   Graphing the function:

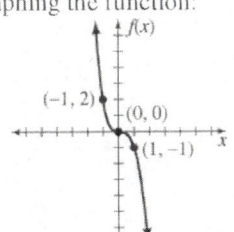

39.   Graphing the function:

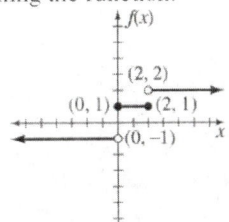

41.   Graphing the function $f(x) = x + |x|$:

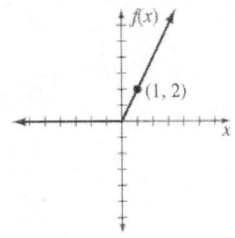

43.   Graphing the function $f(x) = x - |x|$:

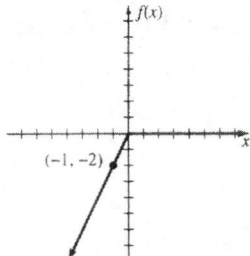

3.5 Combining Functions

1. Finding the functions:
$$f(x) + g(x) = (3x - 4) + (5x + 2) = 8x - 2$$
$$f(x) - g(x) = (3x - 4) - (5x + 2) = -2x - 6$$
$$f(x) \cdot g(x) = (3x - 4)(5x + 2) = 15x^2 - 14x - 8$$
$$\frac{f(x)}{g(x)} = \frac{3x - 4}{5x + 2}, \ x \neq -\frac{2}{5}$$

3. Finding the functions:
$$f(x) + g(x) = (x^2 - 6x + 4) + (-x - 1) = x^2 - 7x + 3$$
$$f(x) - g(x) = (x^2 - 6x + 4) - (-x - 1) = x^2 - 5x + 5$$
$$f(x) \cdot g(x) = (x^2 - 6x + 4)(-x - 1) = -x^3 + 5x^2 + 2x - 4$$
$$\frac{f(x)}{g(x)} = \frac{x^2 - 6x + 4}{-x - 1}, \ x \neq -1$$

5. Finding the functions:
$$f(x) + g(x) = (x^2 - x - 1) + (x^2 + 4x - 5) = 2x^2 + 3x - 6$$
$$f(x) - g(x) = (x^2 - x - 1) - (x^2 + 4x - 5) = -5x + 4$$
$$f(x) \cdot g(x) = (x^2 - x - 1)(x^2 + 4x - 5) = x^4 + 3x^3 - 10x^2 + x + 5$$
$$\frac{f(x)}{g(x)} = \frac{x^2 - x - 1}{x^2 + 4x - 5}, \ x \neq -5, 1$$

7. Finding the functions:
$$f(x) + g(x) = \sqrt{x - 1} + \sqrt{x}$$
$$f(x) - g(x) = \sqrt{x - 1} - \sqrt{x}$$
$$f(x) \cdot g(x) = (\sqrt{x - 1})(\sqrt{x}) = \sqrt{x^2 - x}, \ x \geq 1$$
$$\frac{f(x)}{g(x)} = \frac{\sqrt{x - 1}}{\sqrt{x}} = \frac{\sqrt{x^2 - x}}{x}, \ x \geq 1$$

9. Finding the expression: $\dfrac{f(a + h) - f(a)}{h} = \dfrac{3(a + h) + 8 - (3a + 8)}{h} = \dfrac{3a + 3h + 8 - 3a - 8}{h} = \dfrac{3h}{h} = 3$

11. Finding the expression: $\dfrac{f(a + h) - f(a)}{h} = \dfrac{-7(a + h) - 2 - (-7a - 2)}{h} = \dfrac{-7a - 7h - 2 + 7a + 2}{h} = \dfrac{-7h}{h} = -7$

13. Finding the expression:
$$\frac{f(a + h) - f(a)}{h} = \frac{-(a + h)^2 + 4(a + h) - 2 - (-a^2 + 4a - 2)}{h}$$
$$= \frac{-a^2 - 2ah - h^2 + 4a + 4h - 2 + a^2 - 4a + 2}{h}$$
$$= \frac{-2ah - h^2 + 4h}{h}$$
$$= -2a - h + 4$$

15. Finding the expression:

$$\frac{f(a+h)-f(a)}{h} = \frac{3(a+h)^2 - (a+h) - 4 - \left(3a^2 - a - 4\right)}{h}$$

$$= \frac{3a^2 + 6ah + 3h^2 - a - h - 4 - 3a^2 + a + 4}{h}$$

$$= \frac{6ah + 3h^2 - h}{h}$$

$$= 6a + 3h - 1$$

17. Finding the expression:

$$\frac{f(a+h)-f(a)}{h} = \frac{(a+h)^3 - (a+h)^2 + 2(a+h) - 1 - \left(a^3 - a^2 + 2a - 1\right)}{h}$$

$$= \frac{a^3 + 3a^2h + 3ah^2 + h^3 - a^2 - 2ah - h^2 + 2a + 2h - 1 - a^3 + a^2 - 2a + 1}{h}$$

$$= \frac{3a^2h + 3ah^2 + h^3 - 2ah - h^2 + 2h}{h}$$

$$= 3a^2 + 3ah + h^2 - 2a - h + 2$$

19. Finding the expression:

$$\frac{f(a+h)-f(a)}{h} = \frac{\dfrac{2}{a+h-1} - \dfrac{2}{a-1}}{h}$$

$$= \frac{\dfrac{2}{a+h-1} - \dfrac{2}{a-1}}{h} \cdot \frac{(a-1)(a+h-1)}{(a-1)(a+h-1)}$$

$$= \frac{2a - 2 - 2a - 2h + 2}{h(a-1)(a+h-1)}$$

$$= \frac{-2h}{h(a-1)(a+h-1)}$$

$$= -\frac{2}{(a-1)(a+h-1)}$$

21. Finding the expression:

$$\frac{f(a+h)-f(a)}{h} = \frac{\dfrac{1}{(a+h)^2} - \dfrac{1}{a^2}}{h}$$

$$= \frac{\dfrac{1}{(a+h)^2} - \dfrac{1}{a^2}}{h} \cdot \frac{a^2(a+h)^2}{a^2(a+h)^2}$$

$$= \frac{a^2 - (a+h)^2}{ha^2(a+h)^2}$$

$$= \frac{a^2 - a^2 - 2ah - h^2}{ha^2(a+h)^2}$$

$$= \frac{-2ah - h^2}{ha^2(a+h)^2}$$

$$= -\frac{2a+h}{a^2(a+h)^2}$$

23.   Finding each composition:

$$(f \circ g)(x) = f(3x-1) = 2(3x-1) = 6x-2 \qquad (g \circ f)(x) = g(2x) = 3(2x)-1 = 6x-1$$

The domain for each is $\{x \mid x \text{ is a real number}\}$.

25.   Finding each composition:

$$(f \circ g)(x) = f(2x+1) = 5(2x+1)-3 = 10x+2 \qquad (g \circ f)(x) = g(5x-3) = 2(5x-3)+1 = 10x-5$$

The domain for each is $\{x \mid x \text{ is a real number}\}$.

27.   Finding each composition:

$$(f \circ g)(x) = f(x^2+1) = 3(x^2+1)+4 = 3x^2+7$$
$$(g \circ f)(x) = g(3x+4) = (3x+4)^2+1 = 9x^2+24x+17$$

The domain for each is $\{x \mid x \text{ is a real number}\}$.

29.   Finding each composition:

$$(f \circ g)(x) = f(x^2+3x-4) = 3(x^2+3x-4)-4 = 3x^2+9x-16$$
$$(g \circ f)(x) = g(3x-4) = (3x-4)^2+3(3x-4)-4 = 9x^2-24x+16+9x-12-4 = 9x^2-15x$$

The domain for each is $\{x \mid x \text{ is a real number}\}$.

31.   Finding each composition:

$$(f \circ g)(x) = f(2x+7) = \frac{1}{2x+7} \qquad (g \circ f)(x) = g\left(\frac{1}{x}\right) = \frac{2}{x}+7 = \frac{7x+2}{x}$$

The domain of $f \circ g$ is $\left\{x \mid x \neq -\frac{7}{2}\right\}$ and the domain of $g \circ f$ is $\{x \mid x \neq 0\}$.

33.   Finding each composition:

$$(f \circ g)(x) = f(3x-1) = \sqrt{3x-1-2} = \sqrt{3x-3} \qquad (g \circ f)(x) = g\left(\sqrt{x-2}\right) = 3\sqrt{x-2}-1$$

The domain of $f \circ g$ is $\{x \mid x \geq 1\}$ and the domain of $g \circ f$ is $\{x \mid x \geq 2\}$.

35.   Finding each composition:

$$(f \circ g)(x) = f\left(\frac{2}{x}\right) = \frac{1}{\frac{2}{x}-1} = \frac{x}{2-x} \qquad (g \circ f)(x) = g\left(\frac{1}{x-1}\right) = \frac{2}{\frac{1}{x-1}} = 2x-2$$

The domain of $f \circ g$ is $\{x \mid x \neq 0, 2\}$ and the domain of $g \circ f$ is $\{x \mid x \neq 1\}$.

37.   Finding each composition:

$$(f \circ g)(x) = f\left(\sqrt{x-1}\right) = 2\sqrt{x-1}+1 \qquad (g \circ f)(x) = g(2x+1) = \sqrt{2x+1-1} = \sqrt{2x}$$

The domain of $f \circ g$ is $\{x \mid x \geq 1\}$ and the domain of $g \circ f$ is $\{x \mid x \geq 0\}$.

39.   Finding each composition:

$$(f \circ g)(x) = f\left(\frac{x+1}{x}\right) = \frac{1}{\frac{x+1}{x}-1} = \frac{x}{x+1-x} = x \qquad (g \circ f)(x) = g\left(\frac{1}{x-1}\right) = \frac{\frac{1}{x-1}+1}{\frac{1}{x-1}} = \frac{1+x-1}{1} = x$$

The domain of $f \circ g$ is $\{x \mid x \neq 0\}$ and the domain of $g \circ f$ is $\{x \mid x \neq 1\}$.

41.   Finding each composition:

$$(f \circ g)(-1) = f(1+1) = f(2) = 6-2 = 4 \qquad (g \circ f)(3) = g(9-2) = g(7) = 49+1 = 50$$

43.   Finding each composition:

$$(f \circ g)(-2) = f(4+6-4) = f(6) = 12-3 = 9 \qquad (g \circ f)(1) = g(2-3) = g(-1) = 1+3-4 = 0$$

45.   Finding each composition:

$$(f \circ g)(4) = f(12-1) = f(11) = \sqrt{11} \qquad (g \circ f)(4) = g(2) = 6-1 = 5$$

47. Finding each composition:
$$(f \circ g)(x) = f\left(\frac{1}{2}x\right) = 2\left(\frac{1}{2}x\right) = x \qquad\qquad (g \circ f)(x) = g(2x) = \frac{1}{2}(2x) = x$$

49. Finding each composition:
$$(f \circ g)(x) = f(x+2) = x + 2 - 2 = x \qquad\qquad (g \circ f)(x) = g(x-2) = x - 2 + 2 = x$$

51. Finding each composition:
$$(f \circ g)(x) = f\left(\frac{x-4}{3}\right) = 3\left(\frac{x-4}{3}\right) + 4 = x - 4 + 4 = x$$
$$(g \circ f)(x) = g(3x+4) = \frac{3x+4-4}{3} = \frac{3x}{3} = x$$

57. Finding each composition:
$$(f \circ g)(x) = f\left(\sqrt{x}\right) = \left(\sqrt{x}\right)^2 = x \qquad\qquad (g \circ f)(x) = g\left(x^2\right) = \sqrt{x^2} = x \qquad (\text{since } x \geq 0)$$

3.6 Direct and Inverse Variation

1. The variation equation is $y = kx^3$.

3. The variation equation is $A = klw$.

5. The variation equation is $V = \dfrac{k}{P}$.

7. The variation equation is $V = khr^2$.

9. The variation equation is $y = kx$. Substituting $y = 72$ and $x = 3$:
$$72 = k \cdot 3$$
$$k = 24$$

11. The variation equation is $A = kr^2$. Substituting $A = 154$ and $r = 7$:
$$154 = k \cdot 49$$
$$k = \frac{154}{49} = \frac{22}{7}$$

13. The variation equation is $A = kbh$. Substituting $A = 81$, $b = 9$ and $h = 18$:
$$81 = k \cdot 9 \cdot 18$$
$$k = \frac{81}{162} = \frac{1}{2}$$

15. The variation equation is $y = \dfrac{kxz}{w}$. Substituting $y = 154$, $x = 6$, $z = 11$ and $w = 3$:
$$154 = \frac{k \cdot 6 \cdot 11}{3}$$
$$k = \frac{154}{22} = 7$$

17. The variation equation is $y = \dfrac{kx^2}{w^3}$. Substituting $y = 18$, $x = 9$ and $w = 3$:
$$18 = \frac{k \cdot 81}{27}$$
$$k = \frac{18}{3} = 6$$

19. The variation equation is $y = kx$. Substituting $y = 5$ and $x = -15$:
$$5 = k(-15)$$
$$k = -\frac{1}{3}$$
So $y = -\dfrac{1}{3}x$. Substituting $x = -24$: $y = -\dfrac{1}{3}(-24) = 8$

21. The variation equation is $V = kBh$. Substituting $V = 96$, $B = 36$ and $h = 8$:

$$96 = k(36)(8)$$
$$k = \frac{96}{288} = \frac{1}{3}$$

So $y = \frac{1}{3}Bh$. Substituting $B = 48$ and $h = 6$: $y = \frac{1}{3}(48)(6) = 96$

23. The variation equation is $t = \frac{k}{r}$. Substituting $t = 3$ and $r = 50$:

$$3 = \frac{k}{50}$$
$$k = 150$$

So $t = \frac{150}{r}$. Substituting $r = 30$: $t = \frac{150}{30} = 5$ hours

25. The variation equation is $T = k\sqrt{L}$. Substituting $T = 4$ and $L = 12$:

$$4 = k\sqrt{12}$$
$$k = \frac{4}{2\sqrt{3}} = \frac{2\sqrt{3}}{3}$$

So $T = \frac{2\sqrt{3}}{3}\sqrt{L}$. Substituting $L = 3$: $T = \frac{2\sqrt{3}}{3}\sqrt{3} = \frac{2 \cdot 3}{3} = 2$ seconds

27. The variation equation is $d = \frac{km}{p}$. Substituting $p = 4$, $d = 32$ and $m = 16$:

$$32 = \frac{k \cdot 16}{4}$$
$$k = \frac{32}{4} = 8$$

So $d = \frac{8m}{p}$. Substituting $p = 8$ and $m = 24$: $d = \frac{8 \cdot 24}{8} = 24$ days

29. The variation equation is $V = \frac{kT}{P}$. Substituting $V = 48$, $T = 320$ and $P = 20$:

$$48 = \frac{k \cdot 320}{20}$$
$$k = \frac{48}{16} = 3$$

So $V = \frac{3T}{P}$. Substituting $T = 280$ and $P = 30$: $V = \frac{3 \cdot 280}{30} = 28$

31. The variation equation is $C = kwd$. Substituting $C = 900$, $w = 15$ and $d = 5$:

$$900 = k \cdot 15 \cdot 5$$
$$k = \frac{900}{75} = 12$$

So $C = 12wd$. Substituting $w = 20$ and $d = 10$: $C = 12 \cdot 20 \cdot 10 = \$2400$

37. The variation equation is $T = k\sqrt{L}$. Substituting $T = 2.4$ and $L = 9$:

$$2.4 = k\sqrt{9}$$
$$k = \frac{2.4}{3} = 0.8$$

So $T = 0.8\sqrt{L}$. Substituting $L = 12$: $T = 0.8\sqrt{12} \approx 2.8$ seconds

39. The variation equation is $y = \dfrac{kx}{z^2}$. Substituting $y = 0.336$, $x = 6$ and $z = 5$:

$$0.336 = \frac{k \cdot 6}{25}$$
$$k = \frac{8.4}{6} = 1.4$$

Chapter 3 Review Problem Set

1. Finding the function values:

$$f(2) = 3(2)^2 - 2(2) - 1 = 12 - 4 - 1 = 7$$
$$f(-1) = 3(-1)^2 - 2(-1) - 1 = 3 + 2 - 1 = 4$$
$$f(-3) = 3(-3)^2 - 2(-3) - 1 = 27 + 6 - 1 = 32$$

2. **a.** Finding the expression:

$$\frac{f(a+h) - f(a)}{h} = \frac{-5(a+h) + 4 - (-5a + 4)}{h} = \frac{-5a - 5h + 4 + 5a - 4}{h} = \frac{-5h}{h} = -5$$

b. Finding the expression:

$$\frac{f(a+h) - f(a)}{h} = \frac{2(a+h)^2 - (a+h) + 4 - \left(2a^2 - a + 4\right)}{h}$$
$$= \frac{2a^2 + 4ah + 2h^2 - a - h + 4 - 2a^2 + a - 4}{h}$$
$$= \frac{4ah + 2h^2 - h}{h}$$
$$= 4a + 2h - 1$$

c. Finding the expression:

$$\frac{f(a+h) - f(a)}{h} = \frac{-3(a+h)^2 + 2(a+h) - 5 - \left(-3a^2 + 2a - 5\right)}{h}$$
$$= \frac{-3a^2 - 6ah - 3h^2 + 2a + 2h - 5 + 3a^2 - 2a + 5}{h}$$
$$= \frac{-6ah - 3h^2 + 2h}{h}$$
$$= -6a - 3h + 2$$

3. The domain is $\{x \mid x \text{ is a real number}\}$ and the range is $\{f(x) \mid f(x) \ge 5\}$.

4. Factoring the denominator: $2x^2 + 7x - 4 = (2x - 1)(x + 4)$. Thus the domain is $\left\{ x \mid x \ne -4, \dfrac{1}{2} \right\}$.

5. Factoring the inequality:

$$x^2 - 7x + 10 \ge 0$$
$$(x - 5)(x - 2) \ge 0$$

Forming a sign chart:

The domain is $(-\infty, 2] \cup [5, \infty)$.

6.   Graphing the function $f(x) = -2x + 2$:

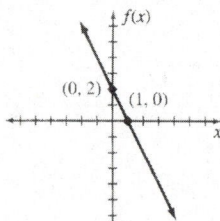

7.   Graphing the function $f(x) = 2x^2 - 1$:

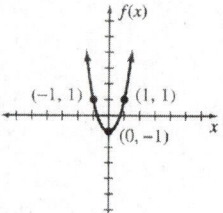

8.   Graphing the function $f(x) = -\sqrt{x-2} + 1$:

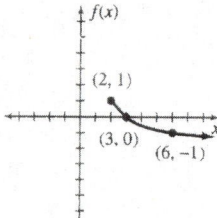

9.   Graphing the function $f(x) = x^2 - 8x + 17$:

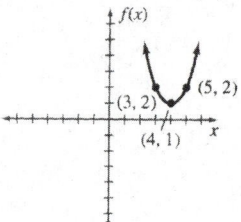

10.   Graphing the function $f(x) = -x^3 + 2$:

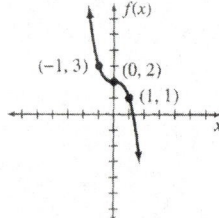

11.   Graphing the function $f(x) = 2|x-1| + 3$:

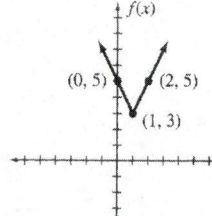

12.   Graphing the function $f(x) = -2x^2 - 12x - 19$:

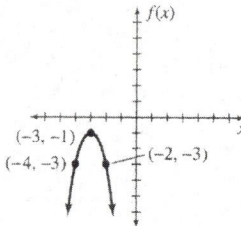

13.   Graphing the function $f(x) = -\dfrac{1}{3}x + 1$:

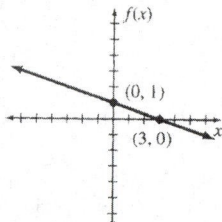

14.   Graphing the function $f(x) = -\dfrac{2}{x^2}$:

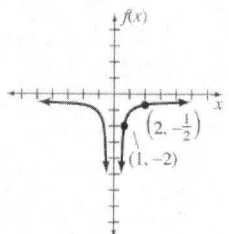

15.   Graphing the function $f(x) = 2|x| - x$:

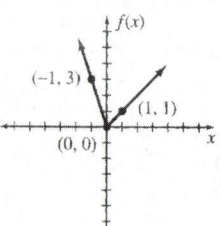

16.   Graphing the function $f(x) = (x-2)^2$:

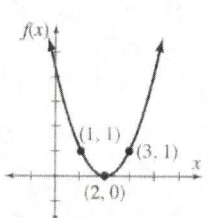

17.   Graphing the function $f(x) = \sqrt{-x+4}$:

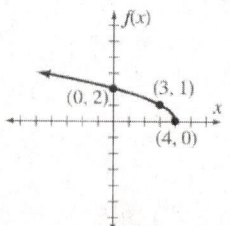

18. Graphing the function $f(x) = -(x+1)^2 - 3$:

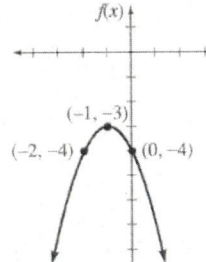

19. Graphing the function $f(x) = \sqrt{x+3} - 2$:

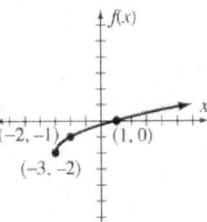

20. Graphing the function $f(x) = -|x| + 4$:

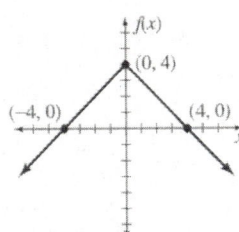

21. Graphing the function $f(x) = (x-2)^3$:

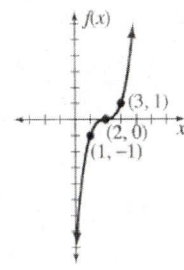

22. Graphing the function $f(x) = \begin{cases} x^2 - 1 & \text{for } x < 0 \\ 3x - 1 & \text{for } x \geq 0 \end{cases}$:

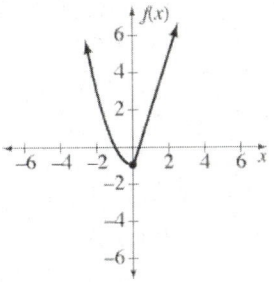

23. Graphing the function $f(x) = \begin{cases} 3 & \text{for } x \leq -3 \\ |x| & \text{for } -3 < x < 3 \\ 2x - 3 & \text{for } x \geq 3 \end{cases}$:

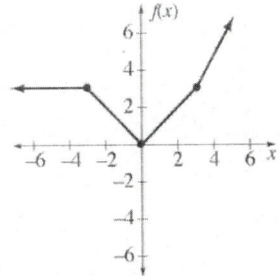

24. Finding the functions:

$$f(x) + g(x) = (2x+3) + (x^2 - 4x - 3) = x^2 - 2x$$

$$f(x) - g(x) = (2x+3) - (x^2 - 4x - 3) = -x^2 + 6x + 6$$

$$f(x) \cdot g(x) = (2x+3)(x^2 - 4x - 3) = 2x^3 - 5x^2 - 18x - 9$$

$$\frac{f(x)}{g(x)} = \frac{2x+3}{x^2 - 4x - 3}, \quad x \neq 2 \pm \sqrt{7}$$

25.   Finding each composition:
$$(f \circ g)(x) = f(-2x+7) = 3(-2x+7) - 9 = -6x + 12$$
$$(g \circ f)(x) = g(3x-9) = -2(3x-9) + 7 = -6x + 25$$

The domain of each is $\{x \mid x \text{ is a real number}\}$.

26.   Finding each composition:
$$(f \circ g)(x) = f(5x-4) = (5x-4)^2 - 5 = 25x^2 - 40x + 11$$
$$(g \circ f)(x) = g(x^2 - 5) = 5(x^2 - 5) - 4 = 5x^2 - 29$$

The domain of each is $\{x \mid x \text{ is a real number}\}$.

27.   Finding each composition:
$$(f \circ g)(x) = f(x+2) = \sqrt{x+2-5} = \sqrt{x-3} \qquad (g \circ f)(x) = g(\sqrt{x-5}) = \sqrt{x-5} + 2$$

The domain of $f \circ g$ is $\{x \mid x \geq 3\}$ and the domain of $g \circ f$ is $\{x \mid x \geq 5\}$.

28.   Finding each composition:
$$(f \circ g)(x) = f(x^2 - x - 6) = \frac{1}{x^2 - x - 6} \qquad (g \circ f)(x) = g\left(\frac{1}{x}\right) = \left(\frac{1}{x}\right)^2 - \frac{1}{x} - 6 = \frac{1}{x^2} - \frac{1}{x} - 6$$

Factoring $x^2 - x - 6 = (x-3)(x+2)$. The domain of $f \circ g$ is $\{x \mid x \neq -2, 3\}$ and the domain of $g \circ f$ is $\{x \mid x \neq 0\}$.

29.   Finding each composition:
$$(f \circ g)(x) = f(\sqrt{x-1}) = (\sqrt{x-1})^2 = x - 1 \qquad (g \circ f)(x) = g(x^2) = \sqrt{x^2 - 1}$$

The domain of $f \circ g$ is $\{x \mid x \geq 1\}$ and the domain of $g \circ f$ is $\{x \mid x \leq -1 \text{ or } x \geq 1\}$.

30.   Finding each composition:
$$(f \circ g)(x) = f\left(\frac{1}{x+2}\right) = \frac{1}{\dfrac{1}{x+2} - 3} = \frac{x+2}{1 - 3x - 6} = \frac{x+2}{-3x-5}$$
$$(g \circ f)(x) = g\left(\frac{1}{x-3}\right) = \frac{1}{\dfrac{1}{x-3} + 2} = \frac{x-3}{1 + 2x - 6} = \frac{x-3}{2x-5}$$

The domain of $f \circ g$ is $\left\{x \mid x \neq -2, -\dfrac{5}{3}\right\}$ and the domain of $g \circ f$ is $\left\{x \mid x \neq \dfrac{5}{2}, 3\right\}$.

31.   Finding the function values:
$$f(5) = 5^2 - 2 = 25 - 2 = 23 \qquad\qquad f(0) = 0^2 - 2 = 0 - 2 = -2$$
$$f(-3) = -3(-3) + 4 = 9 + 4 = 13$$

32.   Finding the function values:
$$f(g(6)) = f(\sqrt{6-2}) = f(2) = -4 - 2 + 4 = -2 \qquad g(f(-2)) = g(-4+2+4) = g(2) = \sqrt{2-2} = 0$$

33.   Finding the function values:
$$f(g(1)) = f(1-1-1) = f(-1) = 1 \qquad\qquad g(f(-3)) = g(3) = 9 - 3 - 1 = 5$$

34.   The parallel slope is $\dfrac{2}{3}$. Using the point-slope formula:
$$f(x) + 2 = \frac{2}{3}(x-5)$$
$$f(x) + 2 = \frac{2}{3}x - \frac{10}{3}$$
$$f(x) = \frac{2}{3}x - \frac{16}{3}$$

35. The perpendicular slope is 2. Using the point-slope formula:
$$f(x) - 3 = 2(x + 6)$$
$$f(x) - 3 = 2x + 12$$
$$f(x) = 2x + 15$$

36. Substituting $h = 120$: $c(120) = 0.006(120) = \$0.72$

37. The function is: $s(p) = p - 0.30p = 0.70p$. Finding the sale prices:
$$s(65) = 0.70(65) = \$45.50 \qquad\qquad s(48) = 0.70(48) = \$33.60$$
$$s(15.50) = 0.70(15.5) = \$10.85$$

38. Finding the x-intercepts:

$$3x^2 + 6x - 24 = 0$$
$$3(x^2 + 2x - 8) = 0$$
$$3(x + 4)(x - 2) = 0$$
$$x = -4, 2$$

The vertex is $(-1, -27)$.

39. Finding the x-intercepts:
$$x^2 - 6x - 5 = 0$$
$$x^2 - 6x + 9 = 5 + 9$$
$$(x - 3)^2 = 14$$
$$x - 3 = \pm\sqrt{14}$$
$$x = 3 \pm \sqrt{14}$$

The vertex is $(3, -14)$.

40. Finding the x-intercepts:
$$2x^2 - 28x + 101 = 0$$
$$x^2 - 14x = -\frac{101}{2}$$
$$x^2 - 14x + 49 = -\frac{101}{2} + 49$$
$$(x - 7)^2 = -\frac{3}{2}$$

Since this solution has non-real complex solutions, there are no x-intercepts. The vertex is $(7, 3)$.

41. Let x and $10 - x$ represent the two numbers. Forming the function: $f(x) = x^2 + 4(10 - x) = x^2 - 4x + 40$

The vertex occurs at $x = \dfrac{4}{2} = 2$, so the two numbers are 2 and 8.

42. Let x represent the number of students above 100 on the flight. The revenue is:
$$R(x) = (100 + x)(496 - 4x) = -4x^2 + 96x + 49600$$

The vertex occurs at $x = \dfrac{-96}{-8} = 12$, so the airline should try to get 112 students on the flight.

43. The variation equation is $y = \dfrac{kx}{w}$. Substituting $y = 27$, $x = 18$ and $w = 6$:

$$27 = \frac{k \cdot 18}{6}$$
$$k = \frac{27}{3} = 9$$

44. The variation equation is $y = kx\sqrt{w}$. Substituting $y = 140$, $x = 5$ and $w = 16$:

$$140 = k(5)\sqrt{16}$$
$$k = \frac{140}{20} = 7$$

So $y = 7x\sqrt{w}$. Substituting $x = 9$ and $w = 49$: $y = 7 \cdot 9\sqrt{49} = 441$

45.   The variation equation is $W = \dfrac{k}{d^2}$. Substituting $W = 200$ and $d = 4000$:

$$200 = \frac{k}{4000^2}$$
$$k = 3.2(10)^9$$

So $W = \dfrac{3.2(10)^9}{d^2}$. Substituting $d = 5000$: $W = \dfrac{3.2(10)^9}{5000^2} = 128$ pounds

46.   The variation equation is $h = \dfrac{kn}{p}$. Substituting $p = 3$, $h = 10$ and $n = 20$:

$$10 = \frac{k \cdot 20}{3}$$
$$k = \frac{30}{20} = \frac{3}{2}$$

So $h = \dfrac{3n}{2p}$. Substituting $p = 4$ and $n = 40$: $h = \dfrac{3 \cdot 40}{2 \cdot 4} = 15$ hours

Chapter 3 Test

1.   Finding the value: $f(-3) = -\dfrac{1}{2}(-3) + \dfrac{1}{3} = \dfrac{3}{2} + \dfrac{1}{3} = \dfrac{11}{6}$

2.   Finding the value: $f(-2) = -(-2)^2 - 6(-2) + 3 = -4 + 12 + 3 = 11$

3.   Simplifying the expression:

$$\frac{f(a+h) - f(a)}{h} = \frac{3(a+h)^2 + 2(a+h) - 5 - \left(3a^2 + 2a - 5\right)}{h}$$
$$= \frac{3a^2 + 6ah + 3h^2 + 2a + 2h - 5 - 3a^2 - 2a + 5}{h}$$
$$= \frac{6ah + 3h^2 + 2h}{h}$$
$$= 6a + 3h + 2$$

4.   Since $2x^2 + 7x - 4 = (2x - 1)(x + 4)$, the domain is $\left\{x \mid x \neq -4, \dfrac{1}{2}\right\}$.

5.   Solving the inequality:
$$5 - 3x \geq 0$$
$$-3x \geq -5$$
$$x \leq \frac{5}{3}$$

The domain is $\left\{x \mid x \leq \dfrac{5}{3}\right\}$.

6.   Finding the functions:

$$f(x) + g(x) = (3x - 1) + \left(2x^2 - x - 5\right) = 2x^2 + 2x - 6$$
$$f(x) - g(x) = (3x - 1) - \left(2x^2 - x - 5\right) = -2x^2 + 4x + 4$$
$$f(x) \bullet g(x) = (3x - 1)\left(2x^2 - x - 5\right) = 6x^3 - 5x^2 - 14x + 5$$

7.   Finding the composition: $(f \circ g)(x) = f(7x + 2) = -3(7x + 2) + 4 = -21x - 2$

8. Finding the composition:
$$(g \circ f)(x) = g(2x+5)$$
$$= 2(2x+5)^2 - (2x+5) + 3$$
$$= 8x^2 + 40x + 50 - 2x - 5 + 3$$
$$= 8x^2 + 38x + 48$$

9. Finding the composition: $(f \circ g)(x) = f\left(\dfrac{2}{x}\right) = \dfrac{3}{\dfrac{2}{x} - 2} = \dfrac{3x}{2 - 2x},\ x \neq 0, 1$

10. Finding the values:
$$f(g(-2)) = f(5) = 25 - 10 - 3 = 12 \qquad g(f(1)) = g(1 - 2 - 3) = g(-4) = 7$$

11. Using the point-slope formula:
$$y - (-8) = -\frac{5}{6}(x - 4)$$
$$y + 8 = -\frac{5}{6}x + \frac{10}{3}$$
$$y = -\frac{5}{6}x - \frac{14}{3}$$
$$f(x) = -\frac{5}{6}x - \frac{14}{3}$$

12. First find the function: $\left(\dfrac{f}{g}\right)(x) = \dfrac{\dfrac{3}{x}}{\dfrac{2}{x-1}} = \dfrac{3x-3}{2x},\ x \neq 0, 1$

    The domain is $\{x \mid x \neq 0, 1\}$.

13. Finding the values:
$$(f+g)(-2) = f(-2) + g(-2) = 11 + 7 = 18$$
$$(f-g)(4) = f(4) - g(4) = 29 - 19 = 10$$
$$(g-f)(-1) = g(-1) - f(-1) = 4 - 4 = 0$$

14. Finding the functions:
$$(f \bullet g)(x) = (x^2 + 5x - 6)(x-1) = x^3 + 4x^2 - 11x + 6$$
$$\left(\frac{f}{g}\right)(x) = \frac{x^2 + 5x - 6}{x - 1} = \frac{(x+6)(x-1)}{x-1} = x + 6,\ x \neq 1$$

15. Let x and $60 - x$ represent the two numbers. The function is: $f(x) = x^2 + 12(60 - x) = x^2 - 12x + 720$

    The vertex is $x = \dfrac{12}{2} = 6$, so the two numbers are 6 and 54.

16. The variation equation is $y = kxz$. Substituting $y = 18$, $x = 8$ and $z = 9$:
$$18 = k(8)(9)$$
$$k = \frac{18}{72} = \frac{1}{4}$$
    So $y = \dfrac{1}{4}xz$. Substituting $x = 5$ and $z = 12$: $y = \dfrac{1}{4}(5)(12) = 15$

17. The variation equation is $y = \dfrac{k}{x}$. Substituting $y = \dfrac{1}{2}$ and $x = -8$:
$$\frac{1}{2} = \frac{k}{-8}$$
$$k = -4$$

18. The variation equation is $I = krt$. Substituting $I = 140$, $r = 0.07$ and $t = 5$:
$$140 = k(0.07)(5)$$
$$k = \frac{140}{0.35} = 400$$

So $I = 400rt$. Substituting $r = 0.08$ and $t = 3$: $I = 400(0.08)(3) = \$96$

19. The function is: $s(c) = c + 0.35c = 1.35c$

Substituting $c = 13$: $s(13) = 1.35(13) = \$17.55$

20. Finding the x-intercepts:
$$4x^2 - 16x - 48 = 0$$
$$4(x^2 - 4x - 12) = 0$$
$$4(x + 2)(x - 6) = 0$$
$$x = -2, 6$$
The vertex is $(2, -64)$.

21. Graphing the function $f(x) = (x - 2)^3 - 3$:

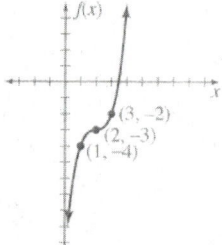

22. Graphing the function $f(x) = -2x^2 - 12x - 14$:

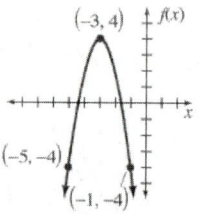

23. Graphing the function $f(x) = 3|x - 2| - 1$:

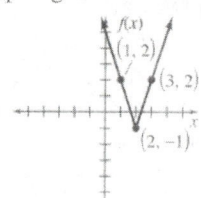

24. Graphing the function $f(x) = \sqrt{-x + 2}$:

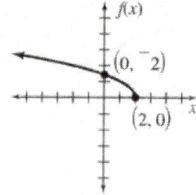

25. Graphing the function $f(x) = -x - 1$:

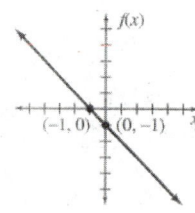

Chapters 0-3 Cumulative Review Problem Set

1.   Evaluating: $\left(3^{-2}\right)^{-1} = 3^2 = 9$

2.   Evaluating: $\left(\dfrac{7}{9}\right)^{-1} = \dfrac{9}{7}$

3.   Evaluating: $\dfrac{1}{\left(\dfrac{1}{2}\right)^{-3}} = \left(\dfrac{1}{2}\right)^3 = \dfrac{1}{8}$

4.   Evaluating: $8^{-1} + 2^{-3} = \dfrac{1}{8} + \dfrac{1}{8} = \dfrac{2}{8} = \dfrac{1}{4}$

5.   Evaluating: $\left(3^{-2} + 2^{-3}\right)^{-1} = \left(\dfrac{1}{9} + \dfrac{1}{8}\right)^{-1} = \left(\dfrac{17}{72}\right)^{-1} = \dfrac{72}{17}$

6.   Evaluating: $-\sqrt{0.16} = -0.4$

7.   Evaluating: $\sqrt[3]{3\dfrac{3}{8}} = \sqrt[3]{\dfrac{27}{8}} = \dfrac{3}{2}$

8.   Evaluating: $9^{3/2} = \left(\sqrt{9}\right)^3 = 3^3 = 27$

9.   Evaluating: $8^{2/3} = \left(\sqrt[3]{8}\right)^2 = 2^2 = 4$

10.   Evaluating: $(-27)^{4/3} = \left(\sqrt[3]{-27}\right)^4 = (-3)^4 = 81$

11.   Evaluating: $-3(-9-1) + 4(-18+3) - (-27+5) = 30 - 60 + 22 = -8$

12.   Evaluating: $\dfrac{3}{-7} - \dfrac{5}{-7} + \dfrac{9}{-7} = \dfrac{7}{-7} = -1$

13.   Evaluating: $\dfrac{4}{6-2} + \dfrac{7}{6+1} = \dfrac{4}{4} + \dfrac{7}{7} = 1 + 1 = 2$

14.   Evaluating: $(10-5)(10+5) = (5)(15) = 75$

15.   Evaluating: $\dfrac{\dfrac{2}{-3} - \dfrac{3}{11}}{\dfrac{1}{-3} + \dfrac{4}{11}} \cdot \dfrac{33}{33} = \dfrac{-22-9}{-11+12} = \dfrac{-31}{1} = -31$

16.   Simplifying: $\dfrac{12x^3 y^2}{27xy} = \dfrac{4x^2 y}{9}$

17.   Simplifying: $\dfrac{6x^2 + 11x - 7}{8x^2 - 22x + 9} = \dfrac{(3x+7)(2x-1)}{(4x-9)(2x-1)} = \dfrac{3x+7}{4x-9}$

18.   Simplifying: $\dfrac{8x^3 + 64}{4x^2 - 16} = \dfrac{8\left(x^3 + 8\right)}{4\left(x^2 - 4\right)} = \dfrac{2(x+2)\left(x^2 - 2x + 4\right)}{(x+2)(x-2)} = \dfrac{2\left(x^2 - 2x + 4\right)}{x-2}$

19.   Simplifying: $\dfrac{xy + 4y - 2x - 8}{x^2 + 4x} = \dfrac{y(x+4) - 2(x+4)}{x(x+4)} = \dfrac{(x+4)(y-2)}{x(x+4)} = \dfrac{y-2}{x}$

20.   Simplifying: $\dfrac{3a^2 b}{4a^3 b^2} \div \dfrac{6a}{27b} = \dfrac{3a^2 b}{4a^3 b^2} \cdot \dfrac{27b}{6a} = \dfrac{27a^2 b^2}{8a^4 b^2} = \dfrac{27}{8a^2}$

21.   Simplifying: $\dfrac{x^2 - x}{x+5} \cdot \dfrac{x^2 + 5x + 4}{x^4 - x^2} = \dfrac{x(x-1)}{x+5} \cdot \dfrac{(x+4)(x+1)}{x^2(x+1)(x-1)} = \dfrac{x+4}{x(x+5)}$

22.   Simplifying: $\dfrac{x+3}{10} + \dfrac{2x+1}{15} - \dfrac{x-2}{18} = \dfrac{9x+27}{90} + \dfrac{12x+6}{90} - \dfrac{5x-10}{90} = \dfrac{9x+27 + 12x+6 - 5x+10}{90} = \dfrac{16x+43}{90}$

23.   Simplifying: $\dfrac{7}{12ab} - \dfrac{11}{15a^2} = \dfrac{35a}{60a^2 b} - \dfrac{44b}{60a^2 b} = \dfrac{35a - 44b}{60a^2 b}$

24.   Simplifying: $\dfrac{8}{x^2 - 4x} + \dfrac{2}{x} = \dfrac{8}{x(x-4)} + \dfrac{2}{x} \cdot \dfrac{x-4}{x-4} = \dfrac{8 + 2x - 8}{x(x-4)} = \dfrac{2x}{x(x-4)} = \dfrac{2}{x-4}$

25.   Simplifying: $\dfrac{\dfrac{2}{x} - 3}{\dfrac{3}{y} + 4} = \dfrac{\dfrac{2}{x} - 3}{\dfrac{3}{y} + 4} \cdot \dfrac{xy}{xy} = \dfrac{2y - 3xy}{3x + 4xy}$

26. Simplifying: $\dfrac{\dfrac{5}{x^2} - \dfrac{3}{x}}{\dfrac{1}{y} + \dfrac{2}{y^2}} = \dfrac{\dfrac{5}{x^2} - \dfrac{3}{x}}{\dfrac{1}{y} + \dfrac{2}{y^2}} \cdot \dfrac{x^2 y^2}{x^2 y^2} = \dfrac{5y^2 - 3xy^2}{x^2 y + 2x^2}$

27. Simplifying: $\dfrac{3a}{2 - \dfrac{1}{a}} - 1 = \dfrac{3a}{2 - \dfrac{1}{a}} \cdot \dfrac{a}{a} - 1 = \dfrac{3a^2}{2a - 1} - 1 = \dfrac{3a^2}{2a - 1} - \dfrac{2a - 1}{2a - 1} = \dfrac{3a^2 - 2a + 1}{2a - 1}$

28. Simplifying: $\left(-3x^{-1}y^2\right)\left(4x^{-2}y^{-3}\right) = -12x^{-3}y^{-1} = -\dfrac{12}{x^3 y}$

29. Simplifying: $\dfrac{48x^{-4}y^2}{6xy} = 8x^{-5}y = \dfrac{8y}{x^5}$

30. Simplifying: $\left(\dfrac{27a^{-4}b^{-3}}{-3a^{-1}b^{-4}}\right)^{-1} = \dfrac{-3a^{-1}b^{-4}}{27a^{-4}b^{-3}} = -\dfrac{a^3 b^{-1}}{9} = -\dfrac{a^3}{9b}$

31. Simplifying: $\sqrt{\dfrac{8}{25}} = \dfrac{\sqrt{8}}{\sqrt{25}} = \dfrac{2\sqrt{2}}{5}$

32. Simplifying: $\dfrac{4\sqrt{3}}{7\sqrt{6}} = \dfrac{4\sqrt{3}}{7\sqrt{6}} \cdot \dfrac{\sqrt{6}}{\sqrt{6}} = \dfrac{4\sqrt{18}}{42} = \dfrac{12\sqrt{2}}{42} = \dfrac{2\sqrt{2}}{7}$

33. Simplifying: $\sqrt{48x^3 y^7} = \sqrt{16x^2 y^6 \cdot 3xy} = 4xy^3 \sqrt{3xy}$

34. Simplifying: $\dfrac{4}{\sqrt{5} - \sqrt{3}} = \dfrac{4}{\sqrt{5} - \sqrt{3}} \cdot \dfrac{\sqrt{5} + \sqrt{3}}{\sqrt{5} + \sqrt{3}} = \dfrac{4\left(\sqrt{5} + \sqrt{3}\right)}{5 - 3} = 2\left(\sqrt{5} + \sqrt{3}\right)$

35. Simplifying: $\sqrt[3]{48x^4 y^5} = \sqrt[3]{8x^3 y^3 \cdot 6xy^2} = 2xy\sqrt[3]{6xy^2}$

36. Simplifying: $\dfrac{\sqrt[3]{4}}{\sqrt[3]{2}} = \dfrac{\sqrt[3]{4}}{\sqrt[3]{2}} \cdot \dfrac{\sqrt[3]{4}}{\sqrt[3]{4}} = \dfrac{\sqrt[3]{16}}{\sqrt[3]{8}} = \dfrac{2\sqrt[3]{2}}{2} = \sqrt[3]{2}$

37. Finding the product: $(5 - 2i)(6 + 5i) = 30 + 13i - 10i^2 = 40 + 13i$

38. Finding the product: $(-3 - i)(-2 - 4i) = 6 + 14i + 4i^2 = 2 + 14i$

39. Finding the quotient: $\dfrac{5}{4i} = \dfrac{5}{4i} \cdot \dfrac{i}{i} = \dfrac{5i}{4i^2} = -\dfrac{5}{4}i = 0 - \dfrac{5}{4}i$

40. Finding the quotient: $\dfrac{6 + 2i}{3 - 4i} = \dfrac{6 + 2i}{3 - 4i} \cdot \dfrac{3 + 4i}{3 + 4i} = \dfrac{18 + 30i + 8i^2}{9 - 16i^2} = \dfrac{10 + 30i}{25} = \dfrac{2}{5} + \dfrac{6}{5}i$

41. Solving the equation:

$$3(2x - 1) - 2(5x + 1) = 4(3x + 4)$$
$$6x - 3 - 10x - 2 = 12x + 16$$
$$-4x - 5 = 12x + 16$$
$$-16x = 21$$
$$x = -\dfrac{21}{16}$$

The solution set is $\left\{-\dfrac{21}{16}\right\}$.

42. Solving the equation:

$$n + \dfrac{3n - 1}{9} - 4 = \dfrac{3n + 1}{3}$$
$$9n + 3n - 1 - 36 = 9n + 3$$
$$12n - 37 = 9n + 3$$
$$3n = 40$$
$$n = \dfrac{40}{3}$$

The solution set is $\left\{\dfrac{40}{3}\right\}$.

43. Solving the equation:
$$0.92 + 0.9(x - 0.3) = 2x - 5.95$$
$$0.92 + 0.9x - 0.27 = 2x - 5.95$$
$$0.9x + 0.65 = 2x - 5.95$$
$$-0.11x = -6.6$$
$$x = 6$$

The solution set is $\{6\}$.

44. Solving the equation:
$$|4x - 1| = 11$$
$$4x - 1 = -11, 11$$
$$4x = -10, 12$$
$$x = -\frac{5}{2}, 3$$

The solution set is $\left\{-\frac{5}{2}, 3\right\}$.

45. Solving the equation:
$$|2x - 1| = |-x + 4|$$
$$2x - 1 = -x + 4 \qquad \text{or} \qquad 2x - 1 = x - 4$$
$$3x = 5 \qquad\qquad\qquad x = -3$$
$$x = \frac{5}{3}$$

The solution set is $\left\{-3, \frac{5}{3}\right\}$.

46. Solving the equation:
$$x^3 = 36x$$
$$x^3 - 36x = 0$$
$$x(x + 6)(x - 6) = 0$$
$$x = -6, 0, 6$$

The solution set is $\{-6, 0, 6\}$.

47. Solving the equation:
$$(3x - 1)^2 = 45$$
$$3x - 1 = \pm\sqrt{45}$$
$$3x = 1 \pm 3\sqrt{5}$$
$$x = \frac{1 \pm 3\sqrt{5}}{3}$$

The solution set is $\left\{\frac{1 \pm 3\sqrt{5}}{3}\right\}$.

48. Solving the equation:
$$(2x + 5)^2 = -32$$
$$2x + 5 = \pm\sqrt{-32}$$
$$2x = -5 \pm 4i\sqrt{2}$$
$$x = \frac{-5 \pm 4i\sqrt{2}}{2}$$

The solution set is $\left\{\frac{-5 \pm 4i\sqrt{2}}{2}\right\}$.

49. Using $a = 2$, $b = -3$, and $c = 4$ in the quadratic formula: $x = \dfrac{3 \pm \sqrt{9 - 4(2)(4)}}{2(2)} = \dfrac{3 \pm \sqrt{-23}}{4} = \dfrac{3 \pm i\sqrt{23}}{4}$

The solution set is $\left\{\dfrac{3 \pm i\sqrt{23}}{4}\right\}$.

50. Solving the equation:
$$(n + 4)(n - 6) = 11$$
$$n^2 - 2n - 24 = 11$$
$$n^2 - 2n - 35 = 0$$
$$(n - 7)(n + 5) = 0$$
$$n = -5, 7$$

The solution set is $\{-5, 7\}$.

51. Solving the equation:
$$(2n - 1)(n + 6) = 11$$
$$n = -6, \frac{1}{2}$$

The solution set is $\left\{-6, \frac{1}{2}\right\}$.

52. Solving the equation:
$$(x+5)(3x-1)=(x+5)(2x+7)$$
$$3x^2+14x-5=2x^2+17x+35$$
$$x^2-3x-40=0$$
$$(x+5)(x-8)=0$$
$$x=-5,8$$
The solution set is $\{-5,8\}$.

53. Solving the equation:
$$(x-4)(2x+9)=(2x-1)(x+2)$$
$$2x^2+x-36=2x^2+3x-2$$
$$-2x=34$$
$$x=-17$$
The solution set is $\{-17\}$.

54. Solving the equation:
$$(3x-1)(x+1)=(2x+1)(x-3)$$
$$3x^2+2x-1=2x^2-5x-3$$
$$x^2+7x+2=0$$

Using $a=1$, $b=7$, and $c=2$ in the quadratic formula: $x=\dfrac{-7\pm\sqrt{49-8}}{2}=\dfrac{-7\pm\sqrt{41}}{2}$

The solution set is $\left\{\dfrac{-7\pm\sqrt{41}}{2}\right\}$.

55. Solving the equation:
$$\sqrt{3x}-x=-6$$
$$\sqrt{3x}=x-6$$
$$3x=(x-6)^2$$
$$3x=x^2-12x+36$$
$$0=x^2-15x+36$$
$$0=(x-12)(x-3)$$
$$x=12 \quad (x=3 \text{ does not check})$$

The solution set is $\{12\}$.

56. Solving the equation:
$$\sqrt{x+19}-\sqrt{x+28}=-1$$
$$\sqrt{x+19}=\sqrt{x+28}-1$$
$$x+19=\left(\sqrt{x+28}-1\right)^2$$
$$x+19=x+28-2\sqrt{x+28}+1$$
$$-10=-2\sqrt{x+28}$$
$$5=\sqrt{x+28}$$
$$25=x+28$$
$$x=-3$$
The solution set is $\{-3\}$.

57. Solving the equation:
$$12x^4-19x^2+5=0$$
$$\left(4x^2-5\right)\left(3x^2-1\right)=0$$
$$x^2=\frac{5}{4},\frac{1}{3}$$
$$x=\pm\frac{\sqrt{5}}{2},\pm\frac{1}{\sqrt{3}}=\pm\frac{\sqrt{3}}{3}$$

The solution set is $\left\{\pm\dfrac{\sqrt{5}}{2},\pm\dfrac{\sqrt{3}}{3}\right\}$.

58. Solving the equation:
$$x^3-4x^2-3x+12=0$$
$$x^2(x-4)-3(x-4)=0$$
$$(x-4)(x^2-3)=0$$
$$x=4,\pm\sqrt{3}$$

The solution set is $\left\{4,\pm\sqrt{3}\right\}$.

59. Solving the inequality:
$$|5x-2|>13$$
$$5x-2<-13 \qquad \text{or} \qquad 5x-2>13$$
$$5x<-11 \qquad\qquad\qquad 5x>15$$
$$x<-\frac{11}{5} \qquad\qquad\qquad x>3$$

The solution set is $\left(-\infty,-\dfrac{11}{5}\right)\cup(3,\infty)$.

60. Forming a sign chart:

The solution set is $[-4,2]$.

61. Solving the inequality:
$$|6x+2| \le 8$$
$$-8 \le 6x+2 \le 8$$
$$-10 \le 6x \le 6$$
$$-\frac{5}{3} \le x \le 1$$

The solution set is $\left[-\frac{5}{3}, 1\right]$.

62. Solving the inequality:
$$x(x+5) < 24$$
$$x^2 + 5x - 24 < 0$$
$$(x+8)(x-3) < 0$$

Forming a sign chart:

The solution set is $(-8, 3)$.

63. Solving the inequality:

$$-5(y-1)+3 > 3y-4-4y$$
$$-5y+5+3 > -y-4$$
$$-4y > -12$$
$$y < 3$$

The solution set is $(-\infty, 3)$.

64. Solving the inequality:
$$\frac{x-2}{5} - \frac{3x-1}{4} \le \frac{3}{10}$$
$$4(x-2)-5(3x-1) \le 6$$
$$4x-8-15x+5 \le 6$$
$$-11x-3 \le 6$$
$$-11x \le 9$$
$$x \ge -\frac{9}{11}$$

The solution set is $\left[-\frac{9}{11}, \infty\right)$.

65. Forming a sign chart:

The solution set is $\left(-5, -\frac{1}{2}\right) \cup (2, \infty)$.

66. Forming a sign chart:

The solution set is $(-\infty, 3] \cup (7, \infty)$.

67. Solving the inequality:
$$\frac{2x}{x+3} > 4$$
$$\frac{2x}{x+3} - 4 > 0$$
$$\frac{2x-4x-12}{x+3} > 0$$
$$\frac{-2x-12}{x+3} > 0$$

Forming a sign chart:

The solution set is $(-6, -3)$.

68. Factoring the inequality:

$$2x^3 + 5x^2 - 3x < 0$$
$$x(2x^2 + 5x - 3) < 0$$
$$x(2x-1)(x+3) < 0$$

Forming a sign chart:

The solution set is $(-\infty, -3) \cup \left(0, \frac{1}{2}\right)$.

69. The point is: $-6 + \frac{3}{4}(16) = -6 + 12 = 6$

70. Finding the coordinates:
$$x = -1 + \frac{2}{3}(9) = -1 + 6 = 5 \qquad\qquad y = 2 + \frac{2}{3}(9) = 2 + 6 = 8$$
The point is $(5,8)$.

71. Solving for y:
$$-2x + 5y = 7$$
$$5y = 2x + 7$$
$$y = \frac{2}{5}x + \frac{7}{5}$$
The slope is $\frac{2}{5}$.

72. Finding the slope: $m = \dfrac{-1+4}{-2-3} = -\dfrac{3}{5}$. Using the point-slope formula:
$$y + 4 = -\frac{3}{5}(x - 3)$$
$$5y + 20 = -3x + 9$$
$$3x + 5y = -11$$

73. Finding the function values:
$$f\big(g(3)\big) = f(9+6) = f(15) = 45 - 2 = 43 \qquad g\big(f(2)\big) = g(6-2) = g(4) = 16 + 8 = 24$$

74. Finding the compositions:
$$f\big(g(x)\big) = f\big(\sqrt{x+2}\big) = 2\sqrt{x+2} - 1 \qquad g\big(f(x)\big) = g(2x-1) = \sqrt{2x-1+2} = \sqrt{2x+1}$$

75. Solving the inequality:
$$x^2 + 7x - 30 \geq 0$$
$$(x+10)(x-3) \geq 0$$
Forming a sign chart:

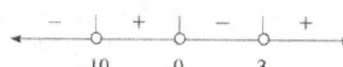

The domain is $(-\infty, -10] \cup [3, \infty)$.

76. Simplifying the expression:
$$\frac{f(a+h) - f(a)}{h} = \frac{-(a+h)^2 + 6(a+h) - 1 - \left(-a^2 + 6a - 1\right)}{h}$$
$$= \frac{-a^2 - 2ah - h^2 + 6a + 6h - 1 + a^2 - 6a + 1}{h}$$
$$= \frac{-2ah - h^2 + 6h}{h}$$
$$= -2a - h + 6$$

77. Graphing the function:

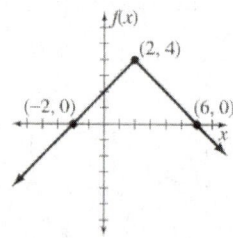

78.   Completing the square: $f(x) = -x^2 - 6x - 10 = -(x^2 + 6x + 9) - 10 + 9 = -(x+3)^2 - 1$. Graphing the function:

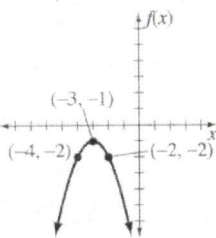

79.   Graphing the function $f(x) = x - 2$:

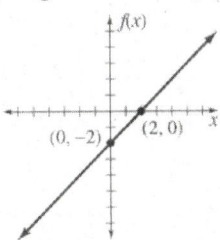

80.   Graphing the function $f(x) = (x-2)^2$:

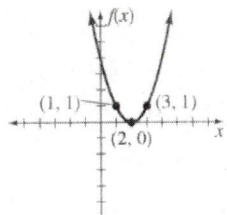

81.   Graphing the function $f(x) = (x-2)^3$:

82.   Graphing the function $f(x) = \sqrt{x-2}$:

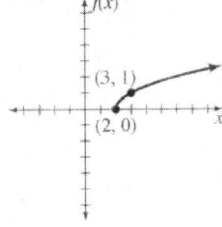

83.   Let x, $x+2$, and $x+4$ represent the three integers. The equation is:
$$x + x + 2 + x + 4 = 57$$
$$3x + 6 = 57$$
$$3x = 51$$
$$x = 17$$
The integers are 17, 19, and 21.

84.   Let n represent the number of nickels, $n+6$ represent the number of dimes, and $2n+1$ represent the number of quarters. The equation is:
$$n + n + 6 + 2n + 1 = 63$$
$$4n + 7 = 63$$
$$4n = 56$$
$$n = 14$$
Eric has 14 nickels, 20 dimes, and 29 quarters.

85.   Let x and $\frac{1}{3}x + 4$ represent the two angles. The equation is:
$$x + \frac{1}{3}x + 4 = 180$$
$$\frac{4}{3}x = 176$$
$$4x = 528$$
$$x = 132$$
The two angles are 132° and 48°.

86.   Let s represent the selling price. The equation is:
$$300 + 0.5s = s$$
$$300 = 0.5s$$
$$s = 600$$
The selling price is $600.

87.   Let x represent the amount invested at 8% and $x + 300$ the amount invested at 9%. The equation is:

$$0.08(x) + 0.09(x + 300) = 316$$
$$0.08x + 0.09x + 27 = 316$$
$$0.17x = 289$$
$$x = 1700$$

Beth invested $1700 at 8% and $2000 at 9%.

88.   Let r and $r + 10$ represent the rates of the two trains. Since their total distance apart is 639 miles, the equation is:

$$4.5r + 4.5(r + 10) = 639$$
$$4.5r + 4.5r + 45 = 639$$
$$9r + 45 = 639$$
$$9r = 594$$
$$r = 66$$

The rates of the two trains are 66 mph and 76 mph.

89.   Let x represent the quantity to be drained out and added. The equation is:

$$0.50(10 - x) + x = 0.70(10)$$
$$5 - 0.5x + x = 7$$
$$0.5x = 2$$
$$x = 4$$

Thus 4 quarts needs to be drained and added.

90.   Let x represent his score on the fourth day. The inequality is:

$$\frac{70 + 73 + 76 + x}{4} \leq 72$$
$$219 + x \leq 288$$
$$x \leq 69$$

Sam must shoot 69 or less on the fourth day.

91.   Let x represent the number. The equation is:

$$x^3 = 9x$$
$$x^3 - 9x = 0$$
$$x(x + 3)(x - 3) = 0$$
$$x = -3, 0, 3$$

The number is either -3, 0 or 3.

92.   Let x represent the width of the strip. The equation is:

$$(8 - 2x)(14 - 2x) = 72$$
$$112 - 44x + 4x^2 = 72$$
$$4x^2 - 44x + 40 = 0$$
$$x^2 - 11x + 10 = 0$$
$$(x - 1)(x - 10) = 0$$
$$x = 1 \quad (x = 10 \text{ is not possible})$$

The width of the strip is 1 inch.

93.   Let $3x$ and $4x$ represent the amount each person receives. The equation is:

$$3x + 4x = 2450$$
$$7x = 2450$$
$$x = 350$$

The amounts each received is $1050 and $1400.

94. Let t represent the time for Sue to complete the task by herself. The equation is:

$$\frac{1}{2} \cdot \frac{6}{5} + \frac{1}{t} \cdot \frac{6}{5} = 1$$

$$\frac{3}{5} + \frac{6}{5t} = 1$$

$$3t + 6 = 5t$$

$$6 = 2t$$

$$t = 3$$

It would take Sue 3 hours to complete the task by herself.

95. Let x represent the number of calories in a serving of meat. Then $\frac{1}{2}x$ represents the calories in a serving of starchy

vegetables, and $\frac{1}{3}x$ represents the calories in a serving of fruit. The equation is:

$$x + \frac{1}{2}x + \frac{1}{3}x = 770$$

$$\frac{11}{6}x = 770$$

$$11x = 4620$$

$$x = 420$$

$$\frac{1}{2}x = 210$$

$$\frac{1}{3}x = 140$$

The meal consists of 420 calories for meat, 210 calories for vegetables, and 140 calories for fruit.

96. Let x represent the cost of a table and $\frac{1}{2}x - 20$ represent the cost of each chair. Each seating arrangement will then

cost $x + 4\left(\frac{1}{2}x - 20\right) = x + 2x - 80 = 3x - 80$. Since there are 24 seating arrangements and the budget is \$27,960, we

have the inequality:

$$24(3x - 80) = 27,960$$

$$72x - 1920 = 27,960$$

$$72x = 29,880$$

$$x = 415$$

Thus the cost of the table is \$415, and the cost of each chair is $\frac{1}{2}(415) - 20 = \$187.50$.

97. Let x, y and z represent the angles in increasing order. Then $x + y = z - 40$. Since the sum of all three angles is $180°$:

$$x + y + z = 180$$

$$z - 40 + z = 180$$

$$2z = 220$$

$$z = 110$$

Thus the largest angle is $110°$. Now $x + z = 2y$, so:

$$x + 110 = 2y$$

$$x = 2y - 110$$

Therefore:

$$2y - 110 + y = 70$$

$$3y = 180$$

$$y = 60$$

$$x = 10$$

The angles are $10°$, $60°$, and $110°$.

Chapter 4
Polynomial and Rational Functions

4.1 Dividing Polynomials and Synthetic Division

1. Using long division:

$$
\require{enclose}
\begin{array}{r}
4x+5 \\
3x-2 \enclose{longdiv}{12x^2+7x-10} \\
\underline{12x^2-8x} \\
15x-10 \\
\underline{15x-10} \\
0
\end{array}
$$

The quotient is $4x+5$ and the remainder is 0.

3. Using long division:

$$
\begin{array}{r}
t^2+2t-4 \\
3t+1 \enclose{longdiv}{3t^3+7t^2-10t-4} \\
\underline{3t^3+\ t^2} \\
6t^2-10t \\
\underline{6t^2+2t} \\
-12t-4 \\
\underline{-12t-4} \\
0
\end{array}
$$

The quotient is t^2+2t-4 and the remainder is 0.

5. Using long division:

$$
\begin{array}{r}
2x+5 \\
3x+2 \enclose{longdiv}{6x^2+19x+11} \\
\underline{6x^2+\ 4x} \\
15x+11 \\
\underline{15x+10} \\
1
\end{array}
$$

The quotient is $2x+5$ and the remainder is 1.

7. Using long division:

$$
\begin{array}{r}
3x-4 \\
x^2+2x \enclose{longdiv}{3x^3+2x^2-5x-1} \\
\underline{3x^3+6x^2} \\
-4x^2-5x \\
\underline{-4x^2-8x} \\
3x-1
\end{array}
$$

The quotient is $3x-4$ and the remainder is $3x-1$.

9. Using long division:

$$
\begin{array}{r}
5y-1 \\
y^2-y \enclose{longdiv}{5y^3-6y^2-7y-2} \\
\underline{5y^3-5y^2} \\
-y^2-7y \\
\underline{-y^2+\ y} \\
-8y-2
\end{array}
$$

The quotient is $5y-1$ and the remainder is $-8y-2$.

11. Using long division:

$$
\begin{array}{r}
4a+6 \\
a^2-2a+3 \enclose{longdiv}{4a^3-2a^2+7a-1} \\
\underline{4a^3-8a^2+12a} \\
6a^2-\ 5a-1 \\
\underline{6a^2-12a+18} \\
7a-19
\end{array}
$$

The quotient is $4a+6$ and the remainder is $7a-19$.

13. Using long division:

$$\begin{array}{r}
3x + 4y \\
x - 2y \overline{\smash{\big)}\ 3x^2 - 2xy - 8y^2} \\
\underline{3x^2 - 6xy} \\
4xy - 8y^2 \\
\underline{4xy - 8y^2} \\
0
\end{array}$$

The quotient is $3x + 4y$ and the remainder is 0.

15. Using synthetic division:

$$\begin{array}{r}
2 \,\big)\overline{\ 4 \quad\ -5 \quad -6} \\
\underline{\quad\ \ 8 \qquad 6} \\
4 \qquad 3 \qquad\ 0
\end{array}$$

The quotient is $4x + 3$ and the remainder is 0.

17. Using synthetic division:

$$\begin{array}{r}
-3 \,\big)\overline{\ 2 \quad\ -1 \quad -21} \\
\underline{\qquad\ -6 \qquad 21} \\
2 \qquad -7 \qquad\ 0
\end{array}$$

The quotient is $2x - 7$ and the remainder is 0.

19. Using synthetic division:

$$\begin{array}{r}
4 \,\big)\overline{\ 3 \quad -16 \quad 17} \\
\underline{\qquad\ 12 \quad -16} \\
3 \qquad -4 \qquad\ 1
\end{array}$$

The quotient is $3x - 4$ and the remainder is 1.

21. Using synthetic division:

$$\begin{array}{r}
-6 \,\big)\overline{\ 4 \quad\ 19 \quad -32} \\
\underline{\qquad -24 \qquad 30} \\
4 \qquad -5 \qquad\ -2
\end{array}$$

The quotient is $4x - 5$ and the remainder is -2.

23. Using synthetic division:

$$\begin{array}{r}
1 \,\big)\overline{\ 1 \quad 2 \quad -7 \quad 4} \\
\underline{\qquad 1 \quad\ 3 \quad -4} \\
1 \quad 3 \quad -4 \quad\ 0
\end{array}$$

The quotient is $x^2 + 3x - 4$ and the remainder is 0.

25. Using synthetic division:

$$\begin{array}{r}
-2 \,\big)\overline{\ 3 \quad 8 \quad 0 \quad -8} \\
\underline{\qquad -6 \quad -4 \quad 8} \\
3 \quad 2 \quad -4 \quad\ 0
\end{array}$$

The quotient is $3x^2 + 2x - 4$ and the remainder is 0.

27. Using synthetic division:

$$\begin{array}{r}
2 \,\big)\overline{\ 5 \quad -9 \quad -3 \quad -2} \\
\underline{\qquad 10 \quad\ 2 \quad -2} \\
5 \quad 1 \quad -1 \quad -4
\end{array}$$

The quotient is $5x^2 + x - 1$ and the remainder is -4.

29. Using synthetic division:

$$\begin{array}{r}
-7 \,\big)\overline{\ 1 \quad 6 \quad -8 \quad 1} \\
\underline{\qquad -7 \quad\ 7 \quad 7} \\
1 \quad -1 \quad -1 \quad 8
\end{array}$$

The quotient is $x^2 - x - 1$ and the remainder is 8.

31. Using synthetic division:

$$\begin{array}{r}
3 \,\big)\overline{\ -1 \quad 7 \quad -14 \quad 6} \\
\underline{\qquad -3 \quad 12 \quad -6} \\
-1 \quad 4 \quad -2 \quad\ 0
\end{array}$$

The quotient is $-x^2 + 4x - 2$ and the remainder is 0.

33. Using synthetic division:

$$\begin{array}{r}
-1 \,\big)\overline{\ -3 \quad 1 \quad 2 \quad 2} \\
\underline{\qquad\ 3 \quad -4 \quad 2} \\
-3 \quad 4 \quad -2 \quad 4
\end{array}$$

The quotient is $-3x^2 + 4x - 2$ and the remainder is 4.

35. Using synthetic division:

$$\begin{array}{r}
2 \,\big)\overline{\ 3 \quad 0 \quad -2 \quad -5} \\
\underline{\qquad 6 \quad 12 \quad 20} \\
3 \quad 6 \quad 10 \quad 15
\end{array}$$

The quotient is $3x^2 + 6x + 10$ and the remainder is 15.

37. Using synthetic division:

$$\begin{array}{r}
-1 \,\big)\overline{\ 2 \quad 1 \quad 3 \quad 2 \quad -2} \\
\underline{\qquad -2 \quad 1 \quad -4 \quad 2} \\
2 \quad -1 \quad 4 \quad -2 \quad\ 0
\end{array}$$

The quotient is $2x^3 - x^2 + 4x - 2$ and the remainder is 0.

39. Using synthetic division:

$$\begin{array}{r}
3 \,\big)\overline{\ 1 \quad 4 \quad 0 \quad -7 \quad -1} \\
\underline{\qquad 3 \quad 21 \quad 63 \quad 168} \\
1 \quad 7 \quad 21 \quad 56 \quad 167
\end{array}$$

The quotient is $x^3 + 7x^2 + 21x + 56$ and the remainder is 167.

41. Using synthetic division:

$$
\begin{array}{r|rrrrr}
-5 & 1 & 5 & -1 & 0 & 25 \\
  &   & -5 & 0 & 5 & -25 \\
\hline
  & 1 & 0 & -1 & 5 & 0
\end{array}
$$

The quotient is $x^3 - x + 5$ and the remainder is 0.

43. Using synthetic division:

$$
\begin{array}{r|rrrrr}
2 & 1 & 0 & 0 & 0 & -16 \\
  &   & 2 & 4 & 8 & 16 \\
\hline
  & 1 & 2 & 4 & 8 & 0
\end{array}
$$

The quotient is $x^3 + 2x^2 + 4x + 8$ and the remainder is 0.

45. Using synthetic division:

$$
\begin{array}{r|rrrrrr}
-1 & 1 & 0 & 0 & 0 & 0 & -1 \\
   &   & -1 & 1 & -1 & 1 & -1 \\
\hline
   & 1 & -1 & 1 & -1 & 1 & -2
\end{array}
$$

The quotient is $x^4 - x^3 + x^2 - x + 1$ and the remainder is -2.

47. Using synthetic division:

$$
\begin{array}{r|rrrrrr}
-1 & 1 & 0 & 0 & 0 & 0 & 1 \\
   &   & -1 & 1 & -1 & 1 & -1 \\
\hline
   & 1 & -1 & 1 & -1 & 1 & 0
\end{array}
$$

The quotient is $x^4 - x^3 + x^2 - x + 1$ and the remainder is 0.

49. Using synthetic division:

$$
\begin{array}{r|rrrrrr}
-4 & 1 & 3 & -5 & -3 & 3 & -4 \\
   &   & -4 & 4 & 4 & -4 & 4 \\
\hline
   & 1 & -1 & -1 & 1 & -1 & 0
\end{array}
$$

The quotient is $x^4 - x^3 - x^2 + x - 1$ and the remainder is 0.

51. Using synthetic division:

$$
\begin{array}{r|rrrrrr}
1 & 4 & -6 & 2 & 2 & -5 & 2 \\
  &   & 4 & -2 & 0 & 2 & -3 \\
\hline
  & 4 & -2 & 0 & 2 & -3 & -1
\end{array}
$$

The quotient is $4x^4 - 2x^3 + 2x - 3$ and the remainder is -1.

53. Using synthetic division:

$$
\begin{array}{r|rrrr}
1/3 & 9 & -6 & 3 & -4 \\
    &   & 3 & -1 & 2/3 \\
\hline
    & 9 & -3 & 2 & -10/3
\end{array}
$$

The quotient is $9x^2 - 3x + 2$ and the remainder is $-10/3$.

55. Using synthetic division:

$$
\begin{array}{r|rrrrr}
-1/3 & 3 & -2 & 5 & -1 & -1 \\
     &   & -1 & 1 & -2 & 1 \\
\hline
     & 3 & -3 & 6 & -3 & 0
\end{array}
$$

The quotient is $3x^3 - 3x^2 + 6x - 3$ and the remainder is 0.

4.2 Remainder and Factor Theorems

1. **a.** Using synthetic division:

$$
\begin{array}{r|rrr}
3 & 1 & 2 & -6 \\
  &   & 3 & 15 \\
\hline
  & 1 & 5 & 9
\end{array}
$$

   b. Evaluating directly: $f(3) = 3^2 + 2(3) - 6 = 9 + 6 - 6 = 9$

3. **a.** Using synthetic division:

$$
\begin{array}{r|rrr}
-1 & 1 & -2 & 3 & -1 \\
   &   & -1 & 3 & -6 \\
\hline
   & 1 & -3 & 6 & -7
\end{array}
$$

   b. Evaluating directly: $f(-1) = (-1)^3 - 2(-1)^2 + 3(-1) - 1 = -1 - 2 - 3 - 1 = -7$

5. a. Using synthetic division:

$$\begin{array}{r|rrrrr} 2 & 2 & -1 & -3 & 4 & -1 \\ & & 4 & 6 & 6 & 20 \\ \hline & 2 & 3 & 3 & 10 & 19 \end{array}$$

   b. Evaluating directly: $f(2)=2(2)^4-(2)^3-3(2)^2+4(2)-1=32-8-12+8-1=19$

7. a. Using synthetic division:

$$\begin{array}{r|rrrr} 6 & 6 & -35 & 8 & -10 \\ & & 36 & 6 & 84 \\ \hline & 6 & 1 & 14 & 74 \end{array}$$

   b. Evaluating directly: $f(6)=6(6)^3-35(6)^2+8(6)-10=1296-1260+48-10=74$

9. a. Using synthetic division:

$$\begin{array}{r|rrrrrr} -2 & 2 & 0 & 0 & 0 & 0 & -1 \\ & & -4 & 8 & -16 & 32 & -64 \\ \hline & 2 & -4 & 8 & -16 & 32 & -65 \end{array}$$

   b. Evaluating directly: $f(-2)=2(-2)^5-1=-64-1=-65$

11. Evaluating: $f(-1)=6(-1)^5-3(-1)^3+2=-6+3+2=-1$

13. Evaluating: $f(8)=2(8)^4-15(8)^3-9(8)^2-2(8)-3=8192-7680-576-16-3=-83$

15. Evaluating: $f(3)=4(3)^7+3=8748+3=8751$

17. Evaluating:

$f(-6)=3(-6)^5+17(-6)^4-4(-6)^3+10(-6)^2-15(-6)+13=-23328+22032+864+360+90+13=31$

19. Evaluating: $f(4)=-4(4)^4-6(4)^2+7=-1024-96+2=-1113$

21. Let $f(x)=5x^2-17x+14$. Evaluating: $f(2)=5(2)^2-17(2)+14=20-34+14=0$

   Thus $x-2$ is a factor of $5x^2-17x+14$.

23. Let $f(x)=6x^2+13x-14$. Evaluating: $f(-3)=6(-3)^2+13(-3)-15=54-39-14=1$

   Thus $x+3$ is not a factor of $6x^2+13x-14$.

25. Let $f(x)=4x^3-13x^2+21x-12$. Evaluating: $f(1)=4(1)^3-13(1)^2+21(1)-12=4-13+21-12=0$

   Thus $x-1$ is a factor of $4x^3-13x^2+21x-12$.

27. Let $f(x)=x^3+7x^2+x-18$. Evaluating: $f(-2)=(-2)^3+7(-2)^2+(-2)-18=-8+28-2-18=0$

   Thus $x+2$ is a factor of x^3+7x^2+x-18.

29. Let $f(x)=3x^3-5x^2-17x+17$. Evaluating: $f(3)=3(3)^3-5(3)^2-17(3)+17=81-45-51+17=2$

   Thus $x-3$ is not a factor of $3x^3-5x^2-17x+17$.

31. Let $f(x)=x^3+8$. Evaluating: $f(-2)=(-2)^3+8=-8+8=0$. Thus $x+2$ is a factor of x^3+8.

33. Let $f(x)=x^4-81$. Evaluating: $f(3)=3^4-81=81-81=0$. Thus $x-3$ is a factor of x^4-81.

35. Using synthetic division:

$$\begin{array}{r|rrrr} 2 & 1 & -6 & -13 & 42 \\ & & 2 & -8 & -42 \\ \hline & 1 & -4 & -21 & 0 \end{array}$$

   Thus: $f(x)=(x-2)(x^2-4x-21)=(x-2)(x-7)(x+3)$

37. Using synthetic division:

$$\begin{array}{r|rrrr} -2 & 12 & 29 & 8 & -4 \\ & & -24 & -10 & 4 \\ \hline & 12 & 5 & -2 & 0 \end{array}$$

   Thus: $f(x)=(x+2)(12x^2+5x-2)=(x+2)(4x-1)(3x+2)$

39. Using synthetic division:

$$
\begin{array}{r|rrr}
-1 & 1 & -2 & -7 & -4 \\
   &   & -1 & 3 & 4 \\
\hline
   & 1 & -3 & -4 & 0
\end{array}
$$

Thus: $f(x) = (x+1)\left(x^2 - 3x - 4\right) = (x+1)(x+1)(x-4) = (x+1)^2 (x-4)$

41. Using synthetic division:

$$
\begin{array}{r|rrrrrr}
6 & 1 & -6 & 0 & 0 & -16 & 96 \\
  &   & 6 & 0 & -30 & 0 & -96 \\
\hline
  & 1 & 0 & 0 & 0 & -16 & 0
\end{array}
$$

Thus: $f(x) = (x-6)\left(x^4 - 16\right) = (x-6)\left(x^2 - 4\right)\left(x^2 + 4\right) = (x-6)(x+2)(x-2)\left(x^2 + 4\right)$

43. Using synthetic division:

$$
\begin{array}{r|rrrr}
-5 & 9 & 21 & -104 & 80 \\
   &   & -45 & 120 & -80 \\
\hline
   & 9 & -24 & 16 & 0
\end{array}
$$

Thus: $f(x) = (x+5)\left(9x^2 - 24x + 16\right) = (x+5)(3x-4)^2$

45. Let $f(x) = k^2 x^4 + 3kx^2 - 4$. Evaluating: $f(1) = k^2 (1)^4 + 3k(1)^2 - 4 = k^2 + 3k - 4$

Now setting $f(1) = 0$:

$$k^2 + 3k - 4 = 0$$
$$(k+4)(k-1) = 0$$
$$k = -4, 1$$

47. Let $f(x) = kx^3 + 19x^2 + x - 6$. Evaluating: $f(-3) = k(-3)^3 + 19(-3)^2 - 3 - 6 = -27k + 171 - 9 = -27k + 162$

Now setting $f(-3) = 0$:

$$-27k + 162 = 0$$
$$-27k = -162$$
$$k = 6$$

49. Since $f(c) = 3c^4 + 2c^2 + 5 > 0$, there is no value of c such that $f(c) = 0$, thus $f(x)$ has no factor of the form $x - c$.

51. Let $f(x) = x^n - 1$, so $f(-1) = (-1)^n - 1 = 1 - 1 = 0$ if n is an even integer. Thus $x + 1$ is a factor of $f(x)$.

53. **a.** Let $f(x) = x^n - y^n$, so $f(y) = y^n - y^n = 0$. Thus $x - y$ is a factor of $x^n - y^n$.

b. Let $f(x) = x^n - y^n$, so $f(-y) = (-y)^n - y^n = y^n - y^n = 0$ if n is an even integer. Thus $x + y$ is a factor of $x^n - y^n$.

c. Let $f(x) = x^n + y^n$, so $f(-y) = (-y)^n + y^n = -y^n + y^n = 0$ if n is an odd integer. Thus $x + y$ is a factor of $x^n + y^n$.

57. Using synthetic division:

$$
\begin{array}{r|rrr}
1+i & 1 & 4 & -2 \\
    &   & 1+i & 4+6i \\
\hline
    & 1 & 5+i & 2+6i
\end{array}
$$

Evaluating: $f(1+i) = (1+i)^2 + 4(1+i) - 2 = 2i + 4 + 4i - 2 = 2 + 6i$

61. **a.** The nested form is: $f(x) = \left[x(x+5) - 2\right]x + 1$. Evaluating the function:

$$f(4) = \left[4(4+5) - 2\right]4 + 1 = 34(4) + 1 = 137$$
$$f(-5) = \left[-5(-5+5) - 2\right](-5) + 1 = 10 + 1 = 11$$
$$f(7) = \left[7(7+5) - 2\right]7 + 1 = 82(7) + 1 = 575$$

b. The nested form is: $f(x) = \left[x(2x-4)-3\right]x+2$. Evaluating the function:

$$f(3) = \left[3(2 \cdot 3 - 4) - 3\right]3 + 2 = 3(3) + 2 = 11$$
$$f(6) = \left[6(2 \cdot 6 - 4) - 3\right]6 + 2 = 45(6) + 2 = 272$$
$$f(-7) = \left[-7(2 \cdot (-7) - 4) - 3\right](-7) + 2 = 123(-7) + 2 = -859$$

c. The nested form is: $f(x) = \left[x(-2x+5)-6\right]x-7$. Evaluating the function:

$$f(4) = \left[4(-2 \cdot 4 + 5) - 6\right]4 - 7 = -18(4) - 7 = -79$$
$$f(5) = \left[5(-2 \cdot 5 + 5) - 6\right]5 - 7 = -31(5) - 7 = -162$$
$$f(-3) = \left[-3(-2 \cdot (-3) + 5) - 6\right](-3) - 7 = -39(-3) - 7 = 110$$

d. The nested form is: $f(x) = \left[\left(x(x+3)-2\right)x+5\right]x-1$. Evaluating the function:

$$f(5) = \left[\left(5(5+3)-2\right)5+5\right]5-1 = \left[38(5)+5\right]5-1 = 195(5)-1 = 974$$
$$f(6) = \left[\left(6(6+3)-2\right)6+5\right]6-1 = \left[52(6)+5\right]6-1 = 317(6)-1 = 1901$$
$$f(-3) = \left[\left(-3(-3+3)-2\right)(-3)+5\right](-3)-1 = 11(-3)-1 = -34$$

4.3 Polynomial Equations

1. The possible rational roots are $\pm 1, \pm 2, \pm 3, \pm 4, \pm 6, \pm 12$. Using synthetic division:

$$
\begin{array}{r|rrrr}
1 & 1 & -2 & -11 & 12 \\
  &   & 1 & -1 & -12 \\
\hline
  & 1 & -1 & -12 & 0
\end{array}
$$

Solving the equation:

$$(x-1)(x^2 - x - 12) = 0$$
$$(x-1)(x-4)(x+3) = 0$$
$$x = -3, 1, 4$$

The solution set is $\{-3, 1, 4\}$.

3. The possible rational roots are $\pm 1, \pm 2, \pm \dfrac{1}{3}, \pm \dfrac{2}{3}, \pm \dfrac{1}{5}, \pm \dfrac{2}{5}, \pm \dfrac{1}{15}, \pm \dfrac{2}{15}$. Using synthetic division:

$$
\begin{array}{r|rrrr}
-1 & 15 & 14 & -3 & -2 \\
   &   & -15 & 1 & 2 \\
\hline
   & 15 & -1 & -2 & 0
\end{array}
$$

Solving the equation:

$$(x+1)(15x^2 - x - 2) = 0$$
$$(x+1)(3x+1)(5x-2) = 0$$
$$x = -1, -\frac{1}{3}, \frac{2}{5}$$

The solution set is $\left\{-1, -\dfrac{1}{3}, \dfrac{2}{5}\right\}$.

5. The possible rational roots are $\pm 1, \pm 2, \pm 5, \pm 10, \pm\frac{1}{2}, \pm\frac{5}{2}, \pm\frac{1}{4}, \pm\frac{5}{4}, \pm\frac{1}{8}, \pm\frac{5}{8}$. Using synthetic division:

$$\begin{array}{r|rrrr} -2) & 8 & -2 & -41 & -10 \\ & & -16 & 36 & 10 \\ \hline & 8 & -18 & -5 & 0 \end{array}$$

Solving the equation:

$$(x+2)\left(8x^2 - 18x - 5\right) = 0$$
$$(x+2)(4x+1)(2x-5) = 0$$
$$x = -2, -\frac{1}{4}, \frac{5}{2}$$

The solution set is $\left\{-2, -\frac{1}{4}, \frac{5}{2}\right\}$.

7. The possible rational roots are $\pm 1, \pm 2, \pm 3, \pm 4, \pm 6, \pm 12$. Using synthetic division:

$$\begin{array}{r|rrrr} -3) & 1 & -1 & -8 & 12 \\ & & -3 & 12 & -12 \\ \hline & 1 & -4 & 4 & 0 \end{array}$$

Solving the equation:

$$(x+3)\left(x^2 - 4x + 4\right) = 0$$
$$(x+3)(x-2)^2 = 0$$
$$x = -3, 2$$

The solution set is $\{-3, 2\}$. Note that 2 has a multiplicity of 2.

9. The possible rational roots are $\pm 1, \pm 2, \pm 4, \pm 8$. Using synthetic division:

$$\begin{array}{r|rrrr} 2) & 1 & -4 & 0 & 8 \\ & & 2 & -4 & -8 \\ \hline & 1 & -2 & -4 & 0 \end{array}$$

Solving the equation:

$$(x-2)\left(x^2 - 2x - 4\right) = 0$$
$$x = 2, \frac{2 \pm \sqrt{4 - 4(-4)}}{2} = \frac{2 \pm \sqrt{20}}{2} = \frac{2 \pm 2\sqrt{5}}{2} = 1 \pm \sqrt{5}$$

The solution set is $\left\{2, 1 \pm \sqrt{5}\right\}$.

11. The possible rational roots are $\pm 1, \pm 2, \pm 3, \pm 4, \pm 6, \pm 12$. Using synthetic division:

$$\begin{array}{r|rrrrr} 2) & 1 & 4 & -1 & -16 & -12 \\ & & 2 & 12 & 22 & 12 \\ \hline & 1 & 6 & 11 & 6 & 0 \end{array}$$

Using synthetic division again:

$$\begin{array}{r|rrrr} -1) & 1 & 6 & 11 & 6 \\ & & -1 & -5 & -6 \\ \hline & 1 & 5 & 6 & 0 \end{array}$$

Solving the equation:

$$(x-2)(x+1)\left(x^2 + 5x + 6\right) = 0$$
$$(x-2)(x+1)(x+2)(x+3) = 0$$
$$x = -3, -2, -1, 2$$

The solution set is $\{-3, -2, -1, 2\}$.

13. The possible rational roots are $\pm 1, \pm 2, \pm 3, \pm 5, \pm 6, \pm 10, \pm 15, \pm 30$. Using synthetic division:

$$
\begin{array}{r|rrrr}
3 & 1 & 1 & -3 & -17 & -30 \\
  &   & 3 & 12 & 27 & 30 \\
\hline
  & 1 & 4 & 9 & 10 & 0
\end{array}
$$

Using synthetic division again:

$$
\begin{array}{r|rrrr}
-2 & 1 & 4 & 9 & 10 \\
   &   & -2 & -4 & -10 \\
\hline
   & 1 & 2 & 5 & 0
\end{array}
$$

Solving the equation:

$$(x-3)(x+2)\left(x^2+2x+5\right)=0$$

$$x = 3, -2, \frac{-2 \pm \sqrt{4-4(5)}}{2} = \frac{-2 \pm \sqrt{-16}}{2} = \frac{-2 \pm 4i}{2} = -1 \pm 2i$$

The solution set is $\{3, -2, -1 \pm 2i\}$.

15. The possible rational roots are ± 1. Using synthetic division:

$$
\begin{array}{r|rrr}
1 & 1 & -1 & 1 & -1 \\
  &   & 1 & 0 & 1 \\
\hline
  & 1 & 0 & 1 & 0
\end{array}
$$

Solving the equation:

$$(x-1)\left(x^2+1\right)=0$$

$$x = 1, \pm i$$

The solution set is $\{1, \pm i\}$.

17. The possible rational roots are $\pm 1, \pm 3, \pm 5, \pm 15, \pm \frac{1}{2}, \pm \frac{3}{2}, \pm \frac{5}{2}, \pm \frac{15}{2}$. Using synthetic division:

$$
\begin{array}{r|rrrr}
1 & 2 & 3 & -11 & -9 & 15 \\
  &   & 2 & 5 & -6 & -15 \\
\hline
  & 2 & 5 & -6 & -15 & 0
\end{array}
$$

Using synthetic division again:

$$
\begin{array}{r|rrrr}
-\frac{5}{2} & 2 & 5 & -6 & -15 \\
             &   & -5 & 0 & 15 \\
\hline
             & 2 & 0 & -6 & 0
\end{array}
$$

Solving the equation:

$$(x-1)\left(x+\frac{5}{2}\right)\left(2x^2-6\right)=0$$

$$(x-1)(2x+5)\left(x^2-3\right)=0$$

$$x = -\frac{5}{2}, 1, \pm \sqrt{3}$$

The solution set is $\left\{-\frac{5}{2}, 1, \pm \sqrt{3}\right\}$.

19. The possible rational roots are $\pm 1, \pm 2, \pm 4, \pm \dfrac{1}{2}$. Using synthetic division:

$$-2 \overline{) \begin{array}{rrrrr} 4 & 12 & 1 & -12 & 4 \\ & -8 & -8 & 14 & -4 \\ \hline 4 & 4 & -7 & 2 & 0 \end{array}}$$

Using synthetic division again:

$$-2 \overline{) \begin{array}{rrrr} 4 & 4 & -7 & 2 \\ & -8 & 8 & -2 \\ \hline 4 & -4 & 1 & 0 \end{array}}$$

Solving the equation:

$$(x+2)^2 \left(4x^2 - 4x + 1\right) = 0$$
$$(x+2)^2 (2x-1)^2 = 0$$
$$x = -2, \frac{1}{2}$$

The solution set is $\left\{ -2, \dfrac{1}{2} \right\}$. Note that -2 has a multiplicity of 2.

21. The possible rational roots are $\pm 1, \pm 2$. None of these produce a remainder of 0, so the equation has no rational solutions.

23. The possible rational roots are $\pm 1, \pm 2, \pm 4, \pm \dfrac{1}{3}, \pm \dfrac{2}{3}, \pm \dfrac{4}{3}$. None of these produce a remainder of 0, so the equation has no rational solutions.

25. The possible rational roots are $\pm 1, \pm 3$. None of these produce a remainder of 0, so the equation has no rational solutions.

27. Multiplying by 10, the equation is $x^3 + 2x^2 - 5x - 6 = 0$. The possible rational roots are $\pm 1, \pm 2, \pm 3, \pm 6$.
Using synthetic division:

$$2 \overline{) \begin{array}{rrrr} 1 & 2 & -5 & -6 \\ & 2 & 8 & 6 \\ \hline 1 & 4 & 3 & 0 \end{array}}$$

Solving the equation:

$$(x-2)\left(x^2 + 4x + 3\right) = 0$$
$$(x-2)(x+3)(x+1) = 0$$
$$x = -3, -1, 2$$

The solution set is $\{-3, -1, 2\}$.

29. Multiplying by 6, the equation is $6x^3 - 5x^2 - 44x + 15 = 0$. The possible rational roots are
$\pm 1, \pm 3, \pm 5, \pm 15, \pm \dfrac{1}{2}, \pm \dfrac{3}{2}, \pm \dfrac{5}{2}, \pm \dfrac{15}{2}, \pm \dfrac{1}{3}, \pm \dfrac{5}{3}, \pm \dfrac{1}{6}, \pm \dfrac{5}{6}$. Using synthetic division:

$$3 \overline{) \begin{array}{rrrr} 6 & -5 & -44 & 15 \\ & 18 & 39 & -15 \\ \hline 6 & 13 & -5 & 0 \end{array}}$$

Solving the equation:

$$(x-3)\left(6x^2 + 13x - 5\right) = 0$$
$$(x-3)(2x+5)(3x-1) = 0$$
$$x = -\frac{5}{2}, \frac{1}{3}, 3$$

The solution set is $\left\{ -\dfrac{5}{2}, \dfrac{1}{3}, 3 \right\}$.

31. The equation has one sign change and the equation with substitution of $-x$ has one sign change, so it has one positive and one negative real solution.

33. The equation has one sign change and the equation with substitution of $-x$ has no sign changes, so it has one positive real solution and two nonreal complex solutions.

35. The equation has two sign changes and the equation with substitution of $-x$ has one sign change, so it has either two positive and one negative real solutions, or one negative real and two nonreal complex solutions.

37. The equation has five sign changes and the equation with substitution of $-x$ has no sign changes, so it has either five positive real solutions, three positive and two nonreal complex solutions, or one positive and four nonreal complex solutions.

39. The equation has no sign changes and the equation with substitution of $-x$ has one sign change, so it has one negative and four nonreal complex solutions.

41. Finding the equation:
$$(x-2)(x-4)(x+3)=0$$
$$(x-2)(x^2-x-12)=0$$
$$x^3-3x^2-10x+24=0$$

43. Finding the equation:
$$(x+2)(2x-1)(3x-2)=0$$
$$(x+2)(6x^2-7x+2)=0$$
$$6x^3+5x^2-12x+4=0$$

45. Finding the equation:
$$(x-1)^5=0$$
$$(x-1)(x^2-2x+1)(x^2-2x+1)=0$$
$$x^5-5x^4+10x^3-10x^2+5x-1=0$$

47. Finding the equation:
$$(x-3)(x-2-3i)(x-2+3i)=0$$
$$(x-3)(x^2-4x+13)=0$$
$$x^3-7x^2+25x-39=0$$

49. Finding the equation:
$$(x-1+i)(x-1-i)(x-2i)(x+2i)=0$$
$$(x^2-2x+2)(x^2+4)=0$$
$$x^4-2x^3+6x^2-8x+8=0$$

53. The possible rational roots of $x^2-2=0$ are $\pm 1, \pm 2$, none of which check. Since the solutions are $x=\pm\sqrt{2}$, then $\sqrt{2}$ is not a rational number.

55. If the polynomial has odd degree, either it has an odd number of sign changes or the equation with substitution of $-x$ has an odd number of sign changes. Thus is must have at least one real number solution.

4.4 Graphing Polynomial Functions

1. Graphing the function $f(x)=-(x-3)^3$:

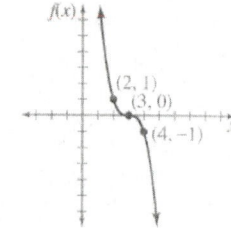

3. Graphing the function $f(x)=(x+1)^3$:

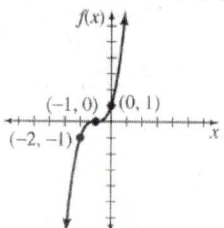

5. Graphing the function $f(x)=(x+3)^4$:

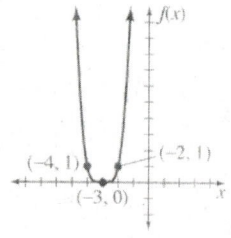

7. Graphing the function $f(x)=-(x-2)^4$:

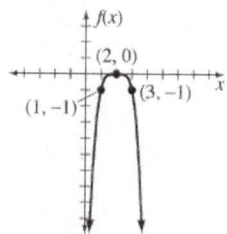

9.   Graphing the function $f(x) = (x+1)^4 + 3$:

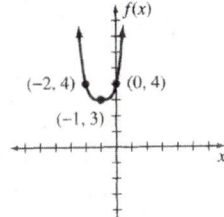

11.   Graphing the function $f(x) = (x-2)(x+1)(x+3)$:

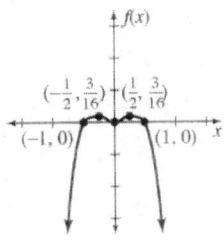

13.   Graphing the function $f(x) = x(x+2)(2-x)$:

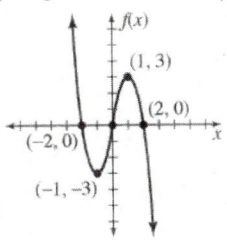

15.   Graphing the function $f(x) = -x^2(x-1)(x+1)$:

17.   Graphing the function $f(x) = (2x-1)(x-2)(x-3)$:

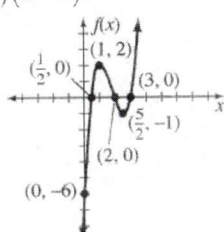

19.   Graphing the function $f(x) = (x-2)(x-1)(x+1)(x+2)$:

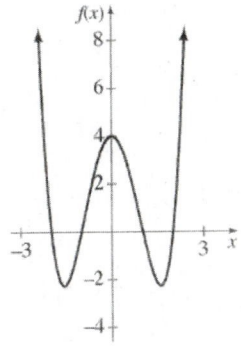

21.   Graphing the function $f(x) = x(x-2)^2(x+1)$:

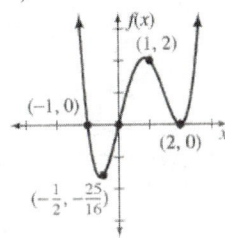

23. Factoring: $f(x) = -x(x^2 + x - 6) = -x(x+3)(x-2)$. Graphing the function:

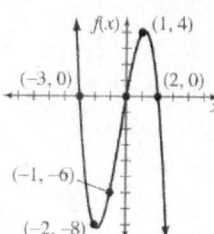

25. Factoring: $f(x) = x^2(x^2 - 5x + 6) = x^2(x-2)(x-3)$. Graphing the function:

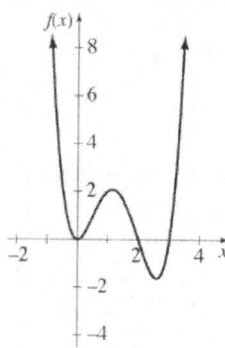

27. Factoring: $f(x) = x^2(x+2) - 1(x+2) = (x+2)(x^2-1) = (x+2)(x+1)(x-1)$. Graphing the function:

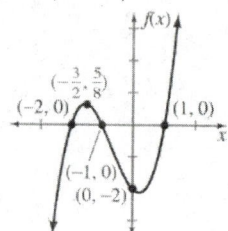

29. Using synthetic division:

$$
\begin{array}{r|rrrr}
1 & 1 & -8 & 19 & -12 \\
  &   & 1 & -7 & 12 \\
\hline
  & 1 & -7 & 12 & 0 \\
\end{array}
$$

Factoring: $f(x) = (x-1)(x^2 - 7x + 12) = (x-1)(x-3)(x-4)$. Graphing the function:

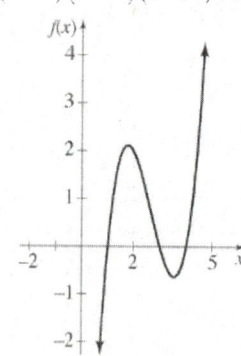

31.   Using synthetic division:

$$
\begin{array}{r|rrrr}
-1 & 2 & -3 & -3 & 2 \\
   &   & -2 & 5 & -2 \\
\hline
   & 2 & -5 & 2 & 0
\end{array}
$$

Factoring: $f(x) = (x+1)\left(2x^2 - 5x + 2\right) = (x+1)(2x-1)(x-2)$. Graphing the function:

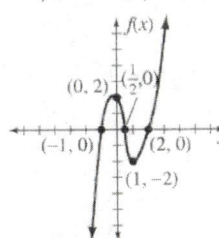

33.   Factoring: $f(x) = \left(x^2 - 1\right)\left(x^2 - 4\right) = (x+1)(x-1)(x+2)(x-2)$. Graphing the function:

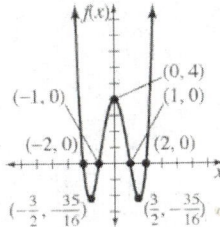

35.   **a.**   The y-intercept is $f(0) = (3)(-6)(8) = -144$.   **b.**   The x-intercepts are -3, 6, and 8.

   c.   Forming a sign chart:

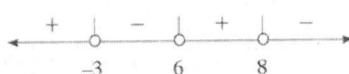

Thus $f(x) > 0$ on the intervals $(-\infty, -3) \cup (6, 8)$, and $f(x) < 0$ on the intervals $(-3, 6) \cup (8, \infty)$.

37.   **a.**   The y-intercept is $f(0) = (3)^4 (-1)^3 = -81$.   **b.**   The x-intercepts are -3 and 1.

   c.   Forming a sign chart:

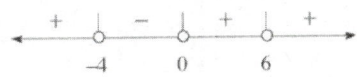

Thus $f(x) > 0$ on the interval $(1, \infty)$, and $f(x) < 0$ on the intervals $(-\infty, -3) \cup (-3, 1)$.

39.   **a.**   The y-intercept is $f(0) = (0)(-6)^2 (4) = 0$.   **b.**   The x-intercepts are -4, 0, and 6.

   c.   Forming a sign chart:

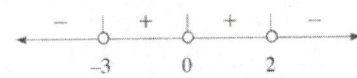

Thus $f(x) > 0$ on the intervals $(-\infty, -4) \cup (0, 6) \cup (6, \infty)$, and $f(x) < 0$ on the interval $(-4, 0)$.

41.   **a.**   The y-intercept is $f(0) = (0)^2 (2)(3) = 0$.   **b.**   The x-intercepts are -3, 0, and 2.

   c.   Forming a sign chart:

$$
\begin{array}{ccccccc}
- & & + & & + & & - \\
\hline
& -3 & & 0 & & 2 &
\end{array}
$$

Thus $f(x) > 0$ on the intervals $(-3, 0) \cup (0, 2)$, and $f(x) < 0$ on the intervals $(-\infty, -3) \cup (2, \infty)$.

47.   **a.**   The approximate solution is 1.6.   **b.**   The approximate solution is 2.8.
   c.   The approximate solution is 6.1.   **d.**   The approximate solution is 1.8.
   e.   The approximate solution is 2.5.   **f.**   The approximate solution is 5.0.

4.5 Graphing Rational Functions

1.   The vertical asymptote is $x = -3$ and the horizontal asymptote is $y = 0$.
3.   The vertical asymptote is $x = 1$ and the horizontal asymptote is $y = 4$.
5.   The vertical asymptotes are $x = -3$ and $x = 4$, and the horizontal asymptote is $y = 0$.
7.   Since $x^2 - 9 = (x + 3)(x - 3)$, the vertical asymptotes are $x = -3$ and $x = 3$, and the horizontal asymptote is $y = 0$.
9.   There are no vertical asymptotes, and the horizontal asymptote is $y = 5$.
11.   The vertical asymptote is $x = 0$ and the horizontal asymptote is $y = 0$:

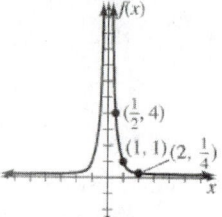

13.   The vertical asymptote is $x = 3$ and the horizontal asymptote is $y = 0$:

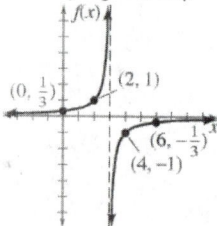

15.   The vertical asymptote is $x = -2$ and the horizontal asymptote is $y = 0$:

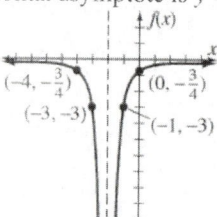

17.   The vertical asymptote is $x = 1$ and the horizontal asymptote is $y = 2$:

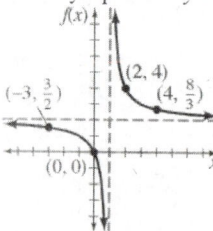

19.   The vertical asymptote is $x = -1$ and the horizontal asymptote is $y = -1$:

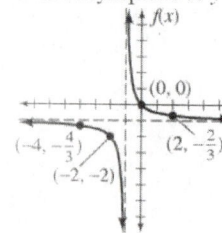

21. Since $x^2 - 4 = (x+2)(x-2)$, the vertical asymptotes are $x = -2$ and $x = 2$, and the horizontal asymptote is $y = 0$:

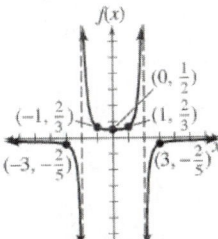

23. The vertical asymptotes are $x = -2$ and $x = 4$, and the horizontal asymptote is $y = 0$:

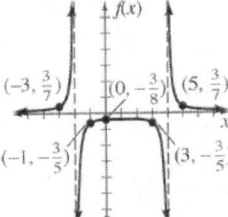

25. Since $x^2 + x - 6 = (x+3)(x-2)$, the vertical asymptotes are $x = -3$ and $x = 2$, and the horizontal asymptote is $y = 0$:

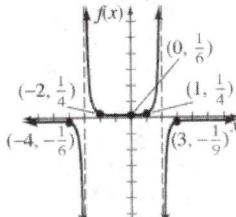

27. The vertical asymptote is $x = 0$ and the horizontal asymptote is $y = 2$:

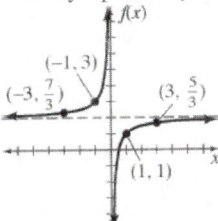

29. There are no vertical asymptotes and the horizontal asymptote is $y = 4$:

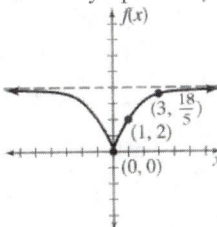

31. There are no vertical asymptotes and the horizontal asymptote is $y = 1$:

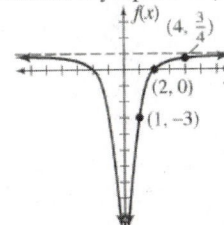

35. **a.** Graphing the function $f(x) = \dfrac{(x+4)(x-1)}{x+4} = x-1$, if $x \neq -4$:

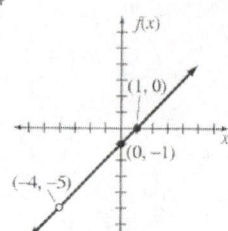

b. Graphing the function $f(x) = \dfrac{x^2 - 5x + 6}{x-2} = \dfrac{(x-2)(x-3)}{x-2} = x-3$, if $x \neq 2$:

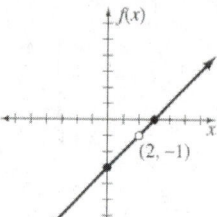

c. Graphing the function $f(x) = \dfrac{x-1}{x^2 - 1} = \dfrac{x-1}{(x+1)(x-1)} = \dfrac{1}{x+1}$, if $x \neq 1$:

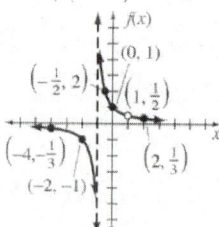

d. Graphing the function $f(x) = \dfrac{x+2}{x^2 + 6x + 8} = \dfrac{x+2}{(x+2)(x+4)} = \dfrac{1}{x+4}$, if $x \neq -2$:

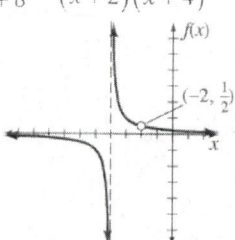

4.6 More on Graphing Rational Functions

1. Using synthetic division:

$$-1 \,\overline{)\begin{array}{ccc} 1 & 0 & 4 \\ & -1 & 1 \\ \hline 1 & -1 & 5 \end{array}}$$

So $f(x) = \dfrac{x^2 + 4}{x+1} = x - 1 + \dfrac{5}{x+1}$. The oblique asymptote is $y = x - 1$.

3. Using synthetic division:

$$-2 \,\overline{)\begin{array}{ccc} 1 & 4 & -6 \\ & -2 & -4 \\ \hline 1 & 2 & -10 \end{array}}$$

So $f(x) = \dfrac{x^2 + 4x - 6}{x+2} = x + 2 - \dfrac{10}{x+2}$. The oblique asymptote is $y = x + 2$.

5. Using synthetic division:

$$-2 \overline{\smash{\big)}\ 3 \quad 1 \quad 2}$$
$$\phantom{-2 \overline{\smash{\big)}\ 3}} \quad -6 \quad 10$$
$$\phantom{-2 \overline{\smash{\big)}}}\ \overline{3 \quad -5 \quad 12}$$

So $f(x) = \dfrac{3x^2 + x + 2}{x + 2} = 3x - 5 + \dfrac{12}{x + 2}$. The oblique asymptote is $y = 3x - 5$.

7. Factor $f(x) = \dfrac{x^2}{x^2 + x - 2} = \dfrac{x^2}{(x + 2)(x - 1)}$. The vertical asymptotes are $x = -2$ and $x = 1$, and the horizontal asymptote is $y = 1$:

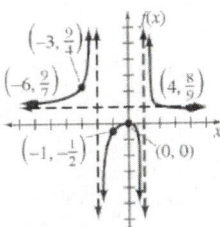

9. Factor $f(x) = \dfrac{2x^2}{x^2 - 2x - 8} = \dfrac{2x^2}{(x + 2)(x - 4)}$. The vertical asymptotes are $x = -2$ and $x = 4$, and the horizontal asymptote is $y = 2$:

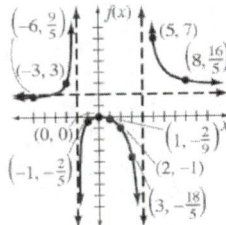

11. Factor $f(x) = \dfrac{-x}{x^2 - 1} = \dfrac{-x}{(x + 1)(x - 1)}$. The vertical asymptotes are $x = -1$ and $x = 1$, and the horizontal asymptote is $y = 0$:

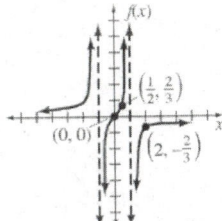

13. Factor $f(x) = \dfrac{x}{x^2 + x - 6} = \dfrac{x}{(x + 3)(x - 2)}$. The vertical asymptotes are $x = -3$ and $x = 2$, and the horizontal asymptote is $y = 0$:

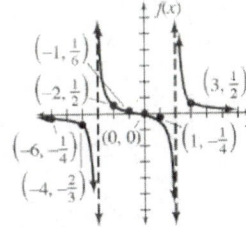

15. Factor $f(x) = \dfrac{x^2}{x^2 - 4x + 3} = \dfrac{x^2}{(x-1)(x-3)}$. The vertical asymptotes are $x = 1$ and $x = 3$, and the horizontal asymptote is $y = 1$:

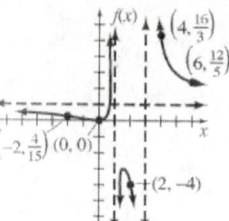

17. There are no vertical asymptotes, and the horizontal asymptote is $y = 0$:

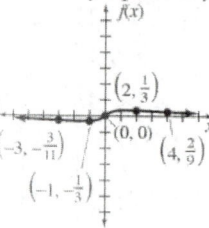

19. There are no vertical asymptotes, and the horizontal asymptote is $y = 0$:

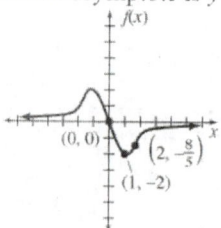

21. Using synthetic division:

$$
\begin{array}{r|rrr}
1 & 1 & 0 & 2 \\
  &   & 1 & 1 \\
\hline
  & 1 & 1 & 3
\end{array}
$$

So $f(x) = \dfrac{x^2 + 2}{x - 1} = x + 1 + \dfrac{3}{x - 1}$. The oblique asymptote is $y = x + 1$.

The vertical asymptote is $x = 1$, and there are no horizontal asymptotes:

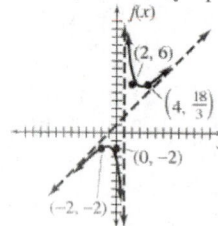

23. Using synthetic division:

$$
\begin{array}{r|rrr}
-1 & 1 & -1 & -6 \\
   &   & -1 & 2 \\
\hline
   & 1 & -2 & -4
\end{array}
$$

So $f(x) = \dfrac{x^2 - x - 6}{x + 1} = x - 2 - \dfrac{4}{x + 1}$. The oblique asymptote is $y = x - 2$.

The vertical asymptote is $x = -1$, and there are no horizontal asymptotes:

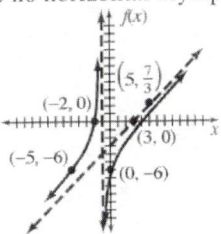

25. First write the function as $f(x) = \dfrac{-x^2 - 1}{x - 1}$. Using synthetic division:

$$
\begin{array}{r|rrr}
1 & -1 & 0 & -1 \\
  &   & -1 & -1 \\
\hline
  & -1 & -1 & -2
\end{array}
$$

So $f(x) = \dfrac{-x^2 - 1}{x - 1} = -x - 1 - \dfrac{2}{x - 1}$. The oblique asymptote is $y = -x - 1$.

The vertical asymptote is $x = 1$, and there are no horizontal asymptotes:

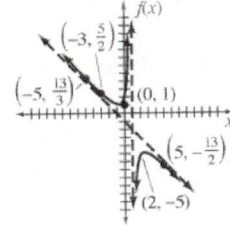

Chapter 4 Review Problem Set

1. Using synthetic division:

$$
\begin{array}{r|rrrr}
1 & 3 & -4 & 6 & -2 \\
  &   & 3 & -1 & 5 \\
\hline
  & 3 & -1 & 5 & 3
\end{array}
$$

The quotient is $3x^2 - x + 5$ and the remainder is 3.

2. Using synthetic division:

$$
\begin{array}{r|rrrr}
-2 & 5 & 7 & -9 & 10 \\
   &   & -10 & 6 & 6 \\
\hline
   & 5 & -3 & -3 & 16
\end{array}
$$

The quotient is $5x^2 - 3x - 3$ and the remainder is 16.

3. Using synthetic division:

$$
\begin{array}{r|rrrrr}
-4 & -2 & 1 & -2 & -1 & -1 \\
   &   & 8 & -36 & 152 & -604 \\
\hline
   & -2 & 9 & -38 & 151 & -605
\end{array}
$$

The quotient is $-2x^3 + 9x^2 - 38x + 151$ and the remainder is -605.

4. Using synthetic division:

$$
\begin{array}{r|rrrrr}
3 & -3 & 0 & -5 & 0 & 9 \\
  &   & -9 & -27 & -96 & -288 \\
\hline
  & -3 & -9 & -32 & -96 & -279
\end{array}
$$

The quotient is $-3x^3 - 9x^2 - 32x - 96$ and the remainder is -279.

5. Evaluating: $f(1) = 4 - 3 + 1 - 1 = 1$

6. Evaluating: $f(-3) = -108 - 63 - 18 - 8 = -197$

7. Evaluating: $f(-2) = -16 + 36 + 2 - 2 = 20$

8. Evaluating: $f(8) = 4096 - 4608 + 576 - 80 + 16 = 0$

9. Let $f(x) = 2x^3 + x^2 - 7x - 2$. Evaluating: $f(-2) = -16 + 4 + 14 - 2 = 0$

Thus $x + 2$ is a factor of $2x^3 + x^2 - 7x - 2$.

10. Let $f(x) = x^4 + 5x^3 - 7x^2 - x + 3$. Evaluating: $f(3) = 81 + 135 - 63 - 3 + 3 = 153$

Thus $x - 3$ is not a factor of $x^4 + 5x^3 - 7x^2 - x + 3$.

11.   Let $f(x) = x^5 - 1024$. Evaluating: $f(4) = 1024 - 1024 = 0$

Thus $x - 4$ is a factor of $x^5 - 1024$.

12.   Let $f(x) = x^5 + 1$. Evaluating: $f(-1) = -1 + 1 = 0$

Thus $x + 1$ is a factor of $x^5 + 1$.

13.   The possible rational roots are $\pm 1, \pm 3, \pm 5, \pm 15$. Using synthetic division:

$$
\begin{array}{r|rrrr}
1) & 1 & -3 & -13 & 15 \\
   &   & 1 & -2 & -15 \\
\hline
   & 1 & -2 & -15 & 0
\end{array}
$$

Solving the equation:

$$(x-1)(x^2 - 2x - 15) = 0$$
$$(x-1)(x-5)(x+3) = 0$$
$$x = -3, 1, 5$$

The solution set is $\{-3, 1, 5\}$.

14.   The possible rational roots are: $\pm 1, \pm 5, \pm 7, \pm 35, \pm \dfrac{1}{2}, \pm \dfrac{5}{2}, \pm \dfrac{7}{2}, \pm \dfrac{35}{2}, \pm \dfrac{1}{4}, \pm \dfrac{5}{4}, \pm \dfrac{7}{4}, \pm \dfrac{35}{4}, \pm \dfrac{1}{8}, \pm \dfrac{5}{8}, \pm \dfrac{7}{8}, \pm \dfrac{35}{8}$

Using synthetic division:

$$
\begin{array}{r|rrrr}
-1) & 8 & 26 & -17 & -35 \\
    &   & -8 & -18 & 35 \\
\hline
    & 8 & 18 & -35 & 0
\end{array}
$$

Solving the equation:

$$(x+1)(8x^2 + 18x - 35) = 0$$
$$(x+1)(2x+7)(4x-5) = 0$$
$$x = -\frac{7}{2}, -1, \frac{5}{4}$$

The solution set is $\left\{-\dfrac{7}{2}, -1, \dfrac{5}{4}\right\}$.

15.   The possible rational roots are $\pm 1, \pm 2, \pm 4, \pm 13, \pm 26, \pm 52$. Using synthetic division:

$$
\begin{array}{r|rrrr}
1) & 1 & -5 & 34 & -82 & 52 \\
   &   & 1 & -4 & 30 & -52 \\
\hline
   & 1 & -4 & 30 & -52 & 0
\end{array}
$$

Using synthetic division again:

$$
\begin{array}{r|rrrr}
2) & 1 & -4 & 30 & -52 \\
   &   & 2 & -4 & 52 \\
\hline
   & 1 & -2 & 26 & 0
\end{array}
$$

Solving the equation:

$$(x-1)(x-2)(x^2 - 2x + 26) = 0$$
$$x = 1, 2, \frac{2 \pm \sqrt{4 - 104}}{2} = \frac{2 \pm \sqrt{-100}}{2} = \frac{2 \pm 10i}{2} = 1 \pm 5i$$

The solution set is $\{1, 2, 1 \pm 5i\}$.

16. The possible rational roots are $\pm 1, \pm 2, \pm 4$. Using synthetic division:

$$-2 \overline{)\begin{array}{rrrr} 1 & -4 & -10 & 4 \\ & -2 & 12 & -4 \\ \hline 1 & -6 & 2 & 0 \end{array}}$$

Solving the equation:

$$(x+2)\left(x^2 - 6x + 2\right) = 0$$

$$x = -2, \frac{6 \pm \sqrt{36-8}}{2} = \frac{6 \pm \sqrt{28}}{2} = \frac{6 \pm 2\sqrt{7}}{2} = 3 \pm \sqrt{7}$$

The solution set is $\left\{-2, 3 \pm \sqrt{7}\right\}$.

17. The equation has two sign changes and the equation with substitution of $-x$ has two sign changes, so it has either two positive real solutions and two negative real solutions, two positive real solutions and two nonreal complex solutions, two negative real solutions and two nonreal complex solutions, or four nonreal complex solutions.

18. The equation has no sign changes and the equation with substitution of $-x$ has one sign change, so it has one negative real solution and four nonreal complex solutions.

19. Graphing the function $f(x) = -(x-2)^3 + 3$:

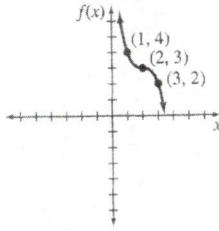

20. Graphing the function $f(x) = (x+3)(x-1)(3-x)$:

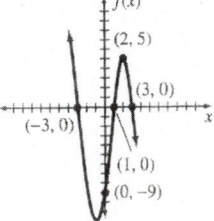

21. Factoring the function: $f(x) = x^4 - 4x^2 = x^2\left(x^2 - 4\right) = x^2(x+2)(x-2)$. Graphing the function:

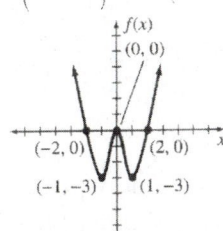

22. Factoring the function using synthetic division: $f(x) = (x+1)(x-2)(x-3)$. Graphing the function:

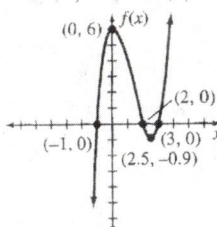

23. The vertical asymptote is $x = 3$ and the horizontal asymptote is $y = 1$:

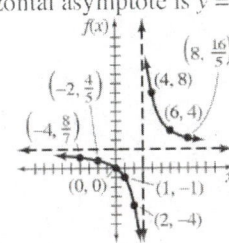

24. There is no vertical asymptote, and the horizontal asymptote is $y = 0$:

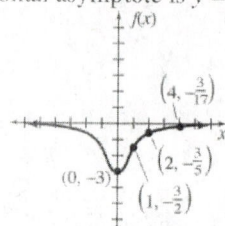

25. Factor $f(x) = \dfrac{-x^2}{x^2 - x - 6} = \dfrac{-x^2}{(x+2)(x-3)}$. The vertical asymptotes are $x = -2$ and $x = 3$, and the horizontal asymptote is $y = -1$:

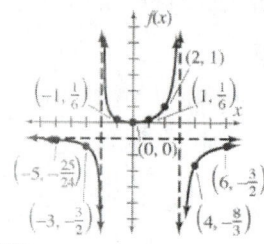

26. Using synthetic division:

$$
\begin{array}{r}
-1\,\big)\,1 \quad\ \ 0 \quad\ \ 3 \\
\underline{\quad\ -1 \quad\ \ 1} \\
1 \quad -1 \quad\ \ 4
\end{array}
$$

So $f(x) = \dfrac{x^2 + 3}{x + 1} = x - 1 + \dfrac{4}{x + 1}$. The oblique asymptote is $y = x - 1$.

The vertical asymptote is $x = -1$, and there is no horizontal asymptote:

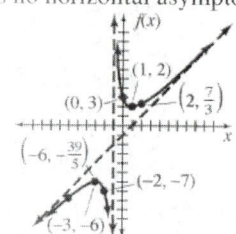

Chapter 4 Test

1. Using synthetic division:

$$
\begin{array}{r}
-3\,\big)\,3 \quad\ \ 5 \quad -14 \quad -6 \\
\underline{\quad\ -9 \quad\ \ 12 \quad\ \ 6} \\
3 \quad -4 \quad\ \ -2 \quad\ \ 0
\end{array}
$$

The quotient is $3x^2 - 4x - 2$ and the remainder is 0.

2. Using synthetic division:

$$
\begin{array}{r}
2\,\big)\,4 \quad\ \ 0 \quad -7 \quad -1 \quad\ \ 4 \\
\underline{\quad\ 8 \quad\ \ 16 \quad\ \ 18 \quad\ \ 34} \\
4 \quad\ \ 8 \quad\ \ 9 \quad\ \ 17 \quad\ \ 38
\end{array}
$$

The quotient is $4x^3 + 8x^2 + 9x + 17$ and the remainder is 38.

3. Evaluating: $f(7) = 7^5 - 8(7)^4 + 9(7)^3 - 13(7)^2 - 9(7) - 10 = -24$

4. Evaluating: $f(-7) = 3(-7)^4 + 20(-7)^3 - 6(-7)^2 + 9(-7) + 19 = 5$

5. Evaluating: $f(6) = (6)^5 - 35(6)^3 - 32(6) + 15 = 39$

6. Let $f(x) = 3x^3 - 11x^2 - 22x - 20$. Evaluating: $f(5) = 3(5)^3 - 11(5)^2 - 22(5) - 20 = -30$

   Thus $x - 5$ is not a factor of $3x^3 - 11x^2 - 22x - 20$.

7.   Let $f(x) = 5x^3 + 9x^2 - 9x - 17$. Evaluating: $f(-2) = 5(-2)^3 + 9(-2)^2 - 9(-2) - 17 = -3$

Thus $x + 2$ is not a factor of $5x^3 + 9x^2 - 9x - 17$.

8.   Let $f(x) = x^4 - 16x^2 - 17x + 12$. Evaluating: $f(-3) = (-3)^4 - 16(-3)^2 - 17(-3) + 12 = 0$

Thus $x + 3$ is a factor of $x^4 - 16x^2 - 17x + 12$.

9.   Let $f(x) = x^4 - 2x^2 + 3x - 12$. Evaluating: $f(6) = (6)^4 - 2(6)^2 + 3(6) - 12 = 1230$

Thus $x - 6$ is not a factor of $x^4 - 2x^2 + 3x - 12$.

10.   The possible rational roots are $\pm 1, \pm 2, \pm 3, \pm 4, \pm 6, \pm 12$. Using synthetic division:

$$
\begin{array}{r|rrrr}
1 & 1 & 0 & -13 & 12 \\
  &   & 1 & 1 & -12 \\
\hline
  & 1 & 1 & -12 & 0
\end{array}
$$

Solving the equation:

$$(x-1)(x^2 + x - 12) = 0$$
$$(x-1)(x+4)(x-3) = 0$$
$$x = -4, 1, 3$$

The solution set is $\{-4, 1, 3\}$.

11.   The possible rational roots are $\pm 1, \pm 2, \pm 4, \pm \dfrac{1}{2}$. Using synthetic division:

$$
\begin{array}{r|rrrr}
-4 & 2 & 5 & -13 & -4 \\
   &   & -8 & 12 & 4 \\
\hline
   & 2 & -3 & -1 & 0
\end{array}
$$

Solving the equation $2x^2 - 3x - 1 = 0$ using the quadratic formula:

$$x = \frac{3 \pm \sqrt{(-3)^2 - 4(2)(-1)}}{2(2)} = \frac{3 \pm \sqrt{9 + 8}}{4} = \frac{3 \pm \sqrt{17}}{4}$$

The solution set is $\left\{-4, \dfrac{3 \pm \sqrt{17}}{4}\right\}$.

12.   The possible rational roots are $\pm 1, \pm 2, \pm 3, \pm 5, \pm 6, \pm 10, \pm 15, \pm 30$. Using synthetic division:

$$
\begin{array}{r|rrrrr}
1 & 1 & -4 & -5 & 38 & -30 \\
  &   & 1 & -3 & -8 & 30 \\
\hline
  & 1 & -3 & -8 & 30 & 0
\end{array}
$$

Using synthetic division again:

$$
\begin{array}{r|rrrr}
-3 & 1 & -3 & -8 & 30 \\
   &   & -3 & 18 & -30 \\
\hline
   & 1 & -6 & 10 & 0
\end{array}
$$

Solving the equation $x^2 - 6x + 10 = 0$ using the quadratic formula:

$$x = \frac{6 \pm \sqrt{(-6)^2 - 4(1)(10)}}{2(1)} = \frac{6 \pm \sqrt{36 - 40}}{2} = \frac{6 \pm \sqrt{-4}}{2} = \frac{6 \pm 2i}{2} = 3 \pm i$$

The solution set is $\{-3, 1, 3 \pm i\}$.

13. The possible rational roots are $\pm 1, \pm 2, \pm 3, \pm 4, \pm 6, \pm 12, \pm \dfrac{1}{2}, \pm \dfrac{3}{2}$. Using synthetic division:

$$
\begin{array}{r|rrrr}
1) & 2 & 3 & -17 & 12 \\
   &   & 2 & 5 & -12 \\
\hline
   & 2 & 5 & -12 & 0
\end{array}
$$

Solving the equation:

$$(x-1)\left(2x^2 + 5x - 12\right) = 0$$
$$(x-1)(2x-3)(x+4) = 0$$
$$x = -4, 1, \frac{3}{2}$$

The solution set is $\left\{ -4, 1, \dfrac{3}{2} \right\}$.

14. The possible rational roots are $\pm 1, \pm 2, \pm 4, \pm 5, \pm 10, \pm 20, \pm \dfrac{1}{3}, \pm \dfrac{2}{3}, \pm \dfrac{4}{3}, \pm \dfrac{5}{3}, \pm \dfrac{10}{3}, \pm \dfrac{20}{3}$. Using synthetic division:

$$
\begin{array}{r|rrrr}
2) & 3 & -7 & -8 & 20 \\
   &   & 6 & -2 & -20 \\
\hline
   & 3 & -1 & -10 & 0
\end{array}
$$

Solving the equation:

$$(x-2)\left(3x^2 - x - 10\right) = 0$$
$$(x-2)(3x+5)(x-2) = 0$$
$$(x-2)^2 (3x+5) = 0$$
$$x = -\frac{5}{3}, 2$$

The solution set is $\left\{ -\dfrac{5}{3}, 2 \right\}$.

15. The equation has one sign change and the equation with substitution of $-x$ has one sign change, so it has one positive real solution, one negative real solution, and two nonreal complex solutions.

16. Solving the equation:

$$3x^3 + 19x^2 - 14x = 0$$
$$x\left(3x^2 + 19x - 14\right) = 0$$
$$x(3x-2)(x+7) = 0$$
$$x = -7, 0, \frac{2}{3}$$

The x-intercepts are $-7, 0, \dfrac{2}{3}$.

17. The equation of the vertical asymptote is $x = -3$.

18. The equation of the horizontal asymptote is $y = 5$.

19. Since $f(-x) = f(x)$, the equation has y-axis symmetry.

20. Since $f(-x) = -f(x)$, the equation has origin symmetry.

21. Graphing the function $f(x) = (2-x)(x-1)(x+1)$:

22. Graphing the function $f(x) = -x(x-3)(x+2)$:

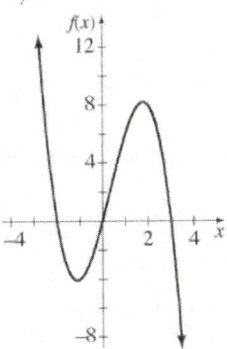

23. The vertical asymptote is $x = 3$, and the horizontal asymptote is $y = -1$:

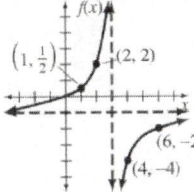

24. Factor $f(x) = \dfrac{-2}{x^2 - 4} = \dfrac{-2}{(x+2)(x-2)}$. The vertical asymptotes are $x = -2$ and $x = 2$, and the horizontal asymptote is $y = 0$:

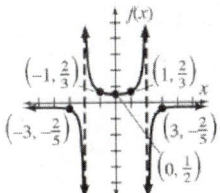

25. Using synthetic division:

$$-1)\overline{)\begin{array}{ccc} 4 & 1 & 1 \\ & -4 & 3 \\ \hline 4 & -3 & 4 \end{array}}$$

So $f(x) = \dfrac{4x^2 + x + 1}{x + 1} = 4x - 3 + \dfrac{4}{x + 1}$. The oblique asymptote is $y = 4x - 3$. The vertical asymptote is $x = -1$, and there is no horizontal asymptote:

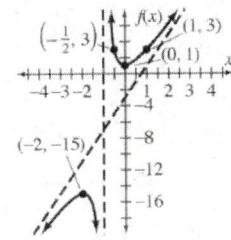

Chapters 0-4 Cumulative Review Problem Set

1. Evaluating: $\left(\dfrac{3}{4}\right)^{-3} = \left(\dfrac{4}{3}\right)^{3} = \dfrac{64}{27}$

2. Evaluating: $\sqrt[3]{-\dfrac{8}{27}} = -\dfrac{2}{3}$

3. Evaluating: $-5^{-2} = -\dfrac{1}{5^2} = -\dfrac{1}{25}$

4. Evaluating: $8^{4/3} = \left(\sqrt[3]{8}\right)^4 = 2^4 = 16$

5. Evaluating: $9^{-3/2} = \left(\sqrt{9}\right)^{-3} = 3^{-3} = \dfrac{1}{3^3} = \dfrac{1}{27}$

6. Evaluating: $\dfrac{1}{\left(\dfrac{2}{3}\right)^{-3}} = \left(\dfrac{2}{3}\right)^3 = \dfrac{8}{27}$

7. Evaluating: $\left(\dfrac{8}{27}\right)^{-2/3} = \left(\dfrac{27}{8}\right)^{2/3} = \left(\sqrt[3]{\dfrac{27}{8}}\right)^2 = \left(\dfrac{3}{2}\right)^2 = \dfrac{9}{4}$

8. Evaluating: $-6^{-2} = -\dfrac{1}{6^2} = -\dfrac{1}{36}$

9. Evaluating: $(-64)^{2/3} = \left(\sqrt[3]{-64}\right)^2 = (-4)^2 = 16$

10. Evaluating: $\sqrt[3]{-\dfrac{64}{27}} = -\dfrac{4}{3}$

11. Factoring the inequality:
    $$2x^2 + 11x - 6 \geq 0$$
    $$(2x - 1)(x + 6) \geq 0$$

    Forming a sign chart:

    The domain is $(-\infty, -6] \cup \left[\dfrac{1}{2}, \infty\right)$.

12. Finding the compositions:
    $$(f \circ g)(-2) = f(9) = 27 - 1 = 26 \qquad (g \circ f)(3) = g(8) = 64 - 8 + 3 = 59$$

13. Finding the compositions:
    $$(f \circ g)(x) = f\left(\dfrac{1}{x-4}\right) = -2(x-4) = -2x + 8 \qquad (g \circ f)(x) = g\left(-\dfrac{2}{x}\right) = \dfrac{1}{-\dfrac{2}{x} - 4} = \dfrac{x}{-2 - 4x} = -\dfrac{x}{4x + 2}$$

    The domain of $f \circ g$ is $\{x \mid x \neq 4\}$ and the domain of $g \circ f$ is $\left\{x \mid x \neq -\dfrac{1}{2}, 0\right\}$.

14. Finding the compositions:
    $$(f \circ g)(x) = f(2x - 1) = (2x - 1)^2 - 4(2x - 1) = 4x^2 - 4x + 1 - 8x + 4 = 4x^2 - 12x + 5$$
    $$(g \circ f)(x) = g\left(x^2 - 4x\right) = 2\left(x^2 - 4x\right) - 1 = 2x^2 - 8x - 1$$

    The domain of $f \circ g$ is $\{x \mid x \text{ is a real number}\}$ and the domain of $g \circ f$ is $\{x \mid x \text{ is a real number}\}$.

15. Simplifying the expression:
    $$\dfrac{f(a+h) - f(a)}{h} = \dfrac{(a+h)^2 + 7(a+h) - 2 - \left(a^2 + 7a - 2\right)}{h}$$
    $$= \dfrac{a^2 + 2ah + h^2 + 7a + 7h - 2 - a^2 - 7a + 2}{h}$$
    $$= \dfrac{2ah + h^2 + 7h}{h}$$
    $$= \dfrac{h(2a + h + 7)}{h}$$
    $$= 2a + h + 7$$

16. Evaluating: $f(9) = 2(9)^4 - 17(9)^3 - 10(9)^2 + 11(9) + 15 = 33$

17. Using synthetic division:

$$\begin{array}{r|rrrrrr} 3 & 3 & 0 & -25 & -7 & 1 & 6 \\ & & 9 & 27 & 6 & -3 & -6 \\ \hline & 3 & 9 & 2 & -1 & -2 & 0 \end{array}$$

The quotient is $3x^4 + 9x^3 + 2x^2 - x - 2$.

18. Let $f(x) = 2x^4 + 3x^3 + x^2 + 2x - 16$. Evaluating: $f(-2) = 2(-2)^4 + 3(-2)^3 + (-2)^2 + 2(-2) - 16 = -8$

Thus $x + 2$ is not a factor of $2x^4 + 3x^3 + x^2 + 2x - 16$.

19. Evaluating: $f(2) = 2^4 + 2(2)^3 - 2^2 + 3(2) - 4 = 30$. By the remainder theorem, the remainder is 30.

20. Completing the square:

$$x^2 + y^2 + 6x - 4y + 4 = 0$$
$$\left(x^2 + 6x + 9\right) + \left(y^2 - 4y + 4\right) = -4 + 9 + 4$$
$$(x + 3)^2 + (y - 2)^2 = 9$$

The center is $(-3, 2)$ and the radius is 3.

21. Finding the slope: $m = \dfrac{-1 - 2}{5 + 4} = \dfrac{-3}{9} = -\dfrac{1}{3}$. Using the point-slope formula:

$$y - 2 = -\frac{1}{3}(x + 4)$$
$$3y - 6 = -x - 4$$
$$x + 3y = 2$$

22. The bisector passes through the midpoint of the line segment, which is $(2, -1)$. The slope of the line segment is $\dfrac{2 + 4}{6 + 2} = \dfrac{3}{4}$, so the perpendicular slope is $-\dfrac{4}{3}$. Using the point-slope formula:

$$y + 1 = -\frac{4}{3}(x - 2)$$
$$3y + 3 = -4x + 8$$
$$4x + 3y = 5$$

23. The standard form of the equation is $\dfrac{x^2}{4} + \dfrac{y^2}{64} = 1$, so $a = 8$ and the length of the major axis is 16.

24. The standard form of the hyperbola is $\dfrac{x^2}{18} - \dfrac{y^2}{2} = 1$, so the slopes of the asymptotes are $\pm\dfrac{\sqrt{2}}{\sqrt{18}} = \pm\dfrac{\sqrt{2}}{3\sqrt{2}} = \pm\dfrac{1}{3}$.

Thus the equations are $y = \pm\dfrac{1}{3}x$.

25. The variation equation is $y = Kx$. Substituting $y = 3$ and $x = 4$:

$$3 = 4K$$
$$K = \frac{3}{4}$$

Thus $y = \dfrac{3}{4}x$. Substituting $x = 16$: $y = \dfrac{3}{4}(16) = 12$

26. The variation equation is $y = \dfrac{K}{\sqrt{x}}$. Substituting $y = \dfrac{2}{5}$ and $x = 25$:

$$\frac{2}{5} = \frac{K}{\sqrt{25}}$$
$$K = 2$$

Thus $y = \dfrac{2}{\sqrt{x}}$. Substituting $x = 49$: $y = \dfrac{2}{\sqrt{49}} = \dfrac{2}{7}$

27.   Let D represent the number of days and n represent the number of people. The variation equation is $D = \dfrac{k}{n}$.

Substituting $n = 8$ and $D = 10$:

$$10 = \frac{k}{8}$$
$$k = 80$$

Thus $D = \dfrac{80}{n}$. Substituting $n = 12$: $D = \dfrac{80}{12} = 6\dfrac{2}{3}$ days.

28.   Using synthetic division:

$$\begin{array}{r|rrr} 1 & 1 & 2 & -1 \\ & & 1 & 3 \\ \hline & 1 & 3 & 2 \end{array}$$

So $f(x) = x + 3 + \dfrac{2}{x-1}$. The oblique asymptote is therefore $y = x + 3$.

29.   Let n represent the number of nickels, $3n + 2$ represent the number of quarters, and $57 - (4n + 2) = 55 - 4n$ represent the number of dimes. The equation is:

$$0.05n + 0.10(55 - 4n) + 0.25(3n + 2) = 10$$
$$0.05n + 5.5 - 0.4n + 0.75n + 0.5 = 10$$
$$0.4n = 4$$
$$n = 10$$

The collection consists of 10 nickels, 32 quarters, and 15 dimes.

30.   Let s represent the selling price. The equation is:

$$75 + 0.4s = s$$
$$75 = 0.6s$$
$$s = 125$$

She should sell the dress for $125.

31.   Let x represent the amount of pure alcohol to add. The equation is:

$$0.30(8) + 1.00(x) = 0.40(8 + x)$$
$$2.4 + x = 3.2 + 0.4x$$
$$0.6x = 0.8$$
$$x = \frac{4}{3} = 1\frac{1}{3}$$

Thus $1\dfrac{1}{3}$ quarts should be added.

32.   Let t represent the time Claire spent riding out, and $\dfrac{15}{2} - t$ represent the time Claire spent riding back. The equation is:

$$15(t) = 10\left(\frac{15}{2} - t\right)$$
$$15t = 75 - 10t$$
$$25t = 75$$
$$t = 3$$

Since she rode for 3 hours at 15 mph, Claire rode 45 miles out before turning around.

33. Let t represent the time for Adam working alone, and $t + 2$ represent the time for Carl. Since 2 hours 24 minutes is $2\dfrac{2}{5}$ hours, the equation is:

$$\frac{1}{t} \cdot \frac{12}{5} + \frac{1}{t+2} \cdot \frac{12}{5} = 1$$

$$\frac{2.4}{t} + \frac{2.4}{t+2} = 1$$

$$2.4(t+2) + 2.4t = t(t+2)$$

$$2.4t + 4.8 + 2.4t = t^2 + 2t$$

$$0 = t^2 - 2.8t - 4.8$$

$$0 = (t-4)(t+1.2)$$

$$t = 4 \qquad (t = -1.2 \text{ does not check})$$

It would take Adam 4 hours working alone.

34. Solving the equation:

$$(2x - 5)(6x + 1) = (3x + 2)(4x - 7)$$

$$12x^2 - 28x - 5 = 12x^2 - 13x - 14$$

$$-15x = -9$$

$$x = \frac{3}{5}$$

The solution set is $\left\{ \dfrac{3}{5} \right\}$.

35. Solving the equation:

$$(2x + 1)(x - 2) = (3x - 2)(x + 4)$$

$$2x^2 - 3x - 2 = 3x^2 + 10x - 8$$

$$0 = x^2 + 13x - 6$$

$$x = \frac{-13 \pm \sqrt{169 + 24}}{2} = \frac{-13 \pm \sqrt{193}}{2}$$

The solution set is $\left\{ \dfrac{-13 \pm \sqrt{193}}{2} \right\}$.

36. Solving the equation:

$$4x^3 + 20x^2 - 56x = 0$$

$$4x(x^2 + 5x - 14) = 0$$

$$4x(x + 7)(x - 2) = 0$$

$$x = -7, 0, 2$$

The solution set is $\{-7, 0, 2\}$.

37. Using synthetic division:

$$
\begin{array}{r|rrrr}
-1 & 6 & 17 & 1 & -10 \\
   &   & -6 & -11 & 10 \\
\hline
   & 6 & 11 & -10 & 0
\end{array}
$$

Solving the equation:

$$(x + 1)(6x^2 + 11x - 10) = 0$$

$$(x + 1)(3x - 2)(2x + 5) = 0$$

$$x = -\frac{5}{2}, -1, \frac{2}{3}$$

The solution set is $\left\{ -\dfrac{5}{2}, -1, \dfrac{2}{3} \right\}$.

38. Solving the equation:

$$|4x - 3| = 7$$
$$4x - 3 = -7, 7$$
$$4x = -4, 10$$
$$x = -1, \frac{5}{2}$$

The solution set is $\left\{-1, \frac{5}{2}\right\}$.

39. Solving the equation:

$$\frac{2x - 1}{3} - \frac{3x + 2}{4} = -\frac{5}{6}$$
$$4(2x - 1) - 3(3x + 2) = -10$$
$$8x - 4 - 9x - 6 = -10$$
$$-x - 10 = -10$$
$$x = 0$$

The solution set is $\{0\}$.

40. Solving the equation:

$$(3x - 2)(3x + 4) = 0$$
$$3x = -4, 2$$
$$x = -\frac{4}{3}, \frac{2}{3}$$

The solution set is $\left\{-\frac{4}{3}, \frac{2}{3}\right\}$.

41. Solving the equation:

$$\sqrt{3x + 1} + 2 = 4$$
$$\sqrt{3x + 1} = 2$$
$$3x + 1 = 4$$
$$3x = 3$$
$$x = 1$$

The solution set is $\{1\}$.

42. Solving the equation:

$$\sqrt{n - 2} - 6 = -3$$
$$\sqrt{n - 2} = 3$$
$$n - 2 = 9$$
$$n = 11$$

The solution set is $\{11\}$.

43. Solving the equation:

$$x^4 + 3x^2 - 54 = 0$$
$$(x^2 + 9)(x^2 - 6) = 0$$
$$x^2 = -9, 6$$
$$x = \pm\sqrt{-9}, \pm\sqrt{6}$$
$$x = \pm 3i, \pm\sqrt{6}$$

The solution set is $\left\{\pm 3i, \pm\sqrt{6}\right\}$.

44. Solving the equation:
$$(2x - 1)(x + 3) = 49$$
$$2x^2 + 5x - 3 = 49$$
$$2x^2 + 5x - 52 = 0$$
$$(2x + 13)(x - 4) = 0$$
$$x = -\frac{13}{2}, 4$$

The solution set is $\left\{-\frac{13}{2}, 4\right\}$.

45. The possible rational roots are $\pm 1, \pm 2, \pm 3, \pm 6$. Using synthetic division:

$$
\begin{array}{r|rrrrr}
1) & 1 & -2 & 2 & -7 & 6 \\
   &   & 1 & -1 & 1 & -6 \\
\hline
   & 1 & -1 & 1 & -6 & 0
\end{array}
$$

Using synthetic division again:

$$
\begin{array}{r|rrrr}
2) & 1 & -1 & 1 & -6 \\
   &   & 2 & 2 & 6 \\
\hline
   & 1 & 1 & 3 & 0
\end{array}
$$

Solving the equation:

$$(x - 1)(x - 2)(x^2 + x + 3) = 0$$
$$x = 1, 2, \frac{-1 \pm \sqrt{1 - 12}}{2} = \frac{-1 \pm \sqrt{-11}}{2} = \frac{-1 \pm i\sqrt{11}}{2}$$

The solution set is $\left\{1, 2, \frac{-1 \pm i\sqrt{11}}{2}\right\}$.

46. Solving the inequality:

$$3(x-1) - 5(x+2) > 3(x+4)$$
$$3x - 3 - 5x - 10 > 3x + 12$$
$$-2x - 13 > 3x + 12$$
$$-5x > 25$$
$$x < -5$$

The solution set is $(-\infty, -5)$.

47. Solving the inequality:

$$\frac{x-1}{2} + \frac{2x+1}{5} \geq \frac{x-2}{3}$$
$$15(x-1) + 6(2x+1) \geq 10(x-2)$$
$$15x - 15 + 12x + 6 \geq 10x - 20$$
$$27x - 9 \geq 10x - 20$$
$$17x \geq -11$$
$$x \geq -\frac{11}{17}$$

The solution set is $\left[-\frac{11}{17}, \infty\right)$.

48. Factoring the inequality:

$$x^2 - 3x < 18$$
$$x^2 - 3x - 18 < 0$$
$$(x-6)(x+3) < 0$$

Forming a sign chart:

The solution set is $(-3, 6)$.

49. Forming a sign chart:

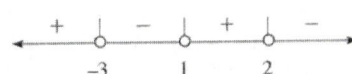

The solution set is $[-3, 1] \cup [2, \infty)$.

50. Solving the inequality:

$$|2x - 1| > 6$$

$$\begin{array}{ccc} 2x - 1 < -6 & \text{or} & 2x - 1 > 6 \\ 2x < -5 & & 2x > 7 \\ x < -\dfrac{5}{2} & & x > \dfrac{7}{2} \end{array}$$

The solution set is $\left(-\infty, -\dfrac{5}{2}\right) \cup \left(\dfrac{7}{2}, \infty\right)$.

51. Solving the inequality:

$$|3x + 2| \leq 8$$
$$-8 \leq 3x + 2 \leq 8$$
$$-10 \leq 3x \leq 6$$
$$-\frac{10}{3} \leq x \leq 2$$

The solution set is $\left[-\dfrac{10}{3}, 2\right]$.

52. Forming a sign chart:

The solution set is $\left(-\infty, \dfrac{3}{4}\right] \cup (2, \infty)$.

53. Solving the inequality:

$$\frac{x+3}{x-4} < 3$$
$$\frac{x+3}{x-4} - 3 < 0$$
$$\frac{x+3-3x+12}{x-4} < 0$$
$$\frac{-2x+15}{x-4} < 0$$

Forming a sign chart:

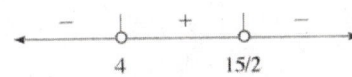

The solution set is $\left(-\infty, 4\right) \cup \left(\dfrac{15}{2}, \infty\right)$.

54. Graphing the function $f(x) = -2x + 4$:

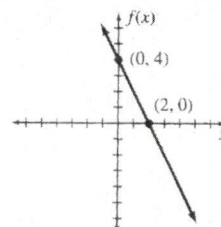

55. Graphing the function $f(x) = 2x^2 - 3$:

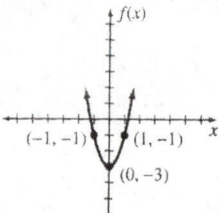

56. Completing the square: $f(x) = 2x^2 + 4x = 2(x^2 + 2x + 1) - 2 = 2(x + 1)^2 - 2$. Graphing the function:

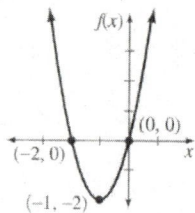

57. Graphing the function $f(x) = 2\sqrt{x + 2} - 1$:

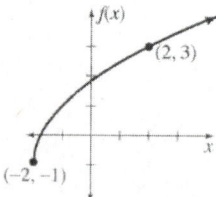

58. Graphing the function $f(x) = \dfrac{2x}{x + 1}$:

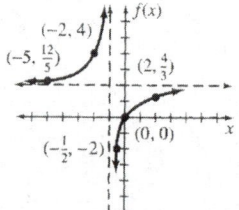

59. Graphing the function $f(x) = -|x - 2| + 1$:

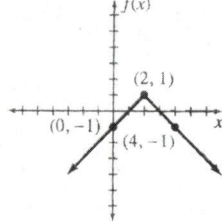

60. Graphing the function $f(x) = 2\sqrt{x} + 1$:

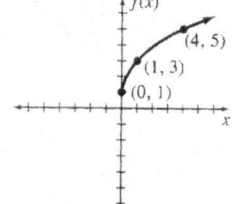

61. Completing the square: $f(x) = 3(x^2 + 4x) + 9 = 3(x^2 + 4x + 4) - 12 + 9 = 3(x + 2)^2 - 3$. Graphing the function:

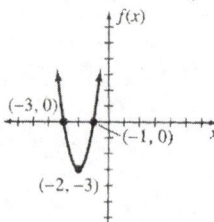

62. Graphing the function $f(x) = -(x - 3)^3 + 1$:

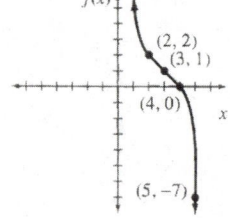

63. Graphing the function $f(x) = (x + 1)(x - 2)(x - 4)$:

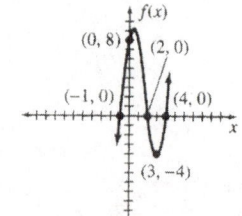

64.   Factoring: $f(x) = x^4 - x^2 = x^2\left(x^2 - 1\right) = x^2\left(x+1\right)\left(x-1\right)$. Graphing the function:

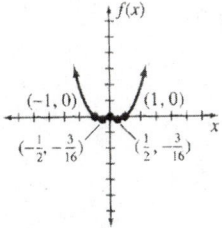

Chapter 5
Exponential and Logarithmic Functions

5.1 Exponents and Exponential Functions

1. Solving the equation:

$$2^x = 64$$
$$2^x = 2^6$$
$$x = 6$$

The solution set is $\{6\}$.

3. Solving the equation:

$$3^{2x} = 27$$
$$3^{2x} = 3^3$$
$$2x = 3$$
$$x = \frac{3}{2}$$

The solution set is $\left\{\dfrac{3}{2}\right\}$.

5. Solving the equation:

$$\left(\frac{1}{2}\right)^x = \frac{1}{128}$$
$$\left(\frac{1}{2}\right)^x = \left(\frac{1}{2}\right)^7$$
$$x = 7$$

The solution set is $\{7\}$.

7. Solving the equation:

$$3^{-x} = \frac{1}{243}$$
$$3^{-x} = 3^{-5}$$
$$x = 5$$

The solution set is $\{5\}$.

9. Solving the equation:

$$6^{3x-1} = 36$$
$$6^{3x-1} = 6^2$$
$$3x - 1 = 2$$
$$3x = 3$$
$$x = 1$$

The solution set is $\{1\}$.

11. Solving the equation:

$$\left(\frac{3}{4}\right)^n = \frac{64}{27}$$
$$\left(\frac{3}{4}\right)^n = \left(\frac{3}{4}\right)^{-3}$$
$$n = -3$$

The solution set is $\{-3\}$.

13.   Solving the equation:

$$16^x = 64$$
$$4^{2x} = 4^3$$
$$2x = 3$$
$$x = \frac{3}{2}$$

The solution set is $\left\{\frac{3}{2}\right\}$.

15.   Solving the equation:

$$27^{4x} = 9^{x+1}$$
$$3^{12x} = 3^{2x+2}$$
$$12x = 2x + 2$$
$$10x = 2$$
$$x = \frac{1}{5}$$

The solution set is $\left\{\frac{1}{5}\right\}$.

17.   Solving the equation:

$$9^{4x-2} = \frac{1}{81}$$
$$9^{4x-2} = 9^{-2}$$
$$4x - 2 = -2$$
$$4x = 0$$
$$x = 0$$

The solution set is $\{0\}$.

19.   Solving the equation:

$$10^x = 0.1$$
$$10^x = 10^{-1}$$
$$x = -1$$

The solution set is $\{-1\}$.

21.   Solving the equation:

$$\left(2^{x+1}\right)\left(2^x\right) = 64$$
$$2^{2x+1} = 2^6$$
$$2x + 1 = 6$$
$$2x = 5$$
$$x = \frac{5}{2}$$

The solution set is $\left\{\frac{5}{2}\right\}$.

23.   Solving the equation:

$$(27)\left(3^x\right) = 9^x$$
$$3^{x+3} = 3^{2x}$$
$$x + 3 = 2x$$
$$-x = -3$$
$$x = 3$$

The solution set is $\{3\}$.

25.   Solving the equation:

$$\left(4^x\right)\left(16^{3x-1}\right) = 8$$
$$\left(2^{2x}\right)\left(2^{12x-4}\right) = 2^3$$
$$2^{14x-4} = 2^3$$
$$14x - 4 = 3$$
$$14x = 7$$
$$x = \frac{1}{2}$$

The solution set is $\left\{\frac{1}{2}\right\}$.

27.   Graphing the function $f(x) = 3^x$:

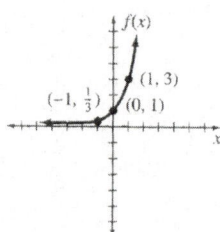

29.   Graphing the function $f(x) = \left(\frac{1}{3}\right)^x$:

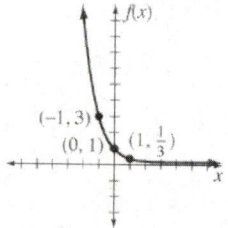

31. Graphing the function $f(x) = \left(\dfrac{3}{2}\right)^x$:

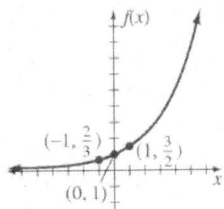

33. Graphing the function $f(x) = 2^x - 3$:

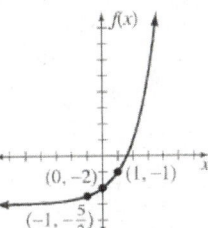

35. Graphing the function $f(x) = 2^{x+2}$:

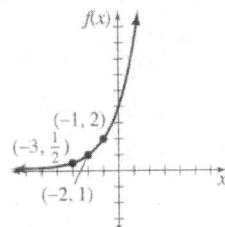

37. Graphing the function $f(x) = -2^x$:

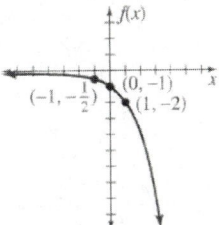

39. Graphing the function $f(x) = 2^{-x-2}$:

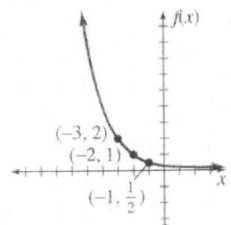

41. Graphing the function $f(x) = 2^{x^2}$:

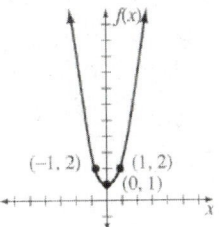

43. Graphing the function $f(x) = 2^{|x|}$:

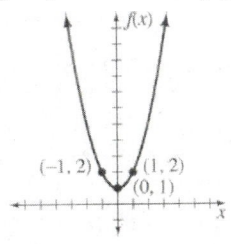

45. Graphing the function $f(x) = 2^x - 2^{-x}$:

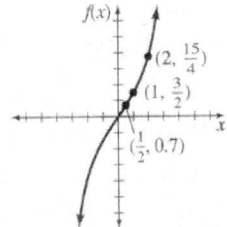

5.2 Applications of Exponential Functions

1. **a.** Computing the price: $P = 1.38(1.04)^3 \approx \$1.55$ **b.** Computing the price: $P = 3.43(1.04)^5 \approx \$4.17$

   c. Computing the price: $P = 1.99(1.04)^4 \approx \$2.33$ **d.** Computing the price: $P = 1.54(1.04)^{10} \approx \$2.28$

   e. Computing the price: $P = 18,000(1.04)^5 \approx \$21,900$

   f. Computing the price: $P = 180,000(1.04)^8 \approx \$246,342$

   g. Computing the price: $P = 500(1.04)^7 \approx \$658$

3. Finding the amount: $A = 200\left(1 + \dfrac{0.06}{1}\right)^{1 \bullet 6} \approx \$283.70$

5. Finding the amount: $A = 500\left(1 + \dfrac{0.04}{2}\right)^{2 \bullet 7} \approx \$659.74$

7.   Finding the amount: $A = 800\left(1 + \dfrac{0.05}{4}\right)^{4 \bullet 9} \approx \$1,251.16$

9.   Finding the amount: $A = 1500\left(1 + \dfrac{0.08}{12}\right)^{12 \bullet 5} \approx \$2,234.77$

11.   Finding the amount: $A = 5000\left(1 + \dfrac{0.045}{1}\right)^{1 \bullet 15} \approx \$9,676.41$

13.   Finding the amount: $A = 8000\left(1 + \dfrac{0.055}{4}\right)^{4 \bullet 10} \approx \$13,814.17$

15.   Finding the amount: $A = 400e^{0.07 \bullet 5} \approx \$567.63$ 17.   Finding the amount: $A = 750e^{0.08 \bullet 8} \approx \$1,422.36$

19.   Finding the amount: $A = 2,000e^{0.07 \bullet 15} \approx \$5,715.30$ 21.   Finding the amount: $A = 7,500e^{0.065 \bullet 10} \approx \$14,366.56$

23.   Finding the balance: $A = 4,830\left(1 + \dfrac{0.109}{12}\right)^{12 \bullet 3} \approx \$6,688.37$

25.   Finding the balance: $A = 3,200\left(1 + \dfrac{0.025}{4}\right)^{4 \bullet 5} \approx \$3,624.66$

27.   Finding the balance: $A = 500e^{0.10 \bullet 3} \approx \$674.93$

29.   Using $A = P\left(1 + \dfrac{r}{n}\right)^{nt}$ with $P = 1500$, $A = 2700$, $n = 4$, and $t = 10$:

$$2700 = 1500\left(1 + \dfrac{r}{4}\right)^{4 \bullet 10}$$

$$\left(1 + \dfrac{r}{4}\right)^{40} = 1.8$$

$$1 + \dfrac{r}{4} = (1.8)^{1/40}$$

$$\dfrac{r}{4} = (1.8)^{1/40} - 1$$

$$r = 4\left((1.8)^{1/40} - 1\right) \approx 0.059$$

The rate is 5.9%.

31.   Solving the equation:

$$P(1 + r) = Pe^{0.0775 \bullet 1}$$

$$1 + r = e^{0.0775}$$

$$r = e^{0.0775} - 1 \approx 0.0806$$

The effective yield is 8.06%.

33.   Finding the effective yield for each:

$$P(1 + r) = P\left(1 + \dfrac{0.0825}{4}\right)^{4 \bullet 1}$$

$$1 + r = \left(1 + \dfrac{0.0825}{4}\right)^{4}$$

$$r = \left(1 + \dfrac{0.0825}{4}\right)^{4} - 1 \approx 0.0851$$

$$P(1 + r) = P\left(1 + \dfrac{0.083}{2}\right)^{2 \bullet 1}$$

$$1 + r = \left(1 + \dfrac{0.083}{2}\right)^{2}$$

$$r = \left(1 + \dfrac{0.083}{2}\right)^{2} - 1 \approx 0.0847$$

The 8.25% compounded quarterly yields the greater return.

35.   Using the half-life formula:

$$Q(87) = 400\left(\frac{1}{2}\right)^{87/29} = 400\left(\frac{1}{2}\right)^{3} = 50 \text{ grams}$$

$$Q(100) = 400\left(\frac{1}{2}\right)^{100/29} \approx 37 \text{ grams}$$

37.   Evaluating the function:

$$Q(2) = 1000e^{0.4 \cdot 2} = 1000e^{0.8} \approx 2,226 \text{ bacteria}$$
$$Q(3) = 1000e^{0.4 \cdot 3} = 1000e^{1.2} \approx 3,320 \text{ bacteria}$$
$$Q(5) = 1000e^{0.4 \cdot 5} = 1000e^{2} \approx 7,389 \text{ bacteria}$$

39.   Solving the equation:

$$6640 = Q_0 e^{0.3 \cdot 4}$$
$$Q_0 = \frac{6640}{e^{1.2}} \approx 2,000 \text{ bacteria}$$

41.   **a.**   Evaluating: $P(3.85) = 14.7e^{-0.21 \cdot 3.85} = 14.7e^{-0.8085} \approx 6.5$ pounds/square inch

  b.   Evaluating: $P(1) = 14.7e^{-0.21 \cdot 1} = 14.7e^{-0.21} \approx 11.9$ pounds/square inch

  c.   Evaluating: $P(1985/5280) = 14.7e^{-0.21 \cdot (1985/5280)} \approx 14.7e^{-0.0789} \approx 13.6$ pounds/square inch

  d.   Evaluating: $P(1090/5280) = 14.7e^{-0.21 \cdot (1090/5280)} \approx 14.7e^{-0.0434} \approx 14.1$ pounds/square inch

43.   Graphing the function $f(x) = e^x + 1$:

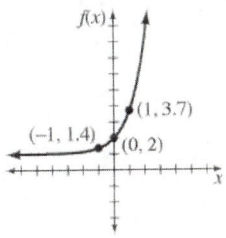

45.   Graphing the function $f(x) = 2e^x$:

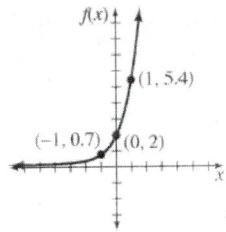

47.   Graphing the function $f(x) = e^{2x}$:

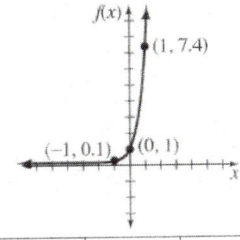

53.   Completing the table:

	4%	5%	6%	7%
5 years	$1,221	$1,284	$1,350	$1,419
10 years	$1,492	$1,649	$1,822	$2,013
15 years	$1,822	$2,117	$2,460	$2,858
20 years	$2,226	$2,718	$3,320	$4,055
25 years	$2,718	$3,490	$4,482	$5,754

	4%	5%	6%	7%
Compounded annually	$1,480	$1,629	$1,791	$1,967
Compounded semiannually	$1,486	$1,639	$1,806	$1,990
Compounded quarterly	$1,489	$1,644	$1,814	$2,002
Compounded monthly	$1,491	$1,647	$1,819	$2,010
Compounded continuously	$1,492	$1,649	$1,822	$2,014

55. Completing the table:

57. Graphing the function $f(x) = \dfrac{e^x + e^{-x}}{2}$:

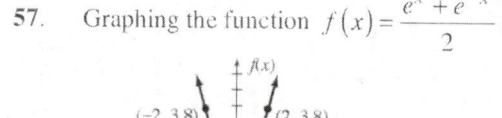

59. Graphing the function $f(x) = \dfrac{e^x - e^{-x}}{2}$:

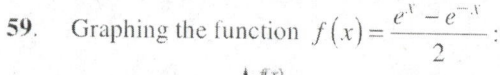

5.3 Inverse Functions

1. Since the graph passes the horizontal line test, this does represent a one-to-one function.

3. Since the graph does not pass the horizontal line test, this does not represent a one-to-one function.

5. Since the graph passes the horizontal line test, this does represent a one-to-one function.

7. Since the graph passes the horizontal line test, this does represent a one-to-one function.

9. Since the graph passes the horizontal line test, this does represent a one-to-one function.

11. Since the graph does not pass the horizontal line test, this does not represent a one-to-one function.

13. Since the graph does not pass the horizontal line test, this does not represent a one-to-one function.

15. The domain of f is $\{1,2,5\}$ and the range of f is $\{5,9,21\}$. Since $f^{-1} = \{(5,1),(9,2),(21,5)\}$, the domain of f^{-1} is $\{5,9,21\}$ and the range of f^{-1} is $\{1,2,5\}$.

17. The domain of f is $\{-2,-1,0,2\}$ and the range of f is $\{-8,-1,0,8\}$. Since $f^{-1} = \{(0,0),(8,2),(-1,-1),(-8,-2)\}$, the domain of f^{-1} is $\{-8,-1,0,8\}$ and the range of f^{-1} is $\{-2,-1,0,2\}$.

19. Verifying the two functions are inverses:
$$f(g(x)) = f\left(\frac{x+9}{5}\right) = 5\left(\frac{x+9}{5}\right) - 9 = x + 9 - 9 = x$$
$$g(f(x)) = g(5x - 9) = \frac{5x - 9 + 9}{5} = \frac{5x}{5} = x$$

21. Verifying the two functions are inverses:
$$f(g(x)) = f\left(-2x + \frac{5}{3}\right) = -\frac{1}{2}\left(-2x + \frac{5}{3}\right) + \frac{5}{6} = x - \frac{5}{6} + \frac{5}{6} = x$$
$$g(f(x)) = g\left(-\frac{1}{2}x + \frac{5}{6}\right) = -2\left(-\frac{1}{2}x + \frac{5}{6}\right) + \frac{5}{3} = x - \frac{5}{3} + \frac{5}{3} = x$$

23. Verifying the two functions are inverses:
$$f(g(x)) = f\left(\frac{x+1}{x}\right) = \frac{1}{\dfrac{x+1}{x} - 1} = \frac{x}{x + 1 - x} = x$$
$$g(f(x)) = g\left(\frac{1}{x-1}\right) = \frac{\dfrac{1}{x-1} + 1}{\dfrac{1}{x-1}} = \frac{1 + x - 1}{1} = x$$

25. Verifying the two functions are inverses:

$$f\big(g(x)\big) = f\left(\frac{x^2+4}{2}\right) = \sqrt{2\left(\frac{x^2+4}{2}\right) - 4} = \sqrt{x^2+4-4} = \sqrt{x^2} = x$$

$$g\big(f(x)\big) = g\left(\sqrt{2x-4}\right) = \frac{\left(\sqrt{2x-4}\right)^2 + 4}{2} = \frac{2x-4+4}{2} = \frac{2x}{2} = x$$

27. Finding the composition: $f\big(g(x)\big) = f\left(-\frac{1}{3}x\right) = 3\left(-\frac{1}{3}x\right) = -x \neq x$

Thus f and g are not inverse functions.

29. Finding the compositions:

$$f\big(g(x)\big) = f\left(\sqrt[3]{x}\right) = \left(\sqrt[3]{x}\right)^3 = x \qquad\qquad g\big(f(x)\big) = g\left(x^3\right) = \sqrt[3]{x^3} = x$$

Thus f and g are inverse functions.

31. Finding the composition: $f\big(g(x)\big) = f\left(\frac{1}{x}\right) = \frac{1}{x} \neq x$. Thus f and g are not inverse functions.

33. Finding the compositions:

$$f\big(g(x)\big) = f\left(\sqrt{x+3}\right) = \left(\sqrt{x+3}\right)^2 - 3 = x+3-3 = x$$

$$g\big(f(x)\big) = g\left(x^2-3\right) = \sqrt{x^2-3+3} = \sqrt{x^2} = x \qquad (\text{since } x \geq 0)$$

Thus f and g are inverse functions.

35. Finding the compositions:

$$f\big(g(x)\big) = f\left(x^2-1\right) = \sqrt{x^2-1+1} = \sqrt{x^2} = x \qquad g\big(f(x)\big) = g\left(\sqrt{x+1}\right) = \left(\sqrt{x+1}\right)^2 - 1 = x+1-1 = x$$

Thus f and g are inverse functions.

37. **a.** Solving the equation:

$$y - 4 = x$$
$$y = x + 4$$
$$f^{-1}(x) = x + 4$$

b. Verifying the compositions:

$$\left(f \circ f^{-1}\right)(x) = f(x+4) = x+4-4 = x \qquad\qquad \left(f^{-1} \circ f\right)(x) = f^{-1}(x-4) = x-4+4 = x$$

39. **a.** Solving the equation:

$$-3y - 4 = x$$
$$-3y = x + 4$$
$$y = -\frac{x+4}{3}$$
$$f^{-1}(x) = \frac{-x-4}{3}$$

b. Verifying the compositions:

$$\left(f \circ f^{-1}\right)(x) = f\left(\frac{-x-4}{3}\right) = -3\left(\frac{-x-4}{3}\right) - 4 = x+4-4 = x$$

$$\left(f^{-1} \circ f\right)(x) = f^{-1}(-3x-4) = \frac{3x+4-4}{3} = \frac{3x}{3} = x$$

41.   **a.**   Solving the equation:

$$\frac{3}{4}y - \frac{5}{6} = x$$

$$\frac{3}{4}y = x + \frac{5}{6}$$

$$y = \frac{4}{3}x + \frac{10}{9}$$

$$f^{-1}(x) = \frac{4}{3}x + \frac{10}{9}$$

b.   Verifying the compositions:

$$\left(f \circ f^{-1}\right)(x) = f\left(\frac{4}{3}x + \frac{10}{9}\right) = \frac{3}{4}\left(\frac{4}{3}x + \frac{10}{9}\right) - \frac{5}{6} = x + \frac{5}{6} - \frac{5}{6} = x$$

$$\left(f^{-1} \circ f\right)(x) = f^{-1}\left(\frac{3}{4}x - \frac{5}{6}\right) = \frac{4}{3}\left(\frac{3}{4}x - \frac{5}{6}\right) + \frac{10}{9} = x - \frac{10}{9} + \frac{10}{9} = x$$

43.   **a.**   Solving the equation:

$$-\frac{2}{3}y = x$$

$$y = -\frac{3}{2}x$$

$$f^{-1}(x) = -\frac{3}{2}x$$

b.   Verifying the compositions:

$$\left(f \circ f^{-1}\right)(x) = f\left(-\frac{3}{2}x\right) = -\frac{2}{3}\left(-\frac{3}{2}x\right) = x$$

$$\left(f^{-1} \circ f\right)(x) = f^{-1}\left(-\frac{2}{3}x\right) = -\frac{3}{2}\left(-\frac{2}{3}x\right) = x$$

45.   **a.**   Solving the equation:

$$\sqrt{y} = x$$

$$y = x^2$$

$$f^{-1}(x) = x^2 \qquad \left(\text{for } x \geq 0\right)$$

b.   Verifying the compositions:

$$\left(f \circ f^{-1}\right)(x) = f\left(x^2\right) = \sqrt{x^2} = x$$

$$\left(f^{-1} \circ f\right)(x) = f^{-1}\left(\sqrt{x}\right) = \left(\sqrt{x}\right)^2 = x$$

47.   **a.**   Solving the equation:

$$\sqrt{y+4} = x$$

$$y + 4 = x^2$$

$$f^{-1}(x) = x^2 - 4 \quad \left(\text{for } x \geq 0\right)$$

b.   Verifying the compositions:

$$\left(f \circ f^{-1}\right)(x) = f\left(x^2 - 4\right) = \sqrt{x^2 - 4 + 4} = \sqrt{x^2} = x$$

$$\left(f^{-1} \circ f\right)(x) = f^{-1}\left(\sqrt{x+4}\right) = \left(\sqrt{x+4}\right)^2 - 4 = x + 4 - 4 = x$$

49.   **a.**   Solving the equation:

$$\frac{1}{y-3} = x$$

$$y - 3 = \frac{1}{x}$$

$$f^{-1}(x) = \frac{1}{x} + 3 = \frac{3x+1}{x}$$

b.   Verifying the compositions:

$$\left(f \circ f^{-1}\right)(x) = f\left(\frac{1}{x} + 3\right) = \frac{1}{\frac{1}{x} + 3 - 3} = \frac{1}{\frac{1}{x}} = x$$

$$\left(f^{-1} \circ f\right)(x) = f^{-1}\left(\frac{1}{x-3}\right) = \frac{1}{\frac{1}{x-3}} + 3 = x - 3 + 3 = x$$

51. a. Solving the equation:

$$y^2 + 4 = x$$
$$y^2 = x - 4$$
$$y = \sqrt{x-4}$$
$$f^{-1}(x) = \sqrt{x-4} \quad (\text{for } x \geq 4)$$

b. Verifying the compositions:

$$\left(f \circ f^{-1}\right)(x) = f\left(\sqrt{x-4}\right) = \left(\sqrt{x-4}\right)^2 + 4 = x - 4 + 4 = x$$
$$\left(f^{-1} \circ f\right)(x) = f^{-1}\left(x^2 + 4\right) = \sqrt{x^2 + 4 - 4} = \sqrt{x^2} = x$$

53. a. Solving the equation:

$$1 + \frac{1}{y} = x$$
$$\frac{1}{y} = x - 1$$
$$y = \frac{1}{x-1}$$
$$f^{-1}(x) = \frac{1}{x-1} \quad (\text{for } x > 1)$$

b. Verifying the compositions:

$$\left(f \circ f^{-1}\right)(x) = f\left(\frac{1}{x-1}\right) = 1 + \frac{1}{\frac{1}{x-1}} = 1 + x - 1 = x$$
$$\left(f^{-1} \circ f\right)(x) = f^{-1}\left(1 + \frac{1}{x}\right) = \frac{1}{1 + \frac{1}{x} - 1} = \frac{1}{\frac{1}{x}} = x$$

55. Solving the equation:

$$3y = x$$
$$y = \frac{1}{3}x$$
$$f^{-1}(x) = \frac{1}{3}x$$

Graphing the curves:

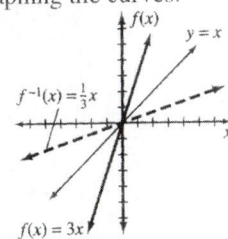

57. Solving the equation:

$$2y + 1 = x$$
$$2y = x - 1$$
$$y = \frac{1}{2}x - \frac{1}{2}$$
$$f^{-1}(x) = \frac{1}{2}x - \frac{1}{2}$$

Graphing the curves:

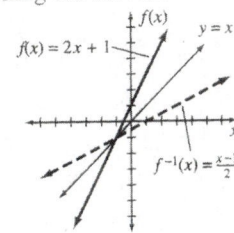

59.   Solving the equation:

$$y^2 - 4 = x$$
$$y^2 = x + 4$$
$$y = \sqrt{x+4}$$
$$f^{-1}(x) = \sqrt{x+4}$$

Graphing the curves:

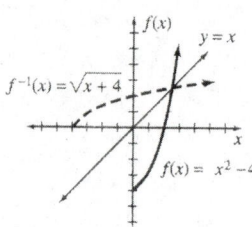

61.   The function is increasing on $[0, \infty)$ and decreasing on $(-\infty, 0]$.

63.   The function is decreasing on $(-\infty, \infty)$.

65.   The function is increasing on $(-\infty, -2]$ and decreasing on $[-2, \infty)$.

The function is increasing on $[1, \infty)$ and decreasing on $(-\infty, 1]$.

67.   Completing the square: $f(x) = -2x^2 - 16x - 35 = -2(x^2 + 8x + 16) - 35 + 32 = -2(x+4)^2 - 3$

The function is increasing on $(-\infty, -4]$ and decreasing on $[-4, \infty)$.

73.   **a.**   Finding the inverse:

$$f\left(f^{-1}(x)\right) = x$$
$$3\left(f^{-1}(x)\right) - 9 = x$$
$$3\left(f^{-1}(x)\right) = x + 9$$
$$f^{-1}(x) = \frac{1}{3}x + 3$$

b.   Finding the inverse:

$$f\left(f^{-1}(x)\right) = x$$
$$-2\left(f^{-1}(x)\right) + 6 = x$$
$$-2\left(f^{-1}(x)\right) = x - 6$$
$$f^{-1}(x) = -\frac{1}{2}x + 3$$

c.   Finding the inverse:

$$f\left(f^{-1}(x)\right) = x$$
$$-\left(f^{-1}(x)\right) + 1 = x$$
$$-\left(f^{-1}(x)\right) = x - 1$$
$$f^{-1}(x) = -x + 1$$

d.   Finding the inverse:

$$f\left(f^{-1}(x)\right) = x$$
$$2\left(f^{-1}(x)\right) = x$$
$$f^{-1}(x) = \frac{1}{2}x$$

e.   Finding the inverse:

$$f\left(f^{-1}(x)\right) = x$$
$$-5\left(f^{-1}(x)\right) = x$$
$$f^{-1}(x) = -\frac{1}{5}x$$

f.   Finding the inverse:

$$f\left(f^{-1}(x)\right) = x$$
$$\left(f^{-1}(x)\right)^2 + 6 = x$$
$$\left(f^{-1}(x)\right)^2 = x - 6$$
$$f^{-1}(x) = \sqrt{x-6} \ \left(\text{for } x \geq 6\right)$$

5.4 Logarithms

1. The logarithmic form is: $\log_2 128 = 7$

3. The logarithmic form is: $\log_5 125 = 3$

5. The logarithmic form is: $\log_{10} 1000 = 3$

7. The logarithmic form is: $\log_2 \frac{1}{4} = -2$

9. The logarithmic form is: $\log_{10} 0.1 = -1$

11. The exponential form is: $3^4 = 81$

13. The exponential form is: $4^3 = 64$

15. The exponential form is: $10^4 = 10,000$

17. The exponential form is: $2^{-4} = \frac{1}{16}$

19. The exponential form is: $10^{-3} = 0.001$

21. Evaluating: $\log_2 16 = \log_2 2^4 = 4$

23. Evaluating: $\log_3 81 = \log_3 3^4 = 4$

25. Evaluating: $\log_6 216 = \log_6 6^3 = 3$

27. Evaluating: $\log_7 \sqrt{7} = \log_7 7^{1/2} = \frac{1}{2}$

29. Evaluating: $\log_{10} 1 = \log_{10} 10^0 = 0$

31. Evaluating: $\log_3 \sqrt{27} = \log_3 3^{3/2} = \frac{3}{2}$

33. Evaluating: $\log_{1/4} 64 = \log_{1/4} \left(\frac{1}{4}\right)^{-3} = -3$

35. Evaluating: $\log_{10} 0.1 = \log_{10} 10^{-1} = -1$

37. Evaluating: $10^{\log_{10} 5} = 5$

39. Evaluating: $\log_2\left(\frac{1}{32}\right) = \log_2\left(2^{-5}\right) = -5$

41. Evaluating: $\log_5\left(\log_2 32\right) = \log_5\left(\log_2 2^5\right) = \log_5 5 = 1$

43. Evaluating: $\log_{10}\left(\log_7 7\right) = \log_{10}(1) = \log_{10} 10^0 = 0$

45. Solving the equation:

$$\log_7 x = 2$$
$$x = 7^2$$
$$x = 49$$

The solution set is $\{49\}$.

47. Solving the equation:

$$\log_8 x = \frac{4}{3}$$
$$x = 8^{4/3}$$
$$x = \left(\sqrt[3]{8}\right)^4$$
$$x = 16$$

The solution set is $\{16\}$.

49. Solving the equation:

$$\log_9 x = -1$$
$$x = 9^{-1}$$
$$x = \frac{1}{9}$$

The solution set is $\left\{\frac{1}{9}\right\}$.

51. Solving the equation:

$$\log_4 x = -\frac{3}{2}$$
$$x = 4^{-3/2}$$
$$x = \left(\sqrt{4}\right)^{-3}$$
$$x = \frac{1}{8}$$

The solution set is $\left\{\frac{1}{8}\right\}$.

53. Solving the equation:

$$\log_{1/2} x = 3$$
$$x = \left(\frac{1}{2}\right)^3$$
$$x = \frac{1}{8}$$

The solution set is $\left\{\frac{1}{8}\right\}$.

55. Solving the equation:

$$\log_x 2 = \frac{1}{2}$$
$$2 = x^{1/2}$$
$$x = 4$$

The solution set is $\{4\}$.

57.   Evaluating: $\log_2 35 = \log_2 5 + \log_2 7 = 2.3219 + 2.8074 = 5.1293$

59.   Evaluating: $\log_2 125 = \log_2 5^3 = 3(2.3219) = 6.9657$

61.   Evaluating: $\log_2 \sqrt{7} = \log_2 7^{1/2} = \frac{1}{2}(2.8074) = 1.4037$

63.   Evaluating: $\log_2 175 = \log_2 \left(7 \cdot 5^2\right) = \log_2 7 + 2\log_2 5 = 2.8074 + 2(2.3219) = 7.4512$

65.   Evaluating: $\log_2 80 = \log_2 \left(5 \cdot 2^4\right) = \log_2 5 + 4\log_2 2 = 2.3219 + 4 = 6.3219$

67.   Evaluating: $\log_8 \dfrac{5}{11} = \log_8 5 - \log_8 11 = 0.7740 - 1.1531 = -0.3791$

69.   Evaluating: $\log_8 \sqrt{11} = \log_8 11^{1/2} = \frac{1}{2}(1.1531) = 0.5766$

71.   Evaluating: $\log_8 88 = \log_8 \left(8 \cdot 11\right) = 1 + 1.1531 = 2.1531$

73.   Evaluating: $\log_8 \dfrac{25}{11} = \log_8 \dfrac{5^2}{11} = 2\log_8 5 - \log_8 11 = 2(0.7740) - 1.1531 = 0.3949$

75.   Expanding the logarithm: $\log_b xyz = \log_b x + \log_b y + \log_b z$

77.   Expanding the logarithm: $\log_b \left(\dfrac{y}{z}\right) = \log_b y - \log_b z$

79.   Expanding the logarithm: $\log_b y^3 z^4 = \log_b y^3 + \log_b z^4 = 3\log_b y + 4\log_b z$

81.   Expanding the logarithm: $\log_b \left(\dfrac{x^{1/2}}{y^{1/3} z^4}\right) = \log_b x^{1/2} - \log_b y^{1/3} - \log_b z^4 = \frac{1}{2}\log_b x - \frac{1}{3}\log_b y - 4\log_b z$

83.   Expanding the logarithm: $\log_b \sqrt[3]{x^2 z} = \log_b x^{2/3} z^{1/3} = \log_b x^{2/3} + \log_b z^{1/3} = \frac{2}{3}\log_b x + \frac{1}{3}\log_b z$

85.   Expanding the logarithm: $\log_b \left[x\sqrt{\dfrac{x}{y}}\right] = \log_b \dfrac{x^{3/2}}{y^{1/2}} = \log_b x^{3/2} - \log_b y^{1/2} = \frac{3}{2}\log_b x - \frac{1}{2}\log_b y$

87.   Expressing as a single logarithm: $2\log_b x - 4\log_b y = \log_b x^2 - \log_b y^4 = \log_b \dfrac{x^2}{y^4}$

89.   Expressing as a single logarithm: $\log_b x - \left(\log_b y - \log_b z\right) = \log_b x - \log_b y + \log_b z = \log_b \dfrac{xz}{y}$

91.   Expressing as a single logarithm: $2\log_b x + 4\log_b y - 3\log_b z = \log_b x^2 + \log_b y^4 - \log_b z^3 = \log_b \dfrac{x^2 y^4}{z^3}$

93.   Expressing as a single logarithm: $\frac{1}{2}\log_b x - \log_b x + 4\log_b y = \log_b \sqrt{x} - \log_b x + \log_b y^4 = \log_b \dfrac{y^4 \sqrt{x}}{x}$

95.   Solving the equation:
$$\log_3 x + \log_3 4 = 2$$
$$\log_3 4x = 2$$
$$4x = 3^2$$
$$4x = 9$$
$$x = \frac{9}{4}$$
The solution set is $\left\{\dfrac{9}{4}\right\}$.

97. Solving the equation:
$$\log_{10} x + \log_{10}(x-21) = 2$$
$$\log_{10}\left(x^2 - 21x\right) = 2$$
$$x^2 - 21x = 10^2$$
$$x^2 - 21x - 100 = 0$$
$$(x-25)(x+4) = 0$$
$$x = 25 \quad (x = -4 \text{ does not check})$$

The solution set is $\{25\}$.

99. Solving the equation:
$$\log_2 x + \log_2(x-3) = 2$$
$$\log_2\left(x^2 - 3x\right) = 2$$
$$x^2 - 3x = 2^2$$
$$x^2 - 3x - 4 = 0$$
$$(x-4)(x+1) = 0$$
$$x = 4 \quad (x = -1 \text{ does not check})$$

The solution set is $\{4\}$.

101. Solving the equation:
$$\log_3(x+3) + \log_3(x+5) = 1$$
$$\log_3\left(x^2 + 8x + 15\right) = 1$$
$$x^2 + 8x + 15 = 3^1$$
$$x^2 + 8x + 12 = 0$$
$$(x+6)(x+2) = 0$$
$$x = -2 \quad (x = -6 \text{ does not check})$$

The solution set is $\{-2\}$.

103. Solving the equation:

$$\log_2 3 + \log_2(x+4) = 3$$
$$\log_2(3x+12) = 3$$
$$3x + 12 = 8$$
$$3x = -4$$
$$x = -\frac{4}{3}$$

The solution set is $\left\{-\dfrac{4}{3}\right\}$.

105. Solving the equation:
$$\log_{10}(2x-1) - \log_{10}(x-2) = 1$$
$$\log_{10}\frac{2x-1}{x-2} = 1$$
$$\frac{2x-1}{x-2} = 10^1$$
$$2x - 1 = 10x - 20$$
$$8x = 19$$
$$x = \frac{19}{8}$$

The solution set is $\left\{\dfrac{19}{8}\right\}$.

107. Solving the equation:
$$\log_5(3x-2) = 1 + \log_5(x-4)$$
$$\log_5(3x-2) - \log_5(x-4) = 1$$
$$\log_5\frac{3x-2}{x-4} = 1$$
$$\frac{3x-2}{x-4} = 5^1$$
$$3x - 2 = 5x - 20$$
$$2x = 18$$
$$x = 9$$

The solution set is $\{9\}$.

109. Solving the equation:

$$\log_2 (x-1) - \log_2 (x+3) = 2$$

$$\log_2 \frac{x-1}{x+3} = 2$$

$$\frac{x-1}{x+3} = 2^2$$

$$x - 1 = 4x + 12$$

$$-3x = 13$$

$$x = -\frac{13}{3}, \text{ which does not check}$$

The solution set is $\varnothing$.

111. Solving the equation:

$$\log_8 (x+7) + \log_8 x = 1$$

$$\log_8 \left(x^2 + 7x\right) = 1$$

$$x^2 + 7x = 8^1$$

$$x^2 + 7x - 8 = 0$$

$$(x+8)(x-1) = 0$$

$$x = 1 \qquad (x = -8 \text{ does not check})$$

The solution set is $\{1\}$.

113. Let $m = \log_b r$ and $n = \log_b s$, so $b^m = r$ and $b^n = s$. Thus $\dfrac{r}{s} = \dfrac{b^m}{b^n} = b^{m-n}$.

Changing to logarithmic form: $\log_b \dfrac{r}{s} = m - n = \log_b r - \log_b s$

5.5 Logarithmic Functions

1. Calculating: $\log 7.24 \approx 0.8597$

3. Calculating: $\log 52.23 \approx 1.7179$

5. Calculating: $\log 3214.1 \approx 3.5071$

7. Calculating: $\log 0.729 \approx -0.1373$

9. Calculating: $\log 0.00034 \approx -3.4685$

11. Calculating: $x = 10^{2.6143} \approx 411.43$

13. Calculating: $x = 10^{4.9547} \approx 90{,}095$

15. Calculating: $x = 10^{1.9006} \approx 79.543$

17. Calculating: $x = 10^{-1.3148} \approx 0.048440$

19. Calculating: $x = 10^{-2.1928} \approx 0.0064150$

21. Calculating: $\ln 5 \approx 1.6094$

23. Calculating: $\ln 32.6 \approx 3.4843$

25. Calculating: $\ln 430 \approx 6.0638$

27. Calculating: $\ln 0.46 \approx -0.7765$

29. Calculating: $\ln 0.0314 \approx -3.4609$

31. Calculating: $x = e^{0.4721} \approx 1.6034$

33. Calculating: $x = e^{1.1425} \approx 3.1346$

35. Calculating: $x = e^{4.6873} \approx 108.56$

37. Calculating: $x = e^{-0.7284} \approx 0.48268$

39. Calculating: $x = e^{-3.3244} \approx 0.035994$

41. **a.** Completing the table:

x	0.1	0.5	1	2	4	8	10
$\log x$	−1	−0.3	0	0.3	0.6	0.9	1

Sketching the graph:

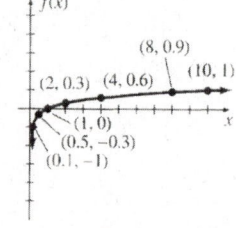

b. Completing the table:

x	−1	−0.3	0	0.3	0.6	0.9	1
10^x	0.1	0.5	1	2	4	8	10

Sketching the graph:

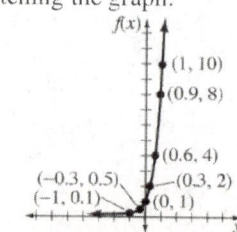

Reflecting the graph results in the graph from part **a.**

43. Graphing the function $y = \log_{1/2} x$:

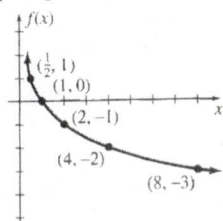

45. Graphing the function $f(x) = \log_3 x$:

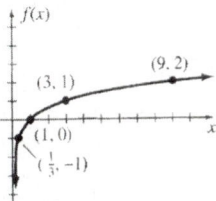

47. Graphing the function $f(x) = 3 + \log_2 x$:

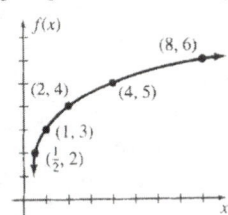

49. Graphing the function $f(x) = \log_2(x + 3)$:

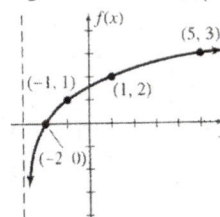

51. Graphing the function $f(x) = \log_2(2x)$:

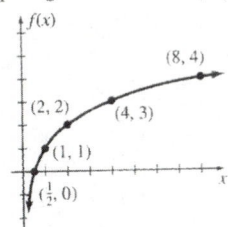

53. Graphing the function $f(x) = 2\log_2 x$:

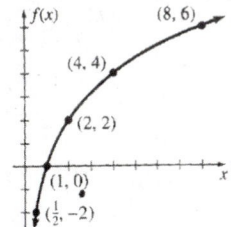

55. Calculating: $\dfrac{\ln 2}{\ln 7} \approx 0.36$

57. Calculating: $\dfrac{\ln 5}{2\ln 3} \approx 0.73$

59. Calculating: $\dfrac{\ln 2}{0.03} \approx 23.10$

61. Calculating: $\dfrac{\log 5}{3\log 1.07} \approx 7.93$

5.6 Exponential Equations, Logarithmic Equations, and Problem Solving

1. Solving the equation:
$$3^x = 13$$
$$\ln(3^x) = \ln 13$$
$$x\ln 3 = \ln 13$$
$$x = \frac{\ln 13}{\ln 3} \approx 2.33$$
The solution set is $\{2.33\}$.

3. Solving the equation:
$$4^n = 35$$
$$\ln(4^n) = \ln 35$$
$$n\ln 4 = \ln 35$$
$$n = \frac{\ln 35}{\ln 4} \approx 2.56$$
The solution set is $\{2.56\}$.

5. Solving the equation:
$$2^x + 7 = 50$$
$$2^x = 43$$
$$\ln(2^x) = \ln 43$$
$$x\ln 2 = \ln 43$$
$$x = \frac{\ln 43}{\ln 2} \approx 5.43$$
The solution set is $\{5.43\}$.

7. Solving the equation:
$$3^{x-2} = 11$$
$$\ln(3^{x-2}) = \ln 11$$
$$(x - 2)\ln 3 = \ln 11$$
$$x - 2 = \frac{\ln 11}{\ln 3}$$
$$x = \frac{\ln 11}{\ln 3} + 2 \approx 4.18$$
The solution set is $\{4.18\}$.

9. Solving the equation:

$$5^{3t+1} = 9$$
$$\ln\left(5^{3t+1}\right) = \ln 9$$
$$(3t+1)\ln 5 = \ln 9$$
$$3t+1 = \frac{\ln 9}{\ln 5}$$
$$3t = \frac{\ln 9}{\ln 5} - 1$$
$$t = \frac{1}{3}\left(\frac{\ln 9}{\ln 5} - 1\right) \approx 0.12$$

The solution set is $\{0.12\}$.

13. Solving the equation:

$$e^{x-2} = 13.1$$
$$x - 2 = \ln 13.1$$
$$x = \ln 13.1 + 2 \approx 4.57$$

The solution set is $\{4.57\}$.

17. Solving the equation:

$$5^{2x+1} = 7^{x+3}$$
$$\ln\left(5^{2x+1}\right) = \ln\left(7^{x+3}\right)$$
$$(2x+1)\ln 5 = (x+3)\ln 7$$
$$2x\ln 5 + \ln 5 = x\ln 7 + 3\ln 7$$
$$x(2\ln 5 - \ln 7) = 3\ln 7 - \ln 5$$
$$x = \frac{3\ln 7 - \ln 5}{2\ln 5 - \ln 7} \approx 3.32$$

The solution set is $\{3.32\}$.

21. Solving the equation:

$$\log x + \log(x + 21) = 2$$
$$\log\left(x^2 + 21x\right) = 2$$
$$x^2 + 21x = 10^2$$
$$x^2 + 21x - 100 = 0$$
$$(x + 25)(x - 4) = 0$$
$$x = 4 \quad (x = -25 \text{ does not check})$$

The solution set is $\{4\}$.

23. Solving the equation:

$$\log(3x - 1) = 1 + \log(5x - 2)$$
$$\log(3x - 1) - \log(5x - 2) = 1$$
$$\log\left(\frac{3x - 1}{5x - 2}\right) = 1$$
$$\frac{3x - 1}{5x - 2} = 10^1$$
$$3x - 1 = 50x - 20$$
$$-47x = -19$$
$$x = \frac{19}{47}$$

The solution set is $\left\{\frac{19}{47}\right\}$.

11. Solving the equation:

$$e^x = 27$$
$$x = \ln 27 \approx 3.30$$

The solution set is $\{3.30\}$.

15. Solving the equation:

$$3e^x - 1 = 17$$
$$3e^x = 18$$
$$e^x = 6$$
$$x = \ln 6 \approx 1.79$$

The solution set is $\{1.79\}$.

19. Solving the equation:

$$3^{2x+1} = 2^{3x+2}$$
$$\ln\left(3^{2x+1}\right) = \ln\left(2^{3x+2}\right)$$
$$(2x+1)\ln 3 = (3x+2)\ln 2$$
$$2x\ln 3 + \ln 3 = 3x\ln 2 + 2\ln 2$$
$$x(2\ln 3 - 3\ln 2) = 2\ln 2 - \ln 3$$
$$x = \frac{2\ln 2 - \ln 3}{2\ln 3 - 3\ln 2} \approx 2.44$$

The solution set is $\{2.44\}$.

25. Solving the equation:

$$\log_3 x + \log_3 (x+4) = 2$$
$$\log_3 (x^2 + 4x) = 2$$
$$x^2 + 4x = 9$$
$$x^2 + 4x - 9 = 0$$
$$x = \frac{-4 \pm \sqrt{16 - 4(1)(-9)}}{2(1)} = \frac{-4 \pm \sqrt{52}}{2} = \frac{-4 \pm 2\sqrt{13}}{2} = -2 \pm \sqrt{13}$$
$$x = -2 + \sqrt{13} \quad \left(x = -2 - \sqrt{13} \text{ does not check} \right)$$

The solution set is $\left\{ -2 + \sqrt{13} \right\}$.

27. Solving the equation:

$$\log_2 (x+5) + \log_2 (x-1) = 4$$
$$\log_2 (x^2 + 4x - 5) = 4$$
$$x^2 + 4x - 5 = 16$$
$$x^2 + 4x - 21 = 0$$
$$(x+7)(x-3) = 0$$
$$x = 3 \quad (x = -7 \text{ does not check})$$

The solution set is $\{3\}$.

29. Solving the equation:

$$\log (x+1) = \log 3 - \log (2x-1)$$
$$\log (x+1) + \log (2x-1) = \log 3$$
$$\log (2x^2 + x - 1) = \log 3$$
$$2x^2 + x - 1 = 3$$
$$2x^2 + x - 4 = 0$$
$$x = \frac{-1 \pm \sqrt{1 - 4(2)(-4)}}{2(2)} = \frac{-1 \pm \sqrt{33}}{4}$$
$$x = \frac{-1 + \sqrt{33}}{4} \quad \left(x = \frac{-1 - \sqrt{33}}{4} \text{ does not check} \right)$$

The solution set is $\left\{ \dfrac{-1 + \sqrt{33}}{4} \right\}$.

31. Solving the equation:

$$\log (x+2) - \log (2x+1) = \log x$$
$$\log \left(\frac{x+2}{2x+1} \right) = \log x$$
$$\frac{x+2}{2x+1} = x$$
$$x + 2 = 2x^2 + x$$
$$1 = x^2$$
$$x = 1 \quad (x = -1 \text{ does not check})$$

The solution set is $\{1\}$.

33. Solving the equation:

$$\ln(2t+5) = \ln 3 + \ln(t-1)$$
$$\ln(2t+5) - \ln(t-1) = \ln 3$$
$$\ln\left(\frac{2t+5}{t-1}\right) = \ln 3$$
$$\frac{2t+5}{t-1} = 3$$
$$2t+5 = 3t-3$$
$$t = 8$$

The solution set is $\{8\}$.

37. Evaluating: $\log_2 14 = \dfrac{\log 40}{\log 2} \approx 5.322$

41. Evaluating: $\log_4 1.6 = \dfrac{\log 1.6}{\log 4} \approx 0.339$

45. Evaluating: $\log_7 500 = \dfrac{\log 500}{\log 7} \approx 3.194$

47. Solving the equation:

$$750\left(1 + \frac{0.06}{4}\right)^{4t} = 1000$$
$$(1.015)^{4t} = \frac{4}{3}$$
$$\ln(1.015)^{4t} = \ln\frac{4}{3}$$
$$4t \ln 1.015 = \ln\frac{4}{3}$$
$$t = \frac{\ln\dfrac{4}{3}}{4 \ln 1.015} \approx 4.8$$

It will take 4.8 years.

51. Solving the equation:

$$500 e^{10r} = 900$$
$$e^{10r} = 1.8$$
$$10r = \ln 1.8$$
$$r = \frac{\ln 1.8}{10} \approx 0.059$$

The rate is 5.9%.

55. Solving the equation:

$$14.7 e^{-0.21t} = 11.53$$
$$e^{-0.21t} = \frac{11.53}{14.7}$$
$$-0.21t = \ln\left(\frac{11.53}{14.7}\right)$$
$$t = \frac{\ln\left(\dfrac{11.53}{14.7}\right)}{-0.21} \approx 1.157 \text{ miles}$$
$$t = 1.157(5280) \approx 6100 \text{ feet}$$

The altitude is approximately 6,100 feet.

35. Solving the equation:

$$\log \sqrt{x} = \sqrt{\log x}$$
$$\frac{1}{2}\log x = \sqrt{\log x}$$
$$\frac{1}{4}(\log x)^2 = \log x$$
$$(\log x)^2 = 4\log x$$
$$(\log x)^2 - 4\log x = 0$$
$$(\log x)(\log x - 4) = 0$$
$$\log x = 0, 4$$
$$x = 10^0, 10^4$$
$$x = 1, 10000$$

The solution set is $\{1, 10000\}$.

39. Evaluating: $\log_3 16 = \dfrac{\log 16}{\log 3} \approx 2.524$

43. Evaluating: $\log_5 0.26 = \dfrac{\log 0.26}{\log 5} \approx -0.837$

49. Solving the equation:

$$2000 e^{0.04t} = 4000$$
$$e^{0.04t} = 2$$
$$0.04t = \ln 2$$
$$t = \frac{\ln 2}{0.04} \approx 17.3$$

It will take 17.3 years.

53. Solving the equation:

$$400 e^{0.34t} = 4000$$
$$e^{0.34t} = 10$$
$$0.34t = \ln 10$$
$$t = \frac{\ln 10}{0.34} \approx 6.8$$

It will take approximately 6.8 hours.

57. Solving the equation:

$$500 e^{0.4t} = 2000$$
$$e^{0.4t} = 4$$
$$0.4t = \ln 4$$
$$t = \frac{\ln 4}{0.4} \approx 3.5$$

It will take approximately 3.5 hours.

59. The Richter number is: $R = \log \dfrac{5{,}000{,}000 I_0}{I_0} = \log 5{,}000{,}000 \approx 6.7$

61. Calculating each intensity:

$$7.3 = \log \frac{I_2}{I_0}$$

$$\frac{I_2}{I_0} = 10^{7.3}$$

$$I_2 = 10^{7.3} I_0$$

$$6.4 = \log \frac{I_1}{I_0}$$

$$\frac{I_1}{I_0} = 10^{6.4}$$

$$I_1 = 10^{6.4} I_0$$

Comparing the two intensities: $\dfrac{I_2}{I_1} = \dfrac{10^{7.3} I_0}{10^{6.4} I_0} = 10^{0.9} \approx 7.9$. The earthquake is approximately 8 times as intense.

69. Solving the equation:

$$\frac{5^x - 5^{-x}}{2} = 3$$

$$5^x - 5^{-x} = 6$$

$$5^x \left(5^x - 5^{-x} \right) = 5^x (6)$$

$$5^{2x} - 1 = 6 \bullet 5^x$$

$$5^{2x} - 6 \bullet 5^x - 1 = 0$$

Using the quadratic formula:

$$5^x = \frac{6 \pm \sqrt{(-6)^2 - 4(1)(-1)}}{2} = \frac{6 \pm \sqrt{36+4}}{2} = \frac{6 \pm \sqrt{40}}{2} = \frac{6 \pm 2\sqrt{10}}{2} = 3 \pm \sqrt{10}$$

Since $5^x > 0$:

$$\ln \left(5^x \right) = \ln \left(3 + \sqrt{10} \right)$$

$$x \ln 5 = \ln \left(3 + \sqrt{10} \right)$$

$$x = \frac{\ln \left(3 + \sqrt{10} \right)}{\ln 5} \approx 1.13$$

71. Solving the equation:

$$y = \frac{e^x - e^{-x}}{2}$$

$$e^x - e^{-x} = 2y$$

$$e^x \left(e^x - e^{-x} \right) = e^x (2y)$$

$$e^{2x} - 1 = 2y \bullet e^x$$

$$e^{2x} - 2y \bullet e^x - 1 = 0$$

Using the quadratic formula:

$$e^x = \frac{2y \pm \sqrt{(-2y)^2 - 4(1)(-1)}}{2} = \frac{2y \pm \sqrt{4y^2 + 4}}{2} = \frac{2y \pm 2\sqrt{y^2 + 1}}{2} = y \pm \sqrt{y^2 + 1}$$

Since $e^x > 0$:

$$\ln \left(e^x \right) = \ln \left(y + \sqrt{y^2 + 1} \right)$$

$$x = \ln \left(y + \sqrt{y^2 + 1} \right)$$

Chapter 5 Review Problem Set

1.   Evaluating: $8^{5/3} = \left(\sqrt[3]{8}\right)^5 = 2^5 = 32$

3.   Evaluating: $(-27)^{4/3} = \left(\sqrt[3]{-27}\right)^4 = (-3)^4 = 81$

5.   Evaluating: $\log_7\left(\dfrac{1}{49}\right) = \log_7 7^{-2} = -2$

7.   Evaluating: $\log_2\left(\dfrac{\sqrt[4]{32}}{2}\right) = \log_2\left(\dfrac{2^{5/4}}{2}\right) = \log_2 2^{1/4} = \dfrac{1}{4}$

9.   Evaluating: $\ln e = 1$

11. Solving the equation:

$$\log_{10} 2 + \log_{10} x = 1$$
$$\log_{10} 2x = 1$$
$$2x = 10$$
$$x = 5$$

The solution set is $\{5\}$.

13. Solving the equation:

$$4^x = 128$$
$$2^{2x} = 2^7$$
$$2x = 7$$
$$x = \dfrac{7}{2}$$

The solution set is $\left\{\dfrac{7}{2}\right\}$.

15. Solving the equation:

$$\log_2 x = 3$$
$$x = 2^3$$
$$x = 8$$

The solution set is $\{8\}$.

17. Solving the equation:

$$2e^x = 14$$
$$e^x = 7$$
$$x = \ln 7$$
$$x \approx 1.95$$

The solution set is $\{1.95\}$.

19. Solving the equation:

$$\ln(x+4) - \ln(x+2) = \ln x$$
$$\ln\left(\dfrac{x+4}{x+2}\right) = \ln x$$
$$\dfrac{x+4}{x+2} = x$$
$$x+4 = x^2 + 2x$$
$$0 = x^2 + x - 4$$
$$x = \dfrac{-1 \pm \sqrt{1 - 4(-4)}}{2} = \dfrac{-1 \pm \sqrt{17}}{2} \approx 1.56 \qquad (x = -2.56 \text{ does not check})$$

The solution set is $\{1.56\}$.

21. Solving the equation:

$$\log(\log x) = 2$$
$$\log x = 10^2 = 100$$
$$x = 10^{100}$$

The solution set is $\left\{10^{100}\right\}$.

23. Solving the equation:

$$\ln(2t-1) = \ln 4 + \ln(t-3)$$
$$\ln(2t-1) - \ln(t-3) = \ln 4$$
$$\ln\left(\dfrac{2t-1}{t-3}\right) = \ln 4$$
$$\dfrac{2t-1}{t-3} = 4$$
$$2t-1 = 4t-12$$
$$-2t = -11$$
$$t = \dfrac{11}{2}$$

The solution set is $\left\{\dfrac{11}{2}\right\}$.

25. Evaluating: $\log\left(\dfrac{7}{3}\right) = \log 7 - \log 3 = 0.8451 - 0.4771 = 0.3680$

27. Evaluating: $\log 27 = \log 3^3 = 3\log 3 = 3(0.4771) = 1.4313$

29. **a.** Expanding: $\log_b\left(\dfrac{x}{y^2}\right) = \log_b x - \log_b y^2 = \log_b x - 2\log_b y$

   b. Expanding: $\log_b \sqrt[4]{xy^2} = \log_b\left(x^{1/4}y^{1/2}\right) = \log_b x^{1/4} + \log_b y^{1/2} = \dfrac{1}{4}\log_b x + \dfrac{1}{2}\log_b y$

   c. Expanding: $\log_b\left(\dfrac{\sqrt{x}}{y^3}\right) = \log_b x^{1/2} - \log_b y^3 = \dfrac{1}{2}\log_b x - 3\log_b y$

31. Calculating: $\log_2 3 = \dfrac{\log 3}{\log 2} \approx 1.585$

33. Calculating: $\log_4 191 = \dfrac{\log 191}{\log 4} \approx 3.789$

35. **a.** Graphing the function $f(x) = \left(\dfrac{3}{4}\right)^x$:

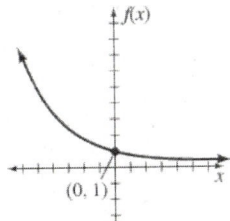

37. **a.** Graphing the function $f(x) = e^{x-1}$:

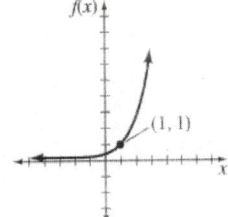

   b. Graphing the function $f(x) = \left(\dfrac{3}{4}\right)^x + 2$:

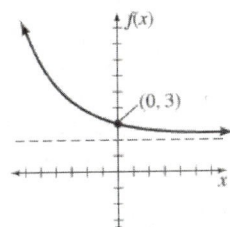

   b. Graphing the function $f(x) = e^x - 1$:

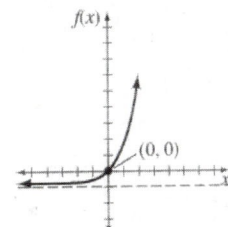

   c. Graphing the function $f(x) = \left(\dfrac{3}{4}\right)^{-x}$:

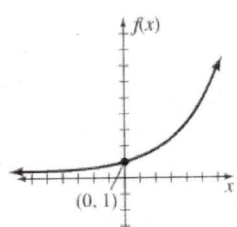

   c. Graphing the function $f(x) = e^{-x+1}$:

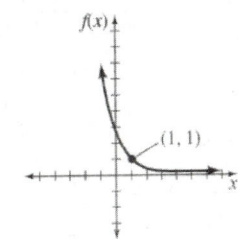

39. Graphing the function $f(x) = 3^x - 3^{-x}$:

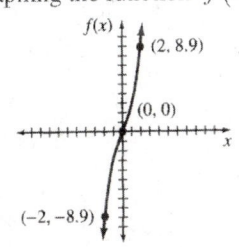

41. Graphing the function $f(x) = \log_2(x-3)$:

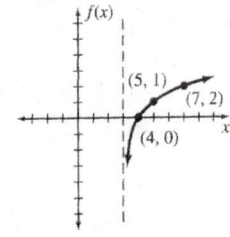

43. Calculating the amount: $A = 7500\left(1 + \dfrac{0.04}{4}\right)^{4 \cdot 10} \approx \$11,166.48$

45. Calculating the amount: $A = 2500\left(1 + \dfrac{0.065}{2}\right)^{2 \cdot 20} \approx \$8,985.50$

47. Finding the compositions:

$(f \circ g)(x) = f\left(\dfrac{3}{2}x\right) = -\dfrac{2}{3}\left(\dfrac{3}{2}x\right) = -x$ $\qquad\qquad (g \circ f)(x) = g\left(-\dfrac{2}{3}x\right) = \dfrac{3}{2}\left(-\dfrac{2}{3}x\right) = -x$

Thus f and g are not inverse functions.

49. Finding the compositions:

$(f \circ g)(x) = f\left(\sqrt{2 - x}\right) = 2 - \left(\sqrt{2 - x}\right)^2 = 2 - 2 + x = x$

$(g \circ f)(x) = g\left(2 - x^2\right) = \sqrt{2 - \left(2 - x^2\right)} = \sqrt{2 - 2 + x^2} = \sqrt{x^2} = x$

Thus f and g are inverse functions.

51. Solving the equation:

$-3y - 7 = x$

$\qquad -3y = x + 7$

$\qquad\quad y = \dfrac{-x - 7}{3}$

$\quad f^{-1}(x) = \dfrac{-x - 7}{3}$

Verifying the compositions:

$\left(f \circ f^{-1}\right)(x) = f\left(\dfrac{-x - 7}{3}\right) = -3\left(\dfrac{-x - 7}{3}\right) - 7 = x + 7 - 7 = x$

$\left(f^{-1} \circ f\right)(x) = f^{-1}(-3x - 7) = \dfrac{-(-3x - 7) - 7}{3} = \dfrac{3x + 7 - 7}{3} = \dfrac{3x}{3} = x$

53. Solving the equation:

$-2 - y^2 = x$

$\qquad y^2 = -x - 2$

$\qquad\; y = \sqrt{-x - 2}$

$\quad f^{-1}(x) = \sqrt{-x - 2}$

Verifying the compositions:

$\left(f \circ f^{-1}\right)(x) = f\left(\sqrt{-x - 2}\right) = -2 - \left(\sqrt{-x - 2}\right)^2 = -2 + x + 2 = x$

$\left(f^{-1} \circ f\right)(x) = f^{-1}\left(-2 - x^2\right) = \sqrt{-\left(-2 - x^2\right) - 2} = \sqrt{2 + x^2 - 2} = \sqrt{x^2} = x$

55. The function is increasing on $[3, \infty)$ and it is never decreasing.

57. Solving the equation:

$1000\left(1 + \dfrac{0.045}{4}\right)^{4t} = 3500$

$\qquad (1.01125)^{4t} = 3.5$

$\qquad \ln(1.01125)^{4t} = \ln 3.5$

$\qquad 4t \ln 1.01125 = \ln 3.5$

$\qquad\qquad t = \dfrac{\ln 3.5}{4 \ln 1.01125} \approx 28.0$

It will take approximately 28.0 years.

59.   Finding the populations:
$$P(10) = 50000e^{0.02 \cdot 10} = 50000e^{0.2} \approx 61,070$$
$$P(15) = 50000e^{0.02 \cdot 15} = 50000e^{0.3} \approx 67,493$$
$$P(20) = 50000e^{0.02 \cdot 20} = 50000e^{0.4} \approx 74,591$$

61.   Evaluating: $Q(100) = 750\left(\frac{1}{2}\right)^{100/40} = 750(0.5)^{2.5} \approx 133$ grams

Chapter 5 Test

1.   Evaluating: $\log_3 \sqrt{3} = \log_3 3^{1/2} = \dfrac{1}{2}$

2.   Evaluating: $\log_2\left(\log_2 4\right) = \log_2\left(\log_2 2^2\right) = \log_2 2 = 1$

3.   Evaluating: $-2 + \ln e^3 = -2 + 3 = 1$

4.   Evaluating: $\log_2(0.5) = \log_2 2^{-1} = -1$

5.   Solving the equation:
$$4^x = \frac{1}{64}$$
$$4^x = 4^{-3}$$
$$x = -3$$

The solution set is $\{-3\}$.

6.   Solving the equation:
$$9^x = \frac{1}{27}$$
$$3^{2x} = 3^{-3}$$
$$2x = -3$$
$$x = -\frac{3}{2}$$

The solution set is $\left\{-\dfrac{3}{2}\right\}$.

7.   Solving the equation:
$$2^{3x-1} = 128$$
$$2^{3x-1} = 2^7$$
$$3x - 1 = 7$$
$$3x = 8$$
$$x = \frac{8}{3}$$

The solution set is $\left\{\dfrac{8}{3}\right\}$.

8.   Solving the equation:
$$\log_9 x = \frac{5}{2}$$
$$x = 9^{5/2}$$
$$x = \left(\sqrt{9}\right)^5$$
$$x = 3^5 = 243$$

The solution set is $\{243\}$.

9.   Solving the equation:

$$\log x + \log(x + 48) = 2$$
$$\log\left(x^2 + 48x\right) = 2$$
$$x^2 + 48x = 10^2$$
$$x^2 + 48x - 100 = 0$$
$$(x + 50)(x - 2) = 0$$
$$x = 2 \quad (x = -50 \text{ does not check})$$

The solution set is $\{2\}$.

10.   Solving the equation:
$$\ln x = \ln 2 + \ln(3x - 1)$$
$$\ln x - \ln(3x - 1) = \ln 2$$
$$\ln \frac{x}{3x - 1} = \ln 2$$
$$\frac{x}{3x - 1} = 2$$
$$x = 6x - 2$$
$$-5x = -2$$
$$x = \frac{2}{5}$$

The solution set is $\left\{\dfrac{2}{5}\right\}$.

11.   Evaluating: $\log_3 100 = \log_3\left(4 \cdot 5^2\right) = \log_3 4 + 2\log_3 5 = 1.2619 + 2(1.4650) = 4.1919$

12.   Evaluating: $\log_3 \dfrac{5}{4} = \log_3 5 - \log_3 4 = 1.4650 - 1.2619 = 0.2031$

13.   Evaluating: $\log_3 \sqrt{5} = \log_3 5^{1/2} = \frac{1}{2}\log_3 5 = \frac{1}{2}(1.4650) = 0.7325$

14. Finding the inverse:
$$-3y - 6 = x$$
$$-3y = x + 6$$
$$y = -\frac{1}{3}x - 2$$
$$f^{-1}(x) = -\frac{1}{3}x - 2$$

15. Solving the equation:

$$e^x = 176$$
$$x = \ln 176$$
$$x \approx 5.17$$

The solution set is $\{5.17\}$.

16. Solving the equation:
$$2^{x-2} = 314$$
$$\ln 2^{x-2} = \ln 314$$
$$(x-2)\ln 2 = \ln 314$$
$$x - 2 = \frac{\ln 314}{\ln 2}$$
$$x = 2 + \frac{\ln 314}{\ln 2} \approx 10.29$$

The solution set is $\{10.29\}$.

17. Calculating the logarithm: $\log_5 632 = \dfrac{\ln 632}{\ln 5} \approx 4.0069$

18. Finding the inverse:
$$\frac{2}{3}y - \frac{3}{5} = x$$
$$10y - 9 = 15x$$
$$10y = 15x + 9$$
$$y = \frac{3}{2}x + \frac{9}{10}$$
$$f^{-1}(x) = \frac{3}{2}x + \frac{9}{10}$$

19. Computing the amount: $A = 3500\left(1 + \dfrac{0.075}{4}\right)^{4 \cdot 8} = 3500(1.01875)^{32} \approx \$6,342.08$

20. Solving the equation:
$$5000(1.07)^t = 12500$$
$$1.07^t = 2.5$$
$$\ln 1.07^t = \ln 2.5$$
$$t \ln 1.07 = \ln 2.5$$
$$t = \frac{\ln 2.5}{\ln 1.07} \approx 13.5$$

It will take approximately 13.5 years.

21. Solving the equation:
$$400e^{0.23t} = 2400$$
$$e^{0.23t} = 6$$
$$0.23t = \ln 6$$
$$t = \frac{\ln 6}{0.23} \approx 7.8$$

It will take approximately 7.8 hours.

22. Finding the amount: $Q(32) = 7500\left(\dfrac{1}{2}\right)^{32/50} = 7500(0.5)^{0.64} \approx 4,813$ grams

23. Graphing the function $f(x) = e^x - 2$:

24. Graphing the function $f(x) = -3^{-x}$:

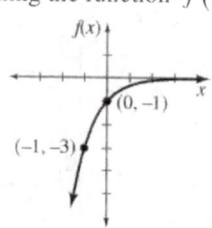

25.   Graphing the function $f(x) = \log_2 (x-2)$:

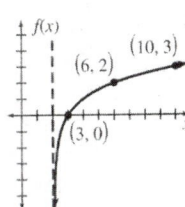

Chapters 0-5 Cumulative Review Problem Set

1.   Evaluating when $x = -2$: $-5(-2-1) - 3(2(-2)+4) + 3(3(-2)-1) = -5(-3) - 3(0) + 3(-7) = 15 - 0 - 21 = -6$

2.   Evaluating when $a = -1$ and $b = 4$: $\dfrac{14(-1)^3 (4)^2}{7(-1)^2 (4)} = \dfrac{-224}{28} = -8$

3.   Evaluating when $n = 4$: $\dfrac{2}{4} - \dfrac{3}{2(4)} + \dfrac{5}{3(4)} = \dfrac{1}{2} - \dfrac{3}{8} + \dfrac{5}{12} = \dfrac{12}{24} - \dfrac{9}{24} + \dfrac{10}{24} = \dfrac{13}{24}$

4.   Evaluating when $x = 16$ and $y = 16$:

$4\sqrt{2(16)-16} + 5\sqrt{3(16)+16} = 4\sqrt{32-16} + 5\sqrt{48+16} = 4\sqrt{16} + 5\sqrt{64} = 16 + 40 = 56$

5.   Evaluating when $x = 3$: $\dfrac{3}{3-2} - \dfrac{5}{3+3} = 3 - \dfrac{5}{6} = \dfrac{18}{6} - \dfrac{5}{6} = \dfrac{13}{6}$

6.   Simplifying: $\left(-5\sqrt{6}\right)\left(3\sqrt{12}\right) = -15\sqrt{72} = -15 \cdot 6\sqrt{2} = -90\sqrt{2}$

7.   Simplifying: $\left(2\sqrt{x} - 3\right)\left(\sqrt{x} + 4\right) = 2x - 3\sqrt{x} + 8\sqrt{x} - 12 = 2x + 5\sqrt{x} - 12$

8.   Simplifying: $\left(3\sqrt{2} - \sqrt{6}\right)\left(\sqrt{2} + 4\sqrt{6}\right) = 6 + 12\sqrt{12} - \sqrt{12} - 24 = -18 + 11\sqrt{12} = -18 + 11 \cdot 2\sqrt{3} = -18 + 22\sqrt{3}$

9.   Simplifying: $(2x-1)\left(x^2 + 6x - 4\right) = 2x^3 + 12x^2 - 8x - x^2 - 6x + 4 = 2x^3 + 11x^2 - 14x + 4$

10.   Simplifying: $\dfrac{x^2 - x}{x+5} \cdot \dfrac{x^2 + 5x + 4}{x^4 - x^2} = \dfrac{x(x-1)}{x+5} \cdot \dfrac{(x+4)(x+1)}{x^2(x+1)(x-1)} = \dfrac{x+4}{x(x+5)}$

11.   Simplifying: $\dfrac{16x^2 y}{24xy^3} \div \dfrac{9xy}{8x^2 y^2} = \dfrac{16x^2 y}{24xy^3} \cdot \dfrac{8x^2 y^2}{9xy} = \dfrac{16x^4 y^3}{27x^2 y^4} = \dfrac{16x^2}{27y}$

12.   Simplifying: $\dfrac{x+3}{10} + \dfrac{2x+1}{15} - \dfrac{x-2}{18} = \dfrac{x+3}{10} \cdot \dfrac{9}{9} + \dfrac{2x+1}{15} \cdot \dfrac{6}{6} - \dfrac{x-2}{18} \cdot \dfrac{5}{5} = \dfrac{9x+27}{90} + \dfrac{12x+6}{90} - \dfrac{5x-10}{90} = \dfrac{16x+43}{90}$

13.   Simplifying: $\dfrac{7}{12ab} - \dfrac{11}{15a^2} = \dfrac{7}{12ab} \cdot \dfrac{5a}{5a} - \dfrac{11}{15a^2} \cdot \dfrac{4b}{4b} = \dfrac{35a}{60a^2 b} - \dfrac{44b}{60a^2 b} = \dfrac{35a - 44b}{60a^2 b}$

14.   Simplifying: $\dfrac{8}{x^2 - 4x} + \dfrac{2}{x} = \dfrac{8}{x(x-4)} + \dfrac{2}{x} \cdot \dfrac{x-4}{x-4} = \dfrac{8}{x(x-4)} + \dfrac{2x-8}{x(x-4)} = \dfrac{2x}{x(x-4)} = \dfrac{2}{x-4}$

15.   Using long division:

$$
\begin{array}{r}
2x^2 - x - 4 \\
4x-1 \overline{\smash{\big)}\ 8x^3 - 6x^2 - 15x + 4} \\
\underline{8x^3 - 2x^2} \\
-4x^2 - 15x \\
\underline{-4x^2 + x} \\
-16x + 4 \\
\underline{-4x + 2} \\
0
\end{array}
$$

The quotient is $2x^2 - x - 4$.

16.   Simplifying: $\dfrac{\dfrac{5}{x^2}-\dfrac{3}{x}}{\dfrac{1}{y}+\dfrac{2}{y^2}} = \dfrac{\dfrac{5}{x^2}-\dfrac{3}{x}}{\dfrac{1}{y}+\dfrac{2}{y^2}} \cdot \dfrac{x^2y^2}{x^2y^2} = \dfrac{5y^2-3xy^2}{x^2y+2x^2}$

17.   Simplifying: $\dfrac{\dfrac{2}{x}-3}{\dfrac{3}{y}+4} = \dfrac{\dfrac{2}{x}-3}{\dfrac{3}{y}+4} \cdot \dfrac{xy}{xy} = \dfrac{2y-3xy}{3x+4xy}$

18.   Simplifying:

$$\dfrac{2-\dfrac{1}{n-2}}{3+\dfrac{4}{n+3}} = \dfrac{2-\dfrac{1}{n-2}}{3+\dfrac{4}{n+3}} \cdot \dfrac{(n-2)(n+3)}{(n-2)(n+3)} = \dfrac{2(n-2)(n+3)-(n+3)}{3(n-2)(n+3)+4(n-2)} = \dfrac{(n+3)\left[2(n-2)-1\right]}{(n-2)\left[3(n+3)+4\right]} = \dfrac{(n+3)(2n-5)}{(n-2)(3n+13)}$$

19.   Simplifying: $\dfrac{3a}{2-\dfrac{1}{a}}-1 = \dfrac{3a}{2-\dfrac{1}{a}} \cdot \dfrac{a}{a}-1 = \dfrac{3a^2}{2a-1}-1 \cdot \dfrac{2a-1}{2a-1} = \dfrac{3a^2-2a+1}{2a-1}$

20.   Factoring: $20x^2+7x-6 = (5x-2)(4x+3)$

21.   Factoring: $16x^3+54 = 2\left(8x^3+27\right) = 2(2x+3)\left(4x^2-6x+9\right)$

22.   Factoring: $4x^4-25x^2+36 = \left(4x^2-9\right)\left(x^2-4\right) = (2x+3)(2x-3)(x+2)(x-2)$

23.   Factoring: $12x^3-52x^2-40x = 4x\left(3x^2-13x-10\right) = 4x(3x+2)(x-5)$

24.   Factoring: $xy-6x+3y-18 = x(y-6)+3(y-6) = (y-6)(x+3)$

25.   Factoring: $10+9x-9x^2 = (5-3x)(2+3x)$

26.   Evaluating: $\left(\dfrac{2}{3}\right)^{-4} = \left(\dfrac{3}{2}\right)^{4} = \dfrac{3^4}{2^4} = \dfrac{81}{16}$

27.   Evaluating: $\dfrac{3}{\left(\dfrac{4}{3}\right)^{-1}} = 3\left(\dfrac{4}{3}\right) = 4$

28.   Evaluating: $\sqrt[3]{-\dfrac{27}{64}} = -\dfrac{3}{4}$

29.   Evaluating: $-\sqrt{0.09} = -0.3$

30.   Evaluating: $(27)^{-4/3} = \left(\sqrt[3]{27}\right)^{-4} = 3^{-4} = \dfrac{1}{81}$

31.   Evaluating: $4^0+4^{-1}+4^{-2} = 1+\dfrac{1}{4}+\dfrac{1}{16} = \dfrac{16}{16}+\dfrac{4}{16}+\dfrac{1}{16} = \dfrac{21}{16}$

32.   Evaluating: $\left(\dfrac{3^{-1}}{2^{-3}}\right)^{-2} = \dfrac{3^2}{2^6} = \dfrac{9}{64}$

33.   Evaluating: $\left(2^{-3}-3^{-2}\right)^{-1} = \left(\dfrac{1}{8}-\dfrac{1}{9}\right)^{-1} = \left(\dfrac{1}{72}\right)^{-1} = 72$

34.   Evaluating: $\log_2 64 = \log_2 2^6 = 6$

35.   Evaluating: $\log_3\left(\dfrac{1}{9}\right) = \log_3 3^{-2} = -2$

36.   Simplifying: $\left(-3x^{-1}y^2\right)\left(4x^{-2}y^{-3}\right) = -12x^{-3}y^{-1} = \dfrac{-12}{x^3y}$

37.   Simplifying: $\dfrac{48x^{-4}y^2}{6xy} = \dfrac{48y^2}{6xy \cdot x^4} = \dfrac{8y}{x^5}$

38.   Simplifying: $\left(\dfrac{27a^{-4}b^{-3}}{-3a^{-1}b^{-4}}\right)^{-1} = \dfrac{-3a^4b^3}{27ab^4} = -\dfrac{a^3}{9b}$

39.   Simplifying: $\sqrt{80} = \sqrt{16 \cdot 5} = 4\sqrt{5}$

40.   Simplifying: $-2\sqrt{54} = -2\sqrt{9 \cdot 6} = -6\sqrt{6}$

41. Simplifying: $\dfrac{\sqrt{75}}{\sqrt{81}} = \dfrac{\sqrt{25 \cdot 3}}{9} = \dfrac{5\sqrt{3}}{9}$

42. Simplifying: $\dfrac{4\sqrt{6}}{3\sqrt{8}} = \dfrac{4\sqrt{6}}{3\sqrt{8}} \cdot \dfrac{\sqrt{2}}{\sqrt{2}} = \dfrac{4\sqrt{12}}{12} = \dfrac{8\sqrt{3}}{12} = \dfrac{2\sqrt{3}}{3}$

43. Simplifying: $\sqrt[3]{56} = \sqrt[3]{8 \cdot 7} = 2\sqrt[3]{7}$

44. Simplifying: $\dfrac{\sqrt[3]{3}}{\sqrt[3]{4}} = \dfrac{\sqrt[3]{3}}{\sqrt[3]{4}} \cdot \dfrac{\sqrt[3]{2}}{\sqrt[3]{2}} = \dfrac{\sqrt[3]{6}}{2}$

45. Simplifying: $4\sqrt{52x^3 y^2} = 4\sqrt{4x^2 y^2 \cdot 13x} = 8xy\sqrt{13x}$

46. Simplifying: $\sqrt{\dfrac{2x}{3y}} = \dfrac{\sqrt{2x}}{\sqrt{3y}} \cdot \dfrac{\sqrt{3y}}{\sqrt{3y}} = \dfrac{\sqrt{6xy}}{3y}$

47. Simplifying: $-3\sqrt{24} + 6\sqrt{54} - \sqrt{6} = -3\sqrt{4 \cdot 6} + 6\sqrt{9 \cdot 6} - \sqrt{6} = -6\sqrt{6} + 18\sqrt{6} - \sqrt{6} = 11\sqrt{6}$

48. Simplifying:

$$\dfrac{\sqrt{8}}{3} - \dfrac{3\sqrt{18}}{4} - \dfrac{5\sqrt{50}}{2} = \dfrac{\sqrt{4 \cdot 2}}{3} - \dfrac{3\sqrt{9 \cdot 2}}{4} - \dfrac{5\sqrt{25 \cdot 2}}{2}$$

$$= \dfrac{2\sqrt{2}}{3} - \dfrac{9\sqrt{2}}{4} - \dfrac{25\sqrt{2}}{2}$$

$$= \dfrac{2\sqrt{2}}{3} \cdot \dfrac{4}{4} - \dfrac{9\sqrt{2}}{4} \cdot \dfrac{3}{3} - \dfrac{25\sqrt{2}}{2} \cdot \dfrac{6}{6}$$

$$= \dfrac{8\sqrt{2}}{12} - \dfrac{27\sqrt{2}}{12} - \dfrac{150\sqrt{2}}{12}$$

$$= -\dfrac{169\sqrt{2}}{12}$$

49. Simplifying: $8\sqrt[3]{3} - 6\sqrt[3]{24} - 4\sqrt[3]{81} = 8\sqrt[3]{3} - 6\sqrt[3]{8 \cdot 3} - 4\sqrt[3]{27 \cdot 3} = 8\sqrt[3]{3} - 12\sqrt[3]{3} - 12\sqrt[3]{3} = -16\sqrt[3]{3}$

50. Simplifying: $\dfrac{\sqrt{3}}{\sqrt{6} - 2\sqrt{2}} = \dfrac{\sqrt{3}}{\sqrt{6} - 2\sqrt{2}} \cdot \dfrac{\sqrt{6} + 2\sqrt{2}}{\sqrt{6} + 2\sqrt{2}} = \dfrac{\sqrt{18} + 2\sqrt{6}}{6 - 8} = \dfrac{3\sqrt{2} + 2\sqrt{6}}{-2} = \dfrac{-3\sqrt{2} - 2\sqrt{6}}{2}$

51. Simplifying: $\dfrac{3\sqrt{5} - \sqrt{3}}{2\sqrt{3} + \sqrt{7}} = \dfrac{3\sqrt{5} - \sqrt{3}}{2\sqrt{3} + \sqrt{7}} \cdot \dfrac{2\sqrt{3} - \sqrt{7}}{2\sqrt{3} - \sqrt{7}} = \dfrac{6\sqrt{15} - 3\sqrt{35} - 6 + \sqrt{21}}{12 - 7} = \dfrac{6\sqrt{15} - 3\sqrt{35} - 6 + \sqrt{21}}{5}$

52. Using scientific notation:

$$\dfrac{(0.00016)(300)(0.028)}{0.064} = \dfrac{\left(1.6 \times 10^{-4}\right)\left(3 \times 10^2\right)\left(2.8 \times 10^{-2}\right)}{6.4 \times 10^{-2}} = \dfrac{13.44 \times 10^{-4}}{6.4 \times 10^{-2}} = 2.1 \times 10^{-2} = 0.021$$

53. Using scientific notation: $\dfrac{0.00072}{0.0000024} = \dfrac{7.2 \times 10^{-4}}{2.4 \times 10^{-6}} = 3.0 \times 10^2 = 300$

54. Using scientific notation: $\sqrt{0.00000009} = \sqrt{9 \times 10^{-8}} = 3 \times 10^{-4} = 0.0003$

55. Simplifying: $(5 - 2i)(4 + 6i) = 20 + 30i - 8i - 12i^2 = 20 + 22i + 12 = 32 + 22i$

56. Simplifying: $(-3 - i)(5 - 2i) = -15 + 6i - 5i + 2i^2 = -15 + i - 2 = -17 + i$

57. Simplifying: $\dfrac{5}{4i} = \dfrac{5}{4i} \cdot \dfrac{i}{i} = \dfrac{5i}{4i^2} = \dfrac{5i}{-4} = -\dfrac{5}{4}i = 0 - \dfrac{5}{4}i$

58. Simplifying: $\dfrac{-1 + 6i}{7 - 2i} = \dfrac{-1 + 6i}{7 - 2i} \cdot \dfrac{7 + 2i}{7 + 2i} = \dfrac{-7 - 2i + 42i + 12i^2}{49 - 4i^2} = \dfrac{-7 + 40i - 12}{49 + 4} = \dfrac{-19 + 40i}{53} = -\dfrac{19}{53} + \dfrac{40}{53}i$

59. Finding the slope: $m = \dfrac{7 - (-3)}{-1 - 2} = \dfrac{10}{-3} = -\dfrac{10}{3}$

60. Solving for y:
$$4x - 7y = 9$$
$$-7y = -4x + 9$$
$$y = \frac{4}{7}x - \frac{9}{7}$$

The slope is $\frac{4}{7}$.

61. Using the distance formula: $\sqrt{(-2-4)^2 + (1-5)^2} = \sqrt{(-6)^2 + (-4)^2} = \sqrt{36+16} = \sqrt{52} = 2\sqrt{13}$

62. Finding the slope: $m = \dfrac{4-(-1)}{7-3} = \dfrac{5}{4}$. Using the point-slope formula:
$$y - 4 = \frac{5}{4}(x - 7)$$
$$4y - 16 = 5x - 35$$
$$-5x + 4y = -19$$
$$5x - 4y = 19$$

63. The slope of $3x - 4y = 6$ is $\dfrac{3}{4}$, so the perpendicular slope is $-\dfrac{4}{3}$. Using the point-slope formula:
$$y + 2 = -\frac{4}{3}(x + 3)$$
$$3y + 6 = -4x - 12$$
$$4x + 3y = -18$$

64. Finding the slope: $m = \dfrac{10-4}{6-2} = \dfrac{6}{4} = \dfrac{3}{2}$. So the perpendicular slope is $-\dfrac{2}{3}$. Using the midpoint $(4,7)$ in the point-slope formula:
$$y - 7 = -\frac{2}{3}(x - 4)$$
$$3y - 21 = -2x + 8$$
$$2x + 3y = 29$$

65. The slope is $-\dfrac{3}{4}$. Using the point-slope formula
$$y - 0 = -\frac{3}{4}(x + 4)$$
$$4y = -3x - 12$$
$$y = -\frac{3}{4}x - 3$$

66. Since the line must be a vertical line, its equation is $x = -2$.

67. Sketching the graph:

68. Sketching the graph:

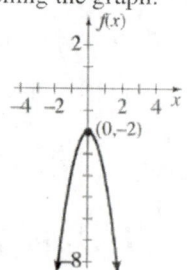

69. Sketching the graph of $f(x) = \left(x^2 - 2x + 1\right) - 3 = (x-1)^2 - 3$:

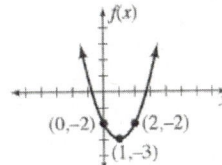

70. Sketching the graph:

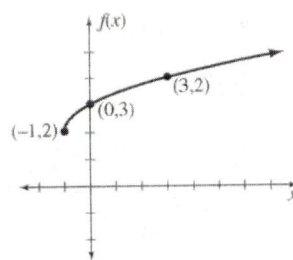

71. Sketching the graph of $f(x) = 2\left(x^2 + 4x\right) + 9 = 2\left(x^2 + 4x + 4\right) + 9 - 8 = 2(x+2)^2 + 1$:

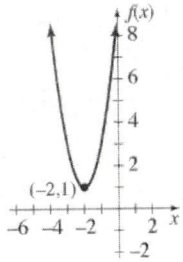

72. Sketching the graph:

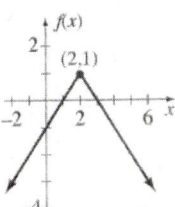

73. Sketching the graph:

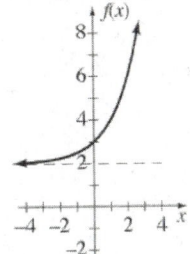

74. Sketching the graph:

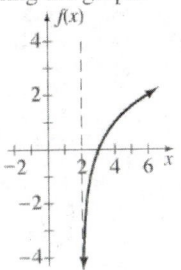

75. Sketching the graph:

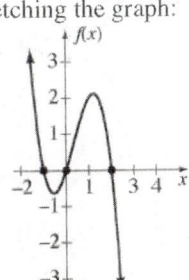

76. Sketching the graph:

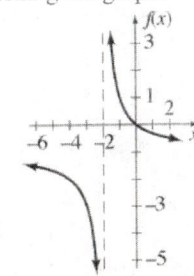

77.   Finding the compositions:

$$g\big(f(x)\big) = g(x-3) = 2(x-3)^2 - (x-3) - 1 = 2x^2 - 12x + 18 - x + 3 - 1 = 2x^2 - 13x + 20$$

$$f\big(g(x)\big) = f\big(2x^2 - x - 1\big) = 2x^2 - x - 1 - 3 = 2x^2 - x - 4$$

78.   Finding the inverse:

$$3y - 7 = x$$
$$3y = x + 7$$
$$y = \frac{1}{3}x + \frac{7}{3}$$
$$f^{-1}(x) = \frac{1}{3}x + \frac{7}{3}$$

79.   Finding the inverse:

$$-\frac{1}{2}y + \frac{2}{3} = x$$
$$-\frac{1}{2}y = x - \frac{2}{3}$$
$$y = -2x + \frac{4}{3}$$
$$f^{-1}(x) = -2x + \frac{4}{3}$$

80.   The variation equation is $y = kx$. Substituting $y = 2$ and $x = -\frac{2}{3}$:

$$2 = k\left(-\frac{2}{3}\right)$$
$$6 = -2k$$
$$k = -3$$

81.   The variation equation is $y = \dfrac{k}{x^2}$. Substituting $y = 4$ and $x = 3$:

$$4 = \frac{k}{3^2}$$
$$4 = \frac{k}{9}$$
$$k = 36$$

So $y = \dfrac{36}{x^2}$. Substituting $x = 6$: $y = \dfrac{36}{6^2} = \dfrac{36}{36} = 1$

82.   The variation equation is $V = \dfrac{k}{P}$. Substituting $V = 15$ and $P = 20$:

$$15 = \frac{k}{20}$$
$$k = 300$$

So $V = \dfrac{300}{P}$. Substituting $P = 25$: $V = \dfrac{300}{25} = 12$ cubic centimeters

83.   Solving the equation:

$$3(2x-1) - 2(5x+1) = 4(3x+4)$$
$$6x - 3 - 10x - 2 = 12x + 16$$
$$-4x - 5 = 12x + 16$$
$$-16x = 21$$
$$x = -\frac{21}{16}$$

The solution set is $\left\{-\dfrac{21}{16}\right\}$.

84.   Solving the equation:

$$n + \frac{3n-1}{9} - 4 = \frac{3n+1}{3}$$
$$9\left(n + \frac{3n-1}{9} - 4\right) = 9\left(\frac{3n+1}{3}\right)$$
$$9n + 3n - 1 - 36 = 9n + 3$$
$$12n - 37 = 9n + 3$$
$$3n = 40$$
$$n = \frac{40}{3}$$

The solution set is $\left\{\dfrac{40}{3}\right\}$.

85. Solving the equation:
$$0.92 + 0.9(x - 0.3) = 2x - 5.95$$
$$0.92 + 0.9x - 0.27 = 2x - 5.95$$
$$0.9x + 0.65 = 2x - 5.95$$
$$-1.1x = -6.6$$
$$x = 6$$

The solution set is $\{6\}$.

86. Solving the equation:
$$|4x - 1| = 11$$
$$4x - 1 = -11, 11$$
$$4x = -10, 12$$
$$x = -\frac{5}{2}, 3$$

The solution set is $\left\{-\frac{5}{2}, 3\right\}$.

87. Solving the equation:
$$3x^2 = 7x$$
$$3x^2 - 7x = 0$$
$$x(3x - 7) = 0$$
$$x = 0, \frac{7}{3}$$

The solution set is $\left\{0, \frac{7}{3}\right\}$.

88. Solving the equation:
$$x^3 - 36x = 0$$
$$x(x^2 - 36) = 0$$
$$x(x + 6)(x - 6) = 0$$
$$x = -6, 0, 6$$

The solution set is $\{-6, 0, 6\}$.

89. Solving the equation:
$$30x^2 + 13x - 10 = 0$$
$$(6x + 5)(5x - 2) = 0$$
$$x = -\frac{5}{6}, \frac{2}{5}$$

The solution set is $\left\{-\frac{5}{6}, \frac{2}{5}\right\}$.

90. Solving the equation:
$$8x^3 + 12x^2 - 36x = 0$$
$$4x(2x^2 + 3x - 9) = 0$$
$$4x(2x - 3)(x + 3) = 0$$
$$x = -3, 0, \frac{3}{2}$$

The solution set is $\left\{-3, 0, \frac{3}{2}\right\}$.

91. Solving the equation:
$$x^4 + 8x^2 - 9 = 0$$
$$(x^2 - 1)(x^2 + 9) = 0$$
$$x^2 = 1, -9$$
$$x = \pm 1, \pm 3i$$

The solution set is $\{\pm 1, \pm 3i\}$.

92. Solving the equation:
$$(n + 4)(n - 6) = 11$$
$$n^2 - 2n - 24 = 11$$
$$n^2 - 2n - 35 = 0$$
$$(n + 5)(n - 7) = 0$$
$$n = -5, 7$$

The solution set is $\{-5, 7\}$.

93. Solving the equation:
$$2 - \frac{3x}{x - 4} = \frac{14}{x + 7}$$
$$2(x - 4)(x + 7) - 3x(x + 7) = 14(x - 4)$$
$$2x^2 + 6x - 56 - 3x^2 - 21x = 14x - 56$$
$$-x^2 - 29x = 0$$
$$-x(x + 29) = 0$$
$$x = -29, 0$$

The solution set is $\{-29, 0\}$.

94.   Solving the equation:

$$\frac{2n}{6n^2+7n-3}-\frac{n-3}{3n^2+11n-4}=\frac{5}{2n^2+11n+12}$$

$$\frac{2n}{(3n-1)(2n+3)}-\frac{n-3}{(3n-1)(n+4)}=\frac{5}{(2n+3)(n+4)}$$

$$\frac{2n(n+4)}{(3n-1)(2n+3)(n+4)}-\frac{(n-3)(2n+3)}{(3n-1)(2n+3)(n+4)}=\frac{5(3n-1)}{(3n-1)(2n+3)(n+4)}$$

$$2n(n+4)-(n-3)(2n+3)=5(3n-1)$$

$$2n^2+8n-2n^2+3n+9=15n-5$$

$$11n+9=15n-5$$

$$-4n=-14$$

$$n=\frac{7}{2}$$

The solution set is $\left\{\dfrac{7}{2}\right\}$.

95.   Solving the equation:

$$\sqrt{3y}-y=-6$$

$$\sqrt{3y}=y-6$$

$$3y=(y-6)^2$$

$$3y=y^2-12y+36$$

$$y^2-15y+36=0$$

$$(y-3)(y-12)=0$$

$$y=12 \quad (y=3 \text{ does not check})$$

The solution set is $\{12\}$.

96.   Solving the equation:

$$\sqrt{x+19}-\sqrt{x+28}=-1$$

$$\sqrt{x+19}=\sqrt{x+28}-1$$

$$x+19=\left(\sqrt{x+28}-1\right)^2$$

$$x+19=x+28-2\sqrt{x+28}+1$$

$$-10=-2\sqrt{x+28}$$

$$5=\sqrt{x+28}$$

$$25=x+28$$

$$x=-3$$

The solution set is $\{-3\}$.

97.   Solving the equation:

$$(3x-1)^2=45$$

$$3x-1=\pm\sqrt{45}$$

$$3x-1=\pm3\sqrt{5}$$

$$3x=1\pm3\sqrt{5}$$

$$x=\frac{1\pm3\sqrt{5}}{3}$$

The solution set is $\left\{\dfrac{1\pm3\sqrt{5}}{3}\right\}$.

98.   Solving the equation:

$$(2x+5)^2=-32$$

$$2x+5=\pm\sqrt{-32}$$

$$2x+5=\pm4i\sqrt{2}$$

$$2x=-5\pm4i\sqrt{2}$$

$$x=\frac{-5\pm4i\sqrt{2}}{2}$$

The solution set is $\left\{\dfrac{-5\pm4i\sqrt{2}}{2}\right\}$.

99.   Solving the equation:

$$2x^2-3x+4=0$$

$$x=\frac{3\pm\sqrt{(-3)^2-4(2)(4)}}{2(2)}=\frac{3\pm\sqrt{9-32}}{4}=\frac{3\pm\sqrt{-23}}{4}=\frac{3\pm i\sqrt{23}}{4}$$

The solution set is $\left\{\dfrac{3\pm i\sqrt{23}}{4}\right\}$.

100. Solving the equation:

$$3n^2 - 6n + 2 = 0$$

$$n = \frac{6 \pm \sqrt{(-6)^2 - 4(3)(2)}}{2(3)} = \frac{6 \pm \sqrt{36-24}}{6} = \frac{6 \pm \sqrt{12}}{6} = \frac{6 \pm 2\sqrt{3}}{6} = \frac{3 \pm \sqrt{3}}{3}$$

The solution set is $\left\{ \dfrac{3 \pm \sqrt{3}}{3} \right\}$.

101. Solving the equation:

$$\frac{5}{n-3} - \frac{3}{n+3} = 1$$
$$5(n+3) - 3(n-3) = (n+3)(n-3)$$
$$5n + 15 - 3n + 9 = n^2 - 9$$
$$2n + 24 = n^2 - 9$$
$$n^2 - 2n - 33 = 0$$

$$n = \frac{2 \pm \sqrt{(-2)^2 - 4(1)(-33)}}{2(1)} = \frac{2 \pm \sqrt{4+132}}{2} = \frac{2 \pm 2\sqrt{34}}{2} = 1 \pm \sqrt{34}$$

The solution set is $\left\{ 1 \pm \sqrt{34} \right\}$.

102. Solving the equation:

$$12x^4 - 19x^2 + 5 = 0$$
$$\left(4x^2 - 5\right)\left(3x^2 - 1\right) = 0$$

$$4x^2 - 5 = 0 \qquad\qquad\qquad 3x^2 - 1 = 0$$
$$4x^2 = 5 \qquad\qquad\qquad\quad 3x^2 = 1$$
$$x^2 = \frac{5}{4} \qquad\qquad\qquad\quad x^2 = \frac{1}{3}$$
$$x = \pm\frac{\sqrt{5}}{2} \qquad\qquad\quad x = \pm\frac{1}{\sqrt{3}} = \pm\frac{\sqrt{3}}{3}$$

The solution set is $\left\{ \pm\dfrac{\sqrt{5}}{2}, \pm\dfrac{\sqrt{3}}{3} \right\}$.

103. Solving the equation:

$$2x^2 + 5x + 5 = 0$$

$$x = \frac{-5 \pm \sqrt{(5)^2 - 4(2)(5)}}{2(2)} = \frac{-5 \pm \sqrt{25-40}}{4} = \frac{-5 \pm \sqrt{-15}}{4} = \frac{-5 \pm i\sqrt{15}}{4}$$

The solution set is $\left\{ \dfrac{-5 \pm i\sqrt{15}}{4} \right\}$.

104. The possible rational roots are $\pm 1, \pm 2, \pm 4, \pm 7, \pm 14, \pm 28$. Using synthetic division:

```
1 ⌋ 1    -4    -25    28
           1     -3   -28
     1    -3    -28     0
```

Solving the equation:

$$(x-1)\left(x^2 - 3x - 28\right) = 0$$
$$(x-1)(x-7)(x+4) = 0$$
$$x = -4, 1, 7$$

The solution set is $\left\{ -4, 1, 7 \right\}$.

105. The possible rational roots are $\pm 1, \pm 2, \pm 5, \pm 10, \pm \frac{1}{2}, \pm \frac{5}{2}, \pm \frac{1}{3}, \pm \frac{2}{3}, \pm \frac{5}{3}, \pm \frac{10}{3}, \pm \frac{1}{6}, \pm \frac{5}{6}$. Using synthetic division:

$$
\begin{array}{r|rrrr}
2 & 6 & -19 & 9 & 10 \\
  &   & 12 & -14 & -10 \\
\hline
  & 6 & -7 & -5 & 0
\end{array}
$$

Solving the equation:
$$(x-2)\left(6x^2 - 7x - 5\right) = 0$$
$$(x-2)(2x+1)(3x-5) = 0$$
$$x = -\frac{1}{2}, \frac{5}{3}, 2$$

The solution set is $\left\{-\frac{1}{2}, \frac{5}{3}, 2\right\}$.

106. Solving the equation:
$$16^x = 64$$
$$4^{2x} = 4^3$$
$$2x = 3$$
$$x = \frac{3}{2}$$

The solution set is $\left\{\frac{3}{2}\right\}$.

107. Solving the equation:
$$\log_3 x = 4$$
$$x = 3^4$$
$$x = 81$$

The solution set is $\{81\}$.

108. Solving the equation:
$$\log_{10} x + \log_{10} 25 = 2$$
$$\log_{10}(25x) = 2$$
$$25x = 10^2$$
$$25x = 100$$
$$x = 4$$

The solution set is $\{4\}$.

109. Solving the equation:
$$\ln(3x-4) - \ln(x+1) = \ln 2$$
$$\ln\left(\frac{3x-4}{x+1}\right) = \ln 2$$
$$\frac{3x-4}{x+1} = 2$$
$$3x - 4 = 2x + 2$$
$$x = 6$$

The solution set is $\{6\}$.

110. Solving the equation:
$$27^{4x} = 9^{x+1}$$
$$3^{12x} = 3^{2x+2}$$
$$12x = 2x + 2$$
$$10x = 2$$
$$x = \frac{1}{5}$$

The solution set is $\left\{\frac{1}{5}\right\}$.

111. Solving the inequality:
$$-5(y-1) + 3 > 3y - 4 - 4y$$
$$-5y + 8 > -y - 4$$
$$-4y > -12$$
$$y < 3$$

The solution set is $(-\infty, 3)$.

112. Solving the inequality:
$$0.06x + 0.08(250 - x) \geq 19$$
$$0.06x + 20 - 0.08x \geq 19$$
$$-0.02x + 20 \geq 19$$
$$-0.02x \geq -1$$
$$x \leq 50$$

The solution set is $(-\infty, 50]$.

113. Solving the inequality:
$$|5x - 2| > 13$$
$$5x - 2 < -13 \quad \text{or} \quad 5x - 2 > 13$$
$$5x < -11 \qquad\qquad 5x > 15$$
$$x < -\frac{11}{5} \qquad\qquad x > 3$$

The solution set is $\left(-\infty, -\frac{11}{5}\right) \cup (3, \infty)$.

114. Solving the inequality:
$$|6x + 2| < 8$$
$$-8 < 6x + 2 < 8$$
$$-10 < 6x < 6$$
$$-\frac{5}{3} < x < 1$$

The solution set is $\left(-\frac{5}{3}, 1\right)$.

115. Solving the inequality:
$$\frac{x-2}{5} - \frac{3x-1}{4} \le \frac{3}{10}$$
$$20\left(\frac{x-2}{5}\right) - 20\left(\frac{3x-1}{4}\right) \le 20\left(\frac{3}{10}\right)$$
$$4(x-2) - 5(3x-1) \le 2(3)$$
$$4x - 8 - 15x + 5 \le 6$$
$$-11x - 3 \le 6$$
$$-11x \le 9$$
$$x \ge -\frac{9}{11}$$

The solution set is $\left[-\frac{9}{11}, \infty\right)$.

116. Forming a sign chart:

The solution set is $[-4, 2]$.

117. Forming a sign chart:

The solution set is $\left(-\infty, \frac{1}{3}\right) \cup (4, \infty)$.

118. Factoring the inequality:
$$x(x+5) < 24$$
$$x^2 + 5x - 24 < 0$$
$$(x+8)(x-3) < 0$$

Forming a sign chart:

The solution set is $(-8, 3)$.

119. Forming a sign chart:

The solution set is $(-\infty, 3] \cup (7, \infty)$.

120. Solving the inequality:
$$\frac{2x}{x+3} > 4$$
$$\frac{2x}{x+3} - 4 > 0$$
$$\frac{2x - 4(x+3)}{x+3} > 0$$
$$\frac{-2x - 12}{x+3} > 0$$
$$\frac{-2(x+6)}{x+3} > 0$$

Forming a sign chart:

The solution set is $(-6, -3)$.

121. Let x, $x + 2$, and $x + 4$ represent the three odd integers. The equation is:
$$x + (x + 2) + (x + 4) = 57$$
$$3x + 6 = 57$$
$$3x = 51$$
$$x = 17$$
The three integers are 17, 19, and 21.

122. Let n, d, and q represent the number of nickels, dimes, and quarters, respectively. The system of equations is:
$$n + d + q = 63$$
$$d = n + 6$$
$$q = 2n + 1$$
Substituting into the first equation:
$$n + (n + 6) + (2n + 1) = 63$$
$$4n + 7 = 63$$
$$4n = 56$$
$$n = 14$$
$$d = 14 + 6 = 20$$
$$q = 28 + 1 = 29$$
There are 14 nickels, 20 dimes, and 29 quarters in Eric's collection.

123. Let x represent one angle and $\frac{1}{3}x + 4$ represent the second angle. Since the two angles are supplementary:
$$x + \frac{1}{3}x + 4 = 180$$
$$\frac{4}{3}x + 4 = 180$$
$$4x + 12 = 540$$
$$4x = 528$$
$$x = 132$$
$$\frac{1}{3}x + 4 = \frac{1}{3}(132) + 4 = 48$$
The two angles are 48° and 132°.

124. Let p represent the selling price. The equation is:
$$300 + 0.50p = p$$
$$300 = 0.5p$$
$$p = 600$$
The selling price should be $600.

125. Let x represent the amount invested at 8% and $x + 300$ the amount invested at 9%. The interest equation is:
$$0.08x + 0.09(x + 300) = 316$$
$$0.08x + 0.09x + 27 = 316$$
$$0.17x = 289$$
$$x = 1700$$
$$x + 300 = 2000$$
Beth invested $1,700 at 8% and $2,000 at 9%.

126. Let r represent the rate of the first train and $r + 10$ represent the rate of the second train. The distance equation after 4.5 hours is given by:

$$4.5r + 4.5(r + 10) = 639$$
$$4.5r + 4.5r + 45 = 639$$
$$9r = 594$$
$$r = 66$$
$$r + 10 = 76$$

The two trains are traveling 66 mph and 76 mph.

127. Let x represent the amount of fluid to drain out and replace with antifreeze. The equation for antifreeze is:

$$0.50(10 - x) + 1.00(x) = 0.70(10)$$
$$5 - 0.5x + x = 7$$
$$5 + 0.5x = 7$$
$$0.5x = 2$$
$$x = 4$$

Thus 4 quarts needs to be drained out and replaced with antifreeze.

128. Let x represent Sam's score on the fourth day. Solving the inequality:

$$\frac{70 + 73 + 76 + x}{4} \le 72$$
$$219 + x \le 288$$
$$x \le 69$$

Sam must shoot a 69 or less on the fourth day.

129. Let x represent the number. Solving the equation:

$$x^3 = 9x$$
$$x^3 - 9x = 0$$
$$x(x + 3)(x - 3) = 0$$
$$x = -3, 0, 3$$

130. Let x represent the width of the strip. The area equation is:

$$(8 - 2x)(14 - 2x) = 72$$
$$112 - 44x + 4x^2 = 72$$
$$4x^2 - 44x + 40 = 0$$
$$x^2 - 11x + 10 = 0$$
$$(x - 1)(x - 10) = 0$$
$$x = 1, 10$$

Since 10 inch strips cannot be cut off, the width is 1 inch.

131. Let $3x$ and $4x$ represent the amount each person receives. The equation is:

$$3x + 4x = 2450$$
$$7x = 2450$$
$$x = 350$$
$$3x = 1050$$
$$4x = 1400$$

The two people receive $1,050 and $1,400, respectively.

132. Let t represent the time it takes for Sue to complete the task alone. The equation is:

$$\frac{1}{2} + \frac{1}{t} = \frac{1}{6/5}$$
$$\frac{1}{2} + \frac{1}{t} = \frac{5}{6}$$
$$6t\left(\frac{1}{2}\right) + 6t\left(\frac{1}{t}\right) = 6t\left(\frac{5}{6}\right)$$
$$3t + 6 = 5t$$
$$6 = 2t$$
$$t = 3$$

It will take Sue 3 hours to complete the task by herself.

Chapter 6
Systems of Equations

6.1 Systems of Two Linear Equations in Two Variables

1. Graphing the equations:

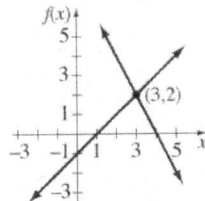

The solution set is $\{(3,2)\}$.

3. Graphing the equations:

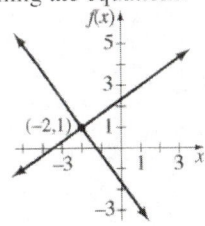

The solution set is $\{(-2,1)\}$.

5. Graphing the equations:

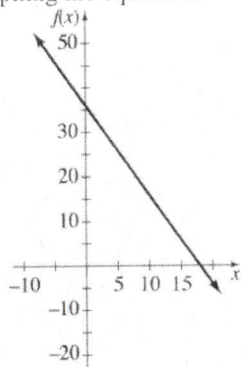

The system is dependent (same line).

7. Graphing the equations:

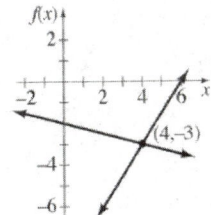

The solution set is $\{(4,-3)\}$.

9. Graphing the equations:

The system is inconsistent (no solution).

211

11.   Substituting into the first equation:
$$x + x + 2 = 16$$
$$2x + 2 = 16$$
$$2x = 14$$
$$x = 7$$
$$y = 9$$

The solution set is $\{(7, 9)\}$.

13.   Substituting into the second equation:
$$4(3y - 25) + 5y = 19$$
$$12y - 100 + 5y = 19$$
$$17y = 119$$
$$y = 7$$
$$x = -4$$

The solution set is $\{(-4, 7)\}$.

15.   Substituting into the second equation:
$$5x - 7\left(\frac{2}{3}x - 1\right) = 9$$
$$5x - \frac{14}{3}x + 7 = 9$$
$$\frac{1}{3}x = 2$$
$$x = 6$$
$$y = 3$$

The solution set is $\{(6, 3)\}$.

17.   Substituting into the second equation:
$$3(4b + 13) + 6b = -33$$
$$12b + 39 + 6b = -33$$
$$18b = -72$$
$$b = -4$$
$$a = -3$$

19.   Substituting into the first equation:
$$2x - 3\left(\frac{2}{3}x - \frac{4}{3}\right) = 4$$
$$2x - 2x + 4 = 4$$
$$4 = 4$$

The system is dependent. The solution set is $\left\{\left(k, \frac{2}{3}k - \frac{4}{3}\right) \middle| k \text{ is a real number}\right\}$.

21.   Substituting into the second equation:
$$t + (t - 2) = 12$$
$$2t - 2 = 12$$
$$2t = 14$$
$$t = 7$$
$$u = 5$$

23.   Solving the second equation for y:
$$3x - 2y = 16$$
$$-2y = -3x + 16$$
$$y = \frac{3}{2}x - 8$$

Substituting into the first equation:
$$4x + 3\left(\frac{3}{2}x - 8\right) = -7$$
$$4x + \frac{9}{2}x - 24 = -7$$
$$\frac{17}{2}x = 17$$
$$x = 2$$
$$y = -5$$

The solution set is $\{(2, -5)\}$.

25.   Substituting into the first equation:
$$5x - (5x + 9) = 4$$
$$5x - 5x - 9 = 4$$
$$-9 = 4$$

The system is inconsistent. The solution set is $\varnothing$.

27. Solving the first equation for x:

$$4x - 5y = 3$$
$$4x = 5y + 3$$
$$x = \frac{5}{4}y + \frac{3}{4}$$

Substituting into the second equation:

$$8\left(\frac{5}{4}y + \frac{3}{4}\right) + 15y = -24$$
$$10y + 6 + 15y = -24$$
$$25y = -30$$
$$y = -\frac{6}{5}$$
$$x = -\frac{3}{4}$$

The solution set is $\left\{\left(-\frac{3}{4}, -\frac{6}{5}\right)\right\}$.

29. Adding the two equations yields:

$$8x = 24$$
$$x = 3$$

Substituting into the first equation:

$$9 + 2y = 1$$
$$2y = -8$$
$$y = -4$$

The solution set is $\left\{(3, -4)\right\}$.

31. Multiply the first equation by -2:

$$-2x + 6y = 44$$
$$2x + 7y = 60$$

Adding yields:

$$13y = 104$$
$$y = 8$$

Substituting into the first equation:

$$x - 3(8) = -22$$
$$x - 24 = -22$$
$$x = 2$$

The solution set is $\left\{(2, 8)\right\}$.

33. Multiply the first equation by -3 and the second equation by 4:

$$-12x + 15y = -63$$
$$12x + 28y = -152$$

Adding yields:

$$43y = -215$$
$$y = -5$$

Substituting into the first equation:

$$4x - 5(-5) = 21$$
$$4x + 25 = 21$$
$$4x = -4$$
$$x = -1$$

The solution set is $\left\{(-1, -5)\right\}$.

35.   Multiply the second equation by -1:
$$5x - 2y = 19$$
$$-5x + 2y = -7$$

Adding yields $0 = 12$, which is false. The system is inconsistent, and the solution set is $\varnothing$.

37.   Multiply the first equation by 2 and the second equation by -5:
$$10a + 12b = 16$$
$$-10a + 75b = -45$$

Adding yields:
$$87b = -29$$
$$b = -\frac{1}{3}$$

Substituting into the first equation:
$$5a + 6\left(-\frac{1}{3}\right) = 8$$
$$5a - 2 = 8$$
$$5a = 10$$
$$a = 2$$

39.   To clear each equation of fractions, multiply the first equation by 12 and the second equation by 6:
$$8s + 3t = -12$$
$$3s - 2t = -42$$

Multiply the first equation by 2 and the second equation by 3:
$$16s + 6t = -24$$
$$9s - 6t = -126$$

Adding yields:
$$25s = -150$$
$$s = -6$$

Substituting into the first equation:
$$8(-6) + 3t = -12$$
$$-48 + 3t = -12$$
$$3t = 36$$
$$t = 12$$

41.   To clear each equation of fractions, multiply the first equation by 60 and the second equation by 12:
$$30x - 24y = -23$$
$$8x + 3y = -3$$

Multiply the second equation by 8:
$$30x - 24y = -23$$
$$64x + 24y = -24$$

Adding yields:
$$94x = -47$$
$$x = -\frac{1}{2}$$

Substituting into the second equation:
$$8\left(-\frac{1}{2}\right) + 3y = -3$$
$$-4 + 3y = -3$$
$$3y = 1$$
$$y = \frac{1}{3}$$

The solution set is $\left\{\left(-\frac{1}{2}, \frac{1}{3}\right)\right\}$.

43. Multiply the first equation by 6 and the second equation by -1:
$$4x + 3y = 1$$
$$-4x - 6y = 1$$

Adding yields:
$$-3y = 2$$
$$y = -\frac{2}{3}$$

Substituting into the second equation:
$$4x + 6\left(-\frac{2}{3}\right) = -1$$
$$4x - 4 = -1$$
$$4x = 3$$
$$x = \frac{3}{4}$$

The solution set is $\left\{\left(\frac{3}{4}, -\frac{2}{3}\right)\right\}$.

45. Multiply the first equation by 3:
$$15x - 3y = -66$$
$$2x + 3y = -2$$

Adding yields:
$$17x = -68$$
$$x = -4$$

Substituting into the second equation:
$$-8 + 3y = -2$$
$$3y = 6$$
$$y = 2$$

The solution set is $\left\{(-4, 2)\right\}$.

47. Substituting into the first equation:
$$-2y + 15 = 3y - 10$$
$$-5y = -25$$
$$y = 5$$
$$x = 5$$

The solution set is $\left\{(5, 5)\right\}$.

49. Multiply the first equation by -2:
$$-6x + 10y = -18$$
$$6x - 10y = -1$$

Adding yields $0 = -19$, which is false. The system is inconsistent, and the solution set is $\varnothing$.

51. To clear each equation of fractions, multiply the first equation by 6 and the second equation by 4:
$$3x - 4y = 132$$
$$2x + y = 0$$

Multiply the second equation by 4:
$$3x - 4y = 132$$
$$8x + 4y = 0$$

Adding yields:
$$11x = 132$$
$$x = 12$$

Substituting into the second equation:
$$24 + y = 0$$
$$y = -24$$

The solution set is $\left\{(12, -24)\right\}$.

53. Substituting into the second equation:
$$9u - 9(2u + 2) = -45$$
$$9u - 18u - 18 = -45$$
$$-9u - 18 = -45$$
$$-9u = -27$$
$$u = 3$$
$$t = 8$$

55. Multiply the first equation by -12 and the second equation by 100:
$$-12x - 12y = -12000$$
$$12x + 14y = 13600$$
Adding yields:
$$2y = 1600$$
$$y = 800$$
Substituting into the first equation:
$$x + 800 = 1000$$
$$x = 200$$
The solution set is $\{(200, 800)\}$.

57. Substituting into the second equation:
$$0.09x + 0.12(2x) = 132$$
$$0.09x + 0.24x = 132$$
$$0.33x = 132$$
$$x = 400$$
$$y = 800$$
The solution set is $\{(400, 800)\}$.

59. Multiply the first equation by -5 and the second equation by 10:
$$-5x - 5y = -52.5$$
$$5x + 8y = 73.5$$
Adding yields:
$$3y = 21$$
$$y = 7$$
Substituting into the first equation:
$$x + 7 = 10.5$$
$$x = 3.5$$
The solution set is $\{(3.5, 7)\}$.

61. Let x and y represent the two numbers. The system of equations is:
$$x + y = 53$$
$$x - y = 19$$
Adding yields:
$$2x = 72$$
$$x = 36$$
$$y = 17$$
The numbers are 17 and 36.

63. Let x and y represent the two angles. The system of equations is:
$$x + y = 90$$
$$y = 4x + 15$$
Substituting into the first equation:
$$x + 4x + 15 = 90$$
$$5x = 75$$
$$x = 15$$
$$y = 75$$
The two angles are $15°$ and $75°$.

65.   Let u represent the units digit and t represent the tens digit. The system of equations is:
$$t = 3u + 1$$
$$t + u = 9$$
Substituting into the second equation:
$$3u + 1 + u = 9$$
$$4u + 1 = 9$$
$$4u = 8$$
$$u = 2$$
$$t = 7$$
The number is 72.

67.   Let s represent the number of sedans rented and c represent the number of convertibles rented.
The system of equations is:
$$45s + 65c = 1680$$
$$s + c = 32$$
Multiplying the second equation by -45:
$$45s + 65c = 1680$$
$$-45s - 45c = -1440$$
Adding yields:
$$20c = 240$$
$$c = 12$$
There were 12 convertibles rented.

69.   Let x represent the number of double rooms rented and y represent the number of single rooms rented. The system of equations is:
$$x + y = 23$$
$$100x + 75y = 2100$$
Multiplying the first equation by -75:
$$-75x - 75y = -1725$$
$$100x + 75y = 2100$$
Adding yields:
$$25x = 375$$
$$x = 15$$
$$y = 8$$
There were 15 double rooms and 8 single rooms rented.

71.   Let s represent the number of student tickets and n the number of non-student tickets. The system of equations is:
$$s + n = 3000$$
$$10s + 15n = 32500$$
Multiplying the first equation by -10:
$$-10s - 10n = -30000$$
$$10s + 15n = 32500$$
Adding yields:
$$5n = 2500$$
$$n = 500$$
$$s = 2500$$
There were 2500 student tickets and 500 non-student tickets sold.

73.   Let x represent the amount invested at 6% and y represent the amount invested at 4%. The system of equations is:
$$x = 3y$$
$$0.06x + 0.04y = 110$$
Substituting into the second equation:
$$0.06(3y) + 0.04y = 110$$
$$0.18y + 0.04y = 110$$
$$0.22y = 110$$
$$y = 500$$
$$x = 1500$$
Melinda invested $500 at 4% and $1500 at 6%.

75. Let x represent the rate he originally paddled and y represent the rate of the current. The system of equations is:
$$4(x-y) = 20$$
$$1(2x+y) = 19$$

Dividing the first equation by 4, the system simplifies to:
$$x - y = 5$$
$$2x + y = 19$$

Adding yields:
$$3x = 24$$
$$x = 8$$
$$y = 3$$

The rate of the current is 3 mph.

77. Let l represent the price for a gallon of latex paint and p the price for a gallon of primer. The system of equations is:
$$4l + 2p = 116$$
$$3l + 1p = 80$$

Multiplying the second equation by -2:
$$4l + 2p = 116$$
$$-6l - 2p = -160$$

Adding yields:
$$-2l = -44$$
$$l = 22$$

The price for a gallon of green latex paint is \$22.

79. Let f represent the number of five-dollar bills and t the number of ten-dollar bills. The system of equations is:
$$f = t + 12$$
$$5f + 10t = 330$$

Substituting into the second equation:
$$5(t + 12) + 10t = 330$$
$$5t + 60 + 10t = 330$$
$$15t = 270$$
$$t = 18$$
$$f = 30$$

The drawer contains 30 five-dollar bills and 18 ten-dollar bills.

85. Adding the two equations yields:
$$\frac{4}{x} = 1$$
$$x = 4$$

Substituting into the first equation:
$$\frac{1}{4} + \frac{2}{y} = \frac{7}{12}$$
$$\frac{2}{y} = \frac{1}{3}$$
$$y = 6$$

The solution set is $\{(4,6)\}$.

87. Multiply the first equation by 3 and the second equation by 2:
$$\frac{9}{x} - \frac{6}{y} = \frac{13}{2}$$
$$\frac{4}{x} + \frac{6}{y} = 0$$

Adding yields:
$$\frac{13}{x} = \frac{13}{2}$$
$$x = 2$$

Substituting into the second equation:

$$1 + \frac{3}{y} = 0$$

$$\frac{3}{y} = -1$$

$$y = -3$$

The solution set is $\left\{ (2, -3) \right\}$.

89. Multiply the first equation by 3 and the second equation by 2:

$$\frac{15}{x} - \frac{6}{y} = 69$$

$$\frac{8}{x} + \frac{6}{y} = 23$$

Adding yields:

$$\frac{23}{x} = 92$$

$$x = \frac{1}{4}$$

Substituting into the second equation:

$$\frac{4}{1/4} + \frac{3}{y} = \frac{23}{2}$$

$$16 + \frac{3}{y} = \frac{23}{2}$$

$$\frac{3}{y} = -\frac{9}{2}$$

$$y = -\frac{2}{3}$$

The solution set is $\left\{ \left(\frac{1}{4}, -\frac{2}{3} \right) \right\}$.

91. **a.** If $\frac{a_1}{a_2} \neq \frac{b_1}{b_2}$, then the lines do not have the same slope, so they will have one intersection point. Thus the system will have exactly one solution.

b. If $\frac{a_1}{a_2} = \frac{b_1}{b_2} \neq \frac{c_1}{c_2}$, then the two lines have the same slope but have different y-intercepts, so they are parallel. Thus the system will have no solutions.

c. If $\frac{a_1}{a_2} = \frac{b_1}{b_2} = \frac{c_1}{c_2}$, then the two lines have the same slope with the same y-intercept, so they are the same line. Thus the system will have infinitely many solutions (dependent).

6.2 Systems of Three Linear Equations in Three Variables

1. Solving the third equation for z yields $z = 3$. Substituting into the second equation:
$$5y - 6 = -16$$
$$5y = -10$$
$$y = -2$$
Substituting into the first equation:
$$2x + 6 + 12 = 10$$
$$2x = -8$$
$$x = -4$$
The solution set is $\{(-4, -2, 3)\}$.

3. Multiply the second equation by 5 and add it to the third equation:
$$15y - 5z = 65$$
$$3y + 5z = 25$$
Adding yields:
$$18y = 90$$
$$y = 5$$
Substituting into the third equation:
$$15 + 5z = 25$$
$$5z = 10$$
$$z = 2$$
Substituting into the first equation:
$$x + 10 - 6 = 2$$
$$x = -2$$
The solution set is $\{(-2, 5, 2)\}$.

5. Multiply the second equation by -2 and add it to the third equation:
$$-2x + 12z = -32$$
$$2x + 5z = -2$$
Adding yields:
$$17z = -34$$
$$z = -2$$
Substituting into the second equation:
$$x + 12 = 16$$
$$x = 4$$
Substituting into the first equation:
$$12 + 2y + 4 = 14$$
$$2y = -2$$
$$y = -1$$
The solution set is $\{(4, -1, -2)\}$.

7. Multiply the second equation by 2 and add it to the first equation:
$$x - 2y + 3z = 7$$
$$4x + 2y + 10z = 34$$
Adding yields $5x + 13z = 41$. Multiply the second equation by 4 and add it to the third equation:
$$8x + 4y + 20z = 68$$
$$3x - 4y - 2z = 1$$
Adding yields $11x + 18z = 69$. So the system becomes:
$$5x + 13z = 41$$
$$11x + 18z = 69$$
Multiply the first equation by 11 and the second equation by -5:
$$55x + 143z = 451$$
$$-55x - 90z = -345$$

Adding yields:
$$53z = 106$$
$$z = 2$$
Substituting:
$$5x + 13(2) = 41$$
$$5x = 15$$
$$x = 3$$
Substituting:
$$2(3) + y + 5(2) = 17$$
$$y = 1$$
The solution set is $\{(3,1,2)\}$.

9.   Adding the first and third equations yields $7x - 5z = -32$. Multiply the third equation by 2 and add it to the second equation:
$$3x - 2y + 4z = 11$$
$$10x + 2y - 12z = -64$$
Adding yields $13x - 8z = -53$. So the system becomes:
$$13x - 8z = -53$$
$$7x - 5z = -32$$
Multiply the first equation by 5 and the second equation by -8:
$$65x - 40z = -265$$
$$-56x + 40z = 256$$
Adding yields:
$$9x = -9$$
$$x = -1$$
Substituting:
$$7(-1) - 5z = -32$$
$$-5z = -25$$
$$z = 5$$
Substituting:
$$2(-1) - y + 5 = 0$$
$$-y + 3 = 0$$
$$y = 3$$
The solution set is $\{(-1,3,5)\}$.

11.   Multiply the third equation by -2 and add it to the first equation:
$$3x + 2y - z = -11$$
$$-10x - 2y + 4z = 34$$
Adding yields $-7x + 3z = 23$. Multiply the third equation by 3 and add it to the second equation:
$$2x - 3y + 4z = 11$$
$$15x + 3y - 6z = -51$$
Adding yields $17x - 2z = -40$. So the system becomes:
$$17x - 2z = -40$$
$$-7x + 3z = 23$$
Multiply the first equation by 3 and the second equation by 2:
$$51x - 6z = -120$$
$$-14x + 6z = 46$$
Adding yields:
$$37x = -74$$
$$x = -2$$

Substituting:
$$-7(-2) + 3z = 23$$
$$3z = 9$$
$$z = 3$$
Substituting:
$$3(-2) + 2y - 3 = -11$$
$$2y = -2$$
$$y = -1$$
The solution set is $\{(-2, -1, 3)\}$.

13.   Multiply the first equation by -2 and add it to the second equation:
$$-4x - 6y + 8z = 20$$
$$4x - 5y + 3z = 2$$
Adding yields $-11y + 11z = 22$, so $y - z = -2$. So the system becomes:
$$y - z = -2$$
$$2y + z = 8$$
Adding yields:
$$3y = 6$$
$$y = 2$$
Substituting:
$$2(2) + z = 8$$
$$z = 4$$
Substituting:
$$4x - 5(2) + 3(4) = 2$$
$$4x = 0$$
$$x = 0$$
The solution set is $\{(0, 2, 4)\}$.

15.   Multiply the first equation by 2 and the second equation by -3:
$$6x + 4y - 4z = 28$$
$$-6x + 15y - 9z = -21$$
Adding yields $19y - 13z = 7$. Multiply the second equation by -2 and add it to the third equation:
$$-4x + 10y - 6z = -14$$
$$4x - 3y + 7z = 5$$
Adding yields $7y + z = -9$. So the system becomes:
$$7y + z = -9$$
$$19y - 13z = 7$$
Multiply the first equation by 13:
$$91y + 13z = -117$$
$$19y - 13z = 7$$
Adding yields:
$$110y = -110$$
$$y = -1$$
Substituting:
$$7(-1) + z = -9$$
$$z = -2$$
Substituting:
$$3x + 2(-1) - 2(-2) = 14$$
$$3x = 12$$
$$x = 4$$
The solution set is $\{(4, -1, -2)\}$.

17.   Multiply the first equation by –2 and add it to the second equation:
$$-4x + 6y - 8z = 24$$
$$4x + 2y - 3z = -13$$
Adding yields $8y - 11z = 11$. Multiply the first equation by –3 and add it to the third equation:
$$-6x + 9y - 12z = 36$$
$$6x - 5y + 7z = -31$$
Adding yields $4y - 5z = 5$. So the system becomes:
$$4y - 5z = 5$$
$$8y - 11z = 11$$
Multiply the first equation by –2:
$$-8y + 10z = -10$$
$$8y - 11z = 11$$
Adding yields:
$$-z = 1$$
$$z = -1$$
Substituting:
$$8y - 11(-1) = 11$$
$$8y = 0$$
$$y = 0$$
Substituting:
$$2x - 3(0) + 4(-1) = -12$$
$$2x = -8$$
$$x = -4$$
The solution set is $\{(-4, 0, -1)\}$.

19.   Multiply the second equation by 6 and add it to the first equation:
$$5x - 3y - 6z = 22$$
$$6x - 6y + 6z = -18$$
Adding yields $11x - 9y = 4$. Multiply the second equation by 5 and add it to the third equation:
$$-3x + 7y - 5z = 23$$
$$5x - 5y + 5z = -15$$
Adding yields $2x + 2y = 8$, so $x + y = 4$. So the system becomes:
$$x + y = 4$$
$$11x - 9y = 4$$
Multiply the first equation by 9:
$$9x + 9y = 36$$
$$11x - 9y = 4$$
Adding yields:
$$20x = 40$$
$$x = 2$$
Substituting:
$$2 + y = 4$$
$$y = 2$$
Substituting:
$$2 - 2 + z = -3$$
$$z = -3$$
The solution set is $\{(2, 2, -3)\}$.

21. Let x, y, and z represent the number of pounds of almonds, pecans, and peanuts used, respectively. The system of equations is:
$$3.5x + 4y + 2z = 2.7(20)$$
$$x + y + z = 20$$
$$z = 3y$$

Substituting the third equation into the other two and simplifying:
$$3.5x + 10y = 54$$
$$x + 4y = 20$$

Multiplying the second equation by -3.5:
$$3.5x + 10y = 54$$
$$-3.5x - 14y = -70$$

Adding yields:
$$-4y = -16$$
$$y = 4, x = 4, z = 12$$

The mixture contains 4 pounds of almonds, 4 pounds of pecans, and 12 pounds of peanuts.

23. Let n, d, and q represent the number of nickels, dimes, and quarters. The system of equations is:
$$0.05n + 0.10d + 0.25q = 7.15$$
$$n + d + q = 42$$
$$n + d = q - 2$$

Substituting into the second equation:
$$q - 2 + q = 42$$
$$2q = 44$$
$$q = 22$$

Substituting into the first and third equations:
$$0.05n + 0.10d + 0.25(22) = 7.15$$
$$n + d = 20$$

Multiplying the second equation by -0.05:
$$0.05n + 0.10d = 1.65$$
$$-0.05n - 0.05d = -1$$

Adding yields:
$$0.05d = 0.65$$
$$d = 13, n = 7$$

The box contains 7 nickels, 13 dimes, and 22 quarters.

25. Let x represent the smallest angle and $2x$ represent the largest angle. If m represents the middle angle, the system of equations is:
$$x + m + 2x = 180$$
$$x + 2x = 2m$$

Substituting into the first equation:
$$m + 2m = 180$$
$$3m = 180$$
$$m = 60, x = 40$$

The angles are $40°$, $60°$, and $80°$.

27. Let x, y, and z represent the amounts invested at 4%, 5%, and 6%, respectively. The system of equations is:
$$x + y + z = 3000$$
$$0.04x + 0.05y + 0.06z = 160$$
$$x + y = z$$

Substituting into the first equation:
$$2z = 3000$$
$$z = 1500$$

The system of equations becomes:
$$x + y = 1500$$
$$0.04x + 0.05y + 0.06(1500) = 160$$

Multiply the first equation by –0.04:

$$-0.04x - 0.04y = -60$$
$$0.04x + 0.05y = 70$$

Adding yields:

$$0.01y = 10$$
$$y = 1000$$
$$x = 500$$

Thus $500 is invested at 4%, $1000 is invested at 5%, and $1500 is invested at 6%.

29.   Let a, b, and c represent the number of each style. The system of equations is:

$$0.1a + 0.2b + 0.1c = 35$$
$$0.4a + 0.4b + 0.3c = 95$$
$$0.2a + 0.1b + 0.3c = 62.5$$

These equations simplify to:

$$a + 2b + c = 350$$
$$4a + 4b + 3c = 950$$
$$2a + b + 3c = 625$$

Multiply the first equation by –3 and add it to the second equation:

$$-3a - 6b - 3c = -1050$$
$$4a + 4b + 3c = 950$$

Adding yields $a - 2b = -100$. Multiply the third equation by –1 and add it to the second equation:

$$4a + 4b + 3c = 950$$
$$-2a - b - 3c = -625$$

Adding yields $2a + 3b = 325$. So the system becomes:

$$a - 2b = -100$$
$$2a + 3b = 325$$

Multiply the first equation by –2:

$$-2a + 4b = 200$$
$$2a + 3b = 325$$

Adding yields:

$$7b = 525$$
$$b = 75$$

Substituting:

$$2a + 3(75) = 325$$
$$2a = 100$$
$$a = 50$$

Substituting:

$$50 + 2(75) + c = 350$$
$$c = 150$$

The company should make 50 Style A, 75 Style B, and 150 Style C birdhouses.

6.3 Matrix Approach to Solving Linear Systems

1. Yes, this is in reduced echelon form.

3. Yes, this is in reduced echelon form.

5. No, since the second row has a 0 rather than 1.

7. No, since the second row has a 2 rather than 0.

9. Yes, this is in reduced echelon form.

11. Form the augmented matrix: $\begin{bmatrix} 1 & -3 & | & 14 \\ 3 & 2 & | & -13 \end{bmatrix}$

    Add -3 times row 1 to row 2: $\begin{bmatrix} 1 & -3 & | & 14 \\ 0 & 11 & | & -55 \end{bmatrix}$

    Divide row 2 by 11: $\begin{bmatrix} 1 & -3 & | & 14 \\ 0 & 1 & | & -5 \end{bmatrix}$

    Add 3 times row 2 to row 1: $\begin{bmatrix} 1 & 0 & | & -1 \\ 0 & 1 & | & -5 \end{bmatrix}$

    The solution set is $\{(-1,-5)\}$.

13. Form the augmented matrix: $\begin{bmatrix} 3 & -4 & | & 33 \\ 1 & 7 & | & -39 \end{bmatrix}$

    Switch row 1 and row 2: $\begin{bmatrix} 1 & 7 & | & -39 \\ 3 & -4 & | & 33 \end{bmatrix}$

    Add -3 times row 1 to row 2: $\begin{bmatrix} 1 & 7 & | & -39 \\ 0 & -25 & | & 150 \end{bmatrix}$

    Divide row 2 by -25: $\begin{bmatrix} 1 & 7 & | & -39 \\ 0 & 1 & | & -6 \end{bmatrix}$

    Add -7 times row 2 to row 1: $\begin{bmatrix} 1 & 0 & | & 3 \\ 0 & 1 & | & -6 \end{bmatrix}$

    The solution set is $\{(3,-6)\}$.

15. Form the augmented matrix: $\begin{bmatrix} 1 & -6 & | & -2 \\ 2 & -12 & | & 5 \end{bmatrix}$

    Add -2 times row 1 to row 2: $\begin{bmatrix} 1 & -6 & | & -2 \\ 0 & 0 & | & 9 \end{bmatrix}$

    Since $0 \neq 9$, there is no solution. The solution set is $\varnothing$.

17. Form the augmented matrix: $\begin{bmatrix} 3 & -5 & | & 39 \\ 2 & 7 & | & -67 \end{bmatrix}$

    Subtract row 2 from row 1: $\begin{bmatrix} 1 & -12 & | & 106 \\ 2 & 7 & | & -67 \end{bmatrix}$

    Add -2 times row 1 to row 2: $\begin{bmatrix} 1 & -12 & | & 106 \\ 0 & 31 & | & -279 \end{bmatrix}$

    Divide row 2 by 31: $\begin{bmatrix} 1 & -12 & | & 106 \\ 0 & 1 & | & -9 \end{bmatrix}$

    Add 12 times row 2 to row 1: $\begin{bmatrix} 1 & 0 & | & -2 \\ 0 & 1 & | & -9 \end{bmatrix}$

    The solution set is $\{(-2,-9)\}$.

19. Form the augmented matrix: $\begin{bmatrix} 1 & -2 & -3 & | & -6 \\ 3 & -5 & -1 & | & 4 \\ 2 & 1 & 2 & | & 2 \end{bmatrix}$

Add –3 times row 1 to row 2, and –2 times row 1 to row 3: $\begin{bmatrix} 1 & -2 & -3 & | & -6 \\ 0 & 1 & 8 & | & 22 \\ 0 & 5 & 8 & | & 14 \end{bmatrix}$

Add –5 times row 2 to row 3: $\begin{bmatrix} 1 & -2 & -3 & | & -6 \\ 0 & 1 & 8 & | & 22 \\ 0 & 0 & -32 & | & -96 \end{bmatrix}$

Divide row 3 by –32: $\begin{bmatrix} 1 & -2 & -3 & | & -6 \\ 0 & 1 & 8 & | & 22 \\ 0 & 0 & 1 & | & 3 \end{bmatrix}$

Add –8 times row 3 to row 2, and 3 times row 3 to row 1: $\begin{bmatrix} 1 & -2 & 0 & | & 3 \\ 0 & 1 & 0 & | & -2 \\ 0 & 0 & 1 & | & 3 \end{bmatrix}$

Add 2 times row 2 to row 1: $\begin{bmatrix} 1 & 0 & 0 & | & -1 \\ 0 & 1 & 0 & | & -2 \\ 0 & 0 & 1 & | & 3 \end{bmatrix}$

The solution set is $\{(-1,-2,3)\}$.

21. Form the augmented matrix: $\begin{bmatrix} -2 & -5 & 3 & | & 11 \\ 1 & 3 & -3 & | & -12 \\ 3 & -2 & 5 & | & 31 \end{bmatrix}$

Switch row 1 and row 2: $\begin{bmatrix} 1 & 3 & -3 & | & -12 \\ -2 & -5 & 3 & | & 11 \\ 3 & -2 & 5 & | & 31 \end{bmatrix}$

Add 2 times row 1 to row 2, and –3 times row 1 to row 3: $\begin{bmatrix} 1 & 3 & -3 & | & -12 \\ 0 & 1 & -3 & | & -13 \\ 0 & -11 & 14 & | & 67 \end{bmatrix}$

Add 11 times row 2 to row 3: $\begin{bmatrix} 1 & 3 & -3 & | & -12 \\ 0 & 1 & -3 & | & -13 \\ 0 & 0 & -19 & | & -76 \end{bmatrix}$

Divide row 3 by –19: $\begin{bmatrix} 1 & 3 & -3 & | & -12 \\ 0 & 1 & -3 & | & -13 \\ 0 & 0 & 1 & | & 4 \end{bmatrix}$

Add 3 times row 3 to row 2 and row 1: $\begin{bmatrix} 1 & 3 & 0 & | & 0 \\ 0 & 1 & 0 & | & -1 \\ 0 & 0 & 1 & | & 4 \end{bmatrix}$

Add –3 times row 2 to row 1: $\begin{bmatrix} 1 & 0 & 0 & | & 3 \\ 0 & 1 & 0 & | & -1 \\ 0 & 0 & 1 & | & 4 \end{bmatrix}$

The solution set is $\{(3,-1,4)\}$.

23.　Form the augmented matrix: $\begin{bmatrix} 1 & -3 & -1 & | & 2 \\ 3 & 1 & -4 & | & -18 \\ -2 & 5 & 3 & | & 2 \end{bmatrix}$

Add -3 times row 1 to row 2, and 2 times row 1 to row 3: $\begin{bmatrix} 1 & -3 & -1 & | & 2 \\ 0 & 10 & -1 & | & -24 \\ 0 & -1 & 1 & | & 6 \end{bmatrix}$

Multiply row 3 by -1, then switch row 2 and row 3: $\begin{bmatrix} 1 & -3 & -1 & | & 2 \\ 0 & 1 & -1 & | & -6 \\ 0 & 10 & -1 & | & -24 \end{bmatrix}$

Add -10 times row 2 to row 3: $\begin{bmatrix} 1 & -3 & -1 & | & 2 \\ 0 & 1 & -1 & | & -6 \\ 0 & 0 & 9 & | & 36 \end{bmatrix}$

Divide row 3 by 9: $\begin{bmatrix} 1 & -3 & -1 & | & 2 \\ 0 & 1 & -1 & | & -6 \\ 0 & 0 & 1 & | & 4 \end{bmatrix}$

Add row 3 to row 1 and row 2: $\begin{bmatrix} 1 & -3 & 0 & | & 6 \\ 0 & 1 & 0 & | & -2 \\ 0 & 0 & 1 & | & 4 \end{bmatrix}$

Add 3 times row 2 to row 1: $\begin{bmatrix} 1 & 0 & 0 & | & 0 \\ 0 & 1 & 0 & | & -2 \\ 0 & 0 & 1 & | & 4 \end{bmatrix}$

The solution set is $\left\{ (0, -2, 4) \right\}$.

25.　Form the augmented matrix: $\begin{bmatrix} 1 & -1 & 2 & | & 1 \\ -3 & 4 & -1 & | & 4 \\ -1 & 2 & 3 & | & 6 \end{bmatrix}$

Add 3 times row 1 to row 2, and row 1 to row 3: $\begin{bmatrix} 1 & -1 & 2 & | & 1 \\ 0 & 1 & 5 & | & 7 \\ 0 & 1 & 5 & | & 7 \end{bmatrix}$

Add -1 times row 2 to row 3, and row 2 to row 1: $\begin{bmatrix} 1 & 0 & 7 & | & 8 \\ 0 & 1 & 5 & | & 7 \\ 0 & 0 & 0 & | & 0 \end{bmatrix}$

Let $z = k$. The first two rows correspond to:

$$x + 7k = 8 \qquad\qquad y + 5k = 7$$
$$x = -7k + 8 \qquad\qquad y = -5k + 7$$

The solution set is $\left\{ (-7k + 8, -5k + 7, k) \mid k \text{ is any real number} \right\}$.

27. Form the augmented matrix:
$$\begin{bmatrix} -2 & 1 & 5 & | & -5 \\ 3 & 8 & -1 & | & -34 \\ 1 & 2 & 1 & | & -12 \end{bmatrix}$$

Switch row 1 and row 3:
$$\begin{bmatrix} 1 & 2 & 1 & | & -12 \\ 3 & 8 & -1 & | & -34 \\ -2 & 1 & 5 & | & -5 \end{bmatrix}$$

Add -3 times row 1 to row 2, and 2 times row 1 to row 3:
$$\begin{bmatrix} 1 & 2 & 1 & | & -12 \\ 0 & 2 & -4 & | & 2 \\ 0 & 5 & 7 & | & -29 \end{bmatrix}$$

Divide row 2 by 2:
$$\begin{bmatrix} 1 & 2 & 1 & | & -12 \\ 0 & 1 & -2 & | & 1 \\ 0 & 5 & 7 & | & -29 \end{bmatrix}$$

Add -5 times row 2 to row 3:
$$\begin{bmatrix} 1 & 2 & 1 & | & -12 \\ 0 & 1 & -2 & | & 1 \\ 0 & 0 & 17 & | & -34 \end{bmatrix}$$

Divide row 3 by 17:
$$\begin{bmatrix} 1 & 2 & 1 & | & -12 \\ 0 & 1 & -2 & | & 1 \\ 0 & 0 & 1 & | & -2 \end{bmatrix}$$

Add 2 times row 3 to row 2, and -1 times row 3 to row 1:
$$\begin{bmatrix} 1 & 2 & 0 & | & -10 \\ 0 & 1 & 0 & | & -3 \\ 0 & 0 & 1 & | & -2 \end{bmatrix}$$

Add -2 times row 2 to row 1:
$$\begin{bmatrix} 1 & 0 & 0 & | & -4 \\ 0 & 1 & 0 & | & -3 \\ 0 & 0 & 1 & | & -2 \end{bmatrix}$$

The solution set is $\{(-4, -3, -2)\}$.

29. Form the augmented matrix:
$$\begin{bmatrix} 2 & 3 & -1 & | & 7 \\ 3 & 4 & 5 & | & -2 \\ 5 & 1 & 3 & | & 13 \end{bmatrix}$$

Subtract row 1 from row 2, and use the resulting line as row 1:
$$\begin{bmatrix} 1 & 1 & 6 & | & -9 \\ 3 & 4 & 5 & | & -2 \\ 5 & 1 & 3 & | & 13 \end{bmatrix}$$

Add -3 times row 1 to row 2, and -5 times row 1 to row 3:
$$\begin{bmatrix} 1 & 1 & 6 & | & -9 \\ 0 & 1 & -13 & | & 25 \\ 0 & -4 & -27 & | & 58 \end{bmatrix}$$

Add 4 times row 2 to row 3:
$$\begin{bmatrix} 1 & 1 & 6 & | & -9 \\ 0 & 1 & -13 & | & 25 \\ 0 & 0 & -79 & | & 158 \end{bmatrix}$$

Divide row 3 by -79:
$$\begin{bmatrix} 1 & 1 & 6 & | & -9 \\ 0 & 1 & -13 & | & 25 \\ 0 & 0 & 1 & | & -2 \end{bmatrix}$$

Add 13 times row 3 to row 2, and –6 times row 3 to row 1:
$\begin{bmatrix} 1 & 1 & 0 & | & 3 \\ 0 & 1 & 0 & | & -1 \\ 0 & 0 & 1 & | & -2 \end{bmatrix}$

Add –1 times row 2 to row 1:
$\begin{bmatrix} 1 & 0 & 0 & | & 4 \\ 0 & 1 & 0 & | & -1 \\ 0 & 0 & 1 & | & -2 \end{bmatrix}$

The solution set is $\{(-4,-1,-2)\}$.

31. Form the augmented matrix:
$\begin{bmatrix} 1 & -3 & -2 & 1 & | & -3 \\ -2 & 7 & 1 & -2 & | & -1 \\ 3 & -7 & -3 & 3 & | & -5 \\ 5 & 1 & 4 & -2 & | & 18 \end{bmatrix}$

Add 2 times row 1 to row 2, –3 times row 1 to row 3, and –5 times row 1 to row 4:
$\begin{bmatrix} 1 & -3 & -2 & 1 & | & -3 \\ 0 & 1 & -3 & 0 & | & -7 \\ 0 & 2 & 3 & 0 & | & 4 \\ 0 & 16 & 14 & -7 & | & 33 \end{bmatrix}$

Add –2 times row 2 to row 3, and –16 times row 2 to row 4:
$\begin{bmatrix} 1 & -3 & -2 & 1 & | & -3 \\ 0 & 1 & -3 & 0 & | & -7 \\ 0 & 0 & 9 & 0 & | & 18 \\ 0 & 0 & 62 & -7 & | & 145 \end{bmatrix}$

Divide row 3 by 9:
$\begin{bmatrix} 1 & -3 & -2 & 1 & | & -3 \\ 0 & 1 & -3 & 0 & | & -7 \\ 0 & 0 & 1 & 0 & | & 2 \\ 0 & 0 & 62 & -7 & | & 145 \end{bmatrix}$

Add –62 times row 3 to row 4:
$\begin{bmatrix} 1 & -3 & -2 & 1 & | & -3 \\ 0 & 1 & -3 & 0 & | & -7 \\ 0 & 0 & 1 & 0 & | & 2 \\ 0 & 0 & 0 & -7 & | & 21 \end{bmatrix}$

Divide row 4 by –7:
$\begin{bmatrix} 1 & -3 & -2 & 1 & | & -3 \\ 0 & 1 & -3 & 0 & | & -7 \\ 0 & 0 & 1 & 0 & | & 2 \\ 0 & 0 & 0 & 1 & | & -3 \end{bmatrix}$

Add –1 times row 4 to row 1:
$\begin{bmatrix} 1 & -3 & -2 & 0 & | & 0 \\ 0 & 1 & -3 & 0 & | & -7 \\ 0 & 0 & 1 & 0 & | & 2 \\ 0 & 0 & 0 & 1 & | & -3 \end{bmatrix}$

Add 3 times row 3 to row 2, and 2 times row 3 to row 1:
$\begin{bmatrix} 1 & -3 & 0 & 0 & | & 4 \\ 0 & 1 & 0 & 0 & | & -1 \\ 0 & 0 & 1 & 0 & | & 2 \\ 0 & 0 & 0 & 1 & | & -3 \end{bmatrix}$

Add 3 times row 2 to row 1:
$$\begin{bmatrix} 1 & 0 & 0 & 0 & | & 1 \\ 0 & 1 & 0 & 0 & | & -1 \\ 0 & 0 & 1 & 0 & | & 2 \\ 0 & 0 & 0 & 1 & | & -3 \end{bmatrix}$$

The solution set is $\{(1,-1,2,-3)\}$.

33. Form the augmented matrix:
$$\begin{bmatrix} 1 & 3 & -1 & 2 & | & -2 \\ 2 & 7 & 2 & -1 & | & 19 \\ -3 & -8 & 3 & 1 & | & -7 \\ 4 & 11 & -2 & -3 & | & 19 \end{bmatrix}$$

Add –2 times row 1 to row 2, 3 times row 1 to row 3, and –4 times row 1 to row 4:
$$\begin{bmatrix} 1 & 3 & -1 & 2 & | & -2 \\ 0 & 1 & 4 & -5 & | & 23 \\ 0 & 1 & 0 & 7 & | & -13 \\ 0 & -1 & 2 & -11 & | & 27 \end{bmatrix}$$

Add –1 times row 2 to row 3, and row 2 to row 4:
$$\begin{bmatrix} 1 & 3 & -1 & 2 & | & -2 \\ 0 & 1 & 4 & -5 & | & 23 \\ 0 & 0 & -4 & 12 & | & -36 \\ 0 & 0 & 6 & -16 & | & 50 \end{bmatrix}$$

Divide row 3 by –4:
$$\begin{bmatrix} 1 & 3 & -1 & 2 & | & -2 \\ 0 & 1 & 4 & -5 & | & 23 \\ 0 & 0 & 1 & -3 & | & 9 \\ 0 & 0 & 6 & -16 & | & 50 \end{bmatrix}$$

Add –6 times row 3 to row 4:
$$\begin{bmatrix} 1 & 3 & -1 & 2 & | & -2 \\ 0 & 1 & 4 & -5 & | & 23 \\ 0 & 0 & 1 & -3 & | & 9 \\ 0 & 0 & 0 & 2 & | & -4 \end{bmatrix}$$

Divide row 4 by 2:
$$\begin{bmatrix} 1 & 3 & -1 & 2 & | & -2 \\ 0 & 1 & 4 & -5 & | & 23 \\ 0 & 0 & 1 & -3 & | & 9 \\ 0 & 0 & 0 & 1 & | & -2 \end{bmatrix}$$

Add 3 times row 4 to row 3, 5 times row 4 to row 2, and –2 times row 4 to row 1:
$$\begin{bmatrix} 1 & 3 & -1 & 0 & | & 2 \\ 0 & 1 & 4 & 0 & | & 13 \\ 0 & 0 & 1 & 0 & | & 3 \\ 0 & 0 & 0 & 1 & | & -2 \end{bmatrix}$$

Add –4 times row 3 to row 2, and row 3 to row 1:
$$\begin{bmatrix} 1 & 3 & 0 & 0 & | & 5 \\ 0 & 1 & 0 & 0 & | & 1 \\ 0 & 0 & 1 & 0 & | & 3 \\ 0 & 0 & 0 & 1 & | & -2 \end{bmatrix}$$

Add –3 times row 2 to row 1:
$$\begin{bmatrix} 1 & 0 & 0 & 0 & | & 2 \\ 0 & 1 & 0 & 0 & | & 1 \\ 0 & 0 & 1 & 0 & | & 3 \\ 0 & 0 & 0 & 1 & | & -2 \end{bmatrix}$$

The solution set is $\{(2,1,3,-2)\}$.

35.   The solution set is $\left\{(-2,4,-3,0)\right\}$.

37.   Since $0 \neq 1$, there is no solution. The solution set is $\varnothing$.

39.   Let $x_4 = k$. From the first and third rows:

$$x_1 + 3k = 5 \qquad\qquad\qquad x_3 + 4k = 2$$
$$x_1 = -3k + 5 \qquad\qquad\qquad x_3 = -4k + 2$$

The solution set is $\left\{(-3k+5,-1,-4k+2,k)\,\middle|\, k \text{ is any real number}\right\}$.

41.   Let $x_2 = k$. From the first row:

$$x_1 + 3k = 9$$
$$x_1 = -3k + 9$$

The solution set is $\left\{(-3k+9,k,2,-3)\,\middle|\, k \text{ is any real number}\right\}$.

45.   Form the augmented matrix: $\begin{bmatrix} 1 & -2 & 3 & | & 4 \\ 3 & -5 & -1 & | & 7 \end{bmatrix}$

Add -3 times row 1 to row 2: $\begin{bmatrix} 1 & -2 & 3 & | & 4 \\ 0 & 1 & -10 & | & -5 \end{bmatrix}$

Add 2 times row 2 to row 1: $\begin{bmatrix} 1 & 0 & -17 & | & -6 \\ 0 & 1 & -10 & | & -5 \end{bmatrix}$

Let $z = k$. From the two rows:

$$x - 17k = -6 \qquad\qquad\qquad y - 10k = -5$$
$$x = 17k - 6 \qquad\qquad\qquad y = 10k - 5$$

The solution set is $\left\{(17k-6,10k-5,k)\,\middle|\, k \text{ is any real number}\right\}$.

47.   Form the augmented matrix: $\begin{bmatrix} 2 & -4 & 3 & | & 8 \\ 3 & 5 & -1 & | & 7 \end{bmatrix}$

Subtract row 1 from row 2, and use the result in row 1: $\begin{bmatrix} 1 & 9 & -4 & | & -1 \\ 3 & 5 & -1 & | & 7 \end{bmatrix}$

Add -3 times row 1 to row 2: $\begin{bmatrix} 1 & 9 & -4 & | & -1 \\ 0 & -22 & 11 & | & 10 \end{bmatrix}$

Divide row 2 by -22: $\begin{bmatrix} 1 & 9 & -4 & | & -1 \\ 0 & 1 & -\dfrac{1}{2} & | & -\dfrac{5}{11} \end{bmatrix}$

Add -9 times row 2 to row 1: $\begin{bmatrix} 1 & 0 & \dfrac{1}{2} & | & \dfrac{34}{11} \\ 0 & 1 & -\dfrac{1}{2} & | & -\dfrac{5}{11} \end{bmatrix}$

Let $z = k$. From the two rows:

$$x + \frac{1}{2}k = \frac{34}{11} \qquad\qquad\qquad y - \frac{1}{2}k = -\frac{5}{11}$$
$$x = -\frac{1}{2}k + \frac{34}{11} \qquad\qquad\qquad x = \frac{1}{2}k - \frac{5}{11}$$

The solution set is $\left\{\left(-\frac{1}{2}k+\frac{34}{11},\frac{1}{2}k-\frac{5}{11},k\right)\,\middle|\, k \text{ is any real number}\right\}$.

49. Form the augmented matrix: $\begin{bmatrix} 1 & -2 & 4 & \vdots & 9 \\ 2 & -4 & 8 & \vdots & 3 \end{bmatrix}$

Add –2 times row 1 to row 2: $\begin{bmatrix} 1 & -2 & 4 & \vdots & 9 \\ 0 & 0 & 0 & \vdots & -15 \end{bmatrix}$

Since $0 \neq -15$, there is no solution. The solution set is $\varnothing$.

6.4 Determinants

1. Evaluating: $\begin{vmatrix} 4 & 3 \\ 2 & 7 \end{vmatrix} = (4)(7) - (2)(3) = 28 - 6 = 22$

3. Evaluating: $\begin{vmatrix} -3 & 2 \\ 7 & 5 \end{vmatrix} = (-3)(5) - (7)(2) = -15 - 14 = -29$

5. Evaluating: $\begin{vmatrix} 2 & -3 \\ 8 & -2 \end{vmatrix} = (2)(-2) - (8)(-3) = -4 + 24 = 20$

7. Evaluating: $\begin{vmatrix} -2 & -3 \\ -1 & -4 \end{vmatrix} = (-2)(-4) - (-1)(-3) = 8 - 3 = 5$

9. Evaluating: $\begin{vmatrix} \frac{1}{2} & \frac{1}{3} \\ -3 & -6 \end{vmatrix} = \left(\frac{1}{2}\right)(-6) - \left(\frac{1}{3}\right)(-3) = -3 + 1 = -2$

11. Evaluating: $\begin{vmatrix} \frac{1}{2} & \frac{2}{3} \\ \frac{3}{4} & -\frac{1}{3} \end{vmatrix} = \left(\frac{1}{2}\right)\left(-\frac{1}{3}\right) - \left(\frac{3}{4}\right)\left(\frac{2}{3}\right) = -\frac{1}{6} - \frac{1}{2} = -\frac{2}{3}$

13. Expanding along the first row:

$\begin{vmatrix} 1 & 2 & -1 \\ 3 & 1 & 2 \\ 2 & 4 & 3 \end{vmatrix} = 1\begin{vmatrix} 1 & 2 \\ 4 & 3 \end{vmatrix} - 2\begin{vmatrix} 3 & 2 \\ 2 & 3 \end{vmatrix} - 1\begin{vmatrix} 3 & 1 \\ 2 & 4 \end{vmatrix} = 1(3 - 8) - 2(9 - 4) - 1(12 - 2) = -5 - 10 - 10 = -25$

15. Expanding along the first row:

$\begin{vmatrix} 1 & -4 & 1 \\ 2 & 5 & -1 \\ 3 & 3 & 4 \end{vmatrix} = 1\begin{vmatrix} 5 & -1 \\ 3 & 4 \end{vmatrix} + 4\begin{vmatrix} 2 & -1 \\ 3 & 4 \end{vmatrix} + 1\begin{vmatrix} 2 & 5 \\ 3 & 3 \end{vmatrix} = 1(20 + 3) + 4(8 + 3) + 1(6 - 15) = 23 + 44 - 9 = 58$

17. Factoring 3 out of the first row, then expanding along the first row:

$\begin{vmatrix} 6 & 12 & 3 \\ -1 & 5 & 1 \\ -3 & 6 & 2 \end{vmatrix} = 3\begin{vmatrix} 2 & 4 & 1 \\ -1 & 5 & 1 \\ -3 & 6 & 2 \end{vmatrix}$

$= 3\left(2\begin{vmatrix} 5 & 1 \\ 6 & 2 \end{vmatrix} - 4\begin{vmatrix} -1 & 1 \\ -3 & 2 \end{vmatrix} + 1\begin{vmatrix} -1 & 5 \\ -3 & 6 \end{vmatrix}\right)$

$= 3\left(2(10 - 6) - 4(-2 + 3) + 1(-6 + 15)\right)$

$= 3(8 - 4 + 9)$

$= 39$

19. Expanding along the first row:

$\begin{vmatrix} 2 & -1 & 3 \\ 0 & 3 & 1 \\ 1 & -2 & -1 \end{vmatrix} = 2\begin{vmatrix} 3 & 1 \\ -2 & -1 \end{vmatrix} + 1\begin{vmatrix} 0 & 1 \\ 1 & -1 \end{vmatrix} + 3\begin{vmatrix} 0 & 3 \\ 1 & -2 \end{vmatrix} = 2(-3 + 2) + 1(0 - 1) + 3(0 - 3) = -2 - 1 - 9 = -12$

21. Expanding along the first row:

$$\begin{vmatrix} -3 & -2 & 1 \\ 5 & 0 & 6 \\ 2 & 1 & -4 \end{vmatrix} = -3\begin{vmatrix} 0 & 6 \\ 1 & -4 \end{vmatrix} + 2\begin{vmatrix} 5 & 6 \\ 2 & -4 \end{vmatrix} + 1\begin{vmatrix} 5 & 0 \\ 2 & 1 \end{vmatrix} = -3(0-6) + 2(-20-12) + 1(5-0) = 18 - 64 + 5 = -41$$

23. Expanding along the third row: $\begin{vmatrix} 3 & -4 & -2 \\ 5 & -2 & 1 \\ 1 & 0 & 0 \end{vmatrix} = 1\begin{vmatrix} -4 & -2 \\ -2 & 1 \end{vmatrix} = 1(-4-4) = -8$

25. Expanding along the third column: $\begin{vmatrix} 24 & -1 & 4 \\ 40 & 2 & 0 \\ -16 & 6 & 0 \end{vmatrix} = 4\begin{vmatrix} 40 & 2 \\ -16 & 6 \end{vmatrix} = 4(240+32) = 1088$

27. First add -2 times the first row to the second row: $\begin{vmatrix} 2 & 3 & -4 \\ 0 & 0 & 7 \\ -6 & 1 & -2 \end{vmatrix}$

Expanding along the second row: $\begin{vmatrix} 2 & 3 & -4 \\ 0 & 0 & 7 \\ -6 & 1 & -2 \end{vmatrix} = -7\begin{vmatrix} 2 & 3 \\ -6 & 1 \end{vmatrix} = -7(2+18) = -140$

29. Add -3 times row 4 to row 1: $\begin{vmatrix} 4 & -5 & 0 & -13 \\ 2 & -1 & 0 & 4 \\ -3 & 4 & 0 & -2 \\ -1 & 1 & 1 & 5 \end{vmatrix}$

Expanding along column 3: $\begin{vmatrix} 4 & -5 & 0 & -13 \\ 2 & -1 & 0 & 4 \\ -3 & 4 & 0 & -2 \\ -1 & 1 & 1 & 5 \end{vmatrix} = -1\begin{vmatrix} 4 & -5 & -13 \\ 2 & -1 & 4 \\ -3 & 4 & -2 \end{vmatrix}$

Adding 2 times column 2 to column 1 and 4 times column 2 to column 3: $-1\begin{vmatrix} -6 & -5 & -33 \\ 0 & -1 & 0 \\ 5 & 4 & 14 \end{vmatrix}$

Expanding along row 2: $-1\begin{vmatrix} -6 & -5 & -33 \\ 0 & -1 & 0 \\ 5 & 4 & 14 \end{vmatrix} = 1\begin{vmatrix} -6 & -33 \\ 5 & 14 \end{vmatrix} = -84 + 165 = 81$

31. Adding -1 times column 1 to column 4, and -2 times column 1 to column 3:

$$\begin{vmatrix} 3 & -1 & -4 & 0 \\ 1 & 0 & 0 & 0 \\ 2 & 3 & -4 & -1 \\ 3 & 2 & -6 & -10 \end{vmatrix} = -1\begin{vmatrix} -1 & -4 & 0 \\ 3 & -4 & -1 \\ 2 & -6 & -10 \end{vmatrix} = -2\begin{vmatrix} -1 & -4 & 0 \\ 3 & -4 & -1 \\ 1 & -3 & -5 \end{vmatrix}$$

Adding 3 times row 1 to row 2, and row 1 to row 3:

$$-2\begin{vmatrix} -1 & -4 & 0 \\ 3 & -4 & -1 \\ 1 & -3 & -5 \end{vmatrix} = -2\begin{vmatrix} -1 & -4 & 0 \\ 0 & -16 & -1 \\ 0 & -7 & -5 \end{vmatrix} = 2\begin{vmatrix} -16 & -1 \\ -7 & -5 \end{vmatrix} = 2(80-7) = 146$$

33. This is property 6.3.

35. This is property 6.2.

37. This is property 6.4.

39. This is property 6.3.

41. This is property 6.5.

6.5 Cramer's Rule

1. Computing the determinants:
$$D = \begin{vmatrix} 2 & -1 \\ 3 & 2 \end{vmatrix} = 4 + 3 = 7$$

$$D_x = \begin{vmatrix} -2 & -1 \\ 11 & 2 \end{vmatrix} = -4 + 11 = 7$$

$$D_y = \begin{vmatrix} 2 & -2 \\ 3 & 11 \end{vmatrix} = 22 + 6 = 28$$

Using Cramer's rule:
$$x = \frac{D_x}{D} = \frac{7}{7} = 1, y = \frac{D_y}{D} = \frac{28}{7} = 4$$

The solution set is $\left\{ (1,4) \right\}$.

3. Computing the determinants:
$$D = \begin{vmatrix} 5 & 2 \\ 3 & -4 \end{vmatrix} = -20 - 6 = -26$$

$$D_x = \begin{vmatrix} 5 & 2 \\ 29 & -4 \end{vmatrix} = -20 - 58 = -78$$

$$D_y = \begin{vmatrix} 5 & 5 \\ 3 & 29 \end{vmatrix} = 145 - 15 = 130$$

Using Cramer's rule:
$$x = \frac{D_x}{D} = \frac{-78}{-26} = 3, y = \frac{D_y}{D} = \frac{130}{-26} = -5$$

The solution set is $\left\{ (3,-5) \right\}$.

5. Computing the determinants:
$$D = \begin{vmatrix} 5 & -4 \\ -1 & 2 \end{vmatrix} = 10 - 4 = 6$$

$$D_x = \begin{vmatrix} 14 & -4 \\ -4 & 2 \end{vmatrix} = 28 - 16 = 12$$

$$D_y = \begin{vmatrix} 5 & 14 \\ -1 & -4 \end{vmatrix} = -20 + 14 = -6$$

Using Cramer's rule: $x = \frac{D_x}{D} = \frac{12}{6} = 2, y = \frac{D_y}{D} = \frac{-6}{6} = -1$. The solution set is $\left\{ (2,-1) \right\}$.

7. First write the system as:
$$-2x + y = -4$$
$$6x - 3y = 1$$

Computing the determinants:
$$D = \begin{vmatrix} -2 & 1 \\ 6 & -3 \end{vmatrix} = 6 - 6 = 0$$

$$D_x = \begin{vmatrix} -4 & 1 \\ 1 & -3 \end{vmatrix} = 12 - 1 = 11$$

Since $D = 0$ and $D_x \neq 0$, the system has no solution. The solution set is $\varnothing$.

9. Computing the determinants:
$$D = \begin{vmatrix} -4 & 3 \\ 4 & -6 \end{vmatrix} = 24 - 12 = 12$$

$$D_x = \begin{vmatrix} 3 & 3 \\ -5 & -6 \end{vmatrix} = -18 + 15 = -3$$

$$D_y = \begin{vmatrix} -4 & 3 \\ 4 & -5 \end{vmatrix} = 20 - 12 = 8$$

Using Cramer's rule: $x = \frac{D_x}{D} = \frac{-3}{12} = -\frac{1}{4}, y = \frac{D_y}{D} = \frac{8}{12} = \frac{2}{3}$. The solution set is $\left\{ \left(-\frac{1}{4}, \frac{2}{3} \right) \right\}$.

11. Computing the determinants:

$$D = \begin{vmatrix} 9 & -1 \\ 8 & 1 \end{vmatrix} = 9 + 8 = 17$$

$$D_x = \begin{vmatrix} -2 & -1 \\ 4 & 1 \end{vmatrix} = -2 + 4 = 2$$

$$D_y = \begin{vmatrix} 9 & -2 \\ 8 & 4 \end{vmatrix} = 36 + 16 = 52$$

Using Cramer's rule: $x = \dfrac{D_x}{D} = \dfrac{2}{17}$, $y = \dfrac{D_y}{D} = \dfrac{52}{17}$. The solution set is $\left\{ \left(\dfrac{2}{17}, \dfrac{52}{17} \right) \right\}$.

13. First clear each equation of fractions by multiplying the first equation by 6 and the second equation by 6:

$$-4x + 3y = -42$$
$$2x - 9y = 36$$

Computing the determinants:

$$D = \begin{vmatrix} -4 & 3 \\ 2 & -9 \end{vmatrix} = 36 - 6 = 30$$

$$D_x = \begin{vmatrix} -42 & 3 \\ 36 & -9 \end{vmatrix} = 378 - 108 = 270$$

$$D_y = \begin{vmatrix} -4 & -42 \\ 2 & 36 \end{vmatrix} = -144 + 84 = -60$$

Using Cramer's rule: $x = \dfrac{D_x}{D} = \dfrac{270}{30} = 9$, $y = \dfrac{D_y}{D} = \dfrac{-60}{30} = -2$. The solution set is $\{(9, -2)\}$.

15. Computing the determinants:

$$D = \begin{vmatrix} 2 & 7 \\ 1 & 0 \end{vmatrix} = 0 - 7 = -7$$

$$D_x = \begin{vmatrix} -1 & 7 \\ 2 & 0 \end{vmatrix} = 0 - 14 = -14$$

$$D_y = \begin{vmatrix} 2 & -1 \\ 1 & 2 \end{vmatrix} = 4 + 1 = 5$$

Using Cramer's rule: $x = \dfrac{D_x}{D} = \dfrac{-14}{-7} = 2$, $y = \dfrac{D_y}{D} = -\dfrac{5}{7}$. The solution set is $\left\{ \left(2, -\dfrac{5}{7} \right) \right\}$.

17. Computing the determinants:

$$D = \begin{vmatrix} 1 & -1 & 2 \\ 2 & 3 & -4 \\ -1 & 2 & -1 \end{vmatrix} = \begin{vmatrix} 1 & -1 & 2 \\ 0 & 5 & -8 \\ 0 & 1 & 1 \end{vmatrix} = 1 \begin{vmatrix} 5 & -8 \\ 1 & 1 \end{vmatrix} = 5 + 8 = 13$$

$$D_x = \begin{vmatrix} -8 & -1 & 2 \\ 18 & 3 & -4 \\ 7 & 2 & -1 \end{vmatrix} = \begin{vmatrix} -8 & -1 & 2 \\ -6 & 0 & 2 \\ -9 & 0 & 3 \end{vmatrix} = 1 \begin{vmatrix} -6 & 2 \\ -9 & 3 \end{vmatrix} = -18 + 18 = 0$$

$$D_y = \begin{vmatrix} 1 & -8 & 2 \\ 2 & 18 & -4 \\ -1 & 7 & -1 \end{vmatrix} = \begin{vmatrix} 1 & -8 & 2 \\ 0 & 34 & -8 \\ 0 & -1 & 1 \end{vmatrix} = 1 \begin{vmatrix} 34 & -8 \\ -1 & 1 \end{vmatrix} = 34 - 8 = 26$$

$$D_z = \begin{vmatrix} 1 & -1 & -8 \\ 2 & 3 & 18 \\ -1 & 2 & 7 \end{vmatrix} = \begin{vmatrix} 1 & -1 & -8 \\ 0 & 5 & 34 \\ 0 & 1 & -1 \end{vmatrix} = 1 \begin{vmatrix} 5 & 34 \\ 1 & -1 \end{vmatrix} = -5 - 34 = -39$$

Using Cramer's rule: $x = \dfrac{D_x}{D} = \dfrac{0}{13} = 0$, $y = \dfrac{D_y}{D} = \dfrac{26}{13} = 2$, $z = \dfrac{D_z}{D} = \dfrac{-39}{13} = -3$. The solution set is $\{(0, 2, -3)\}$.

19. Computing the determinants:

$$D = \begin{vmatrix} 2 & -3 & 1 \\ -3 & 1 & -1 \\ 1 & -2 & -5 \end{vmatrix} = \begin{vmatrix} 0 & 1 & 11 \\ 0 & -5 & -16 \\ 1 & -2 & -5 \end{vmatrix} = 1\begin{vmatrix} 1 & 11 \\ -5 & -16 \end{vmatrix} = -16 + 55 = 39$$

$$D_x = \begin{vmatrix} -7 & -3 & 1 \\ -7 & 1 & -1 \\ -45 & -2 & -5 \end{vmatrix} = \begin{vmatrix} -28 & 0 & -2 \\ -7 & 1 & -1 \\ -59 & 0 & -7 \end{vmatrix} = 1\begin{vmatrix} -28 & -2 \\ -59 & -7 \end{vmatrix} = 196 - 118 = 78$$

$$D_y = \begin{vmatrix} 2 & -7 & 1 \\ -3 & -7 & -1 \\ 1 & -45 & -5 \end{vmatrix} = \begin{vmatrix} 2 & -7 & 1 \\ -1 & -14 & 0 \\ 11 & -80 & 0 \end{vmatrix} = 1\begin{vmatrix} -1 & -14 \\ 11 & -80 \end{vmatrix} = 80 + 154 = 234$$

$$D_z = \begin{vmatrix} 2 & -3 & -7 \\ -3 & 1 & -7 \\ 1 & -2 & -45 \end{vmatrix} = \begin{vmatrix} -7 & 0 & -28 \\ -3 & 1 & -7 \\ -5 & 0 & -59 \end{vmatrix} = 1\begin{vmatrix} -7 & -28 \\ -5 & -59 \end{vmatrix} = 413 - 140 = 273$$

Using Cramer's rule: $x = \dfrac{D_x}{D} = \dfrac{78}{39} = 2, y = \dfrac{D_y}{D} = \dfrac{234}{39} = 6, z = \dfrac{D_z}{D} = \dfrac{273}{39} = 7$. The solution set is $\{(2,6,7)\}$.

21. Computing the determinants:

$$D = \begin{vmatrix} 4 & 5 & -2 \\ 7 & -1 & 2 \\ 3 & 1 & 4 \end{vmatrix} = \begin{vmatrix} 4 & 5 & -2 \\ 11 & 4 & 0 \\ 11 & 11 & 0 \end{vmatrix} = -2\begin{vmatrix} 11 & 4 \\ 11 & 11 \end{vmatrix} = -2(121 - 44) = -154$$

$$D_x = \begin{vmatrix} -14 & 5 & -2 \\ 42 & -1 & 2 \\ 28 & 1 & 4 \end{vmatrix} = \begin{vmatrix} -14 & 5 & -2 \\ 28 & 4 & 0 \\ 0 & 11 & 0 \end{vmatrix} = -2\begin{vmatrix} 28 & 4 \\ 0 & 11 \end{vmatrix} = -2(308) = -616$$

$$D_y = \begin{vmatrix} 4 & -14 & -2 \\ 7 & 42 & 2 \\ 3 & 28 & 4 \end{vmatrix} = \begin{vmatrix} 4 & -14 & -2 \\ 11 & 28 & 0 \\ 11 & 0 & 0 \end{vmatrix} = -2\begin{vmatrix} 11 & 28 \\ 11 & 0 \end{vmatrix} = -2(0 - 308) = 616$$

$$D_z = \begin{vmatrix} 4 & 5 & -14 \\ 7 & -1 & 42 \\ 3 & 1 & 28 \end{vmatrix} = \begin{vmatrix} 39 & 0 & 196 \\ 7 & -1 & 42 \\ 10 & 0 & 70 \end{vmatrix} = -1\begin{vmatrix} 39 & 196 \\ 10 & 70 \end{vmatrix} = -1(2730 - 1960) = -770$$

Using Cramer's rule: $x = \dfrac{D_x}{D} = \dfrac{-616}{-154} = 4, y = \dfrac{D_y}{D} = \dfrac{616}{-154} = -4, z = \dfrac{D_z}{D} = \dfrac{-770}{-154} = 5$

The solution set is $\{(4,-4,5)\}$.

23. Computing the determinants:

$$D = \begin{vmatrix} 2 & -1 & 3 \\ 0 & 3 & 1 \\ 1 & -2 & -1 \end{vmatrix} = \begin{vmatrix} 2 & -10 & 3 \\ 0 & 0 & 1 \\ 1 & 1 & -1 \end{vmatrix} = -1\begin{vmatrix} 2 & -10 \\ 1 & 1 \end{vmatrix} = -1(2 + 10) = -12$$

$$D_x = \begin{vmatrix} -17 & -1 & 3 \\ 5 & 3 & 1 \\ -3 & -2 & -1 \end{vmatrix} = \begin{vmatrix} -32 & -10 & 0 \\ 5 & 3 & 1 \\ 2 & 1 & 0 \end{vmatrix} = -1\begin{vmatrix} -32 & -10 \\ 2 & 1 \end{vmatrix} = -1(-32 + 20) = 12$$

$$D_y = \begin{vmatrix} 2 & -17 & 3 \\ 0 & 5 & 1 \\ 1 & -3 & -1 \end{vmatrix} = \begin{vmatrix} 2 & -32 & 3 \\ 0 & 0 & 1 \\ 1 & 2 & -1 \end{vmatrix} = -1\begin{vmatrix} 2 & -32 \\ 1 & 2 \end{vmatrix} = -1(4 + 32) = -36$$

$$D_z = \begin{vmatrix} 2 & -1 & -17 \\ 0 & 3 & 5 \\ 1 & -2 & -3 \end{vmatrix} = \begin{vmatrix} 0 & 3 & -11 \\ 0 & 3 & 5 \\ 1 & -2 & -3 \end{vmatrix} = 1\begin{vmatrix} 3 & -11 \\ 3 & 5 \end{vmatrix} = 15 + 33 = 48$$

Using Cramer's rule: $x = \dfrac{D_x}{D} = \dfrac{12}{-12} = -1, y = \dfrac{D_y}{D} = \dfrac{-36}{-12} = 3, z = \dfrac{D_z}{D} = \dfrac{48}{-12} = -4$. The solution set is $\{(-1,3,-4)\}$.

25.   Computing the determinants:

$$D = \begin{vmatrix} 1 & 3 & -4 \\ 2 & -1 & 1 \\ 4 & 5 & -7 \end{vmatrix} = \begin{vmatrix} 7 & 3 & -1 \\ 0 & -1 & 0 \\ 14 & 5 & -2 \end{vmatrix} = -1 \begin{vmatrix} 7 & -1 \\ 14 & -2 \end{vmatrix} = -(-14 + 14) = 0$$

$$D_x = \begin{vmatrix} -1 & 3 & -4 \\ 2 & -1 & 1 \\ 0 & 5 & -7 \end{vmatrix} = \begin{vmatrix} -1 & 3 & -4 \\ 0 & 5 & -7 \\ 0 & 5 & -7 \end{vmatrix} = -1 \begin{vmatrix} 5 & -7 \\ 5 & -7 \end{vmatrix} = -(-35 + 35) = 0$$

$$D_y = \begin{vmatrix} 1 & -1 & -4 \\ 2 & 2 & 1 \\ 4 & 0 & -7 \end{vmatrix} = \begin{vmatrix} 1 & -1 & -4 \\ 4 & 0 & -7 \\ 4 & 0 & -7 \end{vmatrix} = 1 \begin{vmatrix} 4 & -7 \\ 4 & -7 \end{vmatrix} = -28 + 28 = 0$$

$$D_z = \begin{vmatrix} 1 & 3 & -1 \\ 2 & -1 & 2 \\ 4 & 5 & 0 \end{vmatrix} = \begin{vmatrix} 1 & 3 & -1 \\ 4 & 5 & 0 \\ 4 & 5 & 0 \end{vmatrix} = -1 \begin{vmatrix} 4 & 5 \\ 4 & 5 \end{vmatrix} = -(20 - 20) = 0$$

Since all of the determinants are 0, the system is dependent. There are infinitely many solutions.

27.   Computing the determinants:

$$D = \begin{vmatrix} 3 & -2 & -3 \\ 1 & 2 & 3 \\ -1 & 4 & -6 \end{vmatrix} = \begin{vmatrix} 4 & 0 & 0 \\ 1 & 2 & 3 \\ -1 & 4 & -6 \end{vmatrix} = 4 \begin{vmatrix} 2 & 3 \\ 4 & -6 \end{vmatrix} = 4(-12 - 12) = -96$$

$$D_x = \begin{vmatrix} -5 & -2 & -3 \\ -3 & 2 & 3 \\ 8 & 4 & -6 \end{vmatrix} = \begin{vmatrix} -8 & 0 & 0 \\ -3 & 2 & 3 \\ 8 & 4 & -6 \end{vmatrix} = -8 \begin{vmatrix} 2 & 3 \\ 4 & -6 \end{vmatrix} = -8(-12 - 12) = 192$$

$$D_y = \begin{vmatrix} 3 & -5 & -3 \\ 1 & -3 & 3 \\ -1 & 8 & -6 \end{vmatrix} = \begin{vmatrix} 0 & 19 & -21 \\ 0 & 5 & -3 \\ -1 & 8 & -6 \end{vmatrix} = -1 \begin{vmatrix} 19 & -21 \\ 5 & -3 \end{vmatrix} = -(-57 + 105) = -48$$

$$D_z = \begin{vmatrix} 3 & -2 & -5 \\ 1 & 2 & -3 \\ -1 & 4 & 8 \end{vmatrix} = \begin{vmatrix} 4 & 0 & -8 \\ 1 & 2 & -3 \\ -3 & 0 & 14 \end{vmatrix} = 2 \begin{vmatrix} 4 & -8 \\ -3 & 14 \end{vmatrix} = 2(56 - 24) = 64$$

Using Cramer's rule: $x = \dfrac{D_x}{D} = \dfrac{192}{-96} = -2, y = \dfrac{D_y}{D} = \dfrac{-48}{-96} = \dfrac{1}{2}, z = \dfrac{D_z}{D} = \dfrac{64}{-96} = -\dfrac{2}{3}$

The solution set is $\left\{ \left(-2, \dfrac{1}{2}, -\dfrac{2}{3} \right) \right\}$.

29.   Computing the determinants:

$$D = \begin{vmatrix} 1 & -2 & 3 \\ -2 & 4 & -3 \\ 5 & -6 & 6 \end{vmatrix} = \begin{vmatrix} 1 & -2 & 3 \\ -1 & 2 & 0 \\ 3 & -2 & 0 \end{vmatrix} = 3 \begin{vmatrix} -1 & 2 \\ 3 & -2 \end{vmatrix} = 3(2 - 6) = -12$$

$$D_x = \begin{vmatrix} 1 & -2 & 3 \\ -3 & 4 & -3 \\ 10 & -6 & 6 \end{vmatrix} = \begin{vmatrix} 1 & -2 & 3 \\ -2 & 2 & 0 \\ 8 & -2 & 0 \end{vmatrix} = 3 \begin{vmatrix} -2 & 2 \\ 8 & -2 \end{vmatrix} = 3(4 - 16) = -36$$

$$D_y = \begin{vmatrix} 1 & 1 & 3 \\ -2 & -3 & -3 \\ 5 & 10 & 6 \end{vmatrix} = \begin{vmatrix} 1 & 1 & 3 \\ -1 & -2 & 0 \\ 3 & 8 & 0 \end{vmatrix} = 3 \begin{vmatrix} -1 & -2 \\ 3 & 8 \end{vmatrix} = 3(-8 + 6) = -6$$

$$D_z = \begin{vmatrix} 1 & -2 & 1 \\ -2 & 4 & -3 \\ 5 & -6 & 10 \end{vmatrix} = \begin{vmatrix} 1 & 0 & 0 \\ -2 & 0 & -1 \\ 5 & 4 & 5 \end{vmatrix} = 1 \begin{vmatrix} 0 & -1 \\ 4 & 5 \end{vmatrix} = 0 + 4 = 4$$

Using Cramer's rule: $x = \dfrac{D_x}{D} = \dfrac{-36}{-12} = 3, y = \dfrac{D_y}{D} = \dfrac{-6}{-12} = \dfrac{1}{2}, z = \dfrac{D_z}{D} = \dfrac{4}{-12} = -\dfrac{1}{3}$. The solution set is $\left\{ \left(3, \dfrac{1}{2}, -\dfrac{1}{3} \right) \right\}$.

31. Computing the determinants:

$$D = \begin{vmatrix} -1 & -1 & 3 \\ -2 & 1 & 7 \\ 3 & 4 & -5 \end{vmatrix} = \begin{vmatrix} -1 & -1 & 3 \\ 0 & 3 & 1 \\ 0 & 1 & 4 \end{vmatrix} = -1\begin{vmatrix} 3 & 1 \\ 1 & 4 \end{vmatrix} = -1(12-1) = -11$$

$$D_x = \begin{vmatrix} -2 & -1 & 3 \\ 14 & 1 & 7 \\ 12 & 4 & -5 \end{vmatrix} = \begin{vmatrix} -2 & -1 & 3 \\ 12 & 0 & 10 \\ 4 & 0 & 7 \end{vmatrix} = 1\begin{vmatrix} 12 & 10 \\ 4 & 7 \end{vmatrix} = 1(84-40) = 44$$

$$D_y = \begin{vmatrix} -1 & -2 & 3 \\ -2 & 14 & 7 \\ 3 & 12 & -5 \end{vmatrix} = \begin{vmatrix} -1 & -2 & 3 \\ 0 & 18 & 1 \\ 0 & 6 & 4 \end{vmatrix} = -1\begin{vmatrix} 18 & 1 \\ 6 & 4 \end{vmatrix} = -1(72-6) = -66$$

$$D_z = \begin{vmatrix} -1 & -1 & -2 \\ -2 & 1 & 14 \\ 3 & 4 & 12 \end{vmatrix} = \begin{vmatrix} -1 & -1 & -2 \\ 0 & 3 & 18 \\ 0 & 1 & 6 \end{vmatrix} = -1\begin{vmatrix} 3 & 18 \\ 1 & 6 \end{vmatrix} = -1(18-18) = 0$$

Using Cramer's rule: $x = \dfrac{D_x}{D} = \dfrac{44}{-11} = -4, y = \dfrac{D_y}{D} = \dfrac{-66}{-11} = 6, z = \dfrac{D_z}{D} = \dfrac{0}{-11} = 0$. The solution set is $\{(-4,6,0)\}$.

37. Computing the determinant: $D = \begin{vmatrix} 2 & -1 & 1 \\ 3 & 2 & 5 \\ 4 & -7 & 1 \end{vmatrix} = \begin{vmatrix} 0 & 0 & 1 \\ -7 & 7 & 5 \\ 2 & -6 & 1 \end{vmatrix} = 1\begin{vmatrix} -7 & 7 \\ 2 & -6 \end{vmatrix} = 1(42-14) = 28$

Since $D \neq 0$, the solution set is $\{(0,0,0)\}$.

39. Computing the determinant: $D = \begin{vmatrix} 2 & -1 & 2 \\ 1 & 2 & 1 \\ 1 & -3 & 1 \end{vmatrix} = \begin{vmatrix} 0 & 5 & 0 \\ 0 & 5 & 0 \\ 1 & -3 & 1 \end{vmatrix} = 1\begin{vmatrix} 5 & 0 \\ 5 & 0 \end{vmatrix} = 1(0-0) = 0$

Since $D = 0$, the system is dependent and has infinitely many solutions.

6.6 Partial Fractions

1. Finding the decomposition:

$$\frac{11x-10}{(x-2)(x+1)} = \frac{A}{x-2} + \frac{B}{x+1}$$
$$11x-10 = A(x+1) + B(x-2)$$

Substituting $x = -1$ and $x = 2$:

$$-21 = B(-3) \qquad\qquad\qquad 12 = A(3)$$
$$B = 7 \qquad\qquad\qquad\qquad A = 4$$

The decomposition is $\dfrac{11x-10}{(x-2)(x+1)} = \dfrac{4}{x-2} + \dfrac{7}{x+1}$.

3. Finding the decomposition:

$$\frac{-2x-8}{x^2-1} = \frac{A}{x+1} + \frac{B}{x-1}$$
$$-2x-8 = A(x-1) + B(x+1)$$

Substituting $x = 1$ and $x = -1$:

$$-10 = B(2) \qquad\qquad\qquad -6 = A(-2)$$
$$B = -5 \qquad\qquad\qquad\qquad A = 3$$

The decomposition is $\dfrac{-2x-8}{x^2-1} = \dfrac{3}{x+1} - \dfrac{5}{x-1}$.

5.   Finding the decomposition:

$$\frac{20x - 3}{6x^2 + 7x - 3} = \frac{A}{3x - 1} + \frac{B}{2x + 3}$$
$$20x - 3 = A(2x + 3) + B(3x - 1)$$

Substituting $x = -\frac{3}{2}$ and $x = \frac{1}{3}$:

$$-33 = B\left(-\frac{11}{2}\right) \qquad\qquad \frac{11}{3} = A\left(\frac{11}{3}\right)$$
$$B = 6 \qquad\qquad\qquad\qquad A = 1$$

The decomposition is $\dfrac{20x - 3}{6x^2 + 7x - 3} = \dfrac{1}{3x - 1} + \dfrac{6}{2x + 3}$.

7.   Finding the decomposition:

$$\frac{x^2 - 18x + 5}{(x - 1)(x + 2)(x - 3)} = \frac{A}{x - 1} + \frac{B}{x + 2} + \frac{C}{x - 3}$$
$$x^2 - 18x + 5 = A(x + 2)(x - 3) + B(x - 1)(x - 3) + C(x - 1)(x + 2)$$

Substituting $x = -2$, $x = 3$, and $x = 1$:

$$45 = B(-3)(-5) \qquad -40 = C(2)(5) \qquad -12 = A(3)(-2)$$
$$B = 3 \qquad\qquad\qquad C = -4 \qquad\qquad A = 2$$

The decomposition is $\dfrac{x^2 - 18x + 5}{(x - 1)(x + 2)(x - 3)} = \dfrac{2}{x - 1} + \dfrac{3}{x + 2} - \dfrac{4}{x - 3}$.

9.   Finding the decomposition:

$$\frac{-6x^2 + 7x + 1}{x(2x - 1)(4x + 1)} = \frac{A}{x} + \frac{B}{2x - 1} + \frac{C}{4x + 1}$$
$$-6x^2 + 7x + 1 = A(2x - 1)(4x + 1) + Bx(4x + 1) + Cx(2x - 1)$$

Substituting $x = 0$, $x = \frac{1}{2}$, and $x = -\frac{1}{4}$:

$$1 = A(-1)(1) \qquad 3 = B\left(\frac{1}{2}\right)(3) \qquad -\frac{9}{8} = C\left(-\frac{1}{4}\right)\left(-\frac{3}{2}\right)$$
$$A = -1 \qquad\qquad\qquad B = 2 \qquad\qquad\qquad C = -3$$

The decomposition is $\dfrac{-6x^2 + 7x + 1}{x(2x - 1)(4x + 1)} = \dfrac{-1}{x} + \dfrac{2}{2x - 1} - \dfrac{3}{4x + 1}$.

11.   Finding the decomposition:

$$\frac{2x + 1}{(x - 2)^2} = \frac{A}{x - 2} + \frac{B}{(x - 2)^2}$$
$$2x + 1 = A(x - 2) + B$$

Substituting $x = 2$ and $x = 0$:

$$5 = B \qquad\qquad\qquad\qquad 1 = A(-2) + 5$$
$$\qquad\qquad\qquad\qquad\qquad A = 2$$

The decomposition is $\dfrac{2x + 1}{(x - 2)^2} = \dfrac{2}{x - 2} + \dfrac{5}{(x - 2)^2}$.

13. Finding the decomposition:

$$\frac{-6x^2 + 19x + 21}{x^2(x+3)} = \frac{A}{x} + \frac{B}{x^2} + \frac{C}{x+3}$$

$$-6x^2 + 19x + 21 = Ax(x+3) + B(x+3) + Cx^2$$

Substituting $x = 0$, $x = -3$, and $x = 1$:

$$21 = B(3) \qquad -90 = C(9) \qquad 34 = A(1)(4) + 7(4) - 10(1)$$
$$B = 7 \qquad\quad C = -10 \qquad\quad 34 = 4A + 18$$
$$4 = A$$

The decomposition is $\dfrac{-6x^2 + 19x + 21}{x^2(x+3)} = \dfrac{4}{x} + \dfrac{7}{x^2} - \dfrac{10}{x+3}$.

15. Finding the decomposition:

$$\frac{-2x^2 - 3x + 10}{(x^2+1)(x-4)} = \frac{Ax+B}{x^2+1} + \frac{C}{x-4}$$

$$-2x^2 - 3x + 10 = (Ax+B)(x-4) + C(x^2+1)$$

Substituting $x = 4$, $x = 0$, and $x = 1$:

$$-34 = C(17) \qquad 10 = B(-4) - 2(1) \qquad 5 = (A-3)(-3) - 2(2)$$
$$-2 = C \qquad\quad 12 = -4B \qquad\qquad 9 = -3A + 9$$
$$-3 = B \qquad\qquad 0 = A$$

The decomposition is $\dfrac{-2x^2 - 3x + 10}{(x^2+1)(x-4)} = \dfrac{-3}{x^2+1} - \dfrac{2}{x-4}$.

17. Finding the decomposition:

$$\frac{3x^2 + 10x + 9}{(x+2)^3} = \frac{A}{x+2} + \frac{B}{(x+2)^2} + \frac{C}{(x+2)^3}$$

$$3x^2 + 10x + 9 = A(x+2)^2 + B(x+2) + C$$

Substituting $x = -2$, $x = 0$, and $x = 1$:

$$1 = C \qquad 9 = 4A + 2B + 1 \qquad 22 = 9A + 3B + 1$$
$$4 = 2A + B \qquad\quad 7 = 3A + B$$

Subtracting the second equation from the third yields $A = 3$, so $B = -2$.

The decomposition is $\dfrac{3x^2 + 10x + 9}{(x+2)^3} = \dfrac{3}{x+2} - \dfrac{2}{(x+2)^2} + \dfrac{1}{(x+2)^3}$.

19. Finding the decomposition:

$$\frac{5x^2 + 3x + 6}{x(x^2 - x + 3)} = \frac{A}{x} + \frac{Bx+C}{x^2 - x + 3}$$

$$5x^2 + 3x + 6 = A(x^2 - x + 3) + (Bx+C)x$$

Substituting $x = 0$, $x = 1$, and $x = -1$:

$$6 = A(3) \qquad 14 = 6 + (B+C) \qquad 8 = 10 + (-B+C)(-1)$$
$$A = 2 \qquad\quad 8 = B + C \qquad\qquad -2 = B - C$$

Adding these last two equations results in:

$$2B = 6$$
$$B = 3$$
$$C = 5$$

The decomposition is $\dfrac{5x^2 + 3x + 6}{x(x^2 - x + 3)} = \dfrac{2}{x} + \dfrac{3x+5}{x^2 - x + 3}$.

21.   Finding the decomposition:

$$\frac{2x^3 + x + 3}{\left(x^2 + 1\right)^2} = \frac{Ax + B}{x^2 + 1} + \frac{Cx + D}{\left(x^2 + 1\right)^2}$$

$$2x^3 + x + 3 = (Ax + B)\left(x^2 + 1\right) + (Cx + D)$$

Substituting $x = 0$, $x = 1$, $x = -1$, and $x = 2$ results in the equations:

$$3 = B + D$$
$$6 = 2(A + B) + (C + D)$$
$$0 = 2(-A + B) + (-C + D)$$
$$21 = 5(2A + B) + (2C + D)$$

These equations simplify to:

$$B + D = 3$$
$$2A + 2B + C + D = 6$$
$$-2A + 2B - C + D = 0$$
$$10A + 5B + 2C + D = 21$$

Adding the second and third equations results in $2B + D = 3$, now subtracting the first equation results in $B = 0$ and $D = 3$. Substituting into the second and fourth equations results in:

$$2A + C = 3$$
$$5A + C = 9$$

Subtracting yields $3A = 6$, so $A = 2$ and $C = -1$. The decomposition is $\dfrac{2x^3 + x + 3}{\left(x^2 + 1\right)^2} = \dfrac{2x}{x^2 + 1} + \dfrac{3 - x}{\left(x^2 + 1\right)^2}$.

Chapter 6 Review Problem Set

1.   Solving the first equation for y:

$$3x - y = 16$$
$$y = 3x - 16$$

Substituting:

$$5x + 7(3x - 16) = -34$$
$$5x + 21x - 112 = -34$$
$$26x = 78$$
$$x = 3$$
$$y = 9 - 16 = -7$$

The solution set is $\{(3, -7)\}$.

2.   Solving the second equation for x:

$$x - 4y = 11$$
$$x = 4y + 11$$

Substituting:

$$6(4y + 11) + 5y = -21$$
$$24y + 66 + 5y = -21$$
$$29y = -87$$
$$y = -3$$
$$x = 4(-3) + 11 = -1$$

The solution set is $\{(-1, -3)\}$.

3.   Solving the first equation for x:

$$2x - 3y = 12$$
$$2x = 3y + 12$$
$$x = \frac{3}{2}y + 6$$

Substituting:

$$3\left(\frac{3}{2}y + 6\right) + 5y = -20$$
$$\frac{9}{2}y + 18 + 5y = -20$$
$$\frac{19}{2}y = -38$$
$$y = -4$$
$$x = \frac{3}{2}(-4) + 6 = 0$$

The solution set is $\{(0, -4)\}$.

4.   Solving the first equation for x:

$$5x + 8y = 1$$
$$5x = -8y + 1$$
$$x = -\frac{8}{5}y + \frac{1}{5}$$

Substituting:

$$4\left(-\frac{8}{5}y + \frac{1}{5}\right) + 7y = -2$$
$$-\frac{32}{5}y + \frac{4}{5} + 7y = -2$$
$$\frac{3}{5}y = -\frac{14}{5}$$
$$y = -\frac{14}{3}$$
$$x = -\frac{8}{5}\left(-\frac{14}{3}\right) + \frac{1}{5} = \frac{112}{15} + \frac{3}{15} = \frac{23}{3}$$

The solution set is $\left\{\left(\frac{23}{3}, -\frac{14}{3}\right)\right\}$.

5.   Multiply the first equation by 2 and the second equation by 3:

$$8x - 6y = 68$$
$$9x + 6y = 0$$

Adding yields:

$$17x = 68$$
$$x = 4$$

Substituting into the second equation:

$$3(4) + 2y = 0$$
$$2y = -12$$
$$y = -6$$

The solution set is $\left\{(4, -6)\right\}$.

6.   To clear each equation of fractions, multiply the first equation by 6 and the second equation by 12:

$$3x - 4y = 6$$
$$9x + 2y = -12$$

Multiply the second equation by 2:

$$3x - 4y = 6$$
$$18x + 4y = -24$$

Adding yields:

$$21x = -18$$
$$x = -\frac{6}{7}$$

Substituting into the first equation:

$$3\left(-\frac{6}{7}\right) - 4y = 6$$
$$-\frac{18}{7} - 4y = 6$$
$$-4y = \frac{60}{7}$$
$$y = -\frac{15}{7}$$

The solution set is $\left\{\left(-\frac{6}{7}, -\frac{15}{7}\right)\right\}$.

7.   Multiply the first equation by 2 and add it to the second equation:
$$4x - 2y + 6z = -38$$
$$3x + 2y - 4z = 21$$

Adding yields $7x + 2z = -17$. Multiply the first equation by -4 and add it to the third equation:
$$-8x + 4y - 12z = 76$$
$$5x - 4y - z = -8$$

Adding yields $-3x - 13z = 68$. So the system becomes:
$$-3x - 13z = 68$$
$$7x + 2z = -17$$

Multiply the first equation by 7 and the second equation by 3:
$$-21x - 91z = 476$$
$$21x + 6z = -51$$

Adding yields:
$$-85z = 425$$
$$z = -5$$

Substituting:
$$7x + 2(-5) = -17$$
$$7x = -7$$
$$x = -1$$

Substituting:
$$2(-1) - y + 3(-5) = -19$$
$$-y - 17 = -19$$
$$y = 2$$

The solution set is $\{(-1, 2, -5)\}$.

8.   Multiply the second equation by -4 and add it to the first equation:
$$3x + 2y - 4z = 4$$
$$-20x - 12y + 4z = -8$$

Adding yields $-17x - 10y = -4$. Multiply the second equation by 3 and add it to the third equation:
$$15x + 9y - 3z = 6$$
$$4x - 2y + 3z = 11$$

Adding yields $19x + 7y = 17$. So the system becomes:
$$19x + 7y = 17$$
$$-17x - 10y = -4$$

Multiply the first equation by 10 and the second equation by 7:
$$190x + 70y = 170$$
$$-119x - 70y = -28$$

Adding yields:
$$71x = 142$$
$$x = 2$$

Substituting:
$$19(2) + 7y = 17$$
$$7y = -21$$
$$y = -3$$

Substituting:
$$5(2) + 3(-3) - z = 2$$
$$1 - z = 2$$
$$z = -1$$

The solution set is $\{(2, -3, -1)\}$.

9. Form the augmented matrix: $\begin{bmatrix} 1 & -3 & | & 17 \\ -3 & 2 & | & -23 \end{bmatrix}$

    Add 3 times row 1 to row 2: $\begin{bmatrix} 1 & -3 & | & 17 \\ 0 & -7 & | & 28 \end{bmatrix}$

    Divide row 2 by –7: $\begin{bmatrix} 1 & -3 & | & 17 \\ 0 & 1 & | & -4 \end{bmatrix}$

    Add 3 times row 2 to row 1: $\begin{bmatrix} 1 & 0 & | & 5 \\ 0 & 1 & | & -4 \end{bmatrix}$

    The solution set is $\{(5,-4)\}$.

10. Form the augmented matrix: $\begin{bmatrix} 2 & 3 & | & 25 \\ 3 & -5 & | & -29 \end{bmatrix}$

    Subtract row 1 from row 2: $\begin{bmatrix} 2 & 3 & | & 25 \\ 1 & -8 & | & -54 \end{bmatrix}$

    Switch row 1 and row 2: $\begin{bmatrix} 1 & -8 & | & -54 \\ 2 & 3 & | & 25 \end{bmatrix}$

    Add –2 times row 1 to row 2: $\begin{bmatrix} 1 & -8 & | & -54 \\ 0 & 19 & | & 133 \end{bmatrix}$

    Divide row 2 by 19: $\begin{bmatrix} 1 & -8 & | & -54 \\ 0 & 1 & | & 7 \end{bmatrix}$

    Add 8 times row 2 to row 1: $\begin{bmatrix} 1 & 0 & | & 2 \\ 0 & 1 & | & 7 \end{bmatrix}$

    The solution set is $\{(2,7)\}$.

11. Form the augmented matrix: $\begin{bmatrix} 1 & -2 & 1 & | & -7 \\ 2 & -3 & 4 & | & -14 \\ -3 & 1 & -2 & | & 10 \end{bmatrix}$

    Add –2 times row 1 to row 2, and 3 times row 1 to row 3: $\begin{bmatrix} 1 & -2 & 1 & | & -7 \\ 0 & 1 & 2 & | & 0 \\ 0 & -5 & 1 & | & -11 \end{bmatrix}$

    Add 5 times row 2 to row 3: $\begin{bmatrix} 1 & -2 & 1 & | & -7 \\ 0 & 1 & 2 & | & 0 \\ 0 & 0 & 11 & | & -11 \end{bmatrix}$

    Divide row 3 by 11: $\begin{bmatrix} 1 & -2 & 1 & | & -7 \\ 0 & 1 & 2 & | & 0 \\ 0 & 0 & 1 & | & -1 \end{bmatrix}$

    Add –2 times row 3 to row 2, and –1 times row 3 to row 1: $\begin{bmatrix} 1 & -2 & 0 & | & -6 \\ 0 & 1 & 0 & | & 2 \\ 0 & 0 & 1 & | & -1 \end{bmatrix}$

    Add 2 times row 2 to row 1: $\begin{bmatrix} 1 & 0 & 0 & | & -2 \\ 0 & 1 & 0 & | & 2 \\ 0 & 0 & 1 & | & -1 \end{bmatrix}$

    The solution set is $\{(-2,2,-1)\}$.

12.   Form the augmented matrix: $\begin{bmatrix} -2 & -7 & 1 & | & 9 \\ 1 & 3 & -4 & | & -11 \\ 4 & 5 & -3 & | & -11 \end{bmatrix}$

Switch row 1 and row 2: $\begin{bmatrix} 1 & 3 & -4 & | & -11 \\ -2 & -7 & 1 & | & 9 \\ 4 & 5 & -3 & | & -11 \end{bmatrix}$

Add 2 times row 1 to row 2, and -4 times row 1 to row 3: $\begin{bmatrix} 1 & 3 & -4 & | & -11 \\ 0 & -1 & -7 & | & -13 \\ 0 & -7 & 13 & | & 33 \end{bmatrix}$

Multiply row 1 by -1: $\begin{bmatrix} 1 & 3 & -4 & | & -11 \\ 0 & 1 & 7 & | & 13 \\ 0 & -7 & 13 & | & 33 \end{bmatrix}$

Add 7 times row 2 to row 3: $\begin{bmatrix} 1 & 3 & -4 & | & -11 \\ 0 & 1 & 7 & | & 13 \\ 0 & 0 & 62 & | & 124 \end{bmatrix}$

Divide row 3 by 62: $\begin{bmatrix} 1 & 3 & -4 & | & -11 \\ 0 & 1 & 7 & | & 13 \\ 0 & 0 & 1 & | & 2 \end{bmatrix}$

Add -7 times row 3 to row 2, and 4 times row 3 to row 1: $\begin{bmatrix} 1 & 3 & 0 & | & -3 \\ 0 & 1 & 0 & | & -1 \\ 0 & 0 & 1 & | & 2 \end{bmatrix}$

Add -3 times row 2 to row 1: $\begin{bmatrix} 1 & 0 & 0 & | & 0 \\ 0 & 1 & 0 & | & -1 \\ 0 & 0 & 1 & | & 2 \end{bmatrix}$

The solution set is $\{(0, -1, 2)\}$.

13.   Computing the determinants:

$$D = \begin{vmatrix} 5 & 3 \\ 4 & -9 \end{vmatrix} = -45 - 12 = -57$$

$$D_x = \begin{vmatrix} -18 & 3 \\ -3 & -9 \end{vmatrix} = 162 + 9 = 171$$

$$D_y = \begin{vmatrix} 5 & -18 \\ 4 & -3 \end{vmatrix} = -15 + 72 = 57$$

Using Cramer's rule:

$$x = \frac{D_x}{D} = \frac{171}{-57} = -3, \, y = \frac{D_y}{D} = \frac{57}{-57} = -1$$

The solution set is $\{(-3, -1)\}$.

14.   Computing the determinants:

$$D = \begin{vmatrix} 0.2 & 0.3 \\ 0.5 & -0.1 \end{vmatrix} = -0.02 - 0.15 = -0.17$$

$$D_x = \begin{vmatrix} 2.6 & 0.3 \\ 1.4 & -0.1 \end{vmatrix} = -0.26 - 0.42 = -0.68$$

$$D_y = \begin{vmatrix} 0.2 & 2.6 \\ 0.5 & 1.4 \end{vmatrix} = 0.28 - 1.3 = -1.02$$

Usign Cramer's rule:

$$x = \frac{D_x}{D} = \frac{-0.68}{-0.17} = 4, \, y = \frac{D_y}{D} = \frac{-1.02}{-0.17} = 6$$

The solution set is $\{(4, 6)\}$.

15. Computing the determinants:

$$D = \begin{vmatrix} 2 & -3 & -3 \\ 3 & 1 & 2 \\ 5 & -2 & -4 \end{vmatrix} = \begin{vmatrix} 11 & 0 & 3 \\ 3 & 1 & 2 \\ 11 & 0 & 0 \end{vmatrix} = 1 \begin{vmatrix} 11 & 3 \\ 11 & 0 \end{vmatrix} = 1(0 - 33) = -33$$

$$D_x = \begin{vmatrix} 25 & -3 & -3 \\ -5 & 1 & 2 \\ 32 & -2 & -4 \end{vmatrix} = \begin{vmatrix} 10 & 0 & 3 \\ -5 & 1 & 2 \\ 22 & 0 & 0 \end{vmatrix} = 1 \begin{vmatrix} 10 & 3 \\ 22 & 0 \end{vmatrix} = 1(0 - 66) = -66$$

$$D_y = \begin{vmatrix} 2 & 25 & -3 \\ 3 & -5 & 2 \\ 5 & 32 & -4 \end{vmatrix} = \begin{vmatrix} 5 & 20 & -1 \\ 3 & -5 & 2 \\ 11 & 22 & 0 \end{vmatrix} = \begin{vmatrix} 5 & 20 & -1 \\ 13 & 35 & 0 \\ 11 & 22 & 0 \end{vmatrix} = -11 \begin{vmatrix} 13 & 35 \\ 1 & 2 \end{vmatrix} = -11(26 - 35) = 99$$

$$D_z = \begin{vmatrix} 2 & -3 & 25 \\ 3 & 1 & -5 \\ 5 & -2 & 32 \end{vmatrix} = \begin{vmatrix} 11 & 0 & 10 \\ 3 & 1 & -5 \\ 11 & 0 & 22 \end{vmatrix} = 11 \begin{vmatrix} 11 & 10 \\ 1 & 2 \end{vmatrix} = 11(22 - 10) = 132$$

Using Cramer's rule: $x = \dfrac{D_x}{D} = \dfrac{-66}{-33} = 2$, $y = \dfrac{D_y}{D} = \dfrac{99}{-33} = -3$, $z = \dfrac{D_z}{D} = \dfrac{132}{-33} = -4$. The solution set is $\{(2, -3, -4)\}$.

16. Computing the determinants:

$$D = \begin{vmatrix} 3 & -1 & 1 \\ 6 & -2 & 5 \\ 7 & 3 & -4 \end{vmatrix} = \begin{vmatrix} 0 & 0 & 1 \\ -9 & 3 & 5 \\ 19 & -1 & -4 \end{vmatrix} = 1 \begin{vmatrix} -9 & 3 \\ 19 & -1 \end{vmatrix} = 1(9 - 57) = -48$$

$$D_x = \begin{vmatrix} -10 & -1 & 1 \\ -35 & -2 & 5 \\ 19 & 3 & -4 \end{vmatrix} = \begin{vmatrix} -10 & -1 & 1 \\ -15 & 0 & 3 \\ -11 & 0 & -1 \end{vmatrix} = 1 \begin{vmatrix} -15 & 3 \\ -11 & -1 \end{vmatrix} = 1(15 + 33) = 48$$

$$D_y = \begin{vmatrix} 3 & -10 & 1 \\ 6 & -35 & 5 \\ 7 & 19 & -4 \end{vmatrix} = \begin{vmatrix} 3 & -10 & 1 \\ -9 & 15 & 0 \\ 19 & -21 & 0 \end{vmatrix} = 1 \begin{vmatrix} -9 & 15 \\ 19 & -21 \end{vmatrix} = 1(189 - 285) = -96$$

$$D_z = \begin{vmatrix} 3 & -1 & -10 \\ 6 & -2 & -35 \\ 7 & 3 & 19 \end{vmatrix} = \begin{vmatrix} 3 & -1 & -10 \\ 0 & 0 & -15 \\ 16 & 0 & -11 \end{vmatrix} = 1 \begin{vmatrix} 0 & -15 \\ 16 & -11 \end{vmatrix} = 1(0 + 240) = 240$$

Using Cramer's rule: $x = \dfrac{D_x}{D} = \dfrac{48}{-48} = -1$, $y = \dfrac{D_y}{D} = \dfrac{-96}{-48} = 2$, $z = \dfrac{D_z}{D} = \dfrac{240}{-48} = -5$. The solution set is $\{(-1, 2, -5)\}$.

17. Form the augmented matrix: $\begin{bmatrix} 4 & 7 & | & -15 \\ 3 & -2 & | & 25 \end{bmatrix}$

Subtract row 2 from row 1: $\begin{bmatrix} 1 & 9 & | & -40 \\ 3 & -2 & | & 25 \end{bmatrix}$

Add -3 times row 1 to row 2: $\begin{bmatrix} 1 & 9 & | & -40 \\ 0 & -29 & | & 145 \end{bmatrix}$

Divide row 2 by -29: $\begin{bmatrix} 1 & 9 & | & -40 \\ 0 & 1 & | & -5 \end{bmatrix}$

Add -9 times row 2 to row 1: $\begin{bmatrix} 1 & 0 & | & 5 \\ 0 & 1 & | & -5 \end{bmatrix}$

The solution set is $\{(5, -5)\}$.

18. Form the augmented matrix: $\begin{bmatrix} \frac{3}{4} & -\frac{1}{2} & \vdots & -15 \\ \frac{2}{3} & \frac{1}{4} & \vdots & -5 \end{bmatrix}$

   Multiply row 1 by 4 and row 2 by 12: $\begin{bmatrix} 3 & -2 & \vdots & -60 \\ 8 & 3 & \vdots & -60 \end{bmatrix}$

   Divide row 1 by 3: $\begin{bmatrix} 1 & -\frac{2}{3} & \vdots & -20 \\ 8 & 3 & \vdots & -60 \end{bmatrix}$

   Add -8 times row 1 to row 2: $\begin{bmatrix} 1 & -\frac{2}{3} & \vdots & -20 \\ 0 & \frac{25}{3} & \vdots & 100 \end{bmatrix}$

   Multiply row 2 by $\frac{3}{25}$: $\begin{bmatrix} 1 & -\frac{2}{3} & \vdots & -20 \\ 0 & 1 & \vdots & 12 \end{bmatrix}$

   Add $\frac{2}{3}$ times row 2 to row 1: $\begin{bmatrix} 1 & 0 & \vdots & -12 \\ 0 & 1 & \vdots & 12 \end{bmatrix}$

   The solution set is $\left\{(-12, 12)\right\}$.

19. Form the augmented matrix: $\begin{bmatrix} 1 & 4 & \vdots & 3 \\ 3 & -2 & \vdots & 1 \end{bmatrix}$

   Add -3 times row 1 to row 2: $\begin{bmatrix} 1 & 4 & \vdots & 3 \\ 0 & -14 & \vdots & -8 \end{bmatrix}$

   Divide row 2 by -14: $\begin{bmatrix} 1 & 4 & \vdots & 3 \\ 0 & 1 & \vdots & \frac{4}{7} \end{bmatrix}$

   Add -4 times row 2 to row 1: $\begin{bmatrix} 1 & 0 & \vdots & \frac{5}{7} \\ 0 & 1 & \vdots & \frac{4}{7} \end{bmatrix}$

   The solution set is $\left\{\left(\frac{5}{7}, \frac{4}{7}\right)\right\}$.

20. Substituting into the first equation:

$$7x - 3\left(\frac{3}{5}x - 1\right) = -49$$
$$7x - \frac{9}{5}x + 3 = -49$$
$$\frac{26}{5}x = -52$$
$$x = -10$$
$$y = \frac{3}{5}(-10) - 1 = -7$$

   The solution set is $\left\{(-10, -7)\right\}$.

21. Form the augmented matrix: $\begin{bmatrix} 1 & -1 & -1 & | & 4 \\ -3 & 2 & 5 & | & -21 \\ 5 & -3 & -7 & | & 30 \end{bmatrix}$

Add 3 times row 1 to row 2, and –5 times row 1 to row 3: $\begin{bmatrix} 1 & -1 & -1 & | & 4 \\ 0 & -1 & 2 & | & -9 \\ 0 & 2 & -2 & | & 10 \end{bmatrix}$

Add 2 times row 2 to row 3, and multiply row 2 by –1: $\begin{bmatrix} 1 & -1 & -1 & | & 4 \\ 0 & 1 & -2 & | & 9 \\ 0 & 0 & 2 & | & -8 \end{bmatrix}$

Divide row 3 by 2: $\begin{bmatrix} 1 & -1 & -1 & | & 4 \\ 0 & 1 & -2 & | & 9 \\ 0 & 0 & 1 & | & -4 \end{bmatrix}$

Add 2 times row 3 to row 2, and row 3 to row 1: $\begin{bmatrix} 1 & -1 & 0 & | & 0 \\ 0 & 1 & 0 & | & 1 \\ 0 & 0 & 1 & | & -4 \end{bmatrix}$

Add row 2 to row 1: $\begin{bmatrix} 1 & 0 & 0 & | & 1 \\ 0 & 1 & 0 & | & 1 \\ 0 & 0 & 1 & | & -4 \end{bmatrix}$

The solution set is $\left\{ (1,1,-4) \right\}$.

22. Form the augmented matrix: $\begin{bmatrix} 2 & -1 & 1 & | & -7 \\ -5 & 2 & -3 & | & 17 \\ 3 & 1 & 7 & | & -5 \end{bmatrix}$

Subtract row 1 from row 3: $\begin{bmatrix} 2 & -1 & 1 & | & -7 \\ -5 & 2 & -3 & | & 17 \\ 1 & 2 & 6 & | & 2 \end{bmatrix}$

Switch row 1 and row 3: $\begin{bmatrix} 1 & 2 & 6 & | & 2 \\ -5 & 2 & -3 & | & 17 \\ 2 & -1 & 1 & | & -7 \end{bmatrix}$

Add 5 times row 1 to row 2, and –2 times row 1 to row 3: $\begin{bmatrix} 1 & 2 & 6 & | & 2 \\ 0 & 12 & 27 & | & 27 \\ 0 & -5 & -11 & | & -11 \end{bmatrix}$

Add 2 times row 3 to row 2: $\begin{bmatrix} 1 & 2 & 6 & | & 2 \\ 0 & 2 & 5 & | & 5 \\ 0 & -5 & -11 & | & -11 \end{bmatrix}$

Divide row 2 by 2: $\begin{bmatrix} 1 & 2 & 6 & | & 2 \\ 0 & 1 & \frac{5}{2} & | & \frac{5}{2} \\ 0 & -5 & -11 & | & -11 \end{bmatrix}$

Add 5 times row 2 to row 3: $\begin{bmatrix} 1 & 2 & 6 & | & 2 \\ 0 & 1 & \frac{5}{2} & | & \frac{5}{2} \\ 0 & 0 & \frac{3}{2} & | & \frac{3}{2} \end{bmatrix}$

Multiply row 3 by $\frac{2}{3}$: $\begin{bmatrix} 1 & 2 & 6 & | & 2 \\ 0 & 1 & \frac{5}{2} & | & \frac{5}{2} \\ 0 & 0 & 1 & | & 1 \end{bmatrix}$

Add $-\frac{5}{2}$ times row 3 to row 2, and -6 times row 3 to row 1: $\begin{bmatrix} 1 & 2 & 0 & | & -4 \\ 0 & 1 & 0 & | & 0 \\ 0 & 0 & 1 & | & 1 \end{bmatrix}$

Add -2 times row 2 to row 1: $\begin{bmatrix} 1 & 0 & 0 & | & -4 \\ 0 & 1 & 0 & | & 0 \\ 0 & 0 & 1 & | & 1 \end{bmatrix}$

The solution set is $\{(-4, 0, 1)\}$.

23. Form the augmented matrix: $\begin{bmatrix} 3 & -2 & -5 & | & 2 \\ -4 & 3 & 11 & | & 3 \\ 2 & -1 & 1 & | & -1 \end{bmatrix}$

Subtract row 3 from row 1: $\begin{bmatrix} 1 & -1 & -6 & | & 3 \\ -4 & 3 & 11 & | & 3 \\ 2 & -1 & 1 & | & -1 \end{bmatrix}$

Add 4 times row 1 to row 2, and -2 times row 1 to row 3: $\begin{bmatrix} 1 & -1 & -6 & | & 3 \\ 0 & -1 & -13 & | & 15 \\ 0 & 1 & 13 & | & -7 \end{bmatrix}$

Add row 2 to row 3, and multiply row 2 by -1: $\begin{bmatrix} 1 & -1 & -6 & | & 3 \\ 0 & 1 & 13 & | & -15 \\ 0 & 0 & 0 & | & 8 \end{bmatrix}$

Since $0 \neq 8$, there is no solution. The solution set is $\varnothing$.

24. Form the augmented matrix: $\begin{bmatrix} 7 & -1 & 1 & | & -4 \\ -2 & 9 & -3 & | & -50 \\ 1 & -5 & 4 & | & 42 \end{bmatrix}$

Switch row 1 and row 3: $\begin{bmatrix} 1 & -5 & 4 & | & 42 \\ -2 & 9 & -3 & | & -50 \\ 7 & -1 & 1 & | & -4 \end{bmatrix}$

Add 2 times row 1 to row 2, and -7 times row 1 to row 3: $\begin{bmatrix} 1 & -5 & 4 & | & 42 \\ 0 & -1 & 5 & | & 34 \\ 0 & 34 & -27 & | & -298 \end{bmatrix}$

Add 34 times row 2 to row 3, and multiply row 2 by -1: $\begin{bmatrix} 1 & -5 & 4 & | & 42 \\ 0 & 1 & -5 & | & -34 \\ 0 & 0 & 143 & | & 858 \end{bmatrix}$

Divide row 3 by 143: $\begin{bmatrix} 1 & -5 & 4 & | & 42 \\ 0 & 1 & -5 & | & -34 \\ 0 & 0 & 1 & | & 6 \end{bmatrix}$

Add 5 times row 3 to row 2, and –4 times row 3 to row 1: $\begin{bmatrix} 1 & -5 & 0 & | & 18 \\ 0 & 1 & 0 & | & -4 \\ 0 & 0 & 1 & | & 6 \end{bmatrix}$

Add 5 times row 2 to row 1: $\begin{bmatrix} 1 & 0 & 0 & | & -2 \\ 0 & 1 & 0 & | & -4 \\ 0 & 0 & 1 & | & 6 \end{bmatrix}$

The solution set is $\{(-2,-4,6)\}$.

25. Evaluating: $\begin{vmatrix} -2 & 6 \\ 3 & 8 \end{vmatrix} = -16 - 18 = -34$

26. Evaluating: $\begin{vmatrix} 5 & -4 \\ 7 & -3 \end{vmatrix} = -15 + 28 = 13$

27. Evaluating: $\begin{vmatrix} 2 & 3 & -1 \\ 3 & 4 & -5 \\ 6 & 4 & 2 \end{vmatrix} = \begin{vmatrix} 0 & 0 & -1 \\ -7 & -11 & -5 \\ 10 & 10 & 2 \end{vmatrix} = -1\begin{vmatrix} -7 & -11 \\ 10 & 10 \end{vmatrix} = -(-70 + 110) = -40$

28. Evaluating: $\begin{vmatrix} 3 & -2 & 4 \\ 1 & 0 & 6 \\ 3 & -3 & 5 \end{vmatrix} = \begin{vmatrix} 3 & -2 & -14 \\ 1 & 0 & 0 \\ 3 & -3 & -13 \end{vmatrix} = -1\begin{vmatrix} -2 & -14 \\ -3 & -13 \end{vmatrix} = -(26 - 42) = 16$

29. Evaluating: $\begin{vmatrix} 5 & 4 & 3 \\ 2 & -7 & 0 \\ 3 & -2 & 0 \end{vmatrix} = 3\begin{vmatrix} 2 & -7 \\ 3 & -2 \end{vmatrix} = 3(-4 + 21) = 51$

30. Evaluating:

$\begin{vmatrix} 5 & -4 & 2 & 1 \\ 3 & 7 & 6 & -2 \\ 2 & 1 & -5 & 0 \\ 3 & -2 & 4 & 0 \end{vmatrix} = \begin{vmatrix} 5 & -4 & 2 & 1 \\ 13 & -1 & 10 & 0 \\ 2 & 1 & -5 & 0 \\ 3 & -2 & 4 & 0 \end{vmatrix} = -1\begin{vmatrix} 13 & -1 & 10 \\ 2 & 1 & -5 \\ 3 & -2 & 4 \end{vmatrix} = -1\begin{vmatrix} 15 & 0 & 5 \\ 2 & 1 & -5 \\ 7 & 0 & -6 \end{vmatrix} = -1\begin{vmatrix} 15 & 5 \\ 7 & -6 \end{vmatrix} = -(-90 - 35) = 125$

31. Let x represent the quarts of 1% milk and y represent the quarts of 4% milk. The system of equation is:
$$x + y = 10$$
$$0.01x + 0.04y = 0.02(10)$$

Multiplying the first equation by –0.01:
$$-0.01x - 0.01y = -0.1$$
$$0.01x + 0.04y = 0.2$$

Adding yields:
$$0.03y = 0.1$$
$$y = \frac{10}{3} = 3\frac{1}{3}$$
$$x = \frac{20}{3} = 6\frac{2}{3}$$

The mixture used $6\frac{2}{3}$ quarts of 1% milk and $3\frac{1}{3}$ quarts of 4% milk.

32. Let w represent the width and l represent the length. The system of equations is:
$$l = 3w$$
$$2w + 2l = 56$$
Substituting into the second equation:
$$2w + 2(3w) = 56$$
$$2w + 6w = 56$$
$$8w = 56$$
$$w = 7$$
$$l = 21$$
The width is 7 cm and the length is 21 cm.

33. Let x represent the debt on the 1% card and y represent the debt on the 1.5% card. The system of equations is:
$$x + y = 4200$$
$$0.01x + 0.015y = 57$$
Multiplying the first equation by -0.01:
$$-0.01x - 0.01y = -42$$
$$0.01x + 0.015y = 57$$
Adding yields:
$$0.005y = 15$$
$$y = 3,000$$
$$x = 1,200$$
Antonio has $1,200 debt on the 1% card and $3,000 debt on the 1.5% card.

34. Let o represent the number of $1 bills and f represent the number of $5 bills. The system of equations is:
$$o + f = 30$$
$$o + 5f = 50$$
Multiplying the first equation by -1:
$$-o - f = -30$$
$$o + 5f = 50$$
Adding yields:
$$4f = 20$$
$$f = 5$$
$$o = 25$$
Kelly had 25 one-dollar bills and 5 five-dollar bills.

35. Let r represent the number of review problems and n represent the number of new material problems. The system of equations is:
$$0.80r + 0.40n = 56$$
$$1.00r + 0.60n = 78$$
Multiplying the first equation 3 and the second equation by -2:
$$2.4r + 1.2n = 168$$
$$-2.0r - 1.2n = -156$$
Adding yields:
$$0.4r = 12$$
$$r = 30$$
Substituting into the first equation:
$$0.80(30) + 0.40n = 56$$
$$24 + 0.4n = 56$$
$$0.4n = 32$$
$$n = 80$$
There are 30 review problems and 80 new material problems.

36. Let x represent the amount invested at 4%, and y represent the amount invested at 6%. The system of equations is:
$$x + y = 2500$$
$$0.06y = 0.04x + 60$$
Substituting into the second equation:
$$0.06y = 0.04(2500 - y) + 60$$
$$0.06y = 100 - 0.04y + 60$$
$$0.10y = 160$$
$$y = 1600$$
$$x = 900$$
Sara invested $900 at 4% and $1600 at 6%.

37. Let n, d, and q represent the number of nickels, dimes, and quarters. The system of equations is:
$$0.05n + 0.10d + 0.25q = 17.70$$
$$d = 2n - 8$$
$$q = n + d + 2$$
Substituting:
$$0.05n + 0.10(2n - 8) + 0.25(3n - 6) = 17.70$$
$$0.05n + 0.20n - 0.80 + 0.75n - 1.50 = 17.70$$
$$n - 2.3 = 17.7$$
$$n = 20$$
$$d = 40 - 8 = 32$$
$$q = 20 + 32 + 2 = 54$$
The box contains 20 nickels, 32 dimes, and 54 quarters.

38. Let x, y, and z represent the number of five, ten, and twenty-dollar bills, respectively. The system of equations is:
$$x + y + z = 64$$
$$5x + 10y + 20z = 620$$
$$y = 3z$$
Substituting for y in the first two equations:
$$x + 3z + z = 64$$
$$5x + 10(3z) + 20z = 620$$
This system simplifies to:
$$x + 4z = 64$$
$$5x + 50z = 620$$
Multiplying the first equation by –5:
$$-5x - 20z = -320$$
$$5x + 50z = 620$$
Adding yields:
$$30z = 300$$
$$z = 10$$
$$y = 30$$
$$x = 24$$
The vendor collected 24 five-dollar bills, 30 five-dollar bills, and 10 twenty-dollar bills.

39. Let x represent the smallest angle, y represent the middle angle, and z represent the largest angle.
The system of equations is:
$$x + y + z = 180$$
$$z = 2x$$
$$x + z = 2y$$
Substituting into the first equation:
$$y + 2y = 180$$
$$3y = 180$$
$$y = 60$$

Substituting into the third equation:

$$x + 2x = 2(60)$$
$$3x = 120$$
$$x = 40$$
$$z = 2(40) = 80$$

The angles are 40°, 60°, and 80°.

40. Let x represent the smallest angle, y represent the middle angle, and z represent the largest angle. The system of equations is:

$$x + y + z = 180$$
$$z = 4x + 10$$
$$x + z = 3y$$

Substituting into the first equation:

$$3y + y = 180$$
$$4y = 180$$
$$y = 45$$

Substituting into the third equation:

$$x + 4x + 10 = 3(45)$$
$$5x = 125$$
$$x = 25$$
$$z = 4(25) + 10 = 110$$

The angles are 25°, 45°, and 110°.

41. Let x represent the charges on the 1% card, y represent the charges on the 1.5% card, and z represent the charges on the 2% card. The system of equations is:

$$x + y + z = 6400$$
$$0.01x + 0.015y + 0.02z = 99$$
$$y = x - 500$$

Substituting for y in the first two equations:

$$x + (x - 500) + z = 6400$$
$$0.01x + 0.015(x - 500) + 0.02z = 99$$

Simplifying the system:

$$2x + z = 6900$$
$$0.025x + 0.02z = 106.5$$

Multiplying the first equation by –0.02:

$$-0.04x - 0.02z = -138$$
$$0.025x + 0.02z = 106.5$$

Adding yields:

$$-0.015x = -31.5$$
$$x = 2,100$$
$$y = 1,600$$
$$z = 2,700$$

Kenisha has $2,100 charged on the Bank of US card, $1,600 charged on the Community Bank card, and $2,700 charged on the First National card.

42. Let x, y, and z represent the shortest, middle, and longest sides, respectively. The system of equations is:

$$x + y + z = 33$$
$$z = 2x + 3$$
$$x + z = y + 9$$

Subsituting into the first equation:

$$y + y + 9 = 33$$
$$2y + 9 = 33$$
$$2y = 24$$
$$y = 12$$

Substituting into the third equation:

$$x + (2x + 3) = 12 + 9$$
$$3x + 3 = 21$$
$$3x = 18$$
$$x = 6$$
$$z = 2(6) + 3 = 15$$

The sides are 6 inches, 12 inches, and 15 inches.

Chapter 6 Test

1. System III represents parallel lines, since the slopes are the same but the y-intercepts are unequal.
2. System I is dependent, since the second equation is a multiple of the first equation.
3. System III has no solution, since it represents parallel lines.
4. System II is consistent and has one unique solution.

5. Evaluating: $\begin{vmatrix} -2 & 4 \\ -5 & 6 \end{vmatrix} = -12 + 20 = 8$

6. Evaluating: $\begin{vmatrix} \dfrac{1}{2} & \dfrac{1}{3} \\ \dfrac{3}{4} & -\dfrac{2}{3} \end{vmatrix} = -\dfrac{1}{3} - \dfrac{1}{4} = -\dfrac{7}{12}$

7. Evaluating: $\begin{vmatrix} -1 & 2 & 1 \\ 3 & 1 & -2 \\ 2 & -1 & 1 \end{vmatrix} = \begin{vmatrix} -1 & 0 & 0 \\ 3 & 7 & 1 \\ 2 & 3 & 3 \end{vmatrix} = -1 \begin{vmatrix} 7 & 1 \\ 3 & 3 \end{vmatrix} = -1(21 - 3) = -18$

8. Evaluating: $\begin{vmatrix} 2 & 4 & -5 \\ -4 & 3 & 0 \\ -2 & 6 & 1 \end{vmatrix} = \begin{vmatrix} -8 & 34 & 0 \\ -4 & 3 & 0 \\ -2 & 6 & 1 \end{vmatrix} = 1 \begin{vmatrix} -8 & 34 \\ -4 & 3 \end{vmatrix} = -24 + 136 = 112$

9. Solving the second equation for y:

$$9x - 3y = 12$$
$$-3y = -9x + 12$$
$$y = 3x - 4$$

Since the two lines are identical, there are infinitely many solutions to the system.

10. Adding the two equations:

$$10x = -20$$
$$x = -2$$

Substituting into the second equation:

$$7(-2) + 2y = -6$$
$$2y = 8$$
$$y = 4$$

The solution set is $\{(-2, 4)\}$.

11. Substituting into the first equation:

$$4x - 5(-3x + 8) = 17$$
$$4x + 15x - 40 = 17$$
$$19x = 57$$
$$x = 3$$
$$y = -9 + 9 = -1$$

The solution set is $\{(3, -1)\}$.

12. Multiply the second equation by 3:
$$\frac{3}{4}x - \frac{1}{2}y = -21$$
$$2x + \frac{1}{2}y = -12$$

Adding yields:
$$\frac{11}{4}x = -33$$
$$x = -12$$

13. Multiply the first equation by 3 and the second equation by –4:
$$12x - 3y = 21$$
$$-12x - 8y = -8$$

Adding yields:
$$-11y = 13$$
$$y = -\frac{13}{11}$$

14. Divide row 3 by 3: $\begin{bmatrix} 1 & 1 & -4 & | & 3 \\ 0 & 1 & 4 & | & 5 \\ 0 & 0 & 1 & | & 2 \end{bmatrix}$

Add –4 times row 3 to row 2, and 4 times row 3 to row 1: $\begin{bmatrix} 1 & 1 & 0 & | & 11 \\ 0 & 1 & 0 & | & -3 \\ 0 & 0 & 1 & | & 2 \end{bmatrix}$

Add –1 times row 2 to row 1: $\begin{bmatrix} 1 & 0 & 0 & | & 14 \\ 0 & 1 & 0 & | & -3 \\ 0 & 0 & 1 & | & 2 \end{bmatrix}$

Thus $x = 14$.

15. Divide row 3 by 2: $\begin{bmatrix} 1 & 2 & -3 & | & 4 \\ 0 & 1 & 2 & | & 5 \\ 0 & 0 & 1 & | & -4 \end{bmatrix}$

Add –2 times row 3 to row 2: $\begin{bmatrix} 1 & 2 & -3 & | & 4 \\ 0 & 1 & 0 & | & 13 \\ 0 & 0 & 1 & | & -4 \end{bmatrix}$

Thus $y = 13$.

16. Form the augmented matrix: $\begin{bmatrix} 1 & 3 & -1 & | & 5 \\ 2 & -1 & -1 & | & 7 \\ 5 & 8 & -4 & | & 22 \end{bmatrix}$

Add –2 times row 1 to row 2, and –5 times row 1 to row 3: $\begin{bmatrix} 1 & 3 & -1 & | & 5 \\ 0 & -7 & 1 & | & -3 \\ 0 & -7 & 1 & | & -3 \end{bmatrix}$

Add –1 times row 2 to row 3: $\begin{bmatrix} 1 & 3 & -1 & | & 5 \\ 0 & -7 & 1 & | & -3 \\ 0 & 0 & 0 & | & 0 \end{bmatrix}$

Since $0 = 0$, the solution is dependent with infinitely many solutions.

17. Form the augmented matrix: $\begin{bmatrix} 3 & -1 & -2 & | & 1 \\ 4 & 2 & 1 & | & 5 \\ 6 & -2 & -4 & | & 9 \end{bmatrix}$

Subtract row 1 from row 2: $\begin{bmatrix} 3 & -1 & -2 & | & 1 \\ 1 & 3 & 3 & | & 4 \\ 6 & -2 & -4 & | & 9 \end{bmatrix}$

Switching row 1 and row 2: $\begin{bmatrix} 1 & 3 & 3 & | & 4 \\ 3 & -1 & -2 & | & 1 \\ 6 & -2 & -4 & | & 9 \end{bmatrix}$

Add –3 times row 1 to row 2, and –6 times row 1 to row 6: $\begin{bmatrix} 1 & 3 & 3 & | & 4 \\ 0 & -10 & -11 & | & -11 \\ 0 & -20 & -22 & | & -15 \end{bmatrix}$

Add –2 times row 2 to row 3: $\begin{bmatrix} 1 & 3 & 3 & | & 4 \\ 0 & -10 & -11 & | & -11 \\ 0 & 0 & 0 & | & 7 \end{bmatrix}$

Since $0 \neq 7$, there are no solutions.

18. Form the augmented matrix: $\begin{bmatrix} 5 & -3 & -2 & | & -1 \\ 0 & 4 & 7 & | & 3 \\ 0 & 0 & 4 & | & -12 \end{bmatrix}$

Dividing row 3 by 3: $\begin{bmatrix} 5 & -3 & -2 & | & -1 \\ 0 & 4 & 7 & | & 3 \\ 0 & 0 & 1 & | & -3 \end{bmatrix}$

Add –7 times row 3 to row 2, and 2 times row 3 to row 1: $\begin{bmatrix} 5 & -3 & 0 & | & -7 \\ 0 & 4 & 0 & | & 24 \\ 0 & 0 & 1 & | & -3 \end{bmatrix}$

Divide row 3 by 4: $\begin{bmatrix} 5 & -3 & 0 & | & -7 \\ 0 & 1 & 0 & | & 6 \\ 0 & 0 & 1 & | & -3 \end{bmatrix}$

Add 3 times row 2 to row 1, and divide row 1 by 5: $\begin{bmatrix} 1 & 0 & 0 & | & \frac{11}{5} \\ 0 & 1 & 0 & | & 6 \\ 0 & 0 & 1 & | & -3 \end{bmatrix}$

The solution set is $\left\{ \left(\frac{11}{5}, 6, -3 \right) \right\}$.

19.   Form the augmented matrix: $\begin{bmatrix} 1 & -2 & 1 & | & 0 \\ 0 & 1 & -3 & | & -1 \\ 0 & 2 & 5 & | & -2 \end{bmatrix}$

Add -2 times row 2 to row 3: $\begin{bmatrix} 1 & -2 & 1 & | & 0 \\ 0 & 1 & -3 & | & -1 \\ 0 & 0 & 11 & | & 0 \end{bmatrix}$

Divide row 3 by 11: $\begin{bmatrix} 1 & -2 & 1 & | & 0 \\ 0 & 1 & -3 & | & -1 \\ 0 & 0 & 1 & | & 0 \end{bmatrix}$

Add 3 times row 3 to row 2, and -1 times row 3 to row 1: $\begin{bmatrix} 1 & -2 & 0 & | & 0 \\ 0 & 1 & 0 & | & -1 \\ 0 & 0 & 1 & | & 0 \end{bmatrix}$

Add 2 times row 2 to row 1: $\begin{bmatrix} 1 & 0 & 0 & | & -2 \\ 0 & 1 & 0 & | & -1 \\ 0 & 0 & 1 & | & 0 \end{bmatrix}$

The solution set is $\{(-2, -1, 0)\}$.

20.   Form the augmented matrix: $\begin{bmatrix} 1 & -4 & 1 & | & 12 \\ -2 & 3 & -1 & | & -11 \\ 5 & -3 & 2 & | & 17 \end{bmatrix}$

Add 2 times row 1 to row 2, and -5 times row 1 to row 3: $\begin{bmatrix} 1 & -4 & 1 & | & 12 \\ 0 & -5 & 1 & | & 13 \\ 0 & 17 & -3 & | & -43 \end{bmatrix}$

Add 3 times row 2 to row 3: $\begin{bmatrix} 1 & -4 & 1 & | & 12 \\ 0 & -5 & 1 & | & 13 \\ 0 & 2 & 0 & | & -4 \end{bmatrix}$

Switch row 2 and row 3: $\begin{bmatrix} 1 & -4 & 1 & | & 12 \\ 0 & 2 & 0 & | & -4 \\ 0 & -5 & 1 & | & 13 \end{bmatrix}$

Divide row 2 by 2: $\begin{bmatrix} 1 & -4 & 1 & | & 12 \\ 0 & 1 & 0 & | & -2 \\ 0 & -5 & 1 & | & 13 \end{bmatrix}$

Add 4 times row 2 to row 1, and 5 times row 2 to row 3: $\begin{bmatrix} 1 & 0 & 1 & | & 4 \\ 0 & 1 & 0 & | & -2 \\ 0 & 0 & 1 & | & 3 \end{bmatrix}$

Add -1 times row 3 to row 1: $\begin{bmatrix} 1 & 0 & 0 & | & 1 \\ 0 & 1 & 0 & | & -2 \\ 0 & 0 & 1 & | & 3 \end{bmatrix}$

Thus $x = 1$.

21. Form the augmented matrix: $\begin{bmatrix} 1 & -3 & 1 & | & -13 \\ 3 & 5 & -1 & | & 17 \\ 5 & -2 & 2 & | & -13 \end{bmatrix}$

Add –3 times row 1 to row 2, and –5 times row 1 to row 3: $\begin{bmatrix} 1 & -3 & 1 & | & -13 \\ 0 & 14 & -4 & | & 56 \\ 0 & 13 & -3 & | & 52 \end{bmatrix}$

Subtracting row 3 from row 2: $\begin{bmatrix} 1 & -3 & 1 & | & -13 \\ 0 & 1 & -1 & | & 4 \\ 0 & 13 & -3 & | & 52 \end{bmatrix}$

Add –13 times row 2 to row 3, and 3 times row 2 to row 1: $\begin{bmatrix} 1 & 0 & -2 & | & -1 \\ 0 & 1 & -1 & | & 4 \\ 0 & 0 & 10 & | & 0 \end{bmatrix}$

Divide row 3 by 10: $\begin{bmatrix} 1 & 0 & -2 & | & -1 \\ 0 & 1 & -1 & | & 4 \\ 0 & 0 & 1 & | & 0 \end{bmatrix}$

Add 2 times row 3 to row 1, and row 3 to row 2: $\begin{bmatrix} 1 & 0 & 0 & | & -1 \\ 0 & 1 & 0 & | & 4 \\ 0 & 0 & 1 & | & 0 \end{bmatrix}$

Thus $y = 4$.

22. Let x represent the amount of 30% solution and y represent the amount of 70% solution used.
The system of equations is:
$$x + y = 8$$
$$0.30x + 0.70y = 0.40(8)$$
Multiplying the first equation by –0.30:
$$-0.3x - 0.3y = -2.4$$
$$0.3x + 0.7y = 3.2$$
Adding yields:
$$0.4y = 0.8$$
$$y = 2$$
$$x = 6$$
Thus 2 liters of the 70% solution should be used.

23. Let e represent the number of express washes and f represent the number of full washes. The system of equations is:
$$e + f = 75$$
$$5e + 15f = 825$$
Multiplying the first equation by –5:
$$-5e - 5f = -375$$
$$5e + 15f = 825$$
Adding yields:
$$10f = 450$$
$$f = 45$$
$$e = 75 - 45 = 30$$
There are 30 express washes.

24.   Let c, e, and d represent the batches of cream puffs, eclairs, and Danish rolls, respectively. The system of equations is:

$$0.2c + 0.5e + 0.4d = 7.0$$
$$0.3c + 0.1e + 0.2d = 3.9$$
$$0.1c + 0.5e + 0.3d = 5.5$$

Multiply each equation by 10:

$$2c + 5e + 4d = 70$$
$$3c + e + 2d = 39$$
$$c + 5e + 3d = 55$$

Form the augmented matrix (using the third equation as row 1): $\begin{bmatrix} 1 & 5 & 3 & | & 55 \\ 2 & 5 & 4 & | & 70 \\ 3 & 1 & 2 & | & 39 \end{bmatrix}$

Add –2 times row 1 to row 2, and –3 times row 1 to row 3: $\begin{bmatrix} 1 & 5 & 3 & | & 55 \\ 0 & -5 & -2 & | & -40 \\ 0 & -14 & -7 & | & -126 \end{bmatrix}$

Dividing the second row by –5: $\begin{bmatrix} 1 & 5 & 3 & | & 55 \\ 0 & 1 & \frac{2}{5} & | & 8 \\ 0 & -14 & -7 & | & -126 \end{bmatrix}$

Add –5 times row 2 to row 1, and 14 times row 2 to row 3: $\begin{bmatrix} 1 & 0 & 1 & | & 15 \\ 0 & 1 & \frac{2}{5} & | & 8 \\ 0 & 0 & -\frac{7}{5} & | & -14 \end{bmatrix}$

Multiply row 3 by $-\frac{5}{7}$: $\begin{bmatrix} 1 & 0 & 1 & | & 15 \\ 0 & 1 & \frac{2}{5} & | & 8 \\ 0 & 0 & 1 & | & 10 \end{bmatrix}$

Add $-\frac{2}{5}$ times row 3 to row 2, and subtract row 3 from row 1: $\begin{bmatrix} 1 & 0 & 0 & | & 5 \\ 0 & 1 & 0 & | & 4 \\ 0 & 0 & 1 & | & 10 \end{bmatrix}$

Thus $c = 5$, $e = 4$, and $d = 10$. They should make 5 batches of cream puffs, 4 batches of eclairs, and 10 batches of Danish rolls.

25.   Let x represent the smallest angle, y represent the middle angle, and z represent the largest angle. The system of equations is:

$$x + y + z = 180$$
$$z = x + y + 20$$
$$z - x = 65$$

Substituting into the first equation:

$$(z - 20) + z = 180$$
$$2z = 200$$
$$z = 100$$

Substituting into the third equation:

$$100 - x = 65$$
$$-x = -35$$
$$x = 35$$
$$y = 180 - 35 - 100 = 45$$

The angles are 35°, 45°, and 100°.

Chapters 0-6 Cumulative Review Problem Set

1. Evaluating: $\left(3^{-2} + 2^{-1}\right)^{-2} = \left(\frac{1}{9} + \frac{1}{2}\right)^{-2} = \left(\frac{11}{18}\right)^{-2} = \left(\frac{18}{11}\right)^{2} = \frac{324}{121}$

2. Evaluating: $\left(\frac{4^{-2}}{2^{-3}}\right)^{-1} = \frac{4^2}{2^3} = \frac{16}{8} = 2$

3. Evaluating: $\ln e^5 = 5 \ln e = 5$

4. Evaluating: $\log 0.001 = \log 10^{-3} = -3$

5. Evaluating: $\sqrt{2\frac{1}{4}} = \sqrt{\frac{9}{4}} = \frac{3}{2}$

6. Evaluating: $(-27)^{4/3} = \left(\sqrt[3]{-27}\right)^4 = (-3)^4 = 81$

7. Evaluating: $\log_2 64 = \log_2 2^6 = 6$

8. Evaluating: $\log_4 64 = \log_4 4^3 = 3$

9. Evaluating: $\left(\frac{3}{2}\right)^{-3} = \left(\frac{2}{3}\right)^3 = \frac{8}{27}$

10. Evaluating: $\sqrt[3]{-0.008} = -0.2$

11. Simplifying: $5(2x-1) - 3(x+4) - 2(3x+11) = 10x - 5 - 3x - 12 - 6x - 22 = x - 39$

    Evaluating when $x = -13$: $x - 39 = -13 - 39 = -52$

12. Simplifying: $\frac{14x^3y^2}{28x^2y^3} = \frac{x}{2y}$. Evaluating when $x = 7$ and $y = -6$: $\frac{x}{2y} = \frac{7}{2(-6)} = -\frac{7}{12}$

13. Simplifying: $\frac{7}{n} - \frac{9}{2n} + \frac{5}{3n} = \frac{42}{6n} - \frac{27}{6n} + \frac{10}{6n} = \frac{25}{6n}$. Evaluating when $n = 15$: $\frac{25}{6n} = \frac{25}{6(15)} = \frac{25}{90} = \frac{5}{18}$

14. Simplifying: $\frac{6x^2 + 7x - 20}{3x^2 - 25x + 28} = \frac{(3x-4)(2x+5)}{(3x-4)(x-7)} = \frac{2x+5}{x-7}$. Evaluating when $x = 8$: $\frac{2x+5}{x-7} = \frac{2(8)+5}{8-7} = 21$

15. Simplifying: $\frac{3a^2b}{7a} \div \frac{9ab^3}{14b^2} = \frac{3a^2b}{7a} \cdot \frac{14b^2}{9ab^3} = \frac{42a^2b^3}{63a^2b^3} = \frac{2}{3}$

16. Simplifying: $\frac{3}{x^2-4} - \frac{2}{x^2+2x} = \frac{3}{(x+2)(x-2)} - \frac{2}{x(x+2)} = \frac{3x - 2(x-2)}{x(x+2)(x-2)} = \frac{x+4}{x(x+2)(x-2)}$

17. Simplifying: $\frac{x^3+1}{x^2+5x+6} \cdot \frac{x^2-9}{x^2-2x-3} = \frac{(x+1)(x^2-x-1)}{(x+2)(x+3)} \cdot \frac{(x+3)(x-3)}{(x-3)(x+1)} = \frac{x^2-x-1}{x+2}$

18. Simplifying: $\frac{2x-1}{4} - \frac{x+2}{6} + \frac{3x-1}{8} = \frac{12x-6}{24} - \frac{4x+8}{24} + \frac{9x-3}{24} = \frac{17x-17}{24}$

19. Simplifying: $\frac{36x^{-2}y^{-4}}{24xy^{-2}} = \frac{3x^{-3}y^{-2}}{2} = \frac{3}{2x^3y^2}$

20. Simplifying: $\left(\frac{28a^2b^{-4}}{7a^{-1}b^3}\right)^{-1} = \frac{7a^{-1}b^3}{28a^2b^{-4}} = \frac{a^{-3}b^7}{4} = \frac{b^7}{4a^3}$

21. Simplifying: $\left(-5x^{-1}y^2\right)\left(6x^{-2}y^{-5}\right) = -30x^{-3}y^{-3} = -\frac{30}{x^3y^3}$

22. Simplifying: $8\sqrt{72} = 8\sqrt{36 \cdot 2} = 48\sqrt{2}$

23. Simplifying: $\sqrt[3]{54x^3y^5} = \sqrt[3]{27x^3y^3 \cdot 2y^2} = 3xy\sqrt[3]{2y^2}$

24. Simplifying: $\frac{3\sqrt{6}}{4\sqrt{27}} = \frac{3\sqrt{6}}{12\sqrt{3}} = \frac{\sqrt{2}}{4}$

25. The slope of the line $2x - y = 4$ is 2, so the perpendicular slope is $-\dfrac{1}{2}$. Using the point-slope formula:

$$y - 2 = -\frac{1}{2}(x - 6)$$
$$2y - 4 = -x + 6$$
$$x + 2y = 10$$

26. Finding the slope: $m = \dfrac{8 + 3}{4 + 2} = \dfrac{11}{6}$. Using the point-slope formula:

$$y - 8 = \frac{11}{6}(x - 4)$$
$$y - 8 = \frac{11}{6}x - \frac{22}{3}$$
$$y = \frac{11}{6}x + \frac{2}{3}$$

The function is $f(x) = \dfrac{11}{6}x + \dfrac{2}{3}$.

27. Finding the compositions:
$$f(g(-2)) = f(13) = -169 - 26 - 4 = -199$$
$$g(f(3)) = g(-9 - 6 - 4) = g(-19) = 76 + 5 = 81$$

28. Finding the compositions:
$$f(g(x)) = f(2x^2 + x - 3) = 2x^2 + x - 3 + 7 = 2x^2 + x + 4$$
$$g(f(x)) = g(x + 7) = 2(x + 7)^2 + x + 7 - 3 = 2x^2 + 28x + 98 + x + 4 = 2x^2 + 29x + 102$$

29. Multiplying: $(4 - 3i)(-2 + 6i) = -8 + 6i + 24i - 18i^2 = 10 + 30i$

30. Solving the equation:
$$-3y + 10 = x$$
$$-3y = x - 10$$
$$y = -\frac{1}{3}x + \frac{10}{3}$$

Thus $f^{-1}(x) = -\dfrac{1}{3}x + \dfrac{10}{3}$.

31. The equation is equivalent to $\dfrac{x^2}{9} + \dfrac{y^2}{27} = 1$, so $b = 3$. The length of the minor axis is 6.

32. Using synthetic division:

$$
\begin{array}{r|rrrr}
2 & 3 & -2 & 1 & -4 \\
  &   & 6 & 8 & 18 \\
\hline
  & 3 & 4 & 9 & 14
\end{array}
$$

The remainder is 14.

33. Using long division:

$$
\begin{array}{r}
x^2 + 4x - 7 \\
2x + 3 \overline{)\; 2x^3 + 11x^2 - 2x - 21} \\
\underline{2x^3 + 3x^2} \\
8x^2 - 2x \\
\underline{8x^2 + 12x} \\
-14x - 21 \\
\underline{-14x - 21} \\
0
\end{array}
$$

The remainder is 0.

34. Using long division:

$$
\begin{array}{r}
x^2 + 7x - 1 \\
3x + 1 \overline{)\; 3x^3 + 22x^2 + 4x - 1} \\
\underline{3x^3 + x^2} \\
21x^2 + 4x \\
\underline{21x^2 + 7x} \\
-3x - 1 \\
\underline{-3x - 1} \\
0
\end{array}
$$

The quotient is $x^2 + 7x - 1$ and the remainder is 0.

35.   Using synthetic division:

$$-1 \overline{\smash{\big)}\ \begin{array}{rrrrr} 2 & -3 & 2 & -1 & 6 \\ & -2 & 5 & -7 & 8 \\ \hline 2 & -5 & 7 & -8 & 14 \end{array}}$$

The quotient is $2x^3 - 5x^2 + 7x - 8$ and the remainder is 14.

36.   Using synthetic division:

$$-3 \overline{\smash{\big)}\ \begin{array}{rrrr} 1 & 9 & 23 & 15 \\ & -3 & -18 & -15 \\ \hline 1 & 6 & 5 & 0 \end{array}}$$

Since the remainder is zero, it is a factor.

37.   Using synthetic division:

$$1 \overline{\smash{\big)}\ \begin{array}{rrrrrrr} 1 & 0 & 0 & 0 & 0 & 0 & 1 \\ & 1 & 1 & 1 & 1 & 1 & 1 \\ \hline 1 & 1 & 1 & 1 & 1 & 1 & 2 \end{array}}$$

Since the remainder is nonzero, it is not a factor.

38.   Using synthetic division:

$$-1 \overline{\smash{\big)}\ \begin{array}{rrrrrrr} 1 & 0 & 0 & 0 & 0 & 0 & 0 & 1 \\ & -1 & 1 & -1 & 1 & -1 & 1 & -1 \\ \hline 1 & -1 & 1 & -1 & 1 & -1 & 1 & 0 \end{array}}$$

Since the remainder is zero, it is a factor.

39.   In standard form the equation is $\dfrac{y^2}{16} - \dfrac{x^2}{16/9} = 1$, so $b = 4$ and $a = \dfrac{4}{3}$.

The equations of the asymptotes are therefore: $y = \pm \dfrac{b}{a} x = \pm 3x$

40.   The variation equation is $y = Kx$. Substituting $y = -2$ and $x = 3$:

$$-2 = K \cdot 3$$
$$K = -\frac{2}{3}$$

So $y = -\dfrac{2}{3}x$. Substituting $x = 9$: $y = -\dfrac{2}{3}(9) = -6$

41.   The variation equation is $y = \dfrac{K}{x}$. Substituting $y = 2$ and $x = -2$:

$$2 = \frac{K}{-2}$$
$$K = -4$$

So $y = \dfrac{-4}{x}$. Substituting $x = 4$: $y = \dfrac{-4}{4} = -1$

42.   Evaluating: $\log_2 101 = \dfrac{\ln 101}{\ln 2} \approx 6.66$

43.   The vertical asymptote is $x = 1$ and the horizontal asymptote is $y = 2$.

44.   Since $(-x, y)$ results in the same equation as (x, y), the curve has y-axis symmetry.

45.   Using synthetic division:

$$-1 \overline{\smash{\big)}\ \begin{array}{rrr} 2 & 0 & 3 \\ & -2 & 2 \\ \hline 2 & -2 & 5 \end{array}}$$

Thus $f(x) = 2x - 2 + \dfrac{5}{x+1}$. The equation of the oblique asymptote is $y = 2x - 2$.

46. Solving the equation:

$$12 \cdot \frac{3}{2}(x-1) - 12 \cdot \frac{2}{3}(x+2) = 12 \cdot \frac{3}{4}(2x-5)$$
$$18(x-1) - 8(x+2) = 9(2x-5)$$
$$18x - 18 - 8x - 16 = 18x - 45$$
$$-8x = -11$$
$$x = \frac{11}{8}$$

The solution set is $\left\{ \frac{11}{8} \right\}$.

47. Solving the equation:

$$\frac{2x-1}{6} + \frac{x+3}{8} = -2$$
$$24\left(\frac{2x-1}{6}\right) + 24\left(\frac{x+3}{8}\right) = -48$$
$$8x - 4 + 3x + 9 = -48$$
$$11x = -53$$
$$x = -\frac{53}{11}$$

The solution set is $\left\{ -\frac{53}{11} \right\}$.

48. Solving the equation:

$$|2x+5| = 6$$
$$2x + 5 = -6, 6$$
$$2x = -11, 1$$
$$x = -\frac{11}{2}, \frac{1}{2}$$

The solution set is $\left\{ -\frac{11}{2}, \frac{1}{2} \right\}$.

49. Multiply the first equation by 5 and the second equation by 2:

$$15x - 10y = -55$$
$$8x + 10y = 32$$

Adding yields:

$$23x = -23$$
$$x = -1$$

Substituting:

$$-3 - 2y = -11$$
$$-2y = -8$$
$$y = 4$$

The solution set is $\left\{ (-1, 4) \right\}$.

50. Solving the equation:

$$14x^3 + 45x^2 - 14x = 0$$
$$x(14x^2 + 45x - 14) = 0$$
$$x(7x-2)(2x+7) = 0$$
$$x = -\frac{7}{2}, 0, \frac{2}{7}$$

The solution set is $\left\{ -\frac{7}{2}, 0, \frac{2}{7} \right\}$.

51. The possible rational roots are $\pm 1, \pm 2, \pm 3, \pm 6, \pm\frac{1}{2}, \pm\frac{3}{2}$. Using synthetic division:

$$
\begin{array}{r|rrrr}
-2 & 2 & 1 & -3 & 6 \\
   &   & -4 & 6 & -6 \\
\hline
   & 2 & -3 & 3 & 0
\end{array}
$$

Now solving the equation:

$$(x+2)(2x^2 - 3x + 3) = 0$$
$$x = -2, \frac{3 \pm \sqrt{9-24}}{4} = \frac{3 \pm i\sqrt{15}}{4}$$

The solution set is $\left\{ -2, \frac{3 \pm i\sqrt{15}}{4} \right\}$.

52. Solving the equation:
$$12x^2 - 14x - 10 = 0$$
$$6x^2 - 7x - 5 = 0$$
$$(3x - 5)(2x + 1) = 0$$
$$x = -\frac{1}{2}, \frac{5}{3}$$

The solution set is $\left\{-\frac{1}{2}, \frac{5}{3}\right\}$.

53. Solving the equation:
$$(x + 1)(3x - 2) = (4x + 5)(x + 1)$$
$$3x^2 + x - 2 = 4x^2 + 9x + 5$$
$$0 = x^2 + 8x + 7$$
$$0 = (x + 7)(x + 1)$$
$$x = -7, -1$$

The solution set is $\{-7, -1\}$.

54. Form the augmented matrix: $\begin{bmatrix} 1 & -1 & 1 & | & -4 \\ 2 & 1 & 3 & | & -5 \\ -3 & -2 & -1 & | & -4 \end{bmatrix}$

Add –2 times row 1 to row 2, and 3 times row 1 to row 3: $\begin{bmatrix} 1 & -1 & 1 & | & -4 \\ 0 & 3 & 1 & | & 3 \\ 0 & -5 & 2 & | & -16 \end{bmatrix}$

Divide row 2 by 3: $\begin{bmatrix} 1 & -1 & 1 & | & -4 \\ 0 & 1 & \frac{1}{3} & | & 1 \\ 0 & -5 & 2 & | & -16 \end{bmatrix}$

Add 5 times row 2 to row 3: $\begin{bmatrix} 1 & -1 & 1 & | & -4 \\ 0 & 1 & \frac{1}{3} & | & 1 \\ 0 & 0 & \frac{11}{3} & | & -11 \end{bmatrix}$

Multiply row 3 by $\frac{3}{11}$: $\begin{bmatrix} 1 & -1 & 1 & | & -4 \\ 0 & 1 & \frac{1}{3} & | & 1 \\ 0 & 0 & 1 & | & -3 \end{bmatrix}$

Add $-\frac{1}{3}$ times row 3 to row 2, and –1 times row 3 to row 1: $\begin{bmatrix} 1 & -1 & 0 & | & -1 \\ 0 & 1 & 0 & | & 2 \\ 0 & 0 & 1 & | & -3 \end{bmatrix}$

Add row 2 to row 1: $\begin{bmatrix} 1 & 0 & 0 & | & 1 \\ 0 & 1 & 0 & | & 2 \\ 0 & 0 & 1 & | & -3 \end{bmatrix}$

The solution set is $\{(1, 2, -3)\}$.

55. Solving the equation:
$$9^{x-2} = 27^{x+3}$$
$$3^{2x-4} = 3^{3x+9}$$
$$2x - 4 = 3x + 9$$
$$-13 = x$$

The solution set is $\{-13\}$.

56. Solving the equation:
$$4^{x+1} = 16^{2x-1}$$
$$4^{x+1} = 4^{4x-2}$$
$$x + 1 = 4x - 2$$
$$-3x = -3$$
$$x = 1$$

The solution set is $\{1\}$.

57. Solving the equation:

$$2^{x+3} = 7^{x-1}$$
$$\ln 2^{x+3} = \ln 7^{x-1}$$
$$(x+3)\ln 2 = (x-1)\ln 7$$
$$x\ln 2 + 3\ln 2 = x\ln 7 - \ln 7$$
$$x(\ln 2 - \ln 7) = -\ln 7 - 3\ln 2$$
$$x = \frac{-\ln 7 - 3\ln 2}{\ln 2 - \ln 7} \approx 3.21$$

The solution set is $\{3.21\}$.

58. Solving the equation:

$$\log(x+3) + \log(x-1) = 1$$
$$\log(x^2 + 2x - 3) = 1$$
$$x^2 + 2x - 3 = 10$$
$$x^2 + 2x - 13 = 0$$
$$x = \frac{-2 \pm \sqrt{4+52}}{2} = \frac{-2 \pm 2\sqrt{14}}{2} = -1 + \sqrt{14}$$
$$\left(x = -1 - \sqrt{14} \text{ does not check}\right)$$

The solution set is $\left\{-1 + \sqrt{14}\right\}$.

59. Solving the inequality:

$$\frac{x-1}{4} - \frac{x+2}{12} \geq 1$$
$$3x - 3 - x - 2 \geq 12$$
$$2x - 5 \geq 12$$
$$2x \geq 17$$
$$x \geq \frac{17}{2}$$

The solution set is $\left[\frac{17}{2}, \infty\right)$.

60. Solving the inequality:

$$|2x - 1| < 4$$
$$-4 < 2x - 1 < 4$$
$$-3 < 2x < 5$$
$$-\frac{3}{2} < x < \frac{5}{2}$$

The solution set is $\left(-\frac{3}{2}, \frac{5}{2}\right)$.

61. Solving the inequality:

$$|3x + 2| > 6$$

$$3x + 2 < -6 \qquad \text{or} \qquad 3x + 2 > 6$$
$$3x < -8 \qquad\qquad 3x > 4$$
$$x < -\frac{8}{3} \qquad\qquad x > \frac{4}{3}$$

The solution set is $\left(-\infty, -\frac{8}{3}\right) \cup \left(\frac{4}{3}, \infty\right)$.

62. Solving the inequality:

$$-3 \leq 2(x-1) - (x+4)$$
$$-3 \leq 2x - 2 - x - 4$$
$$-3 \leq x - 6$$
$$x \geq 3$$

The solution set is $[3, \infty)$.

63. Forming the sign chart:

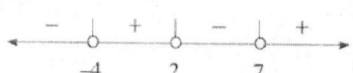

The solution set is $[-4, 2] \cup [7, \infty)$.

64. Solving the inequality:

$$(x-1)(x+5) < (x-1)(2x+3)$$
$$x^2 + 4x - 5 < 2x^2 + x - 3$$
$$0 < x^2 - 3x + 2$$
$$0 < (x-1)(x-2)$$

65. Solving the inequality:

$$\frac{x+1}{x-4} - 2 > 0$$
$$\frac{x+1-2x+8}{x-4} > 0$$
$$\frac{-x+9}{x-4} > 0$$

Forming a sign chart:

The solution set is $(-\infty, 1) \cup (2, \infty)$.

Forming a sign chart:

The solution set is $(4, 9)$.

66.   Solving the inequality:
$$x^2 + 5x \le 24$$
$$x^2 + 5x - 24 \le 0$$
$$(x + 8)(x - 3) \le 0$$
Forming a sign chart:

The solution set is $[-8, 3]$.

67.   Graphing the function $f(x) = 2|x + 3| + 1$:

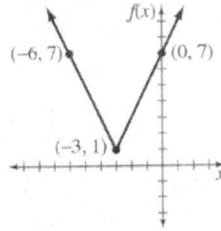

68.   Graphing the function $f(x) = \log_3(x - 1)$:

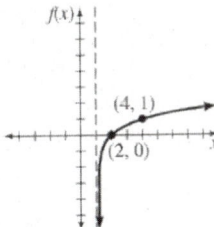

69.   Completing the square: $f(x) = -\left(x^2 - 4x\right) - 5 = -\left(x^2 - 4x + 4\right) + 4 - 5 = -(x - 2)^2 - 1$. Graphing the function:

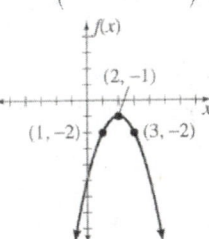

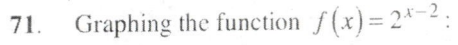

70.   Graphing the function $f(x) = 2\sqrt{x + 1} + 2$:

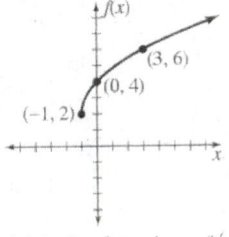

71.   Graphing the function $f(x) = 2^{x-2}$:

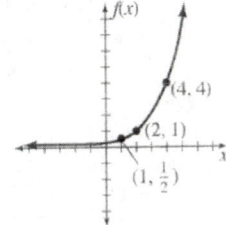

72.   Graphing the function $f(x) = (x - 1)(x + 1)(x - 4)$:

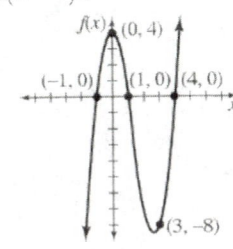

73.   The vertical asymptotes are $x = 3$ and $x = -3$, and the horizontal asymptote is $y = 0$.

Graphing the function $f(x) = \dfrac{-1}{x^2 - 9}$:

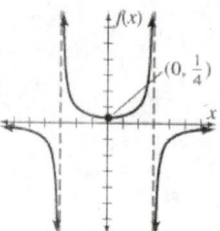

74.   Using synthetic division:

$$1\overline{)\begin{array}{ccc} 1 & 1 & 1 \\ & 1 & 2 \\ \hline 1 & 2 & 3 \end{array}}$$

Thus $f(x) = x + 2 + \dfrac{3}{x+1}$. The equation of the oblique asymptote is $y = x + 2$. The vertical asymptote is $x = 1$.

Graphing the function $f(x) = \dfrac{x^2 + x + 1}{x - 1}$:

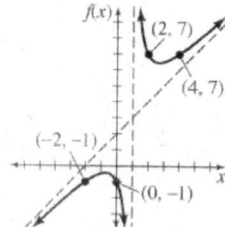

75.   Let $5x$ represent the number of female students and $6x$ the number of male students. The equation is:

$$5x + 6x = 13200$$
$$11x = 13200$$
$$x = 1200$$
$$5x = 6000, 6x = 7200$$

There are 6,000 female students and 7,200 male students.

76.   Let p represent the selling price. The equation is:

$$170 + 0.15p = p$$
$$170 = 0.85p$$
$$p = 200$$

He should price the driver at $200.

77.   The price is: $170 + 0.15(170) = \$195.50$

78.   Let w and l represent the width and length. Since the perimeter is 38 cm:

$$2w + 2l = 38$$
$$w + l = 19$$
$$l = 19 - w$$

Using the area:

$$w(19 - w) = 84$$
$$19w - w^2 = 84$$
$$0 = w^2 - 19w + 84$$
$$0 = (w - 12)(w - 7)$$
$$w = 7, 12$$

The width is 7 cm and the length is 12 cm.

79. Let t represent the time they were working together. The equation is:

$$\frac{1}{30}(10+t)+\frac{1}{20}t = 1$$
$$2(10+t)+3t = 60$$
$$20+5t = 60$$
$$5t = 40$$
$$t = 8$$

It took them 8 minutes to finish the task.

80. Let x represent the amount of 50% acid solution. The equation is:

$$0.50(x)+0.10(5) = 0.30(x+5)$$
$$0.5x+0.5 = 0.3x+1.5$$
$$0.2x = 1$$
$$x = 5$$

Thus 5 liters should be added.

81. Solving the inequality:

$$30+0.15m < 20+0.20m$$
$$10 < 0.05m$$
$$200 < m$$

Agency A is cheaper if the car is driven over 200 miles.

82. Let C represent the cost, w the number of workers, and d the number of days. The variation equation is $C = Kwd$.
Substituting $C = 12500$, $w = 25$, and $d = 5$:

$$12500 = K(25)(5)$$
$$12500 = 125K$$
$$K = 100$$

So $C = 100wd$. Substituting $w = 30$ and $d = 6$: $C = 100(30)(6) = \$18,000$

83. Solving the equation:

$$500\left(1+\frac{0.06}{2}\right)^{2t} = 1000$$
$$(1.03)^{2t} = 2$$
$$\ln(1.03)^{2t} = \ln 2$$
$$2t \ln 1.03 = \ln 2$$
$$t = \frac{\ln 2}{2\ln 1.03} \approx 11.7$$

It will take approximately 11.7 years.

Chapter 7
Algebra of Matrices

7.1 Algebra of 2 x 2 Matrices

1. Computing: $\begin{bmatrix} 1 & -2 \\ 3 & 4 \end{bmatrix} + \begin{bmatrix} 2 & -3 \\ 5 & -1 \end{bmatrix} = \begin{bmatrix} 3 & -5 \\ 8 & 3 \end{bmatrix}$

3. Computing: $3\begin{bmatrix} 0 & 6 \\ -4 & 2 \end{bmatrix} + \begin{bmatrix} -2 & 3 \\ 5 & -4 \end{bmatrix} = \begin{bmatrix} 0 & 18 \\ -12 & 6 \end{bmatrix} + \begin{bmatrix} -2 & 3 \\ 5 & -4 \end{bmatrix} = \begin{bmatrix} -2 & 21 \\ -7 & 2 \end{bmatrix}$

5. Computing: $4\begin{bmatrix} 1 & -2 \\ 3 & 4 \end{bmatrix} - 3\begin{bmatrix} 2 & -3 \\ 5 & -1 \end{bmatrix} = \begin{bmatrix} 4 & -8 \\ 12 & 16 \end{bmatrix} - \begin{bmatrix} 6 & -9 \\ 15 & -3 \end{bmatrix} = \begin{bmatrix} -2 & 1 \\ -3 & 19 \end{bmatrix}$

7. Computing: $\begin{bmatrix} 1 & -2 \\ 3 & 4 \end{bmatrix} - \begin{bmatrix} 2 & -3 \\ 5 & -1 \end{bmatrix} - \begin{bmatrix} 0 & 6 \\ -4 & 2 \end{bmatrix} = \begin{bmatrix} -1 & -5 \\ 2 & 3 \end{bmatrix}$

9. Computing: $2\begin{bmatrix} -2 & 3 \\ 5 & -4 \end{bmatrix} - 4\begin{bmatrix} 2 & 5 \\ 7 & 3 \end{bmatrix} = \begin{bmatrix} -4 & 6 \\ 10 & -8 \end{bmatrix} - \begin{bmatrix} 8 & 20 \\ 28 & 12 \end{bmatrix} = \begin{bmatrix} -12 & -14 \\ -18 & -20 \end{bmatrix}$

11. Computing: $\begin{bmatrix} 2 & -3 \\ 5 & -1 \end{bmatrix} - \begin{bmatrix} -2 & 3 \\ 5 & -4 \end{bmatrix} - \begin{bmatrix} 2 & 5 \\ 7 & 3 \end{bmatrix} = \begin{bmatrix} 2 & -11 \\ -7 & 0 \end{bmatrix}$

13. Computing:

$AB = \begin{bmatrix} 1 & -1 \\ 2 & -2 \end{bmatrix}\begin{bmatrix} 3 & -4 \\ -1 & 2 \end{bmatrix} = \begin{bmatrix} 4 & -6 \\ 8 & -12 \end{bmatrix}$ $\qquad BA = \begin{bmatrix} 3 & -4 \\ -1 & 2 \end{bmatrix}\begin{bmatrix} 1 & -1 \\ 2 & -2 \end{bmatrix} = \begin{bmatrix} -5 & 5 \\ 3 & -3 \end{bmatrix}$

15. Computing:

$AB = \begin{bmatrix} 1 & -3 \\ -4 & 6 \end{bmatrix}\begin{bmatrix} 7 & -3 \\ 4 & 5 \end{bmatrix} = \begin{bmatrix} -5 & -18 \\ -4 & 42 \end{bmatrix}$ $\qquad BA = \begin{bmatrix} 7 & -3 \\ 4 & 5 \end{bmatrix}\begin{bmatrix} 1 & -3 \\ -4 & 6 \end{bmatrix} = \begin{bmatrix} 19 & -39 \\ -16 & 18 \end{bmatrix}$

17. Computing:

$AB = \begin{bmatrix} 2 & -4 \\ 1 & -2 \end{bmatrix}\begin{bmatrix} 1 & -2 \\ -3 & 6 \end{bmatrix} = \begin{bmatrix} 14 & -28 \\ 7 & -14 \end{bmatrix}$ $\qquad BA = \begin{bmatrix} 1 & -2 \\ -3 & 6 \end{bmatrix}\begin{bmatrix} 2 & -4 \\ 1 & -2 \end{bmatrix} = \begin{bmatrix} 0 & 0 \\ 0 & 0 \end{bmatrix}$

19. Computing:

$AB = \begin{bmatrix} -3 & -2 \\ -4 & -1 \end{bmatrix}\begin{bmatrix} 2 & -1 \\ 4 & 5 \end{bmatrix} = \begin{bmatrix} -14 & -7 \\ -12 & -1 \end{bmatrix}$ $\qquad BA = \begin{bmatrix} 2 & -1 \\ 4 & 5 \end{bmatrix}\begin{bmatrix} -3 & -2 \\ -4 & -1 \end{bmatrix} = \begin{bmatrix} -2 & -3 \\ -32 & -13 \end{bmatrix}$

21. Computing:

$$AB = \begin{bmatrix} 2 & -1 \\ -5 & 3 \end{bmatrix}\begin{bmatrix} 3 & 1 \\ 5 & 2 \end{bmatrix} = \begin{bmatrix} 1 & 0 \\ 0 & 1 \end{bmatrix} \qquad BA = \begin{bmatrix} 3 & 1 \\ 5 & 2 \end{bmatrix}\begin{bmatrix} 2 & -1 \\ -5 & 3 \end{bmatrix} = \begin{bmatrix} 1 & 0 \\ 0 & 1 \end{bmatrix}$$

23. Computing:

$$AB = \begin{bmatrix} \dfrac{1}{2} & -\dfrac{1}{3} \\ \dfrac{1}{3} & \dfrac{1}{4} \end{bmatrix}\begin{bmatrix} 4 & -6 \\ 6 & -4 \end{bmatrix} = \begin{bmatrix} 0 & -\dfrac{5}{3} \\ \dfrac{17}{6} & -3 \end{bmatrix} \qquad BA = \begin{bmatrix} 4 & -6 \\ 6 & -4 \end{bmatrix}\begin{bmatrix} \dfrac{1}{2} & -\dfrac{1}{3} \\ \dfrac{1}{3} & \dfrac{1}{4} \end{bmatrix} = \begin{bmatrix} 0 & -\dfrac{17}{6} \\ \dfrac{5}{3} & -3 \end{bmatrix}$$

25. Computing:

$$AB = \begin{bmatrix} 5 & 6 \\ 2 & 3 \end{bmatrix}\begin{bmatrix} 1 & -2 \\ -\dfrac{2}{3} & \dfrac{5}{3} \end{bmatrix} = \begin{bmatrix} 1 & 0 \\ 0 & 1 \end{bmatrix} \qquad BA = \begin{bmatrix} 1 & -2 \\ -\dfrac{2}{3} & \dfrac{5}{3} \end{bmatrix}\begin{bmatrix} 5 & 6 \\ 2 & 3 \end{bmatrix} = \begin{bmatrix} 1 & 0 \\ 0 & 1 \end{bmatrix}$$

27. Computing:

$$AB = \begin{bmatrix} -2 & 3 \\ 5 & 4 \end{bmatrix}\begin{bmatrix} 0 & 1 \\ 1 & 0 \end{bmatrix} = \begin{bmatrix} 3 & -2 \\ 4 & 5 \end{bmatrix} \qquad BA = \begin{bmatrix} 0 & 1 \\ 1 & 0 \end{bmatrix}\begin{bmatrix} -2 & 3 \\ 5 & 4 \end{bmatrix} = \begin{bmatrix} 5 & 4 \\ -2 & 3 \end{bmatrix}$$

29. Computing:

$$AD = \begin{bmatrix} -2 & 3 \\ 5 & 4 \end{bmatrix}\begin{bmatrix} 1 & 1 \\ 1 & 1 \end{bmatrix} = \begin{bmatrix} 1 & 1 \\ 9 & 9 \end{bmatrix} \qquad DA = \begin{bmatrix} 1 & 1 \\ 1 & 1 \end{bmatrix}\begin{bmatrix} -2 & 3 \\ 5 & 4 \end{bmatrix} = \begin{bmatrix} 3 & 7 \\ 3 & 7 \end{bmatrix}$$

31. Computing each matrix:

$$(AB)C = \left(\begin{bmatrix} 2 & 4 \\ 5 & -3 \end{bmatrix}\begin{bmatrix} -2 & 3 \\ -1 & 2 \end{bmatrix}\right)\begin{bmatrix} 2 & 1 \\ 3 & 7 \end{bmatrix} = \begin{bmatrix} -8 & 14 \\ -7 & 9 \end{bmatrix}\begin{bmatrix} 2 & 1 \\ 3 & 7 \end{bmatrix} = \begin{bmatrix} 26 & 90 \\ 13 & 56 \end{bmatrix}$$

$$A(BC) = \begin{bmatrix} 2 & 4 \\ 5 & -3 \end{bmatrix}\left(\begin{bmatrix} -2 & 3 \\ -1 & 2 \end{bmatrix}\begin{bmatrix} 2 & 1 \\ 3 & 7 \end{bmatrix}\right) = \begin{bmatrix} 2 & 4 \\ 5 & -3 \end{bmatrix}\begin{bmatrix} 5 & 19 \\ 4 & 13 \end{bmatrix} = \begin{bmatrix} 26 & 90 \\ 13 & 56 \end{bmatrix}$$

33. Computing each matrix:

$$(A+B)C = \left(\begin{bmatrix} 2 & 4 \\ 5 & -3 \end{bmatrix}+\begin{bmatrix} -2 & 3 \\ -1 & 2 \end{bmatrix}\right)\begin{bmatrix} 2 & 1 \\ 3 & 7 \end{bmatrix} = \begin{bmatrix} 0 & 7 \\ 4 & -1 \end{bmatrix}\begin{bmatrix} 2 & 1 \\ 3 & 7 \end{bmatrix} = \begin{bmatrix} 21 & 49 \\ 5 & -3 \end{bmatrix}$$

$$AC + BC = \begin{bmatrix} 2 & 4 \\ 5 & -3 \end{bmatrix}\begin{bmatrix} 2 & 1 \\ 3 & 7 \end{bmatrix}+\begin{bmatrix} -2 & 3 \\ -1 & 2 \end{bmatrix}\begin{bmatrix} 2 & 1 \\ 3 & 7 \end{bmatrix} = \begin{bmatrix} 16 & 30 \\ 1 & -16 \end{bmatrix}+\begin{bmatrix} 5 & 19 \\ 4 & 13 \end{bmatrix} = \begin{bmatrix} 21 & 49 \\ 5 & -3 \end{bmatrix}$$

35. Computing each matrix:

$$A + B = \begin{bmatrix} a_{11} & a_{12} \\ a_{21} & a_{22} \end{bmatrix}+\begin{bmatrix} b_{11} & b_{12} \\ b_{21} & b_{22} \end{bmatrix} = \begin{bmatrix} a_{11}+b_{11} & a_{12}+b_{12} \\ a_{21}+b_{21} & a_{22}+b_{22} \end{bmatrix}$$

$$B + A = \begin{bmatrix} b_{11} & b_{12} \\ b_{21} & b_{22} \end{bmatrix}+\begin{bmatrix} a_{11} & a_{12} \\ a_{21} & a_{22} \end{bmatrix} = \begin{bmatrix} b_{11}+a_{11} & b_{12}+a_{12} \\ b_{21}+a_{21} & b_{22}+a_{22} \end{bmatrix} = \begin{bmatrix} a_{11}+b_{11} & a_{12}+b_{12} \\ a_{21}+b_{21} & a_{22}+b_{22} \end{bmatrix}$$

37. Computing the sum: $A + (-A) = \begin{bmatrix} a_{11} & a_{12} \\ a_{21} & a_{22} \end{bmatrix}+\begin{bmatrix} -a_{11} & -a_{12} \\ -a_{21} & -a_{22} \end{bmatrix} = \begin{bmatrix} 0 & 0 \\ 0 & 0 \end{bmatrix} = O$

39. Computing each matrix:

$$(k+l)A = (k+l)\begin{bmatrix} a_{11} & a_{12} \\ a_{21} & a_{22} \end{bmatrix} = \begin{bmatrix} ka_{11}+la_{11} & ka_{12}+la_{12} \\ ka_{21}+la_{21} & ka_{22}+la_{22} \end{bmatrix}$$

$$kA + lA = \begin{bmatrix} ka_{11} & ka_{12} \\ ka_{21} & ka_{22} \end{bmatrix}+\begin{bmatrix} lb_{11} & lb_{12} \\ lb_{21} & lb_{22} \end{bmatrix} = \begin{bmatrix} ka_{11}+la_{11} & ka_{12}+la_{12} \\ ka_{21}+la_{21} & ka_{22}+la_{22} \end{bmatrix}$$

49. Computing each power:

$$A^2 = \begin{bmatrix} 1 & -1 \\ 2 & 3 \end{bmatrix}\begin{bmatrix} 1 & -1 \\ 2 & 3 \end{bmatrix} = \begin{bmatrix} -1 & -4 \\ 8 & 7 \end{bmatrix} \qquad A^3 = \begin{bmatrix} -1 & -4 \\ 8 & 7 \end{bmatrix}\begin{bmatrix} 1 & -1 \\ 2 & 3 \end{bmatrix} = \begin{bmatrix} -9 & -11 \\ 22 & 13 \end{bmatrix}$$

7.2 Multiplicative Inverses

1. Finding the inverse: $A^{-1} = \dfrac{1}{15-14}\begin{bmatrix} 3 & -7 \\ -2 & 5 \end{bmatrix} = \begin{bmatrix} 3 & -7 \\ -2 & 5 \end{bmatrix}$

3. Finding the inverse: $A^{-1} = \dfrac{1}{15-16}\begin{bmatrix} 5 & -8 \\ -2 & 3 \end{bmatrix} = -1\begin{bmatrix} 5 & -8 \\ -2 & 3 \end{bmatrix} = \begin{bmatrix} -5 & 8 \\ 2 & -3 \end{bmatrix}$

5. Finding the inverse: $A^{-1} = \dfrac{1}{-4-6}\begin{bmatrix} 4 & -2 \\ -3 & -1 \end{bmatrix} = -\dfrac{1}{10}\begin{bmatrix} 4 & -2 \\ -3 & -1 \end{bmatrix} = \begin{bmatrix} -\dfrac{2}{5} & \dfrac{1}{5} \\ \dfrac{3}{10} & \dfrac{1}{10} \end{bmatrix}$

7. Since the determinant of A is 0, there is no inverse.

9. Finding the inverse: $A^{-1} = \dfrac{1}{-15+8}\begin{bmatrix} 5 & -2 \\ 4 & -3 \end{bmatrix} = -\dfrac{1}{7}\begin{bmatrix} 5 & -2 \\ 4 & -3 \end{bmatrix} = \begin{bmatrix} -\dfrac{5}{7} & \dfrac{2}{7} \\ -\dfrac{4}{7} & \dfrac{3}{7} \end{bmatrix}$

11. Finding the inverse: $A^{-1} = \dfrac{1}{0-5}\begin{bmatrix} 3 & -1 \\ -5 & 0 \end{bmatrix} = -\dfrac{1}{5}\begin{bmatrix} 3 & -1 \\ -5 & 0 \end{bmatrix} = \begin{bmatrix} -\dfrac{3}{5} & \dfrac{1}{5} \\ 1 & 0 \end{bmatrix}$

13. Finding the inverse: $A^{-1} = \dfrac{1}{8-3}\begin{bmatrix} -4 & 3 \\ 1 & -2 \end{bmatrix} = \dfrac{1}{5}\begin{bmatrix} -4 & 3 \\ 1 & -2 \end{bmatrix} = \begin{bmatrix} -\dfrac{4}{5} & \dfrac{3}{5} \\ \dfrac{1}{5} & -\dfrac{2}{5} \end{bmatrix}$

15. Finding the inverse: $A^{-1} = \dfrac{1}{-12+15}\begin{bmatrix} 6 & -5 \\ 3 & -2 \end{bmatrix} = \dfrac{1}{3}\begin{bmatrix} 6 & -5 \\ 3 & -2 \end{bmatrix} = \begin{bmatrix} 2 & -\dfrac{5}{3} \\ 1 & -\dfrac{2}{3} \end{bmatrix}$

17. Finding the inverse: $A^{-1} = \dfrac{1}{-1-1}\begin{bmatrix} -1 & -1 \\ -1 & 1 \end{bmatrix} = -\dfrac{1}{2}\begin{bmatrix} -1 & -1 \\ -1 & 1 \end{bmatrix} = \begin{bmatrix} \dfrac{1}{2} & \dfrac{1}{2} \\ \dfrac{1}{2} & -\dfrac{1}{2} \end{bmatrix}$

19. Computing the product: $AB = \begin{bmatrix} 4 & 3 \\ 2 & 5 \end{bmatrix}\begin{bmatrix} 3 \\ 6 \end{bmatrix} = \begin{bmatrix} 30 \\ 36 \end{bmatrix}$

21. Computing the product: $AB = \begin{bmatrix} -3 & -4 \\ 2 & 1 \end{bmatrix}\begin{bmatrix} 4 \\ -3 \end{bmatrix} = \begin{bmatrix} 0 \\ 5 \end{bmatrix}$

23. Computing the product: $AB = \begin{bmatrix} -4 & 2 \\ 7 & -5 \end{bmatrix}\begin{bmatrix} -1 \\ -4 \end{bmatrix} = \begin{bmatrix} -4 \\ 13 \end{bmatrix}$

25. Computing the product: $AB = \begin{bmatrix} -2 & -3 \\ -5 & -6 \end{bmatrix}\begin{bmatrix} 5 \\ -2 \end{bmatrix} = \begin{bmatrix} -4 \\ -13 \end{bmatrix}$

27. The inverse is: $A^{-1} = \dfrac{1}{4-3}\begin{bmatrix} 2 & -3 \\ -1 & 2 \end{bmatrix} = \begin{bmatrix} 2 & -3 \\ -1 & 2 \end{bmatrix}$

  Thus the solution is: $\begin{bmatrix} x \\ y \end{bmatrix} = \begin{bmatrix} 2 & -3 \\ -1 & 2 \end{bmatrix}\begin{bmatrix} 13 \\ 8 \end{bmatrix} = \begin{bmatrix} 2 \\ 3 \end{bmatrix}$. The solution set is $\{(2,3)\}$.

29. The inverse is: $A^{-1} = \dfrac{1}{8-9}\begin{bmatrix} 2 & 3 \\ 3 & 4 \end{bmatrix} = -1\begin{bmatrix} 2 & 3 \\ 3 & 4 \end{bmatrix} = \begin{bmatrix} -2 & -3 \\ -3 & -4 \end{bmatrix}$

  Thus the solution is: $\begin{bmatrix} x \\ y \end{bmatrix} = \begin{bmatrix} -2 & -3 \\ -3 & -4 \end{bmatrix}\begin{bmatrix} -23 \\ 16 \end{bmatrix} = \begin{bmatrix} -2 \\ 5 \end{bmatrix}$. The solution set is $\{(-2,5)\}$.

31. The inverse is: $A^{-1} = \dfrac{1}{5+42}\begin{bmatrix} 5 & 7 \\ -6 & 1 \end{bmatrix} = \dfrac{1}{47}\begin{bmatrix} 5 & 7 \\ -6 & 1 \end{bmatrix} = \begin{bmatrix} \dfrac{5}{47} & \dfrac{7}{47} \\ -\dfrac{6}{47} & \dfrac{1}{47} \end{bmatrix}$

Thus the solution is: $\begin{bmatrix} x \\ y \end{bmatrix} = \begin{bmatrix} \dfrac{5}{47} & \dfrac{7}{47} \\ -\dfrac{6}{47} & \dfrac{1}{47} \end{bmatrix}\begin{bmatrix} 7 \\ -5 \end{bmatrix} = \begin{bmatrix} 0 \\ -1 \end{bmatrix}$. The solution set is $\{(0,-1)\}$.

33. The inverse is: $A^{-1} = \dfrac{1}{-9+20}\begin{bmatrix} -3 & 5 \\ -4 & 3 \end{bmatrix} = \dfrac{1}{11}\begin{bmatrix} -3 & 5 \\ -4 & 3 \end{bmatrix} = \begin{bmatrix} -\dfrac{3}{11} & \dfrac{5}{11} \\ -\dfrac{4}{11} & \dfrac{3}{11} \end{bmatrix}$

Thus the solution is: $\begin{bmatrix} x \\ y \end{bmatrix} = \begin{bmatrix} -\dfrac{3}{11} & \dfrac{5}{11} \\ -\dfrac{4}{11} & \dfrac{3}{11} \end{bmatrix}\begin{bmatrix} 2 \\ -1 \end{bmatrix} = \begin{bmatrix} -1 \\ -1 \end{bmatrix}$. The solution set is $\{(-1,1)\}$.

35. Write the system as:
$$3x + y = 19$$
$$9x - 5y = 1$$

The inverse is: $A^{-1} = \dfrac{1}{-15-9}\begin{bmatrix} -5 & -1 \\ -9 & 3 \end{bmatrix} = -\dfrac{1}{24}\begin{bmatrix} -5 & -1 \\ -9 & 3 \end{bmatrix} = \begin{bmatrix} \dfrac{5}{24} & \dfrac{1}{24} \\ \dfrac{3}{8} & -\dfrac{1}{8} \end{bmatrix}$

Thus the solution is: $\begin{bmatrix} x \\ y \end{bmatrix} = \begin{bmatrix} \dfrac{5}{24} & \dfrac{1}{24} \\ \dfrac{3}{8} & -\dfrac{1}{8} \end{bmatrix}\begin{bmatrix} 19 \\ 1 \end{bmatrix} = \begin{bmatrix} 4 \\ 7 \end{bmatrix}$. The solution set is $\{(4,7)\}$.

37. The inverse is: $A^{-1} = \dfrac{1}{-54-60}\begin{bmatrix} -18 & -2 \\ -30 & 3 \end{bmatrix} = -\dfrac{1}{114}\begin{bmatrix} -18 & -2 \\ -30 & 3 \end{bmatrix} = \begin{bmatrix} \dfrac{3}{19} & \dfrac{1}{57} \\ \dfrac{5}{19} & -\dfrac{1}{38} \end{bmatrix}$

Thus the solution is: $\begin{bmatrix} x \\ y \end{bmatrix} = \begin{bmatrix} \dfrac{3}{19} & \dfrac{1}{57} \\ \dfrac{5}{19} & -\dfrac{1}{38} \end{bmatrix}\begin{bmatrix} 0 \\ -19 \end{bmatrix} = \begin{bmatrix} -\dfrac{1}{3} \\ \dfrac{1}{2} \end{bmatrix}$. The solution set is $\left\{\left(-\dfrac{1}{3}, \dfrac{1}{2}\right)\right\}$.

39. Multiply the first equation by 12 and the second equation by 15:
$$4x + 9y = 144$$
$$10x + 3y = -30$$

The inverse is: $A^{-1} = \dfrac{1}{12-90}\begin{bmatrix} 3 & -9 \\ -10 & 4 \end{bmatrix} = -\dfrac{1}{78}\begin{bmatrix} 3 & -9 \\ -10 & 4 \end{bmatrix} = \begin{bmatrix} -\dfrac{1}{26} & \dfrac{3}{26} \\ \dfrac{5}{39} & -\dfrac{2}{39} \end{bmatrix}$

Thus the solution is: $\begin{bmatrix} x \\ y \end{bmatrix} = \begin{bmatrix} -\dfrac{1}{26} & \dfrac{3}{26} \\ \dfrac{5}{39} & -\dfrac{2}{39} \end{bmatrix}\begin{bmatrix} 144 \\ -30 \end{bmatrix} = \begin{bmatrix} -9 \\ 20 \end{bmatrix}$. The solution set is $\{(-9,20)\}$.

7.3 *m* x *n* **Matrices**

1. Finding the matrices:

$$A + B = \begin{bmatrix} 2 & -1 & 4 \\ -2 & 0 & 5 \end{bmatrix} + \begin{bmatrix} -1 & 4 & -7 \\ 5 & -6 & 2 \end{bmatrix} = \begin{bmatrix} 1 & 3 & -3 \\ 3 & -6 & 7 \end{bmatrix}$$

$$A - B = \begin{bmatrix} 2 & -1 & 4 \\ -2 & 0 & 5 \end{bmatrix} - \begin{bmatrix} -1 & 4 & -7 \\ 5 & -6 & 2 \end{bmatrix} = \begin{bmatrix} 3 & -5 & 11 \\ -7 & 6 & 3 \end{bmatrix}$$

$$2A + 3B = \begin{bmatrix} 4 & -2 & 8 \\ -4 & 0 & 10 \end{bmatrix} + \begin{bmatrix} -3 & 12 & -21 \\ 15 & -18 & 6 \end{bmatrix} = \begin{bmatrix} 1 & 10 & -13 \\ 11 & -18 & 16 \end{bmatrix}$$

$$4A - 2B = \begin{bmatrix} 8 & -4 & 16 \\ -8 & 0 & 20 \end{bmatrix} - \begin{bmatrix} -2 & 8 & -14 \\ 10 & -12 & 4 \end{bmatrix} = \begin{bmatrix} 10 & -12 & 30 \\ -18 & 12 & 16 \end{bmatrix}$$

3. Finding the matrices:

$$A + B = \begin{bmatrix} 2 & -1 & 4 & 12 \end{bmatrix} + \begin{bmatrix} -3 & -6 & 9 & -5 \end{bmatrix} = \begin{bmatrix} -1 & -7 & 13 & 7 \end{bmatrix}$$

$$A - B = \begin{bmatrix} 2 & -1 & 4 & 12 \end{bmatrix} - \begin{bmatrix} -3 & -6 & 9 & -5 \end{bmatrix} = \begin{bmatrix} 5 & 5 & -5 & 17 \end{bmatrix}$$

$$2A + 3B = \begin{bmatrix} 4 & -2 & 8 & 24 \end{bmatrix} + \begin{bmatrix} -9 & -18 & 27 & -15 \end{bmatrix} = \begin{bmatrix} -5 & -20 & 35 & 9 \end{bmatrix}$$

$$4A - 2B = \begin{bmatrix} 8 & -4 & 16 & 48 \end{bmatrix} - \begin{bmatrix} -6 & -12 & 18 & -10 \end{bmatrix} = \begin{bmatrix} 14 & 8 & -2 & 58 \end{bmatrix}$$

5. Finding the matrices:

$$A + B = \begin{bmatrix} 3 & -2 & 1 \\ -1 & 4 & -7 \\ 0 & 5 & 9 \end{bmatrix} + \begin{bmatrix} 5 & -1 & -3 \\ 10 & -2 & 4 \\ 7 & 0 & 12 \end{bmatrix} = \begin{bmatrix} 8 & -3 & -2 \\ 9 & 2 & -3 \\ 7 & 5 & 21 \end{bmatrix}$$

$$A - B = \begin{bmatrix} 3 & -2 & 1 \\ -1 & 4 & -7 \\ 0 & 5 & 9 \end{bmatrix} - \begin{bmatrix} 5 & -1 & -3 \\ 10 & -2 & 4 \\ 7 & 0 & 12 \end{bmatrix} = \begin{bmatrix} -2 & -1 & 4 \\ -11 & 6 & -11 \\ -7 & 5 & -3 \end{bmatrix}$$

$$2A + 3B = \begin{bmatrix} 6 & -4 & 2 \\ -2 & 8 & -14 \\ 0 & 10 & 18 \end{bmatrix} + \begin{bmatrix} 15 & -3 & -9 \\ 30 & -6 & 12 \\ 21 & 0 & 36 \end{bmatrix} = \begin{bmatrix} 21 & -7 & -7 \\ 28 & 2 & -2 \\ 21 & 10 & 54 \end{bmatrix}$$

$$4A - 2B = \begin{bmatrix} 12 & -8 & 4 \\ -4 & 16 & -28 \\ 0 & 20 & 36 \end{bmatrix} - \begin{bmatrix} 10 & -2 & -6 \\ 20 & -4 & 8 \\ 14 & 0 & 24 \end{bmatrix} = \begin{bmatrix} 2 & -6 & 10 \\ -24 & 20 & -36 \\ -14 & 20 & 12 \end{bmatrix}$$

7. Finding the matrices:

$$A + B = \begin{bmatrix} -1 & 0 \\ 2 & 3 \\ -5 & -4 \\ -7 & 11 \end{bmatrix} + \begin{bmatrix} 1 & 2 \\ -3 & 7 \\ 6 & -5 \\ 9 & -2 \end{bmatrix} = \begin{bmatrix} 0 & 2 \\ -1 & 10 \\ 1 & -9 \\ 2 & 9 \end{bmatrix}$$

$$A - B = \begin{bmatrix} -1 & 0 \\ 2 & 3 \\ -5 & -4 \\ -7 & 11 \end{bmatrix} - \begin{bmatrix} 1 & 2 \\ -3 & 7 \\ 6 & -5 \\ 9 & -2 \end{bmatrix} = \begin{bmatrix} -2 & -2 \\ 5 & -4 \\ -11 & 1 \\ -16 & 13 \end{bmatrix}$$

$$2A + 3B = \begin{bmatrix} -2 & 0 \\ 4 & 6 \\ -10 & -8 \\ -14 & 22 \end{bmatrix} + \begin{bmatrix} 3 & 6 \\ -9 & 21 \\ 18 & -15 \\ 27 & -6 \end{bmatrix} = \begin{bmatrix} 1 & 6 \\ -5 & 27 \\ 8 & -23 \\ 13 & 16 \end{bmatrix}$$

$$4A - 2B = \begin{bmatrix} -4 & 0 \\ 8 & 12 \\ -20 & -16 \\ -28 & 44 \end{bmatrix} - \begin{bmatrix} 2 & 4 \\ -6 & 14 \\ 12 & -10 \\ 18 & -4 \end{bmatrix} = \begin{bmatrix} -6 & -4 \\ 14 & -2 \\ -32 & -6 \\ -46 & 48 \end{bmatrix}$$

9. Finding the products:

$$AB = \begin{bmatrix} 2 & -1 \\ 0 & -4 \\ -5 & 3 \end{bmatrix} \begin{bmatrix} 5 & -2 & 6 \\ -1 & 4 & -2 \end{bmatrix} = \begin{bmatrix} 11 & -8 & 14 \\ 4 & -16 & 8 \\ -28 & 22 & -36 \end{bmatrix}$$

$$BA = \begin{bmatrix} 5 & -2 & 6 \\ -1 & 4 & -2 \end{bmatrix} \begin{bmatrix} 2 & -1 \\ 0 & -4 \\ -5 & 3 \end{bmatrix} = \begin{bmatrix} -20 & 21 \\ 8 & -21 \end{bmatrix}$$

11. Note that BA does not exist. Finding the product: $AB = \begin{bmatrix} 2 & -1 & -3 \\ 0 & -4 & 7 \end{bmatrix} \begin{bmatrix} 2 & 1 & -1 & 4 \\ 0 & -2 & 3 & 5 \\ -6 & 4 & -2 & 0 \end{bmatrix} = \begin{bmatrix} 22 & -8 & 1 & 3 \\ -42 & 36 & -26 & -20 \end{bmatrix}$

13. Finding the products:

$$AB = \begin{bmatrix} 1 & -1 & 2 \\ 0 & 1 & -2 \\ 3 & 1 & 4 \end{bmatrix} \begin{bmatrix} 2 & 3 & -1 \\ 4 & 0 & 2 \\ -5 & 1 & -1 \end{bmatrix} = \begin{bmatrix} -12 & 5 & -5 \\ 14 & -2 & 4 \\ -10 & 13 & -5 \end{bmatrix}$$

$$BA = \begin{bmatrix} 2 & 3 & -1 \\ 4 & 0 & 2 \\ -5 & 1 & -1 \end{bmatrix} \begin{bmatrix} 1 & -1 & 2 \\ 0 & 1 & -2 \\ 3 & 1 & 4 \end{bmatrix} = \begin{bmatrix} -1 & 0 & -6 \\ 10 & -2 & 16 \\ -8 & 5 & -16 \end{bmatrix}$$

15. Finding the products:

$$AB = \begin{bmatrix} 2 & -1 & 3 & 4 \end{bmatrix} \begin{bmatrix} -1 \\ -3 \\ 2 \\ -4 \end{bmatrix} = \begin{bmatrix} -9 \end{bmatrix} \qquad BA = \begin{bmatrix} -1 \\ -3 \\ 2 \\ -4 \end{bmatrix} \begin{bmatrix} 2 & -1 & 3 & 4 \end{bmatrix} = \begin{bmatrix} -2 & 1 & -3 & -4 \\ -6 & 3 & -9 & -12 \\ 4 & -2 & 6 & 8 \\ -8 & 4 & -12 & -16 \end{bmatrix}$$

17. Note that AB does not exist. Finding the product: $BA = \begin{bmatrix} 3 & -2 \\ 1 & 0 \\ -1 & 4 \end{bmatrix} \begin{bmatrix} 2 \\ -7 \end{bmatrix} = \begin{bmatrix} 20 \\ 2 \\ -30 \end{bmatrix}$

19. Note that BA does not exist. Finding the product: $AB = \begin{bmatrix} 3 \\ -4 \\ 2 \end{bmatrix} \begin{bmatrix} 3 & -4 \end{bmatrix} = \begin{bmatrix} 9 & -12 \\ -12 & 16 \\ 6 & -8 \end{bmatrix}$

21. Form the augmented matrix: $\begin{bmatrix} 1 & 3 & | & 1 & 0 \\ 4 & 2 & | & 0 & 1 \end{bmatrix}$

Add -4 times row 1 to row 2: $\begin{bmatrix} 1 & 3 & | & 1 & 0 \\ 0 & -10 & | & -4 & 1 \end{bmatrix}$

Divide row 2 by -10: $\begin{bmatrix} 1 & 3 & | & 1 & 0 \\ 0 & 1 & | & \dfrac{2}{5} & -\dfrac{1}{10} \end{bmatrix}$

Add -3 times row 2 to row 1: $\begin{bmatrix} 1 & 0 & | & -\dfrac{1}{5} & \dfrac{3}{10} \\ 0 & 1 & | & \dfrac{2}{5} & -\dfrac{1}{10} \end{bmatrix}$

The inverse is $\begin{bmatrix} -\dfrac{1}{5} & \dfrac{3}{10} \\ \dfrac{2}{5} & -\dfrac{1}{10} \end{bmatrix}$.

23. Form the augmented matrix: $\begin{bmatrix} 2 & 1 & | & 1 & 0 \\ 7 & 4 & | & 0 & 1 \end{bmatrix}$

Divide row 2 by 1: $\begin{bmatrix} 1 & \frac{1}{2} & | & \frac{1}{2} & 0 \\ 7 & 4 & | & 0 & 1 \end{bmatrix}$

Add -7 times row 1 to row 2: $\begin{bmatrix} 1 & \frac{1}{2} & | & \frac{1}{2} & 0 \\ 0 & \frac{1}{2} & | & -\frac{7}{2} & 1 \end{bmatrix}$

Multiply row 2 by 2: $\begin{bmatrix} 1 & \frac{1}{2} & | & \frac{1}{2} & 0 \\ 0 & 1 & | & -7 & 2 \end{bmatrix}$

Add $-\frac{1}{2}$ times row 2 to row 1: $\begin{bmatrix} 1 & 0 & | & 4 & -1 \\ 0 & 1 & | & -7 & 2 \end{bmatrix}$

The inverse is $\begin{bmatrix} 4 & -1 \\ -7 & 2 \end{bmatrix}$.

25. Form the augmented matrix: $\begin{bmatrix} -2 & 1 & | & 1 & 0 \\ 3 & -4 & | & 0 & 1 \end{bmatrix}$

Add row 2 to row 1: $\begin{bmatrix} 1 & -3 & | & 1 & 1 \\ 3 & -4 & | & 0 & 1 \end{bmatrix}$

Add -3 times row 1 to row 2: $\begin{bmatrix} 1 & -3 & | & 1 & 1 \\ 0 & 5 & | & -3 & -2 \end{bmatrix}$

Divide row 2 by 5: $\begin{bmatrix} 1 & -3 & | & 1 & 1 \\ 0 & 1 & | & -\frac{3}{5} & -\frac{2}{5} \end{bmatrix}$

Add 3 times row 2 to row 1: $\begin{bmatrix} 1 & 0 & | & -\frac{4}{5} & -\frac{1}{5} \\ 0 & 1 & | & -\frac{3}{5} & -\frac{2}{5} \end{bmatrix}$

The inverse is $\begin{bmatrix} -\frac{4}{5} & -\frac{1}{5} \\ -\frac{3}{5} & -\frac{2}{5} \end{bmatrix}$.

27. Form the augmented matrix: $\begin{bmatrix} 1 & 2 & 3 & | & 1 & 0 & 0 \\ 1 & 3 & 4 & | & 0 & 1 & 0 \\ 1 & 4 & 3 & | & 0 & 0 & 1 \end{bmatrix}$

Subtract row 1 from row 2 and row 3: $\begin{bmatrix} 1 & 2 & 3 & | & 1 & 0 & 0 \\ 0 & 1 & 1 & | & -1 & 1 & 0 \\ 0 & 2 & 0 & | & -1 & 0 & 1 \end{bmatrix}$

Add -2 times row 2 to row 3: $\begin{bmatrix} 1 & 2 & 3 & | & 1 & 0 & 0 \\ 0 & 1 & 1 & | & -1 & 1 & 0 \\ 0 & 0 & -2 & | & 1 & -2 & 1 \end{bmatrix}$

Divide row 3 by –2:
$$\begin{bmatrix} 1 & 2 & 3 & | & 1 & 0 & 0 \\ 0 & 1 & 1 & | & -1 & 1 & 0 \\ 0 & 0 & 1 & | & -\dfrac{1}{2} & 1 & -\dfrac{1}{2} \end{bmatrix}$$

Add –1 times row 3 to row 2, and –3 times row 3 to row 1:
$$\begin{bmatrix} 1 & 2 & 0 & | & \dfrac{5}{2} & -3 & \dfrac{3}{2} \\ 0 & 1 & 0 & | & -\dfrac{1}{2} & 0 & \dfrac{1}{2} \\ 0 & 0 & 1 & | & -\dfrac{1}{2} & 1 & -\dfrac{1}{2} \end{bmatrix}$$

Add –2 times row 2 to row 1:
$$\begin{bmatrix} 1 & 0 & 0 & | & \dfrac{7}{2} & -3 & \dfrac{1}{2} \\ 0 & 1 & 0 & | & -\dfrac{1}{2} & 0 & \dfrac{1}{2} \\ 0 & 0 & 1 & | & -\dfrac{1}{2} & 1 & -\dfrac{1}{2} \end{bmatrix}$$

The inverse is
$$\begin{bmatrix} \dfrac{7}{2} & -3 & \dfrac{1}{2} \\ -\dfrac{1}{2} & 0 & \dfrac{1}{2} \\ -\dfrac{1}{2} & 1 & -\dfrac{1}{2} \end{bmatrix}.$$

29.   Form the augmented matrix:
$$\begin{bmatrix} 1 & -2 & 1 & | & 1 & 0 & 0 \\ -2 & 5 & 3 & | & 0 & 1 & 0 \\ 3 & -5 & 7 & | & 0 & 0 & 1 \end{bmatrix}$$

Add 2 times row 1 to row 2, and –3 times row 1 to row 3:
$$\begin{bmatrix} 1 & -2 & 1 & | & 1 & 0 & 0 \\ 0 & 1 & 5 & | & 2 & 1 & 0 \\ 0 & 1 & 4 & | & -3 & 0 & 1 \end{bmatrix}$$

Subtract row 2 from row 3:
$$\begin{bmatrix} 1 & -2 & 1 & | & 1 & 0 & 0 \\ 0 & 1 & 5 & | & 2 & 1 & 0 \\ 0 & 0 & -1 & | & -5 & -1 & 1 \end{bmatrix}$$

Multiply row 3 by –1:
$$\begin{bmatrix} 1 & -2 & 1 & | & 1 & 0 & 0 \\ 0 & 1 & 5 & | & 2 & 1 & 0 \\ 0 & 0 & 1 & | & 5 & 1 & -1 \end{bmatrix}$$

Add –5 times row 3 to row 2, and –1 times row 3 to row 1:
$$\begin{bmatrix} 1 & -2 & 0 & | & -4 & -1 & 1 \\ 0 & 1 & 0 & | & -23 & -4 & 5 \\ 0 & 0 & 1 & | & 5 & 1 & -1 \end{bmatrix}$$

Add 2 times row 2 to row 1:
$$\begin{bmatrix} 1 & 0 & 0 & | & -50 & -9 & 11 \\ 0 & 1 & 0 & | & -23 & -4 & 5 \\ 0 & 0 & 1 & | & 5 & 1 & -1 \end{bmatrix}$$

The inverse is
$$\begin{bmatrix} -50 & -9 & 11 \\ -23 & -4 & 5 \\ 5 & 1 & -1 \end{bmatrix}.$$

31. Form the augmented matrix: $\begin{bmatrix} 2 & 3 & -4 & | & 1 & 0 & 0 \\ 3 & -1 & -2 & | & 0 & 1 & 0 \\ 1 & -4 & 2 & | & 0 & 0 & 1 \end{bmatrix}$

Switch row 1 and row 3: $\begin{bmatrix} 1 & -4 & 2 & | & 0 & 0 & 1 \\ 3 & -1 & -2 & | & 0 & 1 & 0 \\ 2 & 3 & -4 & | & 1 & 0 & 0 \end{bmatrix}$

Add -3 times row 1 to row 2, and -2 times row 1 to row 3: $\begin{bmatrix} 1 & -4 & 2 & | & 0 & 0 & 1 \\ 0 & 11 & -8 & | & 0 & 1 & -3 \\ 0 & 11 & -8 & | & 1 & 0 & -2 \end{bmatrix}$

Subtract row 2 from row 3: $\begin{bmatrix} 1 & -4 & 2 & | & 0 & 0 & 1 \\ 0 & 11 & -8 & | & 0 & 1 & -3 \\ 0 & 0 & 0 & | & 1 & -1 & 1 \end{bmatrix}$

Thus the inverse does not exist.

33. Form the augmented matrix: $\begin{bmatrix} 1 & 2 & 3 & | & 1 & 0 & 0 \\ -3 & -4 & 3 & | & 0 & 1 & 0 \\ 2 & 4 & -1 & | & 0 & 0 & 1 \end{bmatrix}$

Add 3 times row 1 to row 2, and -2 times row 1 to row 3: $\begin{bmatrix} 1 & 2 & 3 & | & 1 & 0 & 0 \\ 0 & 2 & 12 & | & 3 & 1 & 0 \\ 0 & 0 & -7 & | & -2 & 0 & 1 \end{bmatrix}$

Divide row 2 by 2 and row 3 by -7: $\begin{bmatrix} 1 & 2 & 3 & | & 1 & 0 & 0 \\ 0 & 1 & 6 & | & \frac{3}{2} & \frac{1}{2} & 0 \\ 0 & 0 & 1 & | & \frac{2}{7} & 0 & -\frac{1}{7} \end{bmatrix}$

Add -6 times row 3 to row 2, and -3 times row 3 to row 1: $\begin{bmatrix} 1 & 2 & 0 & | & \frac{1}{7} & 0 & \frac{3}{7} \\ 0 & 1 & 0 & | & -\frac{3}{14} & \frac{1}{2} & \frac{6}{7} \\ 0 & 0 & 1 & | & \frac{2}{7} & 0 & -\frac{1}{7} \end{bmatrix}$

Add -2 times row 2 to row 1: $\begin{bmatrix} 1 & 0 & 0 & | & \frac{4}{7} & -1 & -\frac{9}{7} \\ 0 & 1 & 0 & | & -\frac{3}{14} & \frac{1}{2} & \frac{6}{7} \\ 0 & 0 & 1 & | & \frac{2}{7} & 0 & -\frac{1}{7} \end{bmatrix}$

The inverse is $\begin{bmatrix} \frac{4}{7} & -1 & -\frac{9}{7} \\ -\frac{3}{14} & \frac{1}{2} & \frac{6}{7} \\ \frac{2}{7} & 0 & -\frac{1}{7} \end{bmatrix}$.

35.   Form the augmented matrix: $\begin{bmatrix} 2 & 0 & 0 & | & 1 & 0 & 0 \\ 0 & 4 & 0 & | & 0 & 1 & 0 \\ 0 & 0 & 10 & | & 0 & 0 & 1 \end{bmatrix}$

Divide row 1 by 2, row 2 by 4, and row 3 by 10: $\begin{bmatrix} 1 & 0 & 0 & | & \frac{1}{2} & 0 & 0 \\ 0 & 1 & 0 & | & 0 & \frac{1}{4} & 0 \\ 0 & 0 & 1 & | & 0 & 0 & \frac{1}{10} \end{bmatrix}$

The inverse is $\begin{bmatrix} \frac{1}{2} & 0 & 0 \\ 0 & \frac{1}{4} & 0 \\ 0 & 0 & \frac{1}{10} \end{bmatrix}$.

37.   Form the augmented matrix: $\begin{bmatrix} 2 & 1 & | & 1 & 0 \\ 7 & 4 & | & 0 & 1 \end{bmatrix}$

Divide row 1 by 2: $\begin{bmatrix} 1 & \frac{1}{2} & | & \frac{1}{2} & 0 \\ 7 & 4 & | & 0 & 1 \end{bmatrix}$

Add -7 times row 1 to row 2: $\begin{bmatrix} 1 & \frac{1}{2} & | & \frac{1}{2} & 0 \\ 0 & \frac{1}{2} & | & -\frac{7}{2} & 1 \end{bmatrix}$

Subtract row 2 from row 1, and multiply row 2 by 2: $\begin{bmatrix} 1 & 0 & | & 4 & -1 \\ 0 & 1 & | & -7 & 2 \end{bmatrix}$

So the inverse is $\begin{bmatrix} 4 & -1 \\ -7 & 2 \end{bmatrix}$. Now compute the solution: $\begin{bmatrix} x \\ y \end{bmatrix} = \begin{bmatrix} 4 & -1 \\ -7 & 2 \end{bmatrix}\begin{bmatrix} -4 \\ -13 \end{bmatrix} = \begin{bmatrix} -3 \\ 2 \end{bmatrix}$

The solution set is $\{(-3,2)\}$.

39.   Form the augmented matrix: $\begin{bmatrix} -2 & 1 & | & 1 & 0 \\ 3 & -4 & | & 0 & 1 \end{bmatrix}$

Add row 2 to row 1: $\begin{bmatrix} 1 & -3 & | & 1 & 1 \\ 3 & -4 & | & 0 & 1 \end{bmatrix}$

Add -3 times row 1 to row 2: $\begin{bmatrix} 1 & -3 & | & 1 & 1 \\ 0 & 5 & | & -3 & -2 \end{bmatrix}$

Divide row 2 by 5: $\begin{bmatrix} 1 & -3 & | & 1 & 1 \\ 0 & 1 & | & -\frac{3}{5} & -\frac{2}{5} \end{bmatrix}$

Add 3 times row 2 to row 1: $\begin{bmatrix} 1 & 0 & | & -\frac{4}{5} & -\frac{1}{5} \\ 0 & 1 & | & -\frac{3}{5} & -\frac{2}{5} \end{bmatrix}$

So the inverse is $\begin{bmatrix} -\dfrac{4}{5} & -\dfrac{1}{5} \\ -\dfrac{3}{5} & -\dfrac{2}{5} \end{bmatrix}$. Now compute the solution: $\begin{bmatrix} x \\ y \end{bmatrix} = \begin{bmatrix} -\dfrac{4}{5} & -\dfrac{1}{5} \\ -\dfrac{3}{5} & -\dfrac{2}{5} \end{bmatrix} \begin{bmatrix} 1 \\ -14 \end{bmatrix} = \begin{bmatrix} 2 \\ 5 \end{bmatrix}$

The solution set is $\{(2,5)\}$.

41.   Form the augmented matrix: $\begin{bmatrix} 1 & 2 & 3 & | & 1 & 0 & 0 \\ 1 & 3 & 4 & | & 0 & 1 & 0 \\ 1 & 4 & 3 & | & 0 & 0 & 1 \end{bmatrix}$

Subtract row 1 from row 2 and row 3: $\begin{bmatrix} 1 & 2 & 3 & | & 1 & 0 & 0 \\ 0 & 1 & 1 & | & -1 & 1 & 0 \\ 0 & 2 & 0 & | & -1 & 0 & 1 \end{bmatrix}$

Switch row 2 and row 3, and divide row 2 by 2: $\begin{bmatrix} 1 & 2 & 3 & | & 1 & 0 & 0 \\ 0 & 1 & 0 & | & -\dfrac{1}{2} & 0 & \dfrac{1}{2} \\ 0 & 1 & 1 & | & -1 & 1 & 0 \end{bmatrix}$

Subtract row 2 from row 3: $\begin{bmatrix} 1 & 2 & 3 & | & 1 & 0 & 0 \\ 0 & 1 & 0 & | & -\dfrac{1}{2} & 0 & \dfrac{1}{2} \\ 0 & 0 & 1 & | & -\dfrac{1}{2} & 1 & -\dfrac{1}{2} \end{bmatrix}$

Add –3 times row 3 to row 1: $\begin{bmatrix} 1 & 2 & 0 & | & \dfrac{5}{2} & -3 & \dfrac{3}{2} \\ 0 & 1 & 0 & | & -\dfrac{1}{2} & 0 & \dfrac{1}{2} \\ 0 & 0 & 1 & | & -\dfrac{1}{2} & 1 & -\dfrac{1}{2} \end{bmatrix}$

Add –2 times row 2 to row 1: $\begin{bmatrix} 1 & 0 & 0 & | & \dfrac{7}{2} & -3 & \dfrac{1}{2} \\ 0 & 1 & 0 & | & -\dfrac{1}{2} & 0 & \dfrac{1}{2} \\ 0 & 0 & 1 & | & -\dfrac{1}{2} & 1 & -\dfrac{1}{2} \end{bmatrix}$

The inverse is $\begin{bmatrix} \dfrac{7}{2} & -3 & \dfrac{1}{2} \\ -\dfrac{1}{2} & 0 & \dfrac{1}{2} \\ -\dfrac{1}{2} & 1 & -\dfrac{1}{2} \end{bmatrix}$. Now compute the solution: $\begin{bmatrix} x \\ y \\ z \end{bmatrix} = \begin{bmatrix} \dfrac{7}{2} & -3 & \dfrac{1}{2} \\ -\dfrac{1}{2} & 0 & \dfrac{1}{2} \\ -\dfrac{1}{2} & 1 & -\dfrac{1}{2} \end{bmatrix} \begin{bmatrix} -2 \\ -3 \\ -6 \end{bmatrix} = \begin{bmatrix} -1 \\ -2 \\ 1 \end{bmatrix}$

The solution set is $\{(-1,-2,1)\}$.

43. Form the augmented matrix: $\begin{bmatrix} 1 & -2 & 1 & | & 1 & 0 & 0 \\ -2 & 5 & 3 & | & 0 & 1 & 0 \\ 3 & -5 & 7 & | & 0 & 0 & 1 \end{bmatrix}$

Add 2 times row 1 to row 2, and –3 times row 1 to row 3: $\begin{bmatrix} 1 & -2 & 1 & | & 1 & 0 & 0 \\ 0 & 1 & 5 & | & 2 & 1 & 0 \\ 0 & 1 & 4 & | & -3 & 0 & 1 \end{bmatrix}$

Subtract row 2 from row 3: $\begin{bmatrix} 1 & -2 & 1 & | & 1 & 0 & 0 \\ 0 & 1 & 5 & | & 2 & 1 & 0 \\ 0 & 0 & -1 & | & -5 & -1 & 1 \end{bmatrix}$

Multiply row 3 by –1: $\begin{bmatrix} 1 & -2 & 1 & | & 1 & 0 & 0 \\ 0 & 1 & 5 & | & 2 & 1 & 0 \\ 0 & 0 & 1 & | & 5 & 1 & -1 \end{bmatrix}$

Add –5 times row 3 to row 2, and –1 times row 3 to row 1: $\begin{bmatrix} 1 & -2 & 0 & | & -4 & -1 & 1 \\ 0 & 1 & 0 & | & -23 & -4 & 5 \\ 0 & 0 & 1 & | & 5 & 1 & -1 \end{bmatrix}$

Add 2 times row 2 to row 1: $\begin{bmatrix} 1 & 0 & 0 & | & -50 & -9 & 11 \\ 0 & 1 & 0 & | & -23 & -4 & 5 \\ 0 & 0 & 1 & | & 5 & 1 & -1 \end{bmatrix}$

The inverse is $\begin{bmatrix} -50 & -9 & 11 \\ -23 & -4 & 5 \\ 5 & 1 & -1 \end{bmatrix}$. Now compute the solution: $\begin{bmatrix} x \\ y \\ z \end{bmatrix} = \begin{bmatrix} -50 & -9 & 11 \\ -23 & -4 & 5 \\ 5 & 1 & -1 \end{bmatrix} \begin{bmatrix} -3 \\ 34 \\ 14 \end{bmatrix} = \begin{bmatrix} -2 \\ 3 \\ 5 \end{bmatrix}$

The solution set is $\left\{ (-2, 3, 5) \right\}$.

45. Form the augmented matrix: $\begin{bmatrix} 1 & 2 & 3 & | & 1 & 0 & 0 \\ -3 & -4 & 3 & | & 0 & 1 & 0 \\ 2 & 4 & -1 & | & 0 & 0 & 1 \end{bmatrix}$

Add 3 times row 1 to row 2, and –2 times row 1 to row 3: $\begin{bmatrix} 1 & 2 & 3 & | & 1 & 0 & 0 \\ 0 & 2 & 12 & | & 3 & 1 & 0 \\ 0 & 0 & -7 & | & -2 & 0 & 1 \end{bmatrix}$

Divide row 2 by 2 and row 3 by –7: $\begin{bmatrix} 1 & 2 & 3 & | & 1 & 0 & 0 \\ 0 & 1 & 6 & | & \dfrac{3}{2} & \dfrac{1}{2} & 0 \\ 0 & 0 & 1 & | & \dfrac{2}{7} & 0 & -\dfrac{1}{7} \end{bmatrix}$

Add –6 times row 3 to row 2, and –3 times row 3 to row 1: $\begin{bmatrix} 1 & 2 & 0 & | & \dfrac{1}{7} & 0 & \dfrac{3}{7} \\ 0 & 1 & 0 & | & -\dfrac{3}{14} & \dfrac{1}{2} & \dfrac{6}{7} \\ 0 & 0 & 1 & | & \dfrac{2}{7} & 0 & -\dfrac{1}{7} \end{bmatrix}$

Add −2 times row 2 to row 1:
$$\left[\begin{array}{ccc|ccc} 1 & 0 & 0 & \dfrac{4}{7} & -1 & -\dfrac{9}{7} \\ 0 & 1 & 0 & -\dfrac{3}{14} & \dfrac{1}{2} & \dfrac{6}{7} \\ 0 & 0 & 1 & \dfrac{2}{7} & 0 & -\dfrac{1}{7} \end{array}\right]$$

The inverse is $\begin{bmatrix} \dfrac{4}{7} & -1 & -\dfrac{9}{7} \\ -\dfrac{3}{14} & \dfrac{1}{2} & \dfrac{6}{7} \\ \dfrac{2}{7} & 0 & -\dfrac{1}{7} \end{bmatrix}$. Now compute the solution: $\begin{bmatrix} x \\ y \\ z \end{bmatrix} = \begin{bmatrix} \dfrac{4}{7} & -1 & -\dfrac{9}{7} \\ -\dfrac{3}{14} & \dfrac{1}{2} & \dfrac{6}{7} \\ \dfrac{2}{7} & 0 & -\dfrac{1}{7} \end{bmatrix} \begin{bmatrix} 2 \\ 0 \\ 4 \end{bmatrix} = \begin{bmatrix} -4 \\ 3 \\ 0 \end{bmatrix}$

The solution set is $\{(-4,3,0)\}$.

47. From Example 2, the inverse matrix is $\begin{bmatrix} -\dfrac{5}{24} & \dfrac{1}{6} & -\dfrac{7}{24} \\ \dfrac{7}{24} & \dfrac{1}{6} & \dfrac{5}{24} \\ \dfrac{11}{24} & -\dfrac{1}{6} & \dfrac{1}{24} \end{bmatrix}$.

a. Computing the solution: $\begin{bmatrix} x \\ y \\ z \end{bmatrix} = \begin{bmatrix} -\dfrac{5}{24} & \dfrac{1}{6} & -\dfrac{7}{24} \\ \dfrac{7}{24} & \dfrac{1}{6} & \dfrac{5}{24} \\ \dfrac{11}{24} & -\dfrac{1}{6} & \dfrac{1}{24} \end{bmatrix} \begin{bmatrix} 7 \\ 1 \\ -1 \end{bmatrix} = \begin{bmatrix} -1 \\ 2 \\ 3 \end{bmatrix}$. The solution set is $\{(-1,2,3)\}$.

b. Computing the solution: $\begin{bmatrix} x \\ y \\ z \end{bmatrix} = \begin{bmatrix} -\dfrac{5}{24} & \dfrac{1}{6} & -\dfrac{7}{24} \\ \dfrac{7}{24} & \dfrac{1}{6} & \dfrac{5}{24} \\ \dfrac{11}{24} & -\dfrac{1}{6} & \dfrac{1}{24} \end{bmatrix} \begin{bmatrix} -7 \\ 5 \\ 1 \end{bmatrix} = \begin{bmatrix} 2 \\ -1 \\ -4 \end{bmatrix}$. The solution set is $\{(2,-1,-4)\}$.

c. Computing the solution: $\begin{bmatrix} x \\ y \\ z \end{bmatrix} = \begin{bmatrix} -\dfrac{5}{24} & \dfrac{1}{6} & -\dfrac{7}{24} \\ \dfrac{7}{24} & \dfrac{1}{6} & \dfrac{5}{24} \\ \dfrac{11}{24} & -\dfrac{1}{6} & \dfrac{1}{24} \end{bmatrix} \begin{bmatrix} -9 \\ -8 \\ 19 \end{bmatrix} = \begin{bmatrix} -5 \\ 0 \\ -2 \end{bmatrix}$. The solution set is $\{(-5,0,-2)\}$.

d. Computing the solution: $\begin{bmatrix} x \\ y \\ z \end{bmatrix} = \begin{bmatrix} -\dfrac{5}{24} & \dfrac{1}{6} & -\dfrac{7}{24} \\ \dfrac{7}{24} & \dfrac{1}{6} & \dfrac{5}{24} \\ \dfrac{11}{24} & -\dfrac{1}{6} & \dfrac{1}{24} \end{bmatrix} \begin{bmatrix} -1 \\ -13 \\ -17 \end{bmatrix} = \begin{bmatrix} 3 \\ -6 \\ 1 \end{bmatrix}$. The solution set is $\{(3,-6,1)\}$.

e.   Computing the solution: $\begin{bmatrix} x \\ y \\ z \end{bmatrix} = \begin{bmatrix} -\dfrac{5}{24} & \dfrac{1}{6} & -\dfrac{7}{24} \\ \dfrac{7}{24} & \dfrac{1}{6} & \dfrac{5}{24} \\ \dfrac{11}{24} & -\dfrac{1}{6} & \dfrac{1}{24} \end{bmatrix} \begin{bmatrix} -2 \\ 0 \\ -2 \end{bmatrix} = \begin{bmatrix} 1 \\ -1 \\ -1 \end{bmatrix}$. The solution set is $\{(1,-1,-1)\}$.

51.   First find the inverse of the encoding matrix. Form the augmented matrix: $\begin{bmatrix} 2 & 3 & | & 1 & 0 \\ 1 & 2 & | & 0 & 1 \end{bmatrix}$

Switch row 1 and row 2: $\begin{bmatrix} 1 & 2 & | & 0 & 1 \\ 2 & 3 & | & 1 & 0 \end{bmatrix}$

Add –2 times row 1 to row 2: $\begin{bmatrix} 1 & 2 & | & 0 & 1 \\ 0 & -1 & | & 1 & -2 \end{bmatrix}$

Multiply row 2 by –1: $\begin{bmatrix} 1 & 2 & | & 0 & 1 \\ 0 & 1 & | & -1 & 2 \end{bmatrix}$

Add –2 times row 2 to row 1: $\begin{bmatrix} 1 & 0 & | & 2 & -3 \\ 0 & 1 & | & -1 & 2 \end{bmatrix}$

The inverse is $\begin{bmatrix} 2 & -3 \\ -1 & 2 \end{bmatrix}$.

a.   Multiply the inverse by the encoded matrix:

$\begin{bmatrix} 2 & -3 \\ -1 & 2 \end{bmatrix} \begin{bmatrix} 53 & 48 & 39 & 35 & 78 & 56 & 83 \\ 34 & 25 & 22 & 20 & 47 & 37 & 54 \end{bmatrix} = \begin{bmatrix} 4 & 21 & 12 & 10 & 15 & 1 & 4 \\ 15 & 2 & 5 & 5 & 16 & 18 & 25 \end{bmatrix}$

The decoded message is:
4 15 21 2 12 5 10 5 15 16 1 18 4 25
DOUBLE JEOPARDY

b.   Multiply the inverse by the encoded matrix:

$\begin{bmatrix} 2 & -3 \\ -1 & 2 \end{bmatrix} \begin{bmatrix} 62 & 78 & 64 & 19 & 93 & 93 & 88 \\ 40 & 47 & 36 & 11 & 57 & 56 & 57 \end{bmatrix} = \begin{bmatrix} 4 & 15 & 20 & 5 & 15 & 18 & 5 \\ 18 & 16 & 8 & 3 & 21 & 19 & 26 \end{bmatrix}$

The decoded message is:
4 18 15 16 20 8 5 3 15 21 18 19 5 (26)
DROP THE COURSE

c.   Multiply the inverse by the encoded matrix:

$\begin{bmatrix} 2 & -3 \\ -1 & 2 \end{bmatrix} \begin{bmatrix} 64 & 58 & 63 & 21 & 75 & 63 & 38 & 118 \\ 36 & 37 & 36 & 13 & 47 & 36 & 23 & 72 \end{bmatrix} = \begin{bmatrix} 20 & 5 & 18 & 3 & 9 & 18 & 7 & 20 \\ 8 & 16 & 9 & 5 & 19 & 9 & 8 & 26 \end{bmatrix}$

The decoded message is:
20 8 5 16 18 9 3 5 9 19 18 9 7 8 20 (26)
THE PRICE IS RIGHT

d.   Multiply the inverse by the encoded matrix:

$\begin{bmatrix} 2 & -3 \\ -1 & 2 \end{bmatrix} \begin{bmatrix} 29 & 96 & 60 & 75 & 19 & 37 & 70 & 90 & 98 & 72 & 51 & 86 \\ 15 & 58 & 37 & 47 & 10 & 21 & 42 & 55 & 59 & 45 & 28 & 56 \end{bmatrix}$
$= \begin{bmatrix} 13 & 18 & 9 & 9 & 8 & 11 & 14 & 15 & 19 & 9 & 18 & 4 \\ 1 & 20 & 14 & 19 & 1 & 5 & 14 & 20 & 20 & 18 & 5 & 26 \end{bmatrix}$

The decoded message is:
13 1 18 20 9 14 9 19 8 1 11 5 14 14 15 20 19 20 9 18 18 5 4 (26)
MARTINI SHAKEN NOT STIRRED

7.4 Systems of Linear Inequalities: Linear Programming

1.   Graphing the solution set:

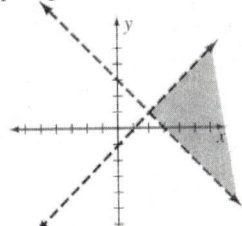

3.   Graphing the solution set:

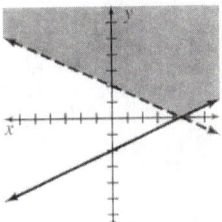

5.   Graphing the solution set:

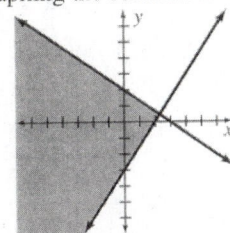

7.   Graphing the solution set:

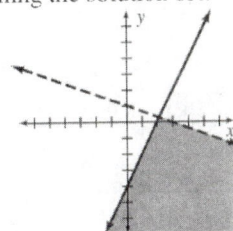

9.   Graphing the solution set:

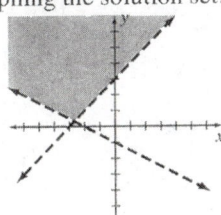

11.   Graphing the solution set:

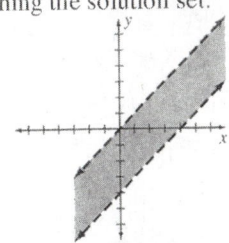

13.   Graphing the solution set:

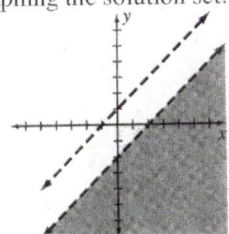

15.   Graphing the solution set:

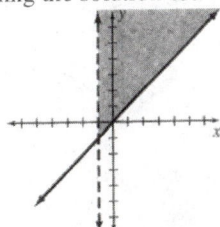

17.   There is no solution set.

19.   Graphing the solution set:

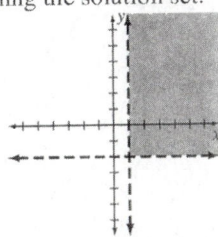

21.   Graphing the solution set:

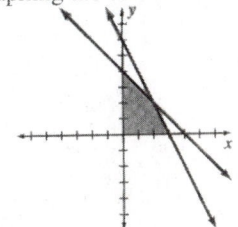

23.   Graphing the solution set:

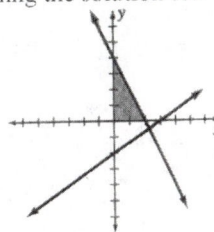

25. Evaluating the function at each vertex:
$$f(1,1) = 3(1) + 5(1) = 8$$
$$f(5,2) = 3(5) + 5(2) = 25$$
$$f(4,8) = 3(4) + 5(8) = 52$$
$$f(2,4) = 3(2) + 5(4) = 26$$
The minimum is 8 and the maximum is 52.

29. Graphing the region:

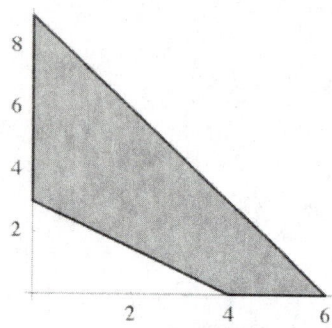

Evaluating the function at each vertex:
$$f(0,3) = 3(0) + 7(3) = 21$$
$$f(0,9) = 3(0) + 7(9) = 63$$
$$f(4,0) = 3(4) + 7(0) = 12$$
$$f(6,0) = 3(6) + 7(0) = 18$$
The maximum is 63.

33. Graphing the region:

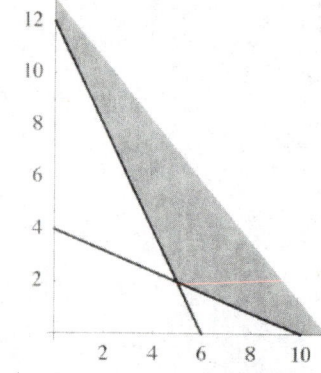

Evaluating the function at each vertex:

$$f(0,12) = 0.2(0) + 0.5(12) = 6$$
$$f(5,2) = 0.2(5) + 0.5(2) = 2$$
$$f(10,0) = 0.2(10) + 0.5(0) = 2$$

The minimum is 2.

27. Evaluating the function at each vertex:
$$f(0,0) = 0 + 4(0) = 0$$
$$f(6,2) = 6 + 4(2) = 14$$
$$f(5,4) = 5 + 4(4) = 21$$
$$f(0,7) = 0 + 4(7) = 28$$
The minimum is 0 and the maximum is 28.

31. Graphing the region:

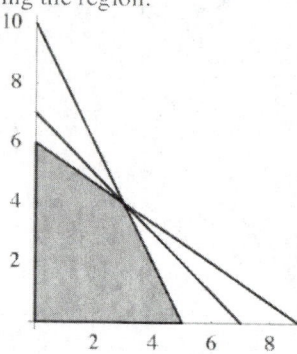

Evaluating the function at each vertex:
$$f(0,0) = 40(0) + 55(0) = 0$$
$$f(0,6) = 40(0) + 55(6) = 330$$
$$f(5,0) = 40(5) + 55(0) = 200$$
$$f(3,4) = 40(3) + 55(4) = 340$$
The maximum is 340.

35. Graphing the region:

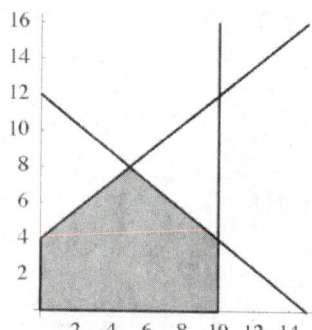

Evaluating the function at each vertex:
$$f(0,0) = 9(0) + 2(0) = 0$$
$$f(0,4) = 9(0) + 2(4) = 8$$
$$f(10,0) = 9(10) + 2(0) = 90$$
$$f(10,4) = 9(10) + 2(4) = 98$$
$$f(5,8) = 9(5) + 2(8) = 61$$
The maximum is 98.

37. Let x represent the amount invested in the speculative stock and y represent the amount invested in the conservative stock. The constraint equations are:

$$x + y \leq 10000$$
$$y \geq 2000$$
$$x \leq 6000$$
$$x \leq y$$
$$x \geq 0$$
$$y \geq 0$$

Graphing the region:

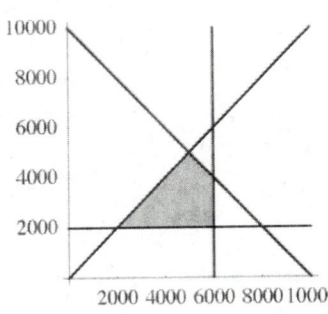

Evaluating $f(x,y) = 0.12x + 0.09y$ at each vertex:

$$f(2000,2000) = 0.12(2000) + 0.09(2000) = 420$$
$$f(6000,2000) = 0.12(6000) + 0.09(2000) = 900$$
$$f(6000,4000) = 0.12(6000) + 0.09(4000) = 1080$$
$$f(5000,5000) = 0.12(5000) + 0.09(5000) = 1050$$

She should invest \$5000 at 9% and \$5000 at 12% to maximize her return.

39. Let x represent the production of model A calculators and y represent the production of model B calculators. The constraint equations are:

$$200 \leq x \leq 300$$
$$100 \leq y \leq 250$$
$$x + y \leq 500$$
$$x \geq 0$$
$$y \geq 0$$

Graphing the region:

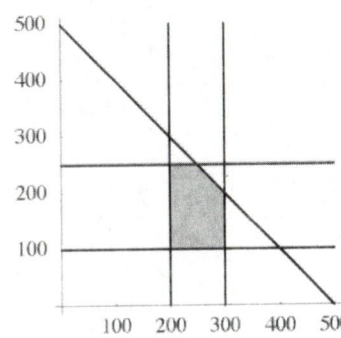

Evaluating $f(x,y) = 3x + 2y$ at each vertex:

$$f(200,100) = 3(200) + 2(100) = 800 \qquad f(300,100) = 3(300) + 2(100) = 1100$$
$$f(200,250) = 3(200) + 2(250) = 1100 \qquad f(250,250) = 3(250) + 2(250) = 1250$$
$$f(300,200) = 3(300) + 2(200) = 1300$$

They should manufacture 300 type A calculators and 200 type B calculators.

41. Let x represent the production of product A and y represent the production of product B. The constraint equations are:
$$x + y \le 40$$
$$2x + y \le 40$$
$$x + 3y \le 60$$
$$x \ge 0$$
$$y \ge 0$$

Graphing the region:

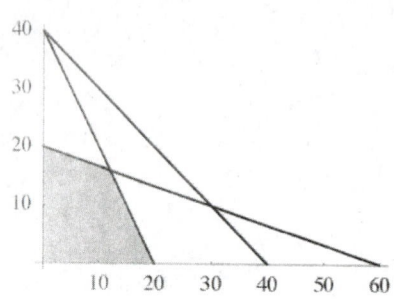

Evaluating $f(x, y) = 2.75x + 3.50y$ at each vertex:

$$f(0, 20) = 2.75(0) + 3.50(20) = 70 \qquad\qquad f(20, 0) = 2.75(20) + 3.50(0) = 55$$
$$f(12, 16) = 2.75(12) + 3.50(16) = 89$$

They should product 12 units of A and 16 units of B.

Chapter 7 Review Problem Set

1. Computing: $A + B = \begin{bmatrix} 2 & -4 \\ -3 & 8 \end{bmatrix} + \begin{bmatrix} 5 & -1 \\ 0 & 2 \end{bmatrix} = \begin{bmatrix} 7 & -5 \\ -3 & 10 \end{bmatrix}$

2. Computing: $B - A = \begin{bmatrix} 5 & -1 \\ 0 & 2 \end{bmatrix} - \begin{bmatrix} 2 & -4 \\ -3 & 8 \end{bmatrix} = \begin{bmatrix} 3 & 3 \\ 3 & -6 \end{bmatrix}$

3. Computing: $C - F = \begin{bmatrix} 3 & -1 \\ -2 & 4 \\ 5 & -6 \end{bmatrix} - \begin{bmatrix} 1 & -2 \\ 4 & -4 \\ 7 & -8 \end{bmatrix} = \begin{bmatrix} 2 & 1 \\ -6 & 8 \\ -2 & 2 \end{bmatrix}$

4. Computing: $2A + 3B = \begin{bmatrix} 4 & -8 \\ -6 & 16 \end{bmatrix} + \begin{bmatrix} 15 & -3 \\ 0 & 6 \end{bmatrix} = \begin{bmatrix} 19 & -11 \\ -6 & 22 \end{bmatrix}$

5. Computing: $3C - 2F = \begin{bmatrix} 9 & -3 \\ -6 & 12 \\ 15 & -18 \end{bmatrix} - \begin{bmatrix} 2 & -4 \\ 8 & -8 \\ 14 & -16 \end{bmatrix} = \begin{bmatrix} 7 & 1 \\ -14 & 20 \\ 1 & -2 \end{bmatrix}$

6. Computing: $CD = \begin{bmatrix} 3 & -1 \\ -2 & 4 \\ 5 & -6 \end{bmatrix} \begin{bmatrix} -2 & -1 & 4 \\ 5 & 0 & -3 \end{bmatrix} = \begin{bmatrix} -11 & -3 & 15 \\ 24 & 2 & -20 \\ -40 & -5 & 38 \end{bmatrix}$

7. Computing: $DC = \begin{bmatrix} -2 & -1 & 4 \\ 5 & 0 & -3 \end{bmatrix} \begin{bmatrix} 3 & -1 \\ -2 & 4 \\ 5 & -6 \end{bmatrix} = \begin{bmatrix} 16 & -26 \\ 0 & 13 \end{bmatrix}$

8. Computing: $DC + AB = \begin{bmatrix} -2 & -1 & 4 \\ 5 & 0 & -3 \end{bmatrix} \begin{bmatrix} 3 & -1 \\ -2 & 4 \\ 5 & -6 \end{bmatrix} + \begin{bmatrix} 2 & -4 \\ -3 & 8 \end{bmatrix} \begin{bmatrix} 5 & -1 \\ 0 & 2 \end{bmatrix} = \begin{bmatrix} 16 & -26 \\ 0 & 13 \end{bmatrix} + \begin{bmatrix} 10 & -10 \\ -15 & 19 \end{bmatrix} = \begin{bmatrix} 26 & -36 \\ -15 & 32 \end{bmatrix}$

9. Computing: $DE = \begin{bmatrix} -2 & -1 & 4 \\ 5 & 0 & -3 \end{bmatrix} \begin{bmatrix} 1 \\ -3 \\ -7 \end{bmatrix} = \begin{bmatrix} -27 \\ 26 \end{bmatrix}$

10. The product EF does not exist.

11. Computing:

$$AB = \begin{bmatrix} 2 & -4 \\ -3 & 8 \end{bmatrix} \begin{bmatrix} 5 & -1 \\ 0 & 2 \end{bmatrix} = \begin{bmatrix} 10 & -10 \\ -15 & 19 \end{bmatrix} \qquad BA = \begin{bmatrix} 5 & -1 \\ 0 & 2 \end{bmatrix} \begin{bmatrix} 2 & -4 \\ -3 & 8 \end{bmatrix} = \begin{bmatrix} 13 & -28 \\ -6 & 16 \end{bmatrix}$$

Note that $AB \neq BA$.

12. Computing:

$$D(C+F) = \begin{bmatrix} -2 & -1 & 4 \\ 5 & 0 & -3 \end{bmatrix} \left(\begin{bmatrix} 3 & -1 \\ -2 & 4 \\ 5 & -6 \end{bmatrix} + \begin{bmatrix} 1 & -2 \\ 4 & -4 \\ 7 & -8 \end{bmatrix} \right) = \begin{bmatrix} -2 & -1 & 4 \\ 5 & 0 & -3 \end{bmatrix} \begin{bmatrix} 4 & -3 \\ 2 & 0 \\ 12 & -14 \end{bmatrix} = \begin{bmatrix} 38 & -50 \\ -16 & 27 \end{bmatrix}$$

$$DC + DF = \begin{bmatrix} -2 & -1 & 4 \\ 5 & 0 & -3 \end{bmatrix} \begin{bmatrix} 3 & -1 \\ -2 & 4 \\ 5 & -6 \end{bmatrix} + \begin{bmatrix} -2 & -1 & 4 \\ 5 & 0 & -3 \end{bmatrix} \begin{bmatrix} 1 & -2 \\ 4 & -4 \\ 7 & -8 \end{bmatrix} = \begin{bmatrix} 16 & -26 \\ 0 & 13 \end{bmatrix} + \begin{bmatrix} 22 & -24 \\ -16 & 14 \end{bmatrix} = \begin{bmatrix} 38 & -50 \\ -16 & 27 \end{bmatrix}$$

So $D(C+F) = DC + DF$.

13. Computing:

$$(C+F)D = \left(\begin{bmatrix} 3 & -1 \\ -2 & 4 \\ 5 & -6 \end{bmatrix} + \begin{bmatrix} 1 & -2 \\ 4 & -4 \\ 7 & -8 \end{bmatrix} \right) \begin{bmatrix} -2 & -1 & 4 \\ 5 & 0 & -3 \end{bmatrix} = \begin{bmatrix} 4 & -3 \\ 2 & 0 \\ 12 & -14 \end{bmatrix} \begin{bmatrix} -2 & -1 & 4 \\ 5 & 0 & -3 \end{bmatrix} = \begin{bmatrix} -23 & -4 & 25 \\ -4 & -2 & 8 \\ -94 & -12 & 90 \end{bmatrix}$$

$$CD + FD = \begin{bmatrix} 3 & -1 \\ -2 & 4 \\ 5 & -6 \end{bmatrix} \begin{bmatrix} -2 & -1 & 4 \\ 5 & 0 & -3 \end{bmatrix} + \begin{bmatrix} 1 & -2 \\ 4 & -4 \\ 7 & -8 \end{bmatrix} \begin{bmatrix} -2 & -1 & 4 \\ 5 & 0 & -3 \end{bmatrix}$$

$$= \begin{bmatrix} -11 & -3 & 15 \\ 24 & 2 & -20 \\ -40 & -5 & 38 \end{bmatrix} + \begin{bmatrix} -12 & -1 & 10 \\ -28 & -4 & 28 \\ -54 & -7 & 52 \end{bmatrix}$$

$$= \begin{bmatrix} -23 & -4 & 25 \\ -4 & -2 & 8 \\ -94 & -12 & 90 \end{bmatrix}$$

So $(C+F)D = CD + FD$.

14. Form the augmented matrix: $\begin{bmatrix} 9 & 5 & | & 1 & 0 \\ 7 & 4 & | & 0 & 1 \end{bmatrix}$

Subtract row 2 from row 1: $\begin{bmatrix} 2 & 1 & | & 1 & -1 \\ 7 & 4 & | & 0 & 1 \end{bmatrix}$

Divide row 2 by 2: $\begin{bmatrix} 1 & \frac{1}{2} & | & \frac{1}{2} & -\frac{1}{2} \\ 7 & 4 & | & 0 & 1 \end{bmatrix}$

Add -7 times row 1 to row 2: $\begin{bmatrix} 1 & \frac{1}{2} & | & \frac{1}{2} & -\frac{1}{2} \\ 0 & \frac{1}{2} & | & -\frac{9}{2} & \frac{9}{2} \end{bmatrix}$

Subtract row 2 from row 1, and multiply row 2 by 2: $\begin{bmatrix} 1 & 0 & | & 4 & -5 \\ 0 & 1 & | & -7 & 9 \end{bmatrix}$

The inverse is $\begin{bmatrix} 4 & -5 \\ -7 & 9 \end{bmatrix}$.

15. Form the augmented matrix: $\begin{bmatrix} 9 & 4 & | & 1 & 0 \\ 7 & 3 & | & 0 & 1 \end{bmatrix}$

Subtract row 2 from row 1: $\begin{bmatrix} 2 & 1 & | & 1 & -1 \\ 7 & 3 & | & 0 & 1 \end{bmatrix}$

Divide row 1 by 2: $\begin{bmatrix} 1 & \frac{1}{2} & | & \frac{1}{2} & -\frac{1}{2} \\ 7 & 3 & | & 0 & 1 \end{bmatrix}$

Add –7 times row 1 to row 2: $\begin{bmatrix} 1 & \frac{1}{2} & | & \frac{1}{2} & -\frac{1}{2} \\ 0 & -\frac{1}{2} & | & -\frac{7}{2} & \frac{9}{2} \end{bmatrix}$

Add row 2 to row 1, and multiply row 2 by –2: $\begin{bmatrix} 1 & 0 & | & -3 & 4 \\ 0 & 1 & | & 7 & -9 \end{bmatrix}$

The inverse is $\begin{bmatrix} -3 & 4 \\ 7 & -9 \end{bmatrix}$.

16. Form the augmented matrix: $\begin{bmatrix} -2 & 1 & | & 1 & 0 \\ 2 & 3 & | & 0 & 1 \end{bmatrix}$

Add row 1 to row 2, and divide row 1 by –2: $\begin{bmatrix} 1 & -\frac{1}{2} & | & -\frac{1}{2} & 0 \\ 0 & 4 & | & 1 & 1 \end{bmatrix}$

Divide row 2 by 4: $\begin{bmatrix} 1 & -\frac{1}{2} & | & -\frac{1}{2} & 0 \\ 0 & 1 & | & \frac{1}{4} & \frac{1}{4} \end{bmatrix}$

Add $\frac{1}{2}$ times row 2 to row 1: $\begin{bmatrix} 1 & 0 & | & -\frac{3}{8} & \frac{1}{8} \\ 0 & 1 & | & \frac{1}{4} & \frac{1}{4} \end{bmatrix}$

The inverse is $\begin{bmatrix} -\frac{3}{8} & \frac{1}{8} \\ \frac{1}{4} & \frac{1}{4} \end{bmatrix}$.

17. Form the augmented matrix: $\begin{bmatrix} 4 & -6 & | & 1 & 0 \\ 2 & -3 & | & 0 & 1 \end{bmatrix}$

Add $-\frac{1}{2}$ times row 1 to row 2: $\begin{bmatrix} 4 & -6 & | & 1 & 0 \\ 0 & 0 & | & -\frac{1}{2} & 1 \end{bmatrix}$

Thus the inverse does not exist.

18. Form the augmented matrix: $\begin{bmatrix} -1 & -3 & | & 1 & 0 \\ -4 & -5 & | & 0 & 1 \end{bmatrix}$

Add −4 times row 1 to row 2, and multiply row 1 by −1: $\begin{bmatrix} 1 & 3 & | & -1 & 0 \\ 0 & 7 & | & -4 & 1 \end{bmatrix}$

Divide row 2 by 7: $\begin{bmatrix} 1 & 3 & | & -1 & 0 \\ 0 & 1 & | & -\dfrac{4}{7} & \dfrac{1}{7} \end{bmatrix}$

Add −3 times row 2 to row 1: $\begin{bmatrix} 1 & 0 & | & \dfrac{5}{7} & -\dfrac{3}{7} \\ 0 & 1 & | & -\dfrac{4}{7} & \dfrac{1}{7} \end{bmatrix}$

The inverse is $\begin{bmatrix} \dfrac{5}{7} & -\dfrac{3}{7} \\ -\dfrac{4}{7} & \dfrac{1}{7} \end{bmatrix}$.

19. Form the augmented matrix: $\begin{bmatrix} 0 & -3 & | & 1 & 0 \\ 7 & 6 & | & 0 & 1 \end{bmatrix}$

Switch row 1 and row 2: $\begin{bmatrix} 7 & 6 & | & 0 & 1 \\ 0 & -3 & | & 1 & 0 \end{bmatrix}$

Divide row 2 by −3: $\begin{bmatrix} 7 & 6 & | & 0 & 1 \\ 0 & 1 & | & -\dfrac{1}{3} & 0 \end{bmatrix}$

Add −6 times row 2 to row 1: $\begin{bmatrix} 7 & 0 & | & 2 & 1 \\ 0 & 1 & | & -\dfrac{1}{3} & 0 \end{bmatrix}$

Divide row 1 by 7: $\begin{bmatrix} 1 & 0 & | & \dfrac{2}{7} & \dfrac{1}{7} \\ 0 & 1 & | & -\dfrac{1}{3} & 0 \end{bmatrix}$

The inverse is $\begin{bmatrix} \dfrac{2}{7} & \dfrac{1}{7} \\ -\dfrac{1}{3} & 0 \end{bmatrix}$.

20. Form the augmented matrix: $\begin{bmatrix} 1 & -2 & 1 & | & 1 & 0 & 0 \\ 2 & -5 & 2 & | & 0 & 1 & 0 \\ -3 & 7 & 5 & | & 0 & 0 & 1 \end{bmatrix}$

Add −2 times row 1 to row 2, and 3 times row 1 to row 3: $\begin{bmatrix} 1 & -2 & 1 & | & 1 & 0 & 0 \\ 0 & -1 & 0 & | & -2 & 1 & 0 \\ 0 & 1 & 8 & | & 3 & 0 & 1 \end{bmatrix}$

Add row 2 to row 3, and multiply row 2 by −1: $\begin{bmatrix} 1 & -2 & 1 & | & 1 & 0 & 0 \\ 0 & 1 & 0 & | & 2 & -1 & 0 \\ 0 & 0 & 8 & | & 1 & 1 & 1 \end{bmatrix}$

Divide row 3 by 8: $\begin{bmatrix} 1 & -2 & 1 & | & 1 & 0 & 0 \\ 0 & 1 & 0 & | & 2 & -1 & 0 \\ 0 & 0 & 1 & | & \dfrac{1}{8} & \dfrac{1}{8} & \dfrac{1}{8} \end{bmatrix}$

Subtract row 3 from row 1: $\begin{bmatrix} 1 & -2 & 0 & | & \dfrac{7}{8} & -\dfrac{1}{8} & -\dfrac{1}{8} \\ 0 & 1 & 0 & | & 2 & -1 & 0 \\ 0 & 0 & 1 & | & \dfrac{1}{8} & \dfrac{1}{8} & \dfrac{1}{8} \end{bmatrix}$

Add 2 times row 2 to row 1: $\begin{bmatrix} 1 & 0 & 0 & | & \dfrac{39}{8} & -\dfrac{17}{8} & -\dfrac{1}{8} \\ 0 & 1 & 0 & | & 2 & -1 & 0 \\ 0 & 0 & 1 & | & \dfrac{1}{8} & \dfrac{1}{8} & \dfrac{1}{8} \end{bmatrix}$

The inverse is $\begin{bmatrix} \dfrac{39}{8} & -\dfrac{17}{8} & -\dfrac{1}{8} \\ 2 & -1 & 0 \\ \dfrac{1}{8} & \dfrac{1}{8} & \dfrac{1}{8} \end{bmatrix}$.

21.   Form the augmented matrix: $\begin{bmatrix} 1 & 3 & -2 & | & 1 & 0 & 0 \\ 4 & 13 & -7 & | & 0 & 1 & 0 \\ 5 & 16 & -8 & | & 0 & 0 & 1 \end{bmatrix}$

Add –4 times row 1 to row 2, and –5 times row 1 to row 3: $\begin{bmatrix} 1 & 3 & -2 & | & 1 & 0 & 0 \\ 0 & 1 & 1 & | & -4 & 1 & 0 \\ 0 & 1 & 2 & | & -5 & 0 & 1 \end{bmatrix}$

Subtract row 2 from row 3: $\begin{bmatrix} 1 & 3 & -2 & | & 1 & 0 & 0 \\ 0 & 1 & 1 & | & -4 & 1 & 0 \\ 0 & 0 & 1 & | & -1 & -1 & 1 \end{bmatrix}$

Add –1 times row 3 to row 2, and 2 times row 3 to row 1: $\begin{bmatrix} 1 & 3 & 0 & | & -1 & -2 & 2 \\ 0 & 1 & 0 & | & -3 & 2 & -1 \\ 0 & 0 & 1 & | & -1 & -1 & 1 \end{bmatrix}$

Add –3 times row 2 to row 1: $\begin{bmatrix} 1 & 0 & 0 & | & 8 & -8 & 5 \\ 0 & 1 & 0 & | & -3 & 2 & -1 \\ 0 & 0 & 1 & | & -1 & -1 & 1 \end{bmatrix}$

The inverse is $\begin{bmatrix} 8 & -8 & 5 \\ -3 & 2 & -1 \\ -1 & -1 & 1 \end{bmatrix}$.

22. Form the augmented matrix:
$$\begin{bmatrix} -2 & 4 & 7 & | & 1 & 0 & 0 \\ 1 & -3 & 5 & | & 0 & 1 & 0 \\ 1 & -5 & 22 & | & 0 & 0 & 1 \end{bmatrix}$$

Switch row 1 and row 2:
$$\begin{bmatrix} 1 & -3 & 5 & | & 0 & 1 & 0 \\ -2 & 4 & 7 & | & 1 & 0 & 0 \\ 1 & -5 & 22 & | & 0 & 0 & 1 \end{bmatrix}$$

Add 2 times row 1 to row 2, and –1 times row 1 to row 3:
$$\begin{bmatrix} 1 & -3 & 5 & | & 0 & 1 & 0 \\ 0 & -2 & 17 & | & 1 & 2 & 0 \\ 0 & -2 & 17 & | & 0 & -1 & 1 \end{bmatrix}$$

Subtract row 2 from row 3:
$$\begin{bmatrix} 1 & -3 & 5 & | & 0 & 1 & 0 \\ 0 & -2 & 17 & | & 1 & 2 & 0 \\ 0 & 0 & 0 & | & -1 & -3 & 1 \end{bmatrix}$$

The inverse does not exist.

23. Form the augmented matrix:
$$\begin{bmatrix} -1 & 2 & 3 & | & 1 & 0 & 0 \\ 2 & -5 & -7 & | & 0 & 1 & 0 \\ -3 & 5 & 11 & | & 0 & 0 & 1 \end{bmatrix}$$

Add 2 times row 1 to row 2, and –3 times row 1 to row 3:
$$\begin{bmatrix} -1 & 2 & 3 & | & 1 & 0 & 0 \\ 0 & -1 & -1 & | & 2 & 1 & 0 \\ 0 & -1 & 2 & | & -3 & 0 & 1 \end{bmatrix}$$

Subtract row 2 from row 3, and multiply row 1 and row 2 by –1:
$$\begin{bmatrix} 1 & -2 & -3 & | & -1 & 0 & 0 \\ 0 & 1 & 1 & | & -2 & -1 & 0 \\ 0 & 0 & 3 & | & -5 & -1 & 1 \end{bmatrix}$$

Divide row 3 by 3:
$$\begin{bmatrix} 1 & -2 & -3 & | & -1 & 0 & 0 \\ 0 & 1 & 1 & | & -2 & -1 & 0 \\ 0 & 0 & 1 & | & -\dfrac{5}{3} & -\dfrac{1}{3} & \dfrac{1}{3} \end{bmatrix}$$

Add –1 times row 3 to row 2, and 3 times row 3 to row 1:
$$\begin{bmatrix} 1 & -2 & 0 & | & -6 & -1 & 1 \\ 0 & 1 & 0 & | & -\dfrac{1}{3} & -\dfrac{2}{3} & -\dfrac{1}{3} \\ 0 & 0 & 1 & | & -\dfrac{5}{3} & -\dfrac{1}{3} & \dfrac{1}{3} \end{bmatrix}$$

Add 2 times row 2 to row 1:
$$\begin{bmatrix} 1 & 0 & 0 & | & -\dfrac{20}{3} & -\dfrac{7}{3} & \dfrac{1}{3} \\ 0 & 1 & 0 & | & -\dfrac{1}{3} & -\dfrac{2}{3} & -\dfrac{1}{3} \\ 0 & 0 & 1 & | & -\dfrac{5}{3} & -\dfrac{1}{3} & \dfrac{1}{3} \end{bmatrix}$$

The inverse is
$$\begin{bmatrix} -\dfrac{20}{3} & -\dfrac{7}{3} & \dfrac{1}{3} \\ -\dfrac{1}{3} & -\dfrac{2}{3} & -\dfrac{1}{3} \\ -\dfrac{5}{3} & -\dfrac{1}{3} & \dfrac{1}{3} \end{bmatrix}.$$

24.   Solving the equation: $\begin{bmatrix} x \\ y \end{bmatrix} = \begin{bmatrix} 4 & -5 \\ -7 & 9 \end{bmatrix}\begin{bmatrix} 12 \\ 10 \end{bmatrix} = \begin{bmatrix} -2 \\ 6 \end{bmatrix}$. The solution set is $\{(-2,6)\}$.

25.   Solving the equation: $\begin{bmatrix} x \\ y \end{bmatrix} = \begin{bmatrix} -\dfrac{3}{8} & \dfrac{1}{8} \\ \dfrac{1}{4} & \dfrac{1}{4} \end{bmatrix}\begin{bmatrix} -9 \\ 5 \end{bmatrix} = \begin{bmatrix} 4 \\ -1 \end{bmatrix}$. The solution set is $\{(4,-1)\}$.

26.   Solving the equation: $\begin{bmatrix} x \\ y \\ z \end{bmatrix} = \begin{bmatrix} \dfrac{39}{8} & -\dfrac{17}{8} & -\dfrac{1}{8} \\ 2 & -1 & 0 \\ \dfrac{1}{8} & \dfrac{1}{8} & \dfrac{1}{8} \end{bmatrix}\begin{bmatrix} 7 \\ 17 \\ -32 \end{bmatrix} = \begin{bmatrix} 2 \\ -3 \\ -1 \end{bmatrix}$. The solution set is $\{(2,-3,-1)\}$.

27.   Solving the equation: $\begin{bmatrix} x \\ y \\ z \end{bmatrix} = \begin{bmatrix} 8 & -8 & 5 \\ -3 & 2 & -1 \\ -1 & -1 & 1 \end{bmatrix}\begin{bmatrix} -7 \\ -21 \\ -23 \end{bmatrix} = \begin{bmatrix} -3 \\ 2 \\ 5 \end{bmatrix}$. The solution set is $\{(-3,2,5)\}$.

28.   Solving the equation: $\begin{bmatrix} x \\ y \\ z \end{bmatrix} = \begin{bmatrix} -\dfrac{20}{3} & -\dfrac{7}{3} & \dfrac{1}{3} \\ -\dfrac{1}{3} & -\dfrac{2}{3} & -\dfrac{1}{3} \\ -\dfrac{5}{3} & -\dfrac{1}{3} & \dfrac{1}{3} \end{bmatrix}\begin{bmatrix} 22 \\ -51 \\ 71 \end{bmatrix} = \begin{bmatrix} -4 \\ 3 \\ 4 \end{bmatrix}$. The solution set is $\{(-4,3,4)\}$.

29.   Graphing the solution set:

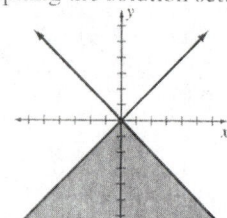

30.   Graphing the solution set:

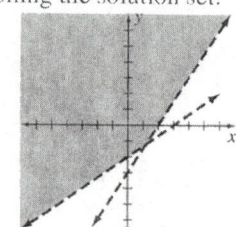

31.   Graphing the solution set:

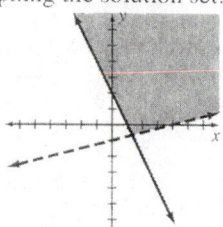

32.   Graphing the solution set:

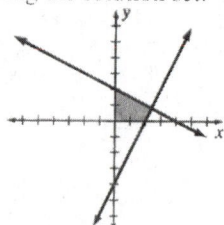

33.   Graphing the region:

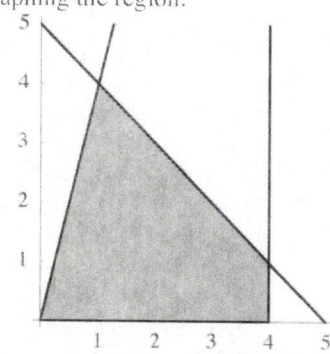

34.   Graphing the region:

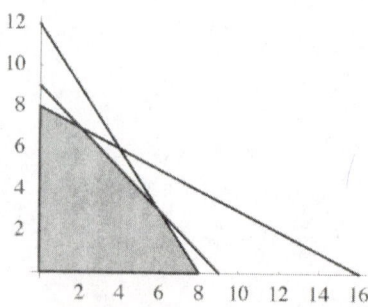

Evaluating the function at each vertex:

$$f(0,0) = 8(0) + 5(0) = 0$$
$$f(1,4) = 8(1) + 5(4) = 28$$
$$f(4,0) = 8(4) + 5(0) = 32$$
$$f(4,1) = 8(4) + 5(1) = 37$$

The maximum value is 37.

Evaluating the function at each vertex:

$$f(0,0) = 2(0) + 7(0) = 0$$
$$f(0,8) = 2(0) + 7(8) = 56$$
$$f(8,0) = 2(8) + 7(0) = 16$$
$$f(2,7) = 2(2) + 7(7) = 53$$
$$f(6,3) = 2(6) + 7(3) = 33$$

The maximum value is 56.

35. Evaluating the function at each vertex:

$$f(0,0) = 7(0) + 5(0) = 0$$
$$f(0,8) = 7(0) + 5(8) = 40$$
$$f(8,0) = 7(8) + 5(0) = 56$$
$$f(2,7) = 7(2) + 5(7) = 49$$
$$f(6,3) = 7(6) + 5(3) = 57$$

The maximum value is 57.

36. Evaluating the function at each vertex:

$$f(0,0) = 150(0) + 200(0) = 0$$
$$f(0,8) = 150(0) + 200(8) = 1600$$
$$f(8,0) = 150(8) + 200(0) = 1200$$
$$f(2,7) = 150(2) + 200(7) = 1700$$
$$f(6,3) = 150(6) + 200(3) = 1500$$

The maximum value is 1700.

37. Let x represent the number of one-gallon freezers and y represent the number of two-gallon freezers. The constraint equations are:

$$x \geq 75$$
$$y \geq 100$$
$$x + y \leq 250$$

Graphing the region:

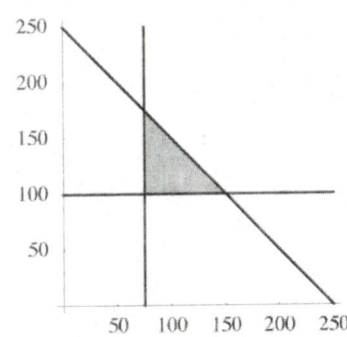

Evaluating $f(x,y) = 4.50x + 5.25y$ at each vertex:

$$f(75,100) = 4.50(75) + 5.25(100) = 862.50 \qquad f(150,100) = 4.50(150) + 5.25(100) = 1200$$
$$f(75,175) = 4.50(75) + 5.25(175) = 1256.25$$

They should produce 75 one-gallon freezers and 175 two-gallon freezers.

Chapter 7 Test

1. Computing: $AB = \begin{bmatrix} -1 & 3 \\ 4 & -2 \end{bmatrix}\begin{bmatrix} 3 & -2 \\ 4 & -1 \end{bmatrix} = \begin{bmatrix} 9 & -1 \\ 4 & -6 \end{bmatrix}$

2. Computing: $BA = \begin{bmatrix} 3 & -2 \\ 4 & -1 \end{bmatrix}\begin{bmatrix} -1 & 3 \\ 4 & -2 \end{bmatrix} = \begin{bmatrix} -11 & 13 \\ -8 & 14 \end{bmatrix}$

3. Computing: $DE = \begin{bmatrix} 2 & -1 \\ 3 & -2 \\ 6 & 5 \end{bmatrix}\begin{bmatrix} 2 & -1 & 4 \\ 5 & 1 & -3 \end{bmatrix} = \begin{bmatrix} -1 & -3 & 11 \\ -4 & -5 & 18 \\ 37 & -1 & 9 \end{bmatrix}$

4. The product BC does not exist.

5. Computing: $EC = \begin{bmatrix} 2 & -1 & 4 \\ 5 & 1 & -3 \end{bmatrix}\begin{bmatrix} -3 \\ 5 \\ -6 \end{bmatrix} = \begin{bmatrix} -35 \\ 8 \end{bmatrix}$

6.   Computing: $2A - B = \begin{bmatrix} -2 & 6 \\ 8 & -4 \end{bmatrix} - \begin{bmatrix} 3 & -2 \\ 4 & -1 \end{bmatrix} = \begin{bmatrix} -5 & 8 \\ 4 & -3 \end{bmatrix}$

7.   Computing: $3D + 2F = \begin{bmatrix} 6 & -3 \\ 9 & -6 \\ 18 & 15 \end{bmatrix} + \begin{bmatrix} -2 & 12 \\ 4 & -10 \\ 6 & 8 \end{bmatrix} = \begin{bmatrix} 4 & 9 \\ 13 & -16 \\ 24 & 23 \end{bmatrix}$

8.   Computing: $-3A - 2B = \begin{bmatrix} 3 & -9 \\ -12 & 6 \end{bmatrix} - \begin{bmatrix} 6 & -4 \\ 8 & -2 \end{bmatrix} = \begin{bmatrix} -3 & -5 \\ -20 & 8 \end{bmatrix}$

9.   Computing: $EF = \begin{bmatrix} 2 & -1 & 4 \\ 5 & 1 & -3 \end{bmatrix} \begin{bmatrix} -1 & 6 \\ 2 & -5 \\ 3 & 4 \end{bmatrix} = \begin{bmatrix} 8 & 33 \\ -12 & 13 \end{bmatrix}$

10.   Computing: $AB - EF = \begin{bmatrix} 9 & -1 \\ 4 & -6 \end{bmatrix} - \begin{bmatrix} 8 & 33 \\ -12 & 13 \end{bmatrix} = \begin{bmatrix} 1 & -34 \\ 16 & -19 \end{bmatrix}$

11.   Form the augmented matrix: $\begin{bmatrix} 3 & -2 & | & 1 & 0 \\ 5 & -3 & | & 0 & 1 \end{bmatrix}$

Add -2 times row 1 to row 2: $\begin{bmatrix} 3 & -2 & | & 1 & 0 \\ -1 & 1 & | & -2 & 1 \end{bmatrix}$

Switch row 1 and row 2, and multiply row 1 by -1: $\begin{bmatrix} 1 & -1 & | & 2 & -1 \\ 3 & -2 & | & 1 & 0 \end{bmatrix}$

Add -3 times row 1 to row 2: $\begin{bmatrix} 1 & -1 & | & 2 & -1 \\ 0 & 1 & | & -5 & 3 \end{bmatrix}$

Add row 2 to row 1: $\begin{bmatrix} 1 & 0 & | & -3 & 2 \\ 0 & 1 & | & -5 & 3 \end{bmatrix}$

The inverse is $\begin{bmatrix} -3 & 2 \\ -5 & 3 \end{bmatrix}$.

12.   Form the augmented matrix: $\begin{bmatrix} -2 & 5 & | & 1 & 0 \\ 3 & -7 & | & 0 & 1 \end{bmatrix}$

Add row 2 to row 1: $\begin{bmatrix} 1 & -2 & | & 1 & 1 \\ 3 & -7 & | & 0 & 1 \end{bmatrix}$

Add -3 times row 1 to row 2: $\begin{bmatrix} 1 & -2 & | & 1 & 1 \\ 0 & -1 & | & -3 & -2 \end{bmatrix}$

Multiply row 2 by -1: $\begin{bmatrix} 1 & -2 & | & 1 & 1 \\ 0 & 1 & | & 3 & 2 \end{bmatrix}$

Add 2 times row 2 to row 1: $\begin{bmatrix} 1 & 0 & | & 7 & 5 \\ 0 & 1 & | & 3 & 2 \end{bmatrix}$

The inverse is $\begin{bmatrix} 7 & 5 \\ 3 & 2 \end{bmatrix}$.

13.   Form the augmented matrix: $\begin{bmatrix} 1 & -3 & | & 1 & 0 \\ -2 & 8 & | & 0 & 1 \end{bmatrix}$

Add 2 times row 1 to row 2: $\begin{bmatrix} 1 & -3 & | & 1 & 0 \\ 0 & 2 & | & 2 & 1 \end{bmatrix}$

Divide row 2 by 2: $\begin{bmatrix} 1 & -3 & | & 1 & 0 \\ 0 & 1 & | & 1 & \frac{1}{2} \end{bmatrix}$

Add 3 times row 2 to row 1: $\begin{bmatrix} 1 & 0 & | & 4 & \frac{3}{2} \\ 0 & 1 & | & 1 & \frac{1}{2} \end{bmatrix}$

The inverse is $\begin{bmatrix} 4 & \frac{3}{2} \\ 1 & \frac{1}{2} \end{bmatrix}$.

14.   Form the augmented matrix: $\begin{bmatrix} 3 & 5 & | & 1 & 0 \\ 1 & 4 & | & 0 & 1 \end{bmatrix}$

Switch row 1 and row 2: $\begin{bmatrix} 1 & 4 & | & 0 & 1 \\ 3 & 5 & | & 1 & 0 \end{bmatrix}$

Add –3 times row 1 to row 2: $\begin{bmatrix} 1 & 4 & | & 0 & 1 \\ 0 & -7 & | & 1 & -3 \end{bmatrix}$

Divide row 2 by –7: $\begin{bmatrix} 1 & 4 & | & 0 & 1 \\ 0 & 1 & | & -\frac{1}{7} & \frac{3}{7} \end{bmatrix}$

Add –4 times row 2 to row 1: $\begin{bmatrix} 1 & 0 & | & \frac{4}{7} & -\frac{5}{7} \\ 0 & 1 & | & -\frac{1}{7} & \frac{3}{7} \end{bmatrix}$

The inverse is $\begin{bmatrix} \frac{4}{7} & -\frac{5}{7} \\ -\frac{1}{7} & \frac{3}{7} \end{bmatrix}$.

15.   Form the augmented matrix: $\begin{bmatrix} -2 & 2 & 3 & | & 1 & 0 & 0 \\ 1 & -1 & 0 & | & 0 & 1 & 0 \\ 0 & 1 & 4 & | & 0 & 0 & 1 \end{bmatrix}$

Switch row 1 and row 2: $\begin{bmatrix} 1 & -1 & 0 & | & 0 & 1 & 0 \\ -2 & 2 & 3 & | & 1 & 0 & 0 \\ 0 & 1 & 4 & | & 0 & 0 & 1 \end{bmatrix}$

Add 2 times row 1 to row 2: $\begin{bmatrix} 1 & -1 & 0 & | & 0 & 1 & 0 \\ 0 & 0 & 3 & | & 1 & 2 & 0 \\ 0 & 1 & 4 & | & 0 & 0 & 1 \end{bmatrix}$

Switch row 2 and row 3: $\begin{bmatrix} 1 & -1 & 0 & | & 0 & 1 & 0 \\ 0 & 1 & 4 & | & 0 & 0 & 1 \\ 0 & 0 & 3 & | & 1 & 2 & 0 \end{bmatrix}$

Divide row 3 by 3: $\begin{bmatrix} 1 & -1 & 0 & | & 0 & 1 & 0 \\ 0 & 1 & 4 & | & 0 & 0 & 1 \\ 0 & 0 & 1 & | & \dfrac{1}{3} & \dfrac{2}{3} & 0 \end{bmatrix}$

Add -4 times row 3 to row 2: $\begin{bmatrix} 1 & -1 & 0 & | & 0 & 1 & 0 \\ 0 & 1 & 0 & | & -\dfrac{4}{3} & -\dfrac{8}{3} & 1 \\ 0 & 0 & 1 & | & \dfrac{1}{3} & \dfrac{2}{3} & 0 \end{bmatrix}$

Add row 2 to row 1: $\begin{bmatrix} 1 & 0 & 0 & | & -\dfrac{4}{3} & -\dfrac{5}{3} & 1 \\ 0 & 1 & 0 & | & -\dfrac{4}{3} & -\dfrac{8}{3} & 1 \\ 0 & 0 & 1 & | & \dfrac{1}{3} & \dfrac{2}{3} & 0 \end{bmatrix}$

The inverse is $\begin{bmatrix} -\dfrac{4}{3} & -\dfrac{5}{3} & 1 \\ -\dfrac{4}{3} & -\dfrac{8}{3} & 1 \\ \dfrac{1}{3} & \dfrac{2}{3} & 0 \end{bmatrix}$.

16. Form the augmented matrix: $\begin{bmatrix} 1 & -2 & 4 & | & 1 & 0 & 0 \\ 0 & 1 & 3 & | & 0 & 1 & 0 \\ 0 & 0 & 1 & | & 0 & 0 & 1 \end{bmatrix}$

Add -3 times row 3 to row 2, and -4 times row 3 to row 1: $\begin{bmatrix} 1 & -2 & 0 & | & 1 & 0 & -4 \\ 0 & 1 & 0 & | & 0 & 1 & -3 \\ 0 & 0 & 1 & | & 0 & 0 & 1 \end{bmatrix}$

Add 2 times row 2 to row 1: $\begin{bmatrix} 1 & 0 & 0 & | & 1 & 2 & -10 \\ 0 & 1 & 0 & | & 0 & 1 & -3 \\ 0 & 0 & 1 & | & 0 & 0 & 1 \end{bmatrix}$

The inverse is $\begin{bmatrix} 1 & 2 & -10 \\ 0 & 1 & -3 \\ 0 & 0 & 1 \end{bmatrix}$.

17. Solving the equation: $\begin{bmatrix} x \\ y \end{bmatrix} = \begin{bmatrix} -3 & 2 \\ -5 & 3 \end{bmatrix}\begin{bmatrix} 48 \\ 76 \end{bmatrix} = \begin{bmatrix} 8 \\ -12 \end{bmatrix}$. The solution set is $\left\{(8, -12)\right\}$.

18. Solving the equation: $\begin{bmatrix} x \\ y \end{bmatrix} = \begin{bmatrix} 4 & \dfrac{3}{2} \\ 1 & \dfrac{1}{2} \end{bmatrix}\begin{bmatrix} 36 \\ -100 \end{bmatrix} = \begin{bmatrix} -6 \\ -14 \end{bmatrix}$. The solution set is $\left\{(-6, -14)\right\}$.

19. Solving the equation: $\begin{bmatrix} x \\ y \end{bmatrix} = \begin{bmatrix} \dfrac{4}{7} & -\dfrac{5}{7} \\ -\dfrac{1}{7} & \dfrac{3}{7} \end{bmatrix} \begin{bmatrix} 92 \\ 61 \end{bmatrix} = \begin{bmatrix} 9 \\ 13 \end{bmatrix}$. The solution set is $\{(9,13)\}$.

20. Solving the equation: $\begin{bmatrix} x \\ y \\ z \end{bmatrix} = \begin{bmatrix} -\dfrac{10}{9} & \dfrac{7}{9} & -\dfrac{5}{9} \\ \dfrac{4}{9} & -\dfrac{1}{9} & \dfrac{2}{9} \\ -\dfrac{13}{9} & \dfrac{10}{9} & -\dfrac{11}{9} \end{bmatrix} \begin{bmatrix} 1 \\ 3 \\ -2 \end{bmatrix} = \begin{bmatrix} \dfrac{7}{3} \\ -\dfrac{1}{3} \\ \dfrac{13}{3} \end{bmatrix}$. The solution set is $\left\{ \left(\dfrac{7}{3}, -\dfrac{1}{3}, \dfrac{13}{3} \right) \right\}$.

21. Solving the equation: $\begin{bmatrix} x \\ y \\ z \end{bmatrix} = \begin{bmatrix} -\dfrac{5}{24} & \dfrac{1}{6} & -\dfrac{7}{24} \\ \dfrac{7}{24} & \dfrac{1}{6} & \dfrac{5}{24} \\ \dfrac{11}{24} & -\dfrac{1}{6} & \dfrac{1}{24} \end{bmatrix} \begin{bmatrix} 3 \\ 3 \\ 3 \end{bmatrix} = \begin{bmatrix} -1 \\ 2 \\ 1 \end{bmatrix}$. The solution set is $\{(-1,2,1)\}$.

22. Graphing the solution set:

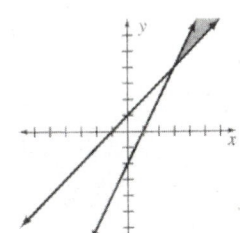

23. Graphing the solution set:

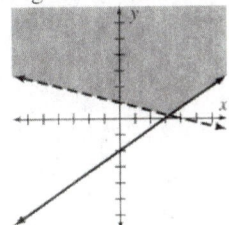

24. Graphing the solution set:

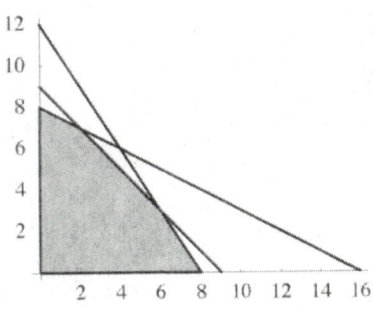

25. Graphing the region:

Evaluating the function at each vertex:

$f(0,0) = 500(0) + 350(0) = 0$

$f(8,0) = 500(8) + 350(0) = 4000$

$f(6,3) = 500(6) + 350(3) = 4050$

$f(0,8) = 500(0) + 350(8) = 2800$

$f(2,7) = 500(2) + 350(7) = 3450$

The maximum value is 4050.

Chapters 0-7 Cumulative Review Problem Set

1. This is the multiplicative inverse property.
2. This is the associative property of multiplication.
3. This is the commutative property of multiplication.
4. This is the identity property of addition.

5. Simplifying: $4(-3)^2 - 2^3 \div 4\left(\dfrac{1}{2}\right) = 4(9) - 8 \div 4\left(\dfrac{1}{2}\right) = 36 - 2\left(\dfrac{1}{2}\right) = 36 - 1 = 35$

6. Simplifying: $\dfrac{1}{2} \div \dfrac{3}{4}(-2)^3 = \dfrac{1}{2} \div \dfrac{3}{4}(-8) = \dfrac{1}{2} \cdot \dfrac{4}{3}(-8) = \dfrac{2}{3}(-8) = -\dfrac{16}{3}$

7. Simplifying: $\dfrac{4}{3} - \left(\dfrac{7}{6}\right)\left(\dfrac{1}{3}\right) \div \dfrac{1}{6} = \dfrac{4}{3} - \dfrac{7}{18} \div \dfrac{1}{6} = \dfrac{4}{3} - \dfrac{7}{18} \cdot \dfrac{6}{1} = \dfrac{4}{3} - \dfrac{7}{3} = -\dfrac{3}{3} = -1$

8. Simplifying: $\dfrac{10}{9} - \dfrac{4}{3} - 3\left(\dfrac{3}{4}\right)\left(\dfrac{1}{6}\right) = \dfrac{10}{9} - \dfrac{4}{3} - \left(\dfrac{9}{4}\right)\left(\dfrac{1}{6}\right) = \dfrac{10}{9} - \dfrac{4}{3} - \dfrac{3}{8} = \dfrac{80}{72} - \dfrac{96}{72} - \dfrac{27}{72} = -\dfrac{43}{72}$

9. Evaluating the expression: $\dfrac{4x^2 - 3y}{2xy} = \dfrac{4(3)^2 - 3(-6)}{2(3)(-6)} = \dfrac{4(9) - 3(-6)}{2(3)(-6)} = \dfrac{36 + 18}{-36} = -\dfrac{54}{36} = -\dfrac{3}{2}$

10. Evaluating the expression:

$\sqrt{3a + b} - 5\sqrt{2a - b} = \sqrt{3(5) + 1} - 5\sqrt{2(5) - 1} = \sqrt{15 + 1} - 5\sqrt{10 - 1} = \sqrt{16} - 5\sqrt{9} = 4 - 15 = -11$

11. Evaluating the expression: $\dfrac{c - d}{\dfrac{1}{c} - \dfrac{1}{d}} = \dfrac{1 - (-1)}{\dfrac{1}{1} - \dfrac{1}{-1}} = \dfrac{1 + 1}{1 + 1} = \dfrac{2}{2} = 1$

12. Evaluating the expression: $\dfrac{1}{x+1} + \dfrac{y}{x-y} = \dfrac{1}{-\dfrac{1}{2}+1} + \dfrac{\dfrac{1}{2}}{-\dfrac{1}{2}-\dfrac{1}{2}} = \dfrac{1}{\dfrac{1}{2}} + \dfrac{\dfrac{1}{2}}{-1} = 2 - \dfrac{1}{2} = \dfrac{3}{2}$

13. Computing the matrix: $A + B = \begin{bmatrix} -4 & 3 \\ -1 & -3 \end{bmatrix} + \begin{bmatrix} 0 & -7 \\ 6 & 1 \end{bmatrix} = \begin{bmatrix} -4 & -4 \\ 5 & -2 \end{bmatrix}$

14. Computing the matrix: $2A - B = 2\begin{bmatrix} -4 & 3 \\ -1 & -3 \end{bmatrix} - \begin{bmatrix} 0 & -7 \\ 6 & 1 \end{bmatrix} = \begin{bmatrix} -8 & 6 \\ -2 & -6 \end{bmatrix} - \begin{bmatrix} 0 & -7 \\ 6 & 1 \end{bmatrix} = \begin{bmatrix} -8 & 13 \\ -8 & -7 \end{bmatrix}$

15. Computing the matrix: $AB = \begin{bmatrix} -4 & 3 \\ -1 & -3 \end{bmatrix}\begin{bmatrix} 0 & -7 \\ 6 & 1 \end{bmatrix} = \begin{bmatrix} 18 & 31 \\ -18 & 4 \end{bmatrix}$

16. Computing the matrix: $BA = \begin{bmatrix} 0 & -7 \\ 6 & 1 \end{bmatrix}\begin{bmatrix} -4 & 3 \\ -1 & -3 \end{bmatrix} = \begin{bmatrix} 7 & 21 \\ -25 & 15 \end{bmatrix}$

17. Evaluating the determinant: $\begin{vmatrix} -5 & -3 \\ 4 & 7 \end{vmatrix} = (-5)(7) - (-3)(4) = -35 + 12 = -23$

18. Evaluating the determinant: $\begin{vmatrix} 1 & -6 & 7 \\ 0 & 2 & 3 \\ 4 & 5 & 2 \end{vmatrix} = 1\begin{vmatrix} 2 & 3 \\ 5 & 2 \end{vmatrix} - 0 + 4\begin{vmatrix} -6 & 7 \\ 2 & 3 \end{vmatrix} = 1(4 - 15) + 4(-18 - 14) = -11 - 128 = -139$

19. First find the slope: $m = \dfrac{-3-6}{2-(-5)} = -\dfrac{9}{7}$. Using the point-slope formula:

$$y-(-3) = -\frac{9}{7}(x-2)$$
$$y+3 = -\frac{9}{7}(x-2)$$
$$7y+21 = -9x+18$$
$$9x+7y = -3$$

20. Using the point-slope formula:

$$y-0 = -\frac{2}{3}(x-1)$$
$$y = -\frac{2}{3}(x-1)$$
$$3y = -2x+2$$
$$2x+3y = 2$$

21. The line $2x-5y = 7$ has a slope of $\dfrac{2}{5}$, so the perpendicular slope is $-\dfrac{5}{2}$. Using the point-slope formula:

$$y-(-4) = -\frac{5}{2}(x-5)$$
$$y+4 = -\frac{5}{2}(x-5)$$
$$2y+8 = -5x+25$$
$$5x+2y = 17$$

22. The line $y = -x$ has a slope of -1, so the perpendicular slope is 1. Using the point-slope formula:

$$y-1 = 1(x-(-2))$$
$$y-1 = x+2$$
$$-x+y = 3$$
$$x-y = -3$$

23. Simplifying: $\left(4a^{-1}b^2\right)^{-3}\left(3^{-1}a^2b^{-3}\right)^{-2} = 4^{-3}a^3b^{-6} \cdot 3^2\, a^{-4}b^6 = \dfrac{9}{64}a^{-1} = \dfrac{9}{64a}$

24. Simplifying: $\left(\dfrac{-6m^3n^{-4}}{4m^{-2}n^{-1}}\right)^{-2} = \left(\dfrac{2m^{-2}n^{-1}}{-3m^3n^{-4}}\right)^2 = \dfrac{4m^{-4}n^{-2}}{9m^6n^{-8}} = \dfrac{4n^6}{9m^{10}}$

25. Simplifying: $\sqrt{68} = \sqrt{4\cdot 17} = 2\sqrt{17}$

26. Simplifying: $\sqrt[3]{81} = \sqrt[3]{27\cdot 3} = 3\sqrt[3]{3}$

27. Simplifying: $\sqrt{\dfrac{16}{5}} = \dfrac{\sqrt{16}}{\sqrt{5}} = \dfrac{4}{\sqrt{5}} \cdot \dfrac{\sqrt{5}}{\sqrt{5}} = \dfrac{4\sqrt{5}}{5}$

28. Simplifying: $\sqrt{\dfrac{7}{18a^3}} = \dfrac{\sqrt{7}}{\sqrt{18a^3}} = \dfrac{\sqrt{7}}{\sqrt{18a^3}} \cdot \dfrac{\sqrt{2a}}{\sqrt{2a}} = \dfrac{\sqrt{14a}}{\sqrt{36a^4}} = \dfrac{\sqrt{14a}}{6a^2}$

29. Simplifying: $\dfrac{3}{10}\sqrt{75x^2 y} = \dfrac{3}{10}\sqrt{25x^2 \cdot 3y} = \dfrac{3}{10} \cdot 5x\sqrt{3y} = \dfrac{3x}{2}\sqrt{3y}$

30. Simplifying: $\dfrac{\sqrt{16a^3b^2}}{\sqrt{2ab}} = \sqrt{\dfrac{16a^3b^2}{2ab}} = \sqrt{8a^2b} = \sqrt{4a^2 \cdot 2b} = 2a\sqrt{2b}$

31. Simplifying: $\sqrt[3]{\dfrac{3}{4}} = \dfrac{\sqrt[3]{3}}{\sqrt[3]{4}} \cdot \dfrac{\sqrt[3]{2}}{\sqrt[3]{2}} = \dfrac{\sqrt[3]{6}}{\sqrt[3]{8}} = \dfrac{\sqrt[3]{6}}{2}$

32. Simplifying: $\sqrt[3]{128c^5d^7} = \sqrt[3]{64c^3d^6 \cdot 2c^2d} = 4cd^2\sqrt[3]{2c^2d}$

33. Rationalizing the denominator: $\dfrac{6}{\sqrt{5}-1} = \dfrac{6}{\sqrt{5}-1} \cdot \dfrac{\sqrt{5}+1}{\sqrt{5}+1} = \dfrac{6\left(\sqrt{5}+1\right)}{5-1} = \dfrac{6\left(\sqrt{5}+1\right)}{4} = \dfrac{3\left(\sqrt{5}+1\right)}{2}$

34. Rationalizing the denominator:

$$\dfrac{8}{3\sqrt{6}+4\sqrt{2}} = \dfrac{8}{3\sqrt{6}+4\sqrt{2}} \cdot \dfrac{3\sqrt{6}-4\sqrt{2}}{3\sqrt{6}-4\sqrt{2}} = \dfrac{8\left(3\sqrt{6}-4\sqrt{2}\right)}{54-32} = \dfrac{8\left(3\sqrt{6}-4\sqrt{2}\right)}{22} = \dfrac{4\left(3\sqrt{6}-4\sqrt{2}\right)}{11}$$

35. Solving for x:

$$2(x-3y) = 4$$
$$2x - 6y = 4$$
$$2x = 6y + 4$$
$$x = 3y + 2$$

36. Solving for x:

$$4y - 9x = 13$$
$$-9x = -4y + 13$$
$$x = \dfrac{4y-13}{9}$$

37. Solving for x:

$$\dfrac{ax+b}{x} = \dfrac{c-d}{y}$$
$$y(ax+b) = x(c-d)$$
$$ayx + by = cx - dx$$
$$by = cx - dx - ayx$$
$$by = x(c-d-ay)$$
$$x = \dfrac{by}{c-d-ay}$$

38. Solving for x:

$$\dfrac{4}{3}x + \dfrac{1}{2}a = c$$
$$6\left(\dfrac{4}{3}x + \dfrac{1}{2}a\right) = 6c$$
$$8x + 3a = 6c$$
$$8x = 6c - 3a$$
$$x = \dfrac{6c-3a}{8}$$

39. Simplifying: $2 - \dfrac{c}{\dfrac{1}{c}-1} = 2 - \dfrac{c}{\dfrac{1}{c}-1} \cdot \dfrac{c}{c} = 2 - \dfrac{c^2}{1-c} = \dfrac{2(1-c)}{1-c} - \dfrac{c^2}{1-c} = \dfrac{2-2c-c^2}{1-c} = \dfrac{c^2+2c-2}{c-1}$

40. Simplifying:

$$\dfrac{\dfrac{1}{a+1}-4}{3+\dfrac{4}{a-2}} = \dfrac{\dfrac{1}{a+1}-4}{3+\dfrac{4}{a-2}} \cdot \dfrac{(a+1)(a-2)}{(a+1)(a-2)}$$

$$= \dfrac{a-2-4(a+1)(a-2)}{3(a+1)(a-2)+4(a+1)}$$

$$= \dfrac{(a-2)(1-4a-4)}{(a+1)(3a-6+4)}$$

$$= \dfrac{(a-2)(-4a-3)}{(a+1)(3a-2)}$$

$$= -\dfrac{(a-2)(4a+3)}{(a+1)(3a-2)}$$

41. Using synthetic division:

$$\begin{array}{r|rrrr} -1 & 2 & -7 & 4 & -1 \\ & & -2 & 9 & -13 \\ \hline & 2 & -9 & 13 & -14 \end{array}$$

So $f(-1) = -14$.

42. Using synthetic division:

$$\begin{array}{r|rrrrr} -2 & 1 & 0 & -1 & 4 & 3 \\ & & -2 & 4 & -6 & 4 \\ \hline & 1 & -2 & 3 & -2 & 7 \end{array}$$

So $f(-2) = 7$.

43. Finding the values:
$$f(0) = -2(0)^2 + 0 - 9 = -9$$
$$f(-1) = -2(-1)^2 + (-1) - 9 = -2 - 1 - 9 = -12$$
$$f(a) = -2a^2 + a - 9$$

44. Finding the values:
$$f(0) = 2(0) = 0 \qquad\qquad f(3) = 2(3) = 6$$
$$f(-1) = -(-1)^2 = -1 \qquad\qquad f(-3) = -(-3)^2 = -9$$

45. Finding the difference quotient: $\dfrac{f(a+h) - f(a)}{h} = \dfrac{-2(a+h) + 7 - (-2a + 7)}{h} = \dfrac{-2a - 2h + 7 + 2a - 7}{h} = \dfrac{-2h}{h} = -2$

46. Finding the difference quotient:
$$\frac{f(a+h) - f(a)}{h} = \frac{3(a+h)^2 - (a+h) + 4 - (3a^2 - a + 4)}{h}$$
$$= \frac{3a^2 + 6ah + 3h^2 - a - h + 4 - 3a^2 + a - 4}{h}$$
$$= \frac{6ah + 3h^2 - h}{h}$$
$$= 6a + 3h - 1$$

47. Setting the denominator equal to 0:
$$3x^2 + x - 2 = 0$$
$$(3x - 2)(x + 1) = 0$$
$$x = -1, \frac{2}{3}$$

So the domain is $\left\{ x \mid x \neq -1, \dfrac{2}{3} \right\}$.

48. Setting the denominator equal to 0:
$$x^2 - 1 = 0$$
$$(x + 1)(x - 1) = 0$$
$$x = -1, 1$$

So the domain is $\left\{ x \mid x \neq -1, 1 \right\}$.

49. The domain is $\left\{ x \mid x \geq -\dfrac{2}{3} \right\}$ and the range is $\left\{ f(x) \mid f(x) \geq 0 \right\}$.

50. The domain is $\left\{ x \mid x \text{ is a real number} \right\}$ and the range is $\left\{ f(x) \mid f(x) \geq 0 \right\}$.

51. The domain is $\left\{ x \mid x \text{ is a real number} \right\}$ and the range is $\left\{ f(x) \mid f(x) \leq -1 \right\}$.

52. Factoring the inequality:
$$x^2 - 25 \geq 0$$
$$(x + 5)(x - 5) \geq 0$$

Forming a sign chart:

The solution set is $(-\infty, -5] \cup [5, \infty)$.

53. Since any real number makes the radicand positive, the domain is $(-\infty, \infty)$.

54. Factoring the inequality:
$$4x^2 - 11x - 3 \geq 0$$
$$(4x + 1)(x - 3) \geq 0$$

Forming a sign chart:

The solution set is $\left(-\infty, -\dfrac{1}{4} \right] \cup [3, \infty)$.

55. Finding the compositions:

$$(f \circ g)(x) = f\left(x^2 + 3x - 4\right) = 3\left(x^2 + 3x - 4\right) - 1 = 3x^2 + 9x - 13$$

$$(g \circ f)(x) = g(3x - 1) = (3x - 1)^2 + 3(3x - 1) - 4 = 9x^2 - 6x + 1 + 9x - 3 - 4 = 9x^2 + 3x - 6$$

The domain is each is $(-\infty, \infty)$.

56. Finding the compositions:

$$(f \circ g)(x) = f\left(\frac{1}{x-1}\right) = \frac{2}{\frac{1}{x-1}} = 2(x-1) = 2x - 2$$

$$(g \circ f)(x) = g\left(\frac{2}{x}\right) = \frac{1}{\frac{2}{x} - 1} \cdot \frac{x}{x} = \frac{x}{2 - x}$$

The domain of $f \circ g$ is $\{x \mid x \neq 1\}$ and the domain of $g \circ f$ is $\{x \mid x \neq 0, 2\}$.

57. Solving the equation:

$$-3(2a - 7) - 6(a + 4) = 8(-3a - 1)$$
$$-6a + 21 - 6a - 24 = -24a - 8$$
$$-12a - 3 = -24a - 8$$
$$12a = -5$$
$$a = -\frac{5}{12}$$

The solution set is $\left\{-\frac{5}{12}\right\}$.

58. Solving the equation:

$$\frac{2x + 9}{3} - \frac{7x - 3}{4} = \frac{11}{2}$$
$$12\left(\frac{2x + 9}{3}\right) - 12\left(\frac{7x - 3}{4}\right) = 12\left(\frac{11}{2}\right)$$
$$4(2x + 9) - 3(7x - 3) = 6(11)$$
$$8x + 36 - 21x + 9 = 66$$
$$-13x + 45 = 66$$
$$-13x = 21$$
$$x = -\frac{21}{13}$$

The solution set is $\left\{-\frac{21}{13}\right\}$.

59. Solving the equation:

$$|4 - 3x| = 16$$
$$4 - 3x = -16, 16$$
$$-3x = -20, 12$$
$$x = \frac{20}{3}, -4$$

The solution set is $\left\{-4, \frac{20}{3}\right\}$.

60. Solving the equation:

$$|-4x + 3| = |-4x - 7|$$

$$-4x + 3 = -4x - 7 \qquad \text{or} \qquad -4x + 3 = 4x + 7$$
$$3 = -7 \qquad\qquad\qquad\qquad -8x = 4$$
$$\text{not possible} \qquad\qquad\qquad x = -\frac{1}{2}$$

The solution set is $\left\{-\frac{1}{2}\right\}$.

61. Solving the equation:

$$x(2x + 5) = 7$$
$$2x^2 + 5x = 7$$
$$2x^2 + 5x - 7 = 0$$
$$(2x + 7)(x - 1) = 0$$
$$x = -\frac{7}{2}, 1$$

The solution set is $\left\{-\frac{7}{2}, 1\right\}$.

62. Solving the equation:

$$9x - 4x^3 = 0$$
$$x\left(9 - 4x^2\right) = 0$$
$$x(3 + 2x)(3 - 2x) = 0$$
$$x = -\frac{3}{2}, 0, \frac{3}{2}$$

The solution set is $\left\{-\frac{3}{2}, 0, \frac{3}{2}\right\}$.

63.   Solving the equation:
$$12x^2 - x - 63 = 0$$
$$(4x + 9)(3x - 7) = 0$$
$$x = -\frac{9}{4}, \frac{7}{3}$$

The solution set is $\left\{ -\dfrac{9}{4}, \dfrac{7}{3} \right\}$.

64.   Solving the equation:
$$\frac{3}{x-2} + \frac{4}{x+3} = \frac{3}{2}$$
$$2(x-2)(x+3)\left(\frac{3}{x-2}\right) + 2(x-2)(x+3)\left(\frac{4}{x+3}\right) = 2(x-2)(x+3)\left(\frac{3}{2}\right)$$
$$6(x+3) + 8(x-2) = 3(x-2)(x+3)$$
$$6x + 18 + 8x - 16 = 3x^2 + 3x - 18$$
$$14x + 2 = 3x^2 + 3x - 18$$
$$3x^2 - 11x - 20 = 0$$
$$(3x+4)(x-5) = 0$$
$$x = -\frac{4}{3}, 5$$

The solution set is $\left\{ -\dfrac{4}{3}, 5 \right\}$.

65.   Solving the equation:
$$\frac{5x}{2x^2 + 3x - 2} - \frac{x+1}{x^2 - 3x - 10} = \frac{-10}{2x^2 - 11x + 5}$$
$$\frac{5x}{(2x-1)(x+2)} - \frac{x+1}{(x-5)(x+2)} = \frac{-10}{(2x-1)(x-5)}$$
$$5x(x-5) - (x+1)(2x-1) = -10(x+2)$$
$$5x^2 - 25x - 2x^2 - x + 1 = -10x - 20$$
$$3x^2 - 16x + 21 = 0$$
$$(3x-7)(x-3) = 0$$
$$x = \frac{7}{3}, 3$$

The solution set is $\left\{ \dfrac{7}{3}, 3 \right\}$.

66.   Solving the equation:
$$\sqrt{3x - 11} = 8$$
$$3x - 11 = 8^2$$
$$3x - 11 = 64$$
$$3x = 75$$
$$x = 25$$

Since this value checks, the solution set is $\{25\}$.

67.   Since $\sqrt{2x + 3}$ cannot be negative, the solution set is $\varnothing$.

68. Solving the equation:

$$\sqrt[3]{x^2 - 8} = 2$$
$$x^2 - 8 = 2^3$$
$$x^2 - 8 = 8$$
$$x^2 = 16$$
$$x = -4, 4$$

Since these values check, the solution set is $\{-4, 4\}$.

69. Solving the equation:

$$5 - \sqrt{x-4} = \sqrt{x+1}$$
$$\left(5 - \sqrt{x-4}\right)^2 = \left(\sqrt{x+1}\right)^2$$
$$25 - 10\sqrt{x-4} + x - 4 = x + 1$$
$$-10\sqrt{x-4} = -20$$
$$\sqrt{x-4} = 2$$
$$x - 4 = 4$$
$$x = 8$$

Since this value checks, the solution set is $\{8\}$.

70. Solving the equation:

$$4\sqrt{x} = x - 5$$
$$\left(4\sqrt{x}\right)^2 = (x-5)^2$$
$$16x = x^2 - 10x + 25$$
$$x^2 - 26x + 25 = 0$$
$$(x-1)(x-25) = 0$$
$$x = 25 \qquad (1 \text{ does not check})$$

The solution set is $\{25\}$.

71. Solving the equation:

$$3n^2 - 7n = 0$$
$$n(3n - 7) = 0$$
$$n = 0, \frac{7}{3}$$

The solution set is $\left\{0, \frac{7}{3}\right\}$.

72. Solving the equation:

$$12x^2 = 15$$
$$x^2 = \frac{15}{12} = \frac{5}{4}$$
$$x = \pm\sqrt{\frac{5}{4}} = \pm\frac{\sqrt{5}}{2}$$

The solution set is $\left\{-\frac{\sqrt{5}}{2}, \frac{\sqrt{5}}{2}\right\}$.

73. Solving the equation $x^2 + 11x - 1 = 0$ using the quadratic formula:

$$x = \frac{-11 \pm \sqrt{11^2 - 4(1)(-1)}}{2(1)} = \frac{-11 \pm \sqrt{121 + 4}}{2} = \frac{-11 \pm \sqrt{125}}{2} = \frac{-11 \pm 5\sqrt{5}}{2}$$

The solution set is $\left\{\frac{-11 - 5\sqrt{5}}{2}, \frac{-11 + 5\sqrt{5}}{2}\right\}$.

74. Solving the equation $2x^2 + x + 7 = 0$ using the quadratic formula:

$$x = \frac{-1 \pm \sqrt{1^2 - 4(2)(7)}}{2(2)} = \frac{-1 \pm \sqrt{1 - 56}}{4} = \frac{-1 \pm \sqrt{-55}}{4} = \frac{-1 \pm i\sqrt{55}}{4}$$

The solution set is $\left\{\frac{-1 - i\sqrt{55}}{4}, \frac{-1 + i\sqrt{55}}{4}\right\}$.

75. Solving the equation:

$$x^4 + 15x^2 - 54 = 0$$
$$\left(x^2 + 18\right)\left(x^2 - 3\right) = 0$$
$$x^2 = -18, 3$$
$$x = \pm 3i\sqrt{2}, \pm\sqrt{3}$$

The solution set is $\left\{\pm 3i\sqrt{2}, \pm\sqrt{3}\right\}$.

76. Solving the equation:

$$\frac{5}{x-3} + \frac{1}{x} = 2$$

$$x(x-3)\left(\frac{5}{x-3}\right) + x(x-3)\left(\frac{1}{x}\right) = 2x(x-3)$$

$$5x + x - 3 = 2x^2 - 6x$$

$$0 = 2x^2 - 12x + 3$$

Solving the equation $2x^2 - 12x + 3 = 0$ using the quadratic formula:

$$x = \frac{12 \pm \sqrt{(-12)^2 - 4(2)(3)}}{2(2)} = \frac{12 \pm \sqrt{144 - 24}}{4} = \frac{12 \pm \sqrt{120}}{4} = \frac{12 \pm 2\sqrt{30}}{4} = \frac{6 \pm \sqrt{30}}{2}$$

The solution set is $\left\{ \dfrac{6 - \sqrt{30}}{2}, \dfrac{6 + \sqrt{30}}{2} \right\}$.

77. The possible rational roots are $\pm 1, \pm 2, \pm 3, \pm 5, \pm 6, \pm 10, \pm 15, \pm 30$. Using synthetic division:

$$\begin{array}{r|rrrr} 2 & 1 & 0 & -19 & 30 \\ & & 2 & 4 & -30 \\ \hline & 1 & 2 & -15 & 0 \end{array}$$

Solving the equation:

$$(x-2)(x^2 + 2x - 15) = 0$$
$$(x-2)(x-3)(x+5) = 0$$
$$x = -5, 2, 3$$

The solution set is $\{-5, 2, 3\}$.

78. The possible rational roots are $\pm 1, \pm 3, \pm 5, \pm 15$. Using synthetic division:

$$\begin{array}{r|rrrrr} -1 & 1 & -6 & 10 & 2 & -15 \\ & & -1 & 7 & -17 & 15 \\ \hline & 1 & -7 & 17 & -15 & 0 \end{array}$$

Using synthetic division again:

$$\begin{array}{r|rrrr} 3 & 1 & -7 & 17 & -15 \\ & & 3 & -12 & 15 \\ \hline & 1 & -4 & 5 & 0 \end{array}$$

Solving the equation:

$$(x+1)(x-3)(x^2 - 4x + 5) = 0$$

$$x = -1, 3, \frac{4 \pm \sqrt{(-4)^2 - 4(1)(5)}}{2(1)} = \frac{4 \pm \sqrt{16 - 20}}{2} = \frac{4 \pm 2i}{2} = 2 \pm i$$

The solution set is $\{-1, 3, 2 - i, 2 + i\}$.

79. Solving the equation:

$$\log_x\left(\frac{9}{4}\right) = 2$$

$$\frac{9}{4} = x^2$$

$$x = \sqrt{\frac{9}{4}} = \frac{3}{2} \quad (\text{since } x > 0)$$

The solution set is $\left\{\dfrac{3}{2}\right\}$.

80. Solving the equation:

$$\log_9 x = -\frac{1}{2}$$

$$x = 9^{-1/2}$$

$$x = \sqrt{\frac{1}{9}} = \frac{1}{3}$$

The solution set is $\left\{\dfrac{1}{3}\right\}$.

81.   Solving the equation:

$$\log_5 x + \log_5 4 = -1$$
$$\log_5 (4x) = -1$$
$$4x = 5^{-1} = \frac{1}{5}$$
$$x = \frac{1}{20}$$

The solution set is $\left\{ \dfrac{1}{20} \right\}$.

82.   Solving the equation:

$$\log_5 (4x - 3) = 1 + \log_5 (x - 2)$$
$$\log_5 (4x - 3) - \log_5 (x - 2) = 1$$
$$\log_5 \frac{4x - 3}{x - 2} = 1$$
$$\frac{4x - 3}{x - 2} = 5$$
$$4x - 3 = 5x - 10$$
$$x = 7$$

The solution set is $\{7\}$.

83.   Solving the equation:

$$\log_6 (3x + 9) + \log_6 (x + 4) = 1$$
$$\log_6 \left[(3x + 9)(x + 4) \right] = 1$$
$$3x^2 + 21x + 36 = 6$$
$$3x^2 + 21x + 30 = 0$$
$$3\left(x^2 + 7x + 10 \right) = 0$$
$$3(x + 5)(x + 2) = 0$$
$$x = -2 \quad (x = -5 \text{ does not check})$$

The solution set is $\{-2\}$.

84.   Solving the equation:

$$16^x = 8^{2x+1}$$
$$2^{4x} = 2^{6x+3}$$
$$4x = 6x + 3$$
$$-2x = 3$$
$$x = -\frac{3}{2}$$

The solution set is $\left\{ -\dfrac{3}{2} \right\}$.

85.   Solving the equation:

$$5^x = 9$$
$$\ln 5^x = \ln 9$$
$$x \ln 5 = \ln 9$$
$$x = \frac{\ln 9}{\ln 5} \approx 1.37$$

The solution set is $\{1.37\}$.

86.   Solving the equation:

$$5^{x+1} = 3^{2x-1}$$
$$\ln 5^{x+1} = \ln 3^{2x-1}$$
$$(x + 1) \ln 5 = (2x - 1) \ln 3$$
$$x \ln 5 + \ln 5 = 2x \ln 3 - \ln 3$$
$$\ln 5 + \ln 3 = 2x \ln 3 - x \ln 5$$
$$\ln 5 + \ln 3 = x(2 \ln 3 - \ln 5)$$
$$x = \frac{\ln 5 + \ln 3}{2 \ln 3 - \ln 5} \approx 4.61$$

The solution set is $\{4.61\}$.

87.   Solving the equation:

$$2e^{x+3} = 7$$
$$e^{x+3} = \frac{7}{2}$$
$$x + 3 = \ln \frac{7}{2}$$
$$x = \ln \frac{7}{2} - 3 \approx -1.75$$

The solution set is $\{-1.75\}$.

88.   Solving the equation:

$$e^{2x-1} = 12.3$$
$$2x - 1 = \ln 12.3$$
$$2x = 1 + \ln 12.3$$
$$x = \frac{1 + \ln 12.3}{2} \approx 1.75$$

The solution set is $\{1.75\}$.

89.   Solving the inequality:

$$\frac{2}{3}x - \frac{6}{7}x > \frac{2}{21}$$
$$21\left(\frac{2}{3}x - \frac{6}{7}x \right) > 21\left(\frac{2}{21} \right)$$
$$14x - 18x > 2$$
$$-4x > 2$$
$$x < -\frac{1}{2}$$

The solution set is $\left(-\infty, -\dfrac{1}{2} \right)$.

90. Solving the inequality:
$$-3 \leq 9 - x < 3$$
$$-12 \leq -x \leq -6$$
$$12 \geq x \geq 6$$
The solution set is $[6, 12]$.

91. Solving the inequality:
$$\left|\frac{3x-2}{5}\right| + 2 < 6$$
$$\left|\frac{3x-2}{5}\right| < 4$$
$$-4 < \frac{3x-2}{5} < 4$$
$$-20 < 3x - 2 < 20$$
$$-18 < 3x < 22$$
$$-6 < x < \frac{22}{3}$$
The solution set is $\left(-6, \frac{22}{3}\right)$.

92. Solving the inequality:
$$|5x - 8| > 0$$
$$\begin{array}{ccc} 5x - 8 < 0 & \text{or} & 5x - 8 > 0 \\ 5x < 8 & & 5x > 8 \\ x < \dfrac{8}{5} & & x > \dfrac{8}{5} \end{array}$$
The solution set is $\left(-\infty, -\frac{8}{5}\right) \cup \left(\frac{8}{5}, \infty\right)$.

93. Factoring the inequality:
$$2x^2 + 7x \geq 0$$
$$x(2x + 7) \geq 0$$
Forming a sign chart:

The solution set is $\left(-\infty, -\frac{7}{2}\right] \cup [0, \infty)$.

94. Solving the inequality:
$$\frac{3x-1}{x+1} < 2$$
$$\frac{3x-1}{x+1} - 2 < 0$$
$$\frac{3x-1}{x+1} - 2 \cdot \frac{x+1}{x+1} < 0$$
$$\frac{3x-1-2x-2}{x+1} < 0$$
$$\frac{x-3}{x+1} < 0$$
Forming a sign chart:

The solution set is $(-1, 3)$.

95.   Multiplying the second equation by 2 yields the system:
$$3x - 2y = -4$$
$$x - 10y = -48$$
Multiplying the first equation by -5 yields the system:
$$-15x + 10y = 20$$
$$x - 10y = -48$$
Adding yields:
$$-14x = -28$$
$$x = 2$$
Substituting into the first equation:
$$3(2) - 2y = -4$$
$$6 - 2y = -4$$
$$-2y = -10$$
$$y = 5$$
The solution set is $\{(2,5)\}$.

96.   Setting the two equations equal:
$$4y - 7 = 7y + 8$$
$$-3y = 15$$
$$y = -5$$
Substituting into the first equation:
$$x = 4(-5) - 7$$
$$x = -20 - 7$$
$$x = -27$$
The solution set is $\{(-27,-5)\}$.

97.   Using the third equation as the first row, form the augmented matrix: $\begin{bmatrix} 1 & 2 & 2 & | & 5 \\ 2 & 6 & 5 & | & 13 \\ 4 & 5 & -1 & | & -7 \end{bmatrix}$

Adding -2 times row 1 to row 2, and -4 times row 1 to row 3: $\begin{bmatrix} 1 & 2 & 2 & | & 5 \\ 0 & 2 & 1 & | & 3 \\ 0 & -3 & -9 & | & -27 \end{bmatrix}$

Dividing row 3 by -3 and switching with row 2: $\begin{bmatrix} 1 & 2 & 2 & | & 5 \\ 0 & 1 & 3 & | & 9 \\ 0 & 2 & 1 & | & 3 \end{bmatrix}$

Adding -2 times row 2 to both row 1 and row 3: $\begin{bmatrix} 1 & 0 & -4 & | & -13 \\ 0 & 1 & 3 & | & 9 \\ 0 & 0 & -5 & | & -15 \end{bmatrix}$

Dividing row 3 by -5: $\begin{bmatrix} 1 & 0 & -4 & | & -13 \\ 0 & 1 & 3 & | & 9 \\ 0 & 0 & 1 & | & 3 \end{bmatrix}$

Adding 4 times row 3 to row 2, and -3 times row 3 to row 2: $\begin{bmatrix} 1 & 0 & 0 & | & -1 \\ 0 & 1 & 0 & | & 0 \\ 0 & 0 & 1 & | & 3 \end{bmatrix}$

The solution set is $\{(-1,0,3)\}$.

98.   Graphing the system:

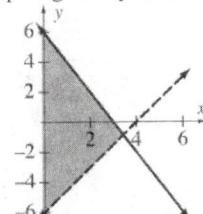

99.   Graphing the system:

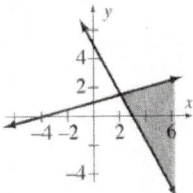

100.   Sketching the graph:

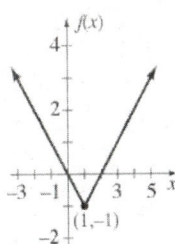

101.   Sketching the graph:

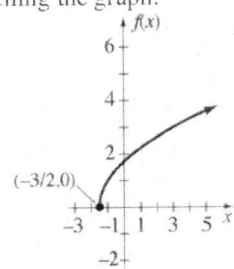

102.   Sketching the graph:

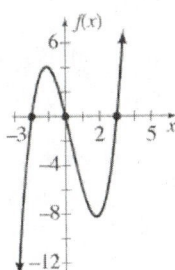

103.   Solving the proportion:

$$\frac{0.8 \text{ in.}}{30 \text{ mi}} = \frac{1.2 \text{ in.}}{x}$$

$$0.8x = 36$$

$$x = 45$$

The rest stations are 45 miles apart.

104.   Let x and $10 - x$ represent the two numbers. The equation is:

$$x(10 - x) = 8$$

$$10x - x^2 = 8$$

$$x^2 - 10x = -8$$

$$x^2 - 10x + 25 = -8 + 25$$

$$(x - 5)^2 = 17$$

$$x - 5 = \pm\sqrt{17}$$

$$x = 5 \pm \sqrt{17}$$

$$10 - x = 5 \pm \sqrt{17}$$

The two numbers are $5 - \sqrt{17}$ and $5 + \sqrt{17}$.

105. Let r represent the radius of the circle and s the side of the square. Then $r = 3s$. Using the perimeter and area formulas for a square and circle, respectively:

$$\pi r^2 = 4s$$
$$\pi (3s)^2 = 4s$$
$$9\pi s^2 = 4s$$
$$9\pi s^2 - 4s = 0$$
$$s(9\pi s - 4) = 0$$
$$s = \frac{4}{9\pi} \quad (\text{since } s > 0)$$
$$r = \frac{4}{3\pi}$$

The side of the square is $\dfrac{4}{9\pi}$ units and the radius of the circle is $\dfrac{4}{3\pi}$ units.

106. Let $r - 10$ represent her rate on the Tallahassee trip, and r represent her rate on the Orlando trip. Setting up the equation involving time:

$$\frac{195}{r-10} = \frac{100}{r} + 1\frac{2}{3}$$
$$3r(r-10)\left(\frac{195}{r-10}\right) = 3r(r-10)\left(\frac{100}{r}\right) + 3r(r-10)\left(\frac{5}{3}\right)$$
$$585r = 300(r-10) + 5r(r-10)$$
$$585r = 300r - 3000 + 5r^2 - 50r$$
$$0 = 5r^2 - 335r - 3000$$
$$0 = 5(r^2 - 67r - 600)$$
$$r = 5(r-75)(r-8)$$
$$r = 75 \quad (r = 8 \text{ is impossible})$$
$$r - 10 = 65$$

Shanna's speed on the Tallahassee trip was 65 mph.

107. Using $P = 5000$, $r = 0.055$, $n = 12$ and $A = 10{,}000$ in the compound interest formula

$$A = P\left(1 + \frac{r}{n}\right)^{nt}$$
$$10{,}000 = 5{,}000\left(1 + \frac{0.055}{12}\right)^{12t}$$
$$2 = \left(1 + \frac{0.055}{12}\right)^{12t}$$
$$\ln 2 = \ln\left(1 + \frac{0.055}{12}\right)^{12t}$$
$$\ln 2 = 12t \ln\left(1 + \frac{0.055}{12}\right)$$
$$t = \frac{\ln 2}{12\ln\left(1 + \dfrac{0.055}{12}\right)} \approx 12.63$$

It will take 12.63 years for the money to double.

108. Let x represent the cups of pure orange juice. Using the percent mixture equation:

$$0.20(6) + 1.00(x) = 0.40(x+6)$$
$$1.2 + x = 0.4x + 2.4$$
$$0.6x = 1.2$$
$$x = 2$$

So 2 cups of orange juice should be added.

109. We use $P = 1000$, $r = 0.04$, and $t = 5$ for this problem.

a. Using $n = 1$: $A = P\left(1+\dfrac{r}{n}\right)^{nt} = 1000\left(1+\dfrac{0.04}{1}\right)^{1(5)} \approx \$1,216.65$

b. Using $n = 2$: $A = P\left(1+\dfrac{r}{n}\right)^{nt} = 1000\left(1+\dfrac{0.04}{2}\right)^{2(5)} \approx \$1,218.99$

c. Using $n = 4$: $A = P\left(1+\dfrac{r}{n}\right)^{nt} = 1000\left(1+\dfrac{0.04}{4}\right)^{4(5)} \approx \$1,220.19$

d. Using $n = 12$: $A = P\left(1+\dfrac{r}{n}\right)^{nt} = 1000\left(1+\dfrac{0.04}{12}\right)^{12(5)} \approx \$1,221.00$

e. Using $A = Pe^{rt}$: $A = Pe^{rt} = 1000e^{0.04(5)} \approx \$1,221.40$

110. Let x, y, and z represent the measures of the smallest, middle, and largest angles. The system of equations is:
$$x + y + z = 180$$
$$x = y - 5$$
$$3y = 2z + 15$$

Substituting into the first equation results in the system:
$$y - 5 + y + z = 180$$
$$3y = 2z + 15$$

This simplifies to:
$$2y + z = 185$$
$$3y - 2z = 15$$

Multiplying the first equation by 2:
$$4y + 2z = 370$$
$$3y - 2z = 15$$

Adding yields:
$$7y = 385$$
$$y = 55$$
$$x = 55 - 5 = 50$$
$$z = 180 - 55 - 50 = 75$$

The angles have measures of 50°, 55°, and 75°.

111. Let x and y represent the quantity of 10% and 80% saline solutions, respectively. The system of equations is:
$$x + y = 7$$
$$0.10x + 0.80y = 0.60(7)$$

Multiplying the first equation by –0.10:
$$-0.10x - 0.10y = -0.7$$
$$0.10x + 0.80y = 4.2$$

Adding yields:
$$0.7y = 3.5$$
$$y = 5$$
$$x = 7 - 5 = 2$$

The pharmacist should use 5 cups of the 80% saline solution.

112. Let r, b, and m represent the number of road, beach, and mountain bicycles to produce. The system of equations is:
$$2r + b + 3m = 196$$
$$4r + 2b + 2m = 264$$
$$r + b + 2m = 124$$

Using the third equation as the first row, form the augmented matrix: $\begin{bmatrix} 1 & 1 & 2 & | & 124 \\ 2 & 1 & 3 & | & 196 \\ 4 & 2 & 2 & | & 264 \end{bmatrix}$

Adding -2 times row 1 to row 2, and -4 times row 1 to row 3: $\begin{bmatrix} 1 & 1 & 2 & | & 124 \\ 0 & -1 & -1 & | & -52 \\ 0 & -2 & -6 & | & -232 \end{bmatrix}$

Dividing row 2 by -1 and row 3 by -2: $\begin{bmatrix} 1 & 1 & 2 & | & 124 \\ 0 & 1 & 1 & | & 52 \\ 0 & 1 & 3 & | & 116 \end{bmatrix}$

Adding -1 times row 2 to both row 1 and row 3: $\begin{bmatrix} 1 & 0 & 1 & | & 72 \\ 0 & 1 & 1 & | & 52 \\ 0 & 0 & 2 & | & 64 \end{bmatrix}$

Dividing row 3 by 2: $\begin{bmatrix} 1 & 0 & 1 & | & 72 \\ 0 & 1 & 1 & | & 52 \\ 0 & 0 & 1 & | & 32 \end{bmatrix}$

Adding -1 times row 3 to both row 2 and row 1: $\begin{bmatrix} 1 & 0 & 0 & | & 40 \\ 0 & 1 & 0 & | & 20 \\ 0 & 0 & 1 & | & 32 \end{bmatrix}$

The company should build 40 road bicycles, 20 beach bicycles, and 32 mountain bicycles.

113. Let o, f, and t represent the number of one-, five-, and twenty-dollar bills, respectively. The system of equations is:
$$o + f + t = 18$$
$$o = t + 1$$
$$o + t = 5f$$

Substituting into the first equation:
$$5f + f = 18$$
$$6f = 18$$
$$f = 3$$

Substituting into the third equation:
$$o = t + 1$$
$$o + t = 5(3)$$

Rewriting the system:
$$o - t = 1$$
$$o + t = 15$$

Adding yields:
$$2o = 16$$
$$o = 8$$
$$t = 18 - 8 - 3 = 7$$

Tianna saved 8 one-dollar bills, 3 five-dollar bills, and 7 twenty-dollar bills.

114. Let s represent Sophie's rate and $\dfrac{1}{2}s$ represent Finn's rate, and let r represent the rate of the river. Adding and subtracting the river's rate for downstream and upstream, respectively, results in the system of equations:

$$3(s+r)=15$$
$$1\left(\dfrac{1}{2}s-r\right)=1$$

Dividing the first equation by 3 and multiplying the second equation by 2 simplifies the system:

$$s+r=5$$
$$s-2r=2$$

Multiplying the first equation by 2:

$$2s+2r=10$$
$$s-2r=2$$

Adding yields:

$$3s=12$$
$$s=4$$
$$\dfrac{1}{2}s=2$$
$$r=5-4=1$$

The rate of the current is 1 mph, Sophie's rate is 4 mph, and Finn's rate is 2 mph.

115. Let x represent the production of crickets and y represent the production of beetles. The constraint equations are:

$$3x+8y\le240$$
$$9x+4y\le360$$
$$x\ge0$$
$$y\ge0$$

Graphing the region:

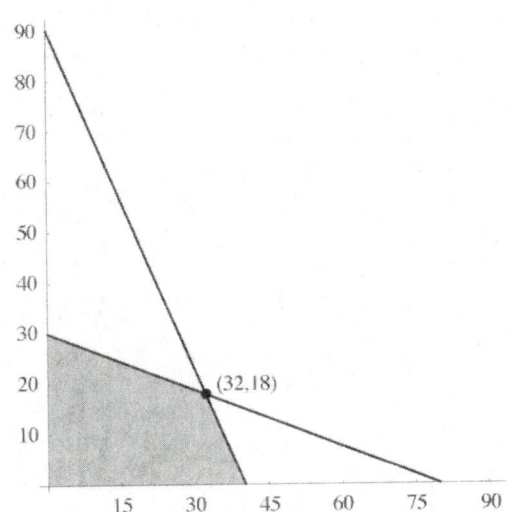

The vertex (32,18) was found by finding the intersection of the two lines:

$$3x+8y=240$$
$$9x+4y=360$$

Multiplying the first equation by -3:

$$-9x-24y=-720$$
$$9x+4y=360$$

Adding yields:

$$-20y=-360$$
$$y=18$$

Substituting into the first equation:

$$3x + 8(18) = 240$$
$$3x + 144 = 240$$
$$3x = 96$$
$$x = 32$$

Evaluating $f(x, y) = 2.90x + 3.50y$ at each vertex:

$$f(0, 30) = 2.9(0) + 3.5(30) = 105 \qquad\qquad f(40, 0) = 2.9(40) + 3.5(0) = 116$$
$$f(32, 18) = 2.9(32) + 3.5(18) = 155.80$$

They should produce 32 cricket action figures and 18 beetle action figures to maximize their profit.

Chapter 8
Conic Sections

8.1 Parabolas

1. The vertex is $(0,0)$, the focus is $(2,0)$, and the directrix is $x = -2$. Graphing the parabola:

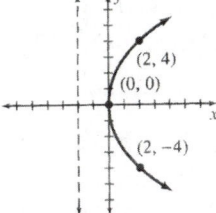

3. The vertex is $(0,0)$, the focus is $(0,-3)$, and the directrix is $y = 3$. Graphing the parabola:

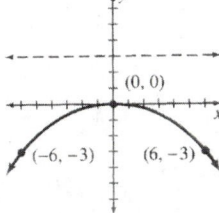

5. The vertex is $(0,0)$, the focus is $\left(-\dfrac{1}{2},0\right)$, and the directrix is $x = \dfrac{1}{2}$. Graphing the parabola:

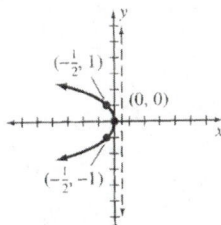

317

7. The vertex is (0,0), the focus is $\left(0, \frac{3}{2}\right)$, and the directrix is $y = -\frac{3}{2}$. Graphing the parabola:

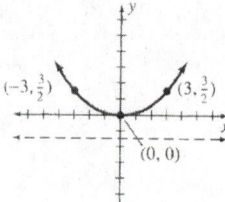

9. The vertex is (0,–1), the focus is (0,2), and the directrix is $y = -4$. Graphing the parabola:

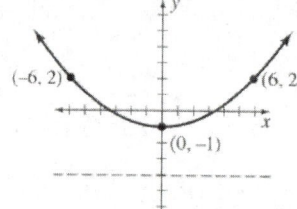

11. The vertex is (3,0), the focus is (1,0), and the directrix is $x = 5$. Graphing the parabola:

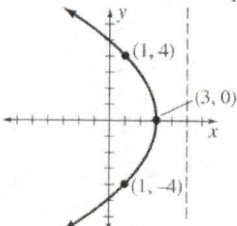

13. Write the equation in standard form:

$$x^2 - 4y + 8 = 0$$
$$x^2 = 4y - 8$$
$$x^2 = 4(y - 2)$$

The vertex is (0,2), the focus is (0,3), and the directrix is $y = 1$. Graphing the parabola:

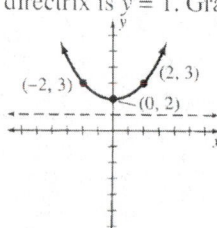

15. Write the equation in standard form:

$$x^2 + 8y + 16 = 0$$
$$x^2 = -8y - 16$$
$$x^2 = -8(y + 2)$$

The vertex is (0,–2), the focus is (0,–4), and the directrix is $y = 0$. Graphing the parabola:

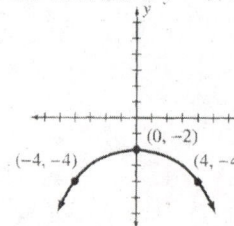

17. Write the equation in standard form:
$$y^2 - 12x + 24 = 0$$
$$y^2 = 12x - 24$$
$$y^2 = 12(x - 2)$$
The vertex is $(2,0)$, the focus is $(5,0)$, and the directrix is $x = -1$. Graphing the parabola:

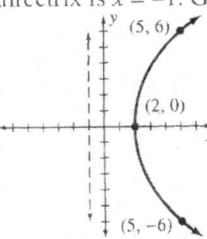

19. The vertex is $(2,-2)$, the focus is $(2,-3)$, and the directrix is $y = -1$. Graphing the parabola:

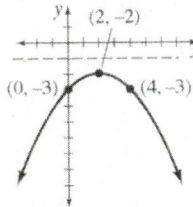

21. The vertex is $(-2,-4)$, the focus is $(-4,-4)$, and the directrix is $x = 0$. Graphing the parabola:

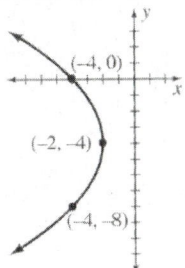

23. Write the equation in standard form:
$$x^2 - 2x - 4y + 9 = 0$$
$$x^2 - 2x + 1 = 4y - 9 + 1$$
$$(x - 1)^2 = 4(y - 2)$$
The vertex is $(1,2)$, the focus is $(1,3)$, and the directrix is $y = 1$. Graphing the parabola:

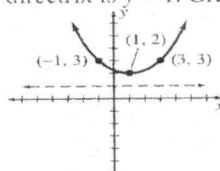

25. Write the equation in standard form:
$$x^2 + 6x + 8y + 1 = 0$$
$$x^2 + 6x + 9 = -8y - 1 + 9$$
$$(x + 3)^2 = -8(y - 1)$$
The vertex is $(-3,1)$, the focus is $(-3,-1)$, and the directrix is $y = 3$. Graphing the parabola:

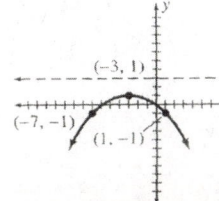

27.   Write the equation in standard form:

$$y^2 - 2y + 12x - 35 = 0$$
$$y^2 - 2y + 1 = -12x + 35 + 1$$
$$(y-1)^2 = -12(x-3)$$

The vertex is $(3,1)$, the focus is $(0,1)$, and the directrix is $x = 6$. Graphing the parabola:

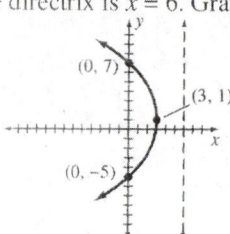

29.   Write the equation in standard form:

$$y^2 + 6y - 4x + 1 = 0$$
$$y^2 + 6y + 9 = 4x - 1 + 9$$
$$(y+3)^2 = 4(x+2)$$

The vertex is $(-2,-3)$, the focus is $(-1,-3)$, and the directrix is $x = -3$. Graphing the parabola:

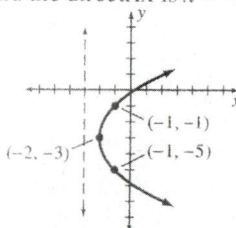

31.   The vertex is $(0,0)$ and $p = 3$, so the equation is:

$$x^2 = 4 \cdot 3y$$
$$x^2 = 12y$$

33.   The vertex is $(0,0)$ and $p = -1$, so the equation is:

$$y^2 = 4 \cdot (-1)y$$
$$y^2 = -4x$$

35.   The vertex is $(0,4)$ and $p = -3$, so the equation is:

$$x^2 = 4 \cdot (-3)(y-4)$$
$$x^2 = -12y + 48$$
$$x^2 + 12y - 48 = 0$$

37.   The vertex is $(3,1)$ and $p = 3$, so the equation is:

$$(x-3)^2 = 4 \cdot 3(y-1)$$
$$x^2 - 6x + 9 = 12y - 12$$
$$x^2 - 6x - 12y + 21 = 0$$

39.   The vertex is $(-2,5)$ and $p = -2$, so the equation is:

$$(y-5)^2 = 4 \cdot (-2)(x+2)$$
$$y^2 - 10y + 25 = -8x - 16$$
$$y^2 - 10y + 8x + 41 = 0$$

41.   The form of the equation is $y^2 = 4px$. Substituting the point:

$$y^2 = 4px$$
$$(5)^2 = 4p(-3)$$
$$-\frac{25}{3} = 4p$$

So the equation is $y^2 = -\dfrac{25}{3}x$, or $3y^2 = -25x$.

43.   The form of the equation is $y^2 = 4px$. Using $p = \dfrac{5}{2}$:

$$y^2 = 4\left(\frac{5}{2}\right)x$$
$$y^2 = 10x$$

45. Since $p = 2$, the equation is:
$$(x-7)^2 = 4 \cdot 2(y-3)$$
$$x^2 - 14x + 49 = 8y - 24$$
$$x^2 - 14x - 8y + 73 = 0$$

47. Since $p = 3$, the equation is:
$$(y+3)^2 = 4 \cdot 3(x-8)$$
$$y^2 + 6y + 9 = 12x - 96$$
$$y^2 + 6y - 12x + 105 = 0$$

49. The form of the equation is $(x+9)^2 = 4p(y-1)$. Substituting the point $(-8,0)$:
$$(-8+9)^2 = 4p(0-1)$$
$$1 = -4p$$
$$-1 = 4p$$

Thus the equation is:
$$(x+9)^2 = -1(y-1)$$
$$x^2 + 18x + 81 = -y + 1$$
$$x^2 + 18x + y + 80 = 0$$

51. The form of the equation is $x^2 = 4p(y-10)$. Substituting the point $(150,40)$:
$$150^2 = 4p(40-10)$$
$$4p = 750$$

The equation is $x^2 = 750(y-10)$.

53. The vertex is $(0,100)$, so the form of the equation is $x^2 = 4p(y-100)$. Substituting $(\pm 10, 0)$:
$$100 = 4p(-100)$$
$$-1 = 4p$$

The equation of the arch is $x^2 = -1(y-100)$. Substituting $y = 50$:
$$x^2 = -1(50-100)$$
$$x^2 = 50$$
$$x = 5\sqrt{2}$$

The width of the arch 50 feet above the ground is $10\sqrt{2}$ feet.

55. The vertex is $(0,h)$, so the form of the equation is $x^2 = 4p(y-h)$. Substituting $(\pm 100, 0)$:
$$(\pm 100)^2 = 4p(-h)$$
$$4p = -\frac{10000}{h}$$

So the equation is $x^2 = -\dfrac{10000}{h}(y-h)$. Substituting the point $(60,40)$:
$$60^2 = -\frac{10000}{h}(40-h)$$
$$3600h = -400000 + 10000h$$
$$-6400h = -400000$$
$$h = 62.5$$

The height of the arch is 62.5 feet.

8.2 Ellipses

1. The foci are $\left(\pm\sqrt{3},0\right)$. The vertices and endpoints are indicated on the graph:

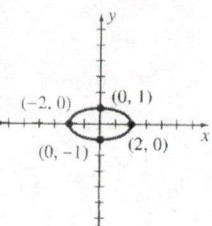

3. The foci are $\left(0,\pm\sqrt{5}\right)$. The vertices and endpoints are indicated on the graph:

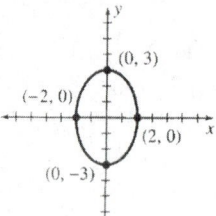

5. The standard form is $\dfrac{x^2}{3}+\dfrac{y^2}{9}=1$. The foci are $\left(0,\pm\sqrt{6}\right)$. The vertices and endpoints are indicated on the graph:

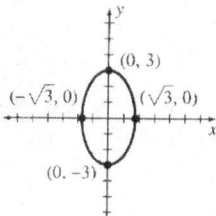

7. The standard form is $\dfrac{x^2}{25}+\dfrac{y^2}{10}=1$. The foci are $\left(\pm\sqrt{15},0\right)$. The vertices and endpoints are indicated on the graph:

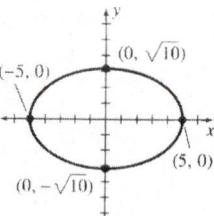

9. The standard form is $\dfrac{x^2}{3}+\dfrac{y^2}{36}=1$. The foci are $\left(0,\pm\sqrt{33}\right)$. The vertices and endpoints are indicated on the graph:

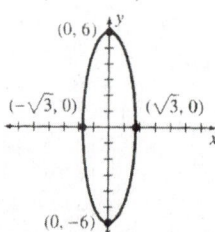

11.   The standard form is $\dfrac{x^2}{11} + \dfrac{y^2}{7} = 1$. The foci are $(\pm 2, 0)$. The vertices and endpoints are indicated on the graph:

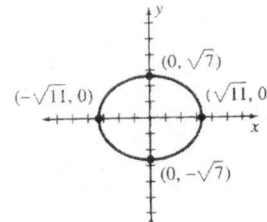

13.   The center is $(2,1)$. The foci are $\left(2 \pm \sqrt{5}, 1\right)$. The vertices and endpoints are indicated on the graph:

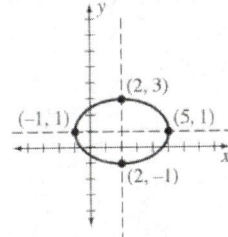

15.   The center is $(-1, -2)$. The foci are $\left(-1, -2 \pm \sqrt{7}\right)$. The vertices and endpoints are indicated on the graph:

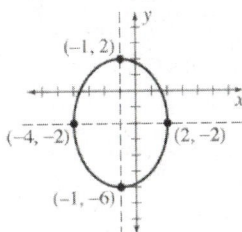

17.   First complete the square:
$$4x^2 - 8x + 9y^2 - 36y + 4 = 0$$
$$4\left(x^2 - 2x + 1\right) + 9\left(y^2 - 4y + 4\right) = -4 + 4 + 36$$
$$4(x-1)^2 + 9(y-2)^2 = 36$$
$$\frac{(x-1)^2}{9} + \frac{(y-2)^2}{4} = 1$$

The foci are $\left(1 \pm \sqrt{5}, 2\right)$. The vertices and endpoints are indicated on the graph:

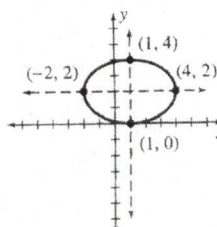

19.   First complete the square:
$$4x^2 + 16x + y^2 + 2y + 1 = 0$$
$$4\left(x^2 + 4x + 4\right) + \left(y^2 + 2y + 1\right) = -1 + 16 + 1$$
$$4(x+2)^2 + (y+1)^2 = 16$$
$$\frac{(x+2)^2}{4} + \frac{(y+1)^2}{16} = 1$$

The foci are $\left(-2,-1\pm 2\sqrt{3}\right)$. The vertices and endpoints are indicated on the graph:

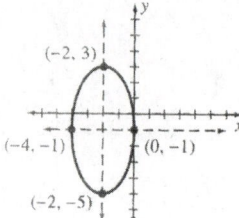

21.   First complete the square:

$$x^2 - 6x + 4y^2 + 5 = 0$$
$$\left(x^2 - 6x + 9\right) + 4y^2 = -5 + 9$$
$$\left(x - 3\right)^2 + 4y^2 = 4$$
$$\frac{\left(x-3\right)^2}{4} + \frac{y^2}{1} = 1$$

The foci are $\left(3 \pm \sqrt{3}, 0\right)$. The vertices and endpoints are indicated on the graph:

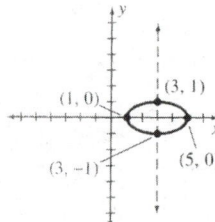

23.   First complete the square:

$$9x^2 - 72x + 2y^2 + 4y + 128 = 0$$
$$9\left(x^2 - 8x + 16\right) + 2\left(y^2 + 2y + 1\right) = -128 + 144 + 2$$
$$9\left(x - 4\right)^2 + 2\left(y + 1\right)^2 = 18$$
$$\frac{\left(x-4\right)^2}{2} + \frac{\left(y+1\right)^2}{9} = 1$$

The foci are $\left(4, -1 \pm \sqrt{7}\right)$. The vertices and endpoints are indicated on the graph:

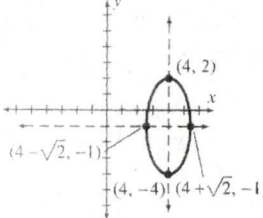

25.   First complete the square:

$$2x^2 + 12x + 11y^2 - 88y + 172 = 0$$
$$2\left(x^2 + 6x + 9\right) + 11\left(y^2 - 8y + 16\right) = -172 + 18 + 176$$
$$2\left(x + 3\right)^2 + 11\left(y - 4\right)^2 = 22$$
$$\frac{\left(x+3\right)^2}{11} + \frac{\left(y-4\right)^2}{2} = 1$$

The foci are $(-6,4)$ and $(0,4)$. The vertices and endpoints are indicated on the graph:

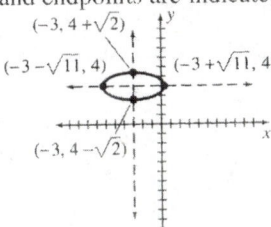

27. Since $a = 5$ and $c = 3$, $b = 4$ and the equation is:

$$\frac{x^2}{25} + \frac{y^2}{16} = 1$$
$$16x^2 + 25y^2 = 400$$

29. Since $a = 6$ and $c = 5$, $b = \sqrt{11}$ and the equation is:

$$\frac{x^2}{11} + \frac{y^2}{36} = 1$$
$$36x^2 + 11y^2 = 396$$

31. Since $a = 3$ and $b = 1$, the equation is:

$$\frac{x^2}{9} + \frac{y^2}{1} = 1$$
$$x^2 + 9y^2 = 9$$

33. Since $c = 2$ and $b = \frac{3}{2}$, $a = \frac{5}{2}$ and the equation is:

$$\frac{x^2}{9/4} + \frac{y^2}{25/4} = 1$$
$$\frac{4x^2}{9} + \frac{4y^2}{25} = 1$$
$$100x^2 + 36y^2 = 225$$

35. The form of the equation is $\frac{x^2}{b^2} + \frac{y^2}{25} = 1$. Substituting the point $(3,2)$:

$$\frac{9}{b^2} + \frac{4}{25} = 1$$
$$\frac{9}{b^2} = \frac{21}{25}$$
$$21b^2 = 225$$
$$b^2 = \frac{75}{7}$$

Thus the equation is:

$$\frac{x^2}{75/7} + \frac{y^2}{25} = 1$$
$$\frac{7x^2}{75} + \frac{y^2}{25} = 1$$
$$7x^2 + 3y^2 = 75$$

37. The center is $(1,1)$, so $a = 4$, $c = 2$, and $b = 2\sqrt{3}$. The equation is:

$$\frac{(x-1)^2}{16} + \frac{(y-1)^2}{12} = 1$$
$$3(x-1)^2 + 4(y-1)^2 = 48$$
$$3x^2 - 6x + 3 + 4y^2 - 8y + 4 = 48$$
$$3x^2 - 6x + 4y^2 - 8y - 41 = 0$$

39. Since $c = 4$ and $b = 3$, $a = 5$ and the equation is:

$$\frac{x^2}{25} + \frac{(y-1)^2}{9} = 1$$
$$9x^2 + 25(y-1)^2 = 225$$
$$9x^2 + 25y^2 - 50y + 25 = 225$$
$$9x^2 + 25y^2 - 50y - 200 = 0$$

41. Since $c = 2$ and $2a = 8$, $a = 4$ and $b = 2\sqrt{3}$. The equation is:

$$\frac{x^2}{16} + \frac{y^2}{12} = 1$$
$$3x^2 + 4y^2 = 48$$

43. The equation of the arch is $\frac{x^2}{15^2} + \frac{y^2}{10^2} = 1$. Substituting $x = 10$:

$$\frac{10^2}{15^2} + \frac{y^2}{10^2} = 1$$
$$\frac{4}{9} + \frac{y^2}{100} = 1$$
$$\frac{y^2}{100} = \frac{5}{9}$$
$$y^2 = \frac{500}{9}$$
$$y = \frac{10\sqrt{5}}{3}$$

The height of the arch 10 feet from the center of the base is $\frac{10\sqrt{5}}{3}$ feet.

8.3 Hyperbolas

1. The foci are $\left(\pm\sqrt{13}, 0\right)$ and the asymptotes are $y = \pm\frac{2}{3}x$. Vertices are labeled on the graph:

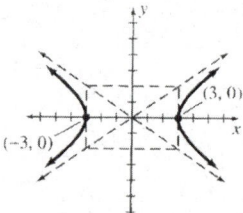

3. The foci are $\left(0, \pm\sqrt{13}\right)$ and the asymptotes are $y = \pm\frac{2}{3}x$. Vertices are labeled on the graph:

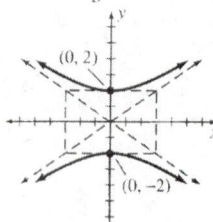

5. The standard form is $\frac{y^2}{16} - \frac{x^2}{9} = 1$. The foci are $\left(0, \pm 5\right)$ and the asymptotes are $y = \pm\frac{4}{3}x$. Vertices are labeled on the graph:

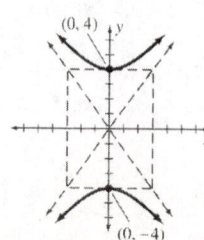

7. The standard form is $\dfrac{x^2}{9} - \dfrac{y^2}{9} = 1$. The foci are $\left(\pm 3\sqrt{2}, 0\right)$ and the asymptotes are $y = \pm x$.

Vertices are labeled on the graph:

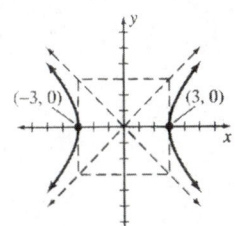

9. The standard form is $\dfrac{y^2}{5} - \dfrac{x^2}{25} = 1$. The foci are $\left(0, \pm\sqrt{30}\right)$ and the asymptotes are $y = \pm \dfrac{\sqrt{5}}{5}x$.

Vertices are labeled on the graph:

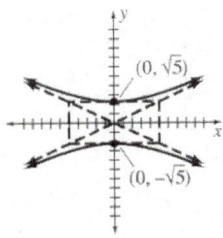

11. The standard form is $\dfrac{x^2}{1} - \dfrac{y^2}{9} = 1$. The foci are $\left(\pm\sqrt{10}, 0\right)$ and the asymptotes are $y = \pm 3x$.

Vertices are labeled on the graph:

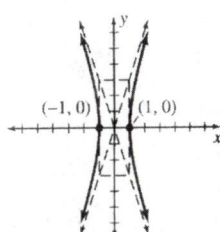

13. The center is $(1, -1)$. The foci are $\left(1 \pm \sqrt{13}, -1\right)$. Finding the asymptotes:

$$y + 1 = -\frac{2}{3}(x - 1) \qquad\qquad y + 1 = \frac{2}{3}(x - 1)$$

$$y + 1 = -\frac{2}{3}x + \frac{2}{3} \qquad\qquad y + 1 = \frac{2}{3}x - \frac{2}{3}$$

$$y = -\frac{2}{3}x - \frac{1}{3} \qquad\qquad y = \frac{2}{3}x - \frac{5}{3}$$

Vertices are labeled on the graph:

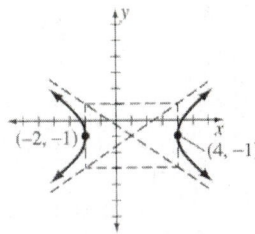

15.   The center is $(1,2)$. The foci are $(1,-3)$ and $(1,7)$. Finding the asymptotes:

$$y - 2 = -\frac{3}{4}(x-1) \qquad\qquad y - 2 = \frac{3}{4}(x-1)$$

$$y - 2 = -\frac{3}{4}x + \frac{3}{4} \qquad\qquad y - 2 = \frac{3}{4}x - \frac{3}{4}$$

$$y = -\frac{3}{4}x + \frac{11}{4} \qquad\qquad y = \frac{3}{4}x + \frac{5}{4}$$

Vertices are labeled on the graph:

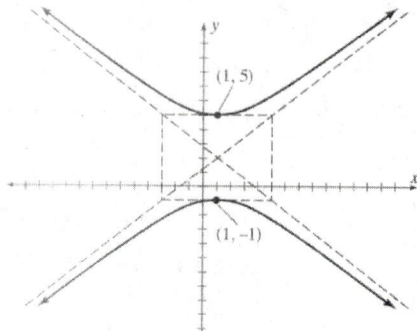

17.   First complete the square:

$$4x^2 - 24x - 9y^2 - 18y - 9 = 0$$

$$4(x^2 - 6x + 9) - 9(y^2 + 2y + 1) = 9 + 36 - 9$$

$$4(x-3)^2 - 9(y+1)^2 = 36$$

$$\frac{(x-3)^2}{9} - \frac{(y+1)^2}{4} = 1$$

The foci are $\left(3 \pm \sqrt{13}, -1\right)$. Finding the asymptotes:

$$y + 1 = -\frac{2}{3}(x-3) \qquad\qquad y + 1 = \frac{2}{3}(x-3)$$

$$3y + 3 = -2x + 6 \qquad\qquad 3y + 3 = 2x - 6$$

$$2x + 3y = 3 \qquad\qquad 2x - 3y = 9$$

Vertices are labeled on the graph:

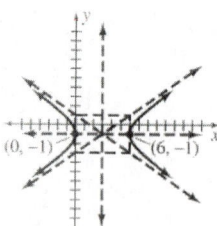

19.   First complete the square:

$$y^2 - 4y - 4x^2 - 24x - 36 = 0$$

$$(y^2 - 4y + 4) - 4(x^2 + 6x + 9) = 36 + 4 - 36$$

$$(y-2)^2 - 4(x+3)^2 = 4$$

$$\frac{(y-2)^2}{4} - \frac{(x+3)^2}{1} = 1$$

The foci are $\left(-3, 2 \pm \sqrt{5}\right)$. Finding the asymptotes:

$$y - 2 = -2(x + 3)$$
$$y - 2 = -2x - 6$$
$$2x + y = -4$$

$$y - 2 = 2(x + 3)$$
$$y - 2 = 2x + 6$$
$$2x - y = -8$$

Vertices are labeled on the graph:

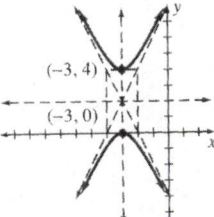

21. First complete the square:

$$2x^2 - 8x - y^2 + 4 = 0$$
$$2\left(x^2 - 4x + 4\right) - y^2 = -4 + 8$$
$$2(x - 2)^2 - y^2 = 4$$
$$\frac{(x - 2)^2}{2} - \frac{y^2}{4} = 1$$

The foci are $\left(2 \pm \sqrt{6}, 0\right)$. Finding the asymptotes:

$$y = -\sqrt{2}(x - 2)$$
$$y = -\sqrt{2}x + 2\sqrt{2}$$
$$\sqrt{2}x + y = 2\sqrt{2}$$

$$y = \sqrt{2}(x - 2)$$
$$y = \sqrt{2}x - 2\sqrt{2}$$
$$\sqrt{2}x - y = 2\sqrt{2}$$

Vertices are labeled on the graph:

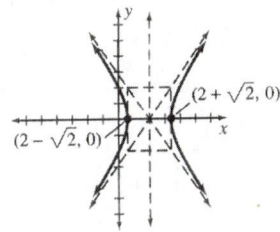

23. First complete the square:

$$y^2 + 10y - 9x^2 + 16 = 0$$
$$\left(y^2 + 10y + 25\right) - 9x^2 = -16 + 25$$
$$(y + 5)^2 - 9x^2 = 9$$
$$\frac{(y + 5)^2}{9} - \frac{x^2}{1} = 1$$

The foci are $\left(0, -5 \pm \sqrt{10}\right)$. Finding the asymptotes:

$$y + 5 = -3x$$
$$3x + y = -5$$

$$y + 5 = 3x$$
$$3x - y = 5$$

Vertices are labeled on the graph:

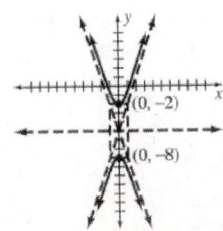

25. First complete the square:

$$x^2 + 4x - y^2 - 4y - 1 = 0$$
$$\left(x^2 + 4x + 4\right) - \left(y^2 + 4y + 4\right) = 1 + 4 - 4$$
$$(x+2)^2 - (y+2)^2 = 1$$

The foci are $\left(-2 \pm \sqrt{2}, -2\right)$. Finding the asymptotes:

$$y + 2 = -1(x+2) \qquad\qquad y + 2 = 1(x+2)$$
$$y + 2 = -x - 2 \qquad\qquad\qquad y + 2 = x + 2$$
$$x + y = -4 \qquad\qquad\qquad\quad x - y = 0$$

Vertices are labeled on the graph:

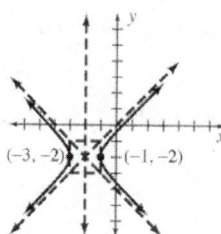

27. Since $a = 2$ and $c = 3$, $b = \sqrt{5}$ and the equation is:

$$\frac{x^2}{4} - \frac{y^2}{5} = 1$$
$$5x^2 - 4y^2 = 20$$

29. Since $a = 3$ and $c = 5$, $b = 4$ and the equation is:

$$\frac{y^2}{9} - \frac{x^2}{16} = 1$$
$$16y^2 - 9x^2 = 144$$

31. Since $a = 1$, the equation has the form $\dfrac{x^2}{1} - \dfrac{y^2}{b^2} = 1$. Substituting the point $(2,3)$:

$$\frac{4}{1} - \frac{9}{b^2} = 1$$
$$\frac{9}{b^2} = 3$$
$$b^2 = 3$$

Thus the equation is:

$$\frac{x^2}{1} - \frac{y^2}{3} = 1$$
$$3x^2 - y^2 = 3$$

33. Since $a = \sqrt{3}$ and $b = 2$, the equation is:

$$\frac{y^2}{3} - \frac{x^2}{4} = 1$$
$$4y^2 - 3x^2 = 12$$

35. Since $c = \sqrt{23}$ and $a = 4$, $b = \sqrt{7}$ and the equation is:

$$\frac{x^2}{16} - \frac{y^2}{7} = 1$$
$$7x^2 - 16y^2 = 112$$

37. The center is $(4, -3)$, so $a = 2$, $c = 3$, and $b = \sqrt{5}$. The equation is:

$$\frac{(x-4)^2}{4} - \frac{(y+3)^2}{5} = 1$$
$$5(x-4)^2 - 4(y+3)^2 = 20$$
$$5x^2 - 40x + 80 - 4y^2 - 24y - 36 = 20$$
$$5x^2 - 40x - 4y^2 - 24y + 24 = 0$$

39.   The center is $(-3,5)$, so $a = 2$, $c = 4$, and $b = 2\sqrt{3}$. The equation is:

$$\frac{(y-5)^2}{4} - \frac{(x+3)^2}{12} = 1$$
$$3(y-5)^2 - 1(x+3)^2 = 12$$
$$3y^2 - 30y + 75 - x^2 - 6x - 9 = 12$$
$$3y^2 - 30y - x^2 - 6x + 54 = 0$$

41.   The center is $(2,0)$, $a = 2$, $c = 3$, and $b = \sqrt{5}$. The equation is:

$$\frac{(x-2)^2}{4} - \frac{y^2}{5} = 1$$
$$5(x-2)^2 - 4y^2 = 20$$
$$5x^2 - 20x + 20 - 4y^2 = 20$$
$$5x^2 - 20x - 4y^2 = 0$$

43.   This is the equation of a circle.

45.   This is the equation of a straight line.

47.   This is the equation of an ellipse.

49.   This is the equation of a hyperbola.

51.   This is the equation of a parabola.

8.4 Systems Involving Nonlinear Equations

1.   Graphing the equations:

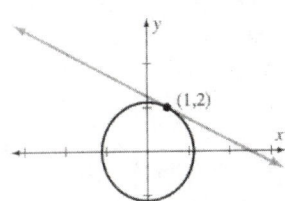

Solving the second equation for x yields $x = 5 - 2y$. Substituting into the first equation:

$$(5 - 2y)^2 + y^2 = 5$$
$$25 - 20y + 4y^2 + y^2 = 5$$
$$5y^2 - 20y + 20 = 0$$
$$y^2 - 4y + 4 = 0$$
$$(y-2)^2 = 0$$
$$y = 2$$
$$x = 5 - 2(2) = 1$$

The solution set is $\{(1,2)\}$.

3.   Graphing the equations:

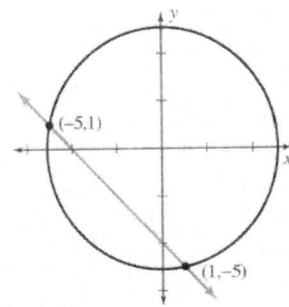

Solving the second equation for y yields $y = -x - 4$. Substituting into the first equation:

$$x^2 + (-x - 4)^2 = 26$$
$$x^2 + x^2 + 8x + 16 = 26$$
$$2x^2 + 8x - 10 = 0$$
$$x^2 + 4x - 5 = 0$$
$$(x + 5)(x - 1) = 0$$
$$x = -5, 1$$
$$y = 1, -5$$

The solution set is $\{(-5, 1), (1, -5)\}$.

5.   Graphing the equations:

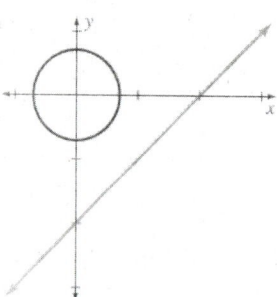

Solving the second equation for x yields $x = y + 4$. Substituting into the first equation:

$$(y + 4)^2 + y^2 = 2$$
$$y^2 + 8y + 16 + y^2 = 2$$
$$2y^2 + 8y + 14 = 0$$
$$y^2 + 4y + 7 = 0$$
$$y = \frac{-4 \pm \sqrt{16 - 28}}{2} = \frac{-4 \pm 2i\sqrt{3}}{2} = -2 \pm i\sqrt{3}$$
$$x = 2 \pm i\sqrt{3}$$

The solution set is $\left\{ \left(2 + i\sqrt{3}, -2 + i\sqrt{3}\right), \left(2 - i\sqrt{3}, -2 - i\sqrt{3}\right) \right\}$.

7.   Graphing the equations:

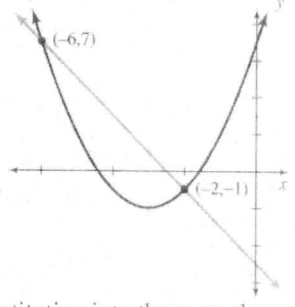

Substituting into the second equation:

$$2x + x^2 + 6x + 7 = -5$$
$$x^2 + 8x + 12 = 0$$
$$(x + 6)(x + 2) = 0$$
$$x = -6, -2$$
$$y = 7, -1$$

The solution set is $\{(-6, 7), (-2, -1)\}$.

9.   Graphing the equations:

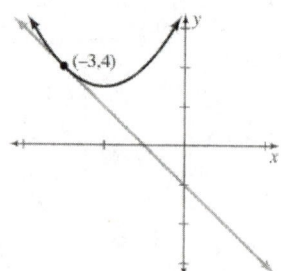

Substituting into the first equation:

$$2x + x^2 + 4x + 7 = -2$$
$$x^2 + 6x + 9 = 0$$
$$(x + 3)^2 = 0$$
$$x = -3$$
$$y = 4$$

The solution set is $\{(-3, 4)\}$.

11.   Graphing the equations:

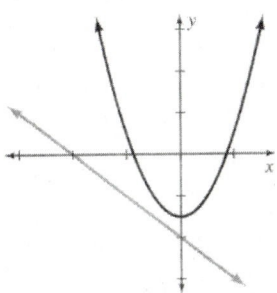

Substituting into the second equation:

$x + x^2 - 3 = -4$

$x^2 + x + 1 = 0$

$$x = \frac{-1 \pm \sqrt{1-4}}{2} = \frac{-1 \pm i\sqrt{3}}{2}$$

$$y = -4 - \frac{-1 \pm i\sqrt{3}}{2} = \frac{-7 \mp i\sqrt{3}}{2}$$

The solution set is $\left\{ \left(\dfrac{-1+i\sqrt{3}}{2}, \dfrac{-7-i\sqrt{3}}{2} \right), \left(\dfrac{-1-i\sqrt{3}}{2}, \dfrac{-7+i\sqrt{3}}{2} \right) \right\}$.

13.   Graphing the equations:

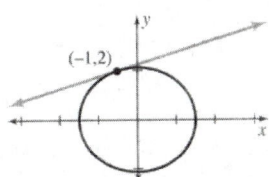

Solving the second equation for x yields $x = 4y - 9$. Substituting into the first equation:

$$(4y - 9)^2 + 2y^2 = 9$$
$$16y^2 - 72y + 81 + 2y^2 = 9$$
$$18y^2 - 72y + 72 = 0$$
$$y^2 - 4y + 4 = 0$$
$$(y - 2)^2 = 0$$
$$y = 2$$
$$x = -1$$

The solution set is $\left\{ (-1, 2) \right\}$.

15.   Graphing the equations:

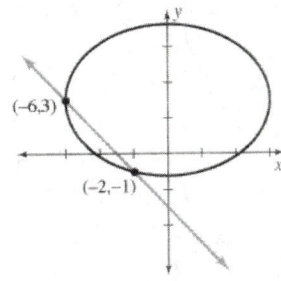

Solving the first equation for x yields $x = -y - 3$. Substituting into the second equation:

$$(-y-3)^2 + 2y^2 - 12y - 18 = 0$$
$$y^2 + 6y + 9 + 2y^2 - 12y - 18 = 0$$
$$3y^2 - 6y - 9 = 0$$
$$y^2 - 2y - 3 = 0$$
$$(y-3)(y+1) = 0$$
$$y = -1, 3$$
$$x = -2, -6$$

The solution set is $\{(-2,-1),(-6,3)\}$.

17. Graphing the equations:

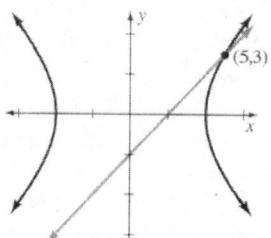

Solving the first equation for x yields $x = y + 2$. Substituting into the second equation:

$$(y+2)^2 - y^2 = 16$$
$$y^2 + 4y + 4 - y^2 = 16$$
$$4y = 12$$
$$y = 3$$
$$x = 5$$

The solution set is $\{(5,3)\}$.

19. Graphing the equations:

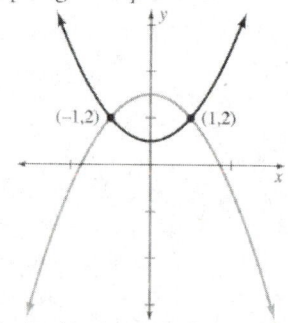

Substituting into the first equation:
$$x^2 + 1 = -x^2 + 3$$
$$2x^2 = 2$$
$$x^2 = 1$$
$$x = \pm 1$$
$$y = 2$$

The solution set is $\{(-1,2),(1,2)\}$.

21. Graphing the equations:

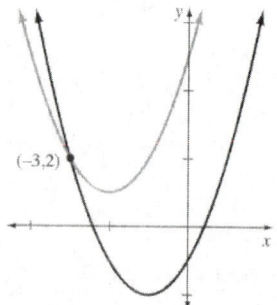

Substituting into the first equation:

$$x^2 + 4x + 5 = x^2 + 2x - 1$$
$$2x = -6$$
$$x = -3$$
$$y = 2$$

The solution set is $\{(-3,2)\}$.

23. Graphing the equations:

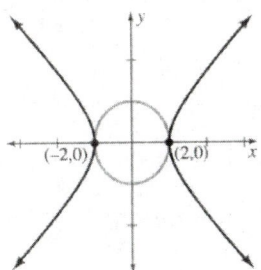

Adding the two equations:

$$2x^2 = 8$$
$$x^2 = 4$$
$$x = \pm 2$$
$$y = 0$$

The solution set is $\{(-2,0),(2,0)\}$.

25. Graphing the equations:

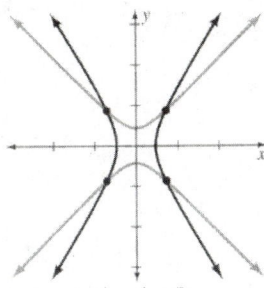

Multiply the first equation by 3 and the second equation by 8:

$$-27x^2 + 24y^2 = 18$$
$$64x^2 - 24y^2 = 56$$

Adding yields:

$$37x^2 = 74$$
$$x^2 = 2$$
$$x = \pm\sqrt{2}$$

Substituting into the first equation:

$$8y^2 - 18 = 6$$
$$8y^2 = 24$$
$$y^2 = 3$$
$$y = \pm\sqrt{3}$$

The solution set is $\left\{\left(-\sqrt{2},-\sqrt{3}\right),\left(-\sqrt{2},\sqrt{3}\right),\left(\sqrt{2},-\sqrt{3}\right),\left(\sqrt{2},\sqrt{3}\right)\right\}$.

27. Graphing the equations:

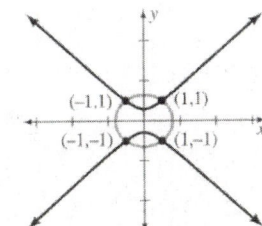

Adding the two equations:

$$4x^2 = 4$$
$$x^2 = 1$$
$$x = \pm 1$$

Substituting into the second equation:

$$2 + 3y^2 = 5$$
$$3y^2 = 3$$
$$y^2 = 1$$
$$y = \pm 1$$

The solution set is $\{(-1,-1),(-1,1),(1,-1),(1,1)\}$.

29.   Graphing the equations:

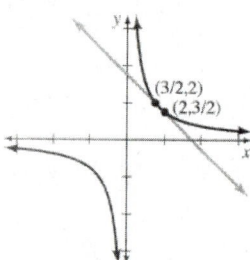

Solving the first equation for y yields $y = \dfrac{3}{x}$. Substituting into the second equation:

$$2x + 2\left(\frac{3}{x}\right) = 7$$
$$2x^2 + 6 = 7x$$
$$2x^2 - 7x + 6 = 0$$
$$(2x - 3)(x - 2) = 0$$
$$x = \frac{3}{2}, 2$$
$$y = 2, \frac{3}{2}$$

The solution set is $\left\{\left(\dfrac{3}{2}, 2\right), \left(2, \dfrac{3}{2}\right)\right\}$.

31.   Substituting into the first equation:

$$-\log_3 x = \log_3 (x - 6) - 3$$
$$\log_3 (x - 6) + \log_3 x = 3$$
$$\log_3 (x^2 - 6x) = 3$$
$$x^2 - 6x = 27$$
$$x^2 - 6x - 27 = 0$$
$$(x - 9)(x + 3) = 0$$
$$x = 9 \quad (x = -3 \text{ does not check})$$
$$y = -2$$

The solution set is $\{(9,-2)\}$.

33. Substituting into the first equation:

$$2e^{-x} = e^x - 1$$
$$2e^{-x} \cdot e^x = \left(e^x - 1\right) \cdot e^x$$
$$2 = \left(e^x\right)^2 - e^x$$
$$\left(e^x\right)^2 - e^x - 2 = 0$$
$$\left(e^x - 2\right)\left(e^x + 1\right) = 0$$
$$e^x = 2 \qquad \left(e^x = -1 \text{ is impossible}\right)$$
$$x = \ln 2$$
$$y = 1$$

The solution set is $\left\{ \left(\ln 2, 1\right) \right\}$.

35. Substituting into the first equation:

$$x^3 + 2x^2 + 5x - 3 = x^3$$
$$2x^2 + 5x - 3 = 0$$
$$\left(2x - 1\right)\left(x + 3\right) = 0$$
$$x = -3, \frac{1}{2}$$
$$y = -27, \frac{1}{8}$$

The solution set is $\left\{ \left(-3, -27\right), \left(\frac{1}{2}, \frac{1}{8}\right) \right\}$.

Chapter 8 Review Problem Set

1. The standard form is $\dfrac{x^2}{32} + \dfrac{y^2}{16} = 1$. This is an ellipse with foci $(\pm 4, 0)$. The vertices and endpoints are on the graph:

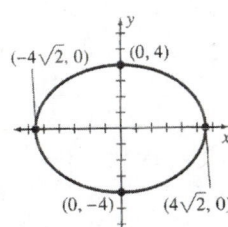

2. This is a parabola with vertex $(0,0)$, focus $(-3,0)$ and directrix $x = 3$:

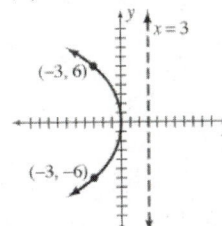

3.   The standard form is $\dfrac{y^2}{3} - \dfrac{x^2}{9} = 1$. This is a hyperbola with foci $\left(0,\pm 2\sqrt{3}\right)$ and asymptotes $y = \pm\dfrac{\sqrt{3}}{3}x$.
The endpoints are on the graph:

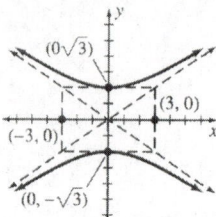

4.   The standard form is $\dfrac{x^2}{9} - \dfrac{y^2}{6} = 1$. This is a hyperbola with foci $\left(\pm\sqrt{15},0\right)$ and asymptotes $y = \pm\dfrac{\sqrt{6}}{3}x$.
The endpoints are on the graph:

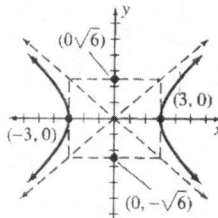

5.   The standard form is $\dfrac{x^2}{4} + \dfrac{y^2}{10} = 1$. This is an ellipse with foci $\left(0,\pm\sqrt{6}\right)$. The vertices and endpoints are on the graph:

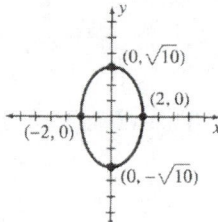

6.   This is a parabola with vertex $(0,0)$, focus $\left(0,\dfrac{1}{2}\right)$, and directrix $y = -\dfrac{1}{2}$:

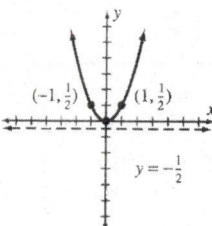

7.   First complete the square:
$$x^2 - 8x - 2y^2 + 4y + 10 = 0$$
$$\left(x^2 - 8x + 16\right) - 2\left(y^2 - 2y + 1\right) = -10 + 16 - 2$$
$$(x-4)^2 - 2(y-1)^2 = 4$$
$$\dfrac{(x-4)^2}{4} - \dfrac{(y-1)^2}{2} = 1$$

This is a hyperbola with center $(4,1)$ and foci $\left(4 \pm \sqrt{6},1\right)$. Finding the asymptotes:

$$y - 1 = -\dfrac{\sqrt{2}}{2}(x-4)$$
$$2y - 2 = -\sqrt{2}x + 4\sqrt{2}$$
$$\sqrt{2}x + 2y = 4\sqrt{2} + 2$$

$$y - 1 = \dfrac{\sqrt{2}}{2}(x-4)$$
$$2y - 2 = \sqrt{2}x - 4\sqrt{2}$$
$$\sqrt{2}x - 2y = 4\sqrt{2} - 2$$

The vertices and endpoints are on the graph:

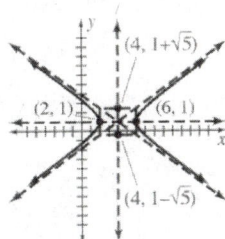

8.   First complete the square:

$$9x^2 - 54x + 2y^2 + 8y + 71 = 0$$
$$9\left(x^2 - 6x + 9\right) + 2\left(y^2 + 4y + 4\right) = -71 + 81 + 8$$
$$9(x-3)^2 + 2(y+2)^2 = 18$$
$$\frac{(x-3)^2}{2} + \frac{(y+2)^2}{9} = 1$$

This is an ellipse with center $(3,-2)$ and foci $\left(3, -2 \pm \sqrt{7}\right)$. The vertices and endpoints are on the graph:

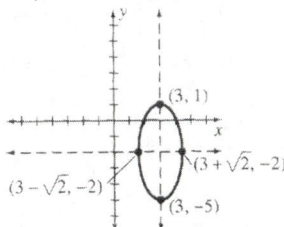

9.   First complete the square:

$$y^2 - 2y + 4x + 9 = 0$$
$$y^2 - 2y + 1 = -4x - 9 + 1$$
$$(y-1)^2 = -4(x+2)$$

This is a parabola with vertex $(-2,1)$, focus $(-3,1)$, and directrix $x = -1$. Sketching the graph:

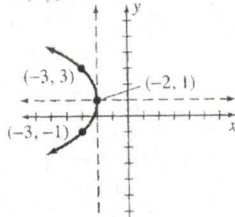

10.   First complete the square:

$$x^2 + 2x + 8y + 25 = 0$$
$$x^2 + 2x + 1 = -8y - 25 + 1$$
$$(x+1)^2 = -8(y+3)$$

This is a parabola with vertex $(-1,-3)$, focus $(-1,-5)$, and directrix $y = -1$. Sketching the graph:

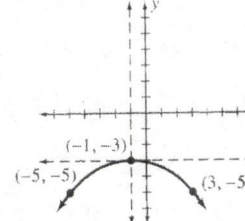

11. First complete the square:

$$x^2 + 10x + 4y^2 - 16y + 25 = 0$$

$$\left(x^2 + 10x + 25\right) + 4\left(y^2 - 4y + 4\right) = -25 + 25 + 16$$

$$(x+5)^2 + 4(y-2)^2 = 16$$

$$\frac{(x+5)^2}{16} + \frac{(y-2)^2}{4} = 1$$

This is an ellipse with center $(-5,2)$ and foci $\left(-5 \pm 2\sqrt{3}, 2\right)$. The vertices and endpoints are on the graph:

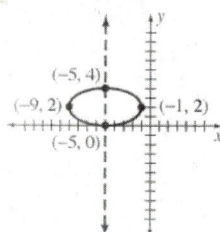

12. First complete the square:

$$3y^2 + 12y - 2x^2 - 8x - 8 = 0$$

$$3\left(y^2 + 4y + 4\right) - 2\left(x^2 + 4x + 4\right) = 8 + 12 - 8$$

$$3(y+2)^2 - 2(x+2)^2 = 12$$

$$\frac{(y+2)^2}{4} - \frac{(x+2)^2}{6} = 1$$

The center is $(-2,-2)$ and the foci are $\left(-2, -2 \pm \sqrt{10}\right)$. Finding the asymptotes:

$$y + 2 = -\frac{\sqrt{6}}{3}(x+2) \qquad\qquad y + 2 = \frac{\sqrt{6}}{3}(x+2)$$

$$3y + 6 = -\sqrt{6}x - 2\sqrt{6} \qquad\qquad 3y + 6 = \sqrt{6}x + 2\sqrt{6}$$

$$\sqrt{6}x + 3y = -2\sqrt{6} - 6 \qquad\qquad \sqrt{6}x - 3y = 6 - 2\sqrt{6}$$

The vertices and endpoints are on the graph:

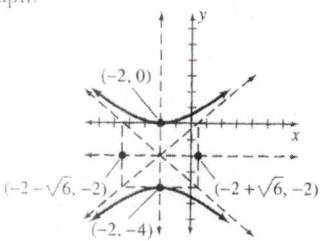

13. Here $p = -5$, so the equation is:

$$y^2 = 4(-5)x$$

$$y^2 = -20x$$

14. Here $a = 4$ and $c = \sqrt{15}$, so $b = 1$ and the equation is:

$$\frac{x^2}{1} + \frac{y^2}{16} = 1$$

$$16x^2 + y^2 = 16$$

15. Here $a = \sqrt{2}$ and $b = 5$, so the equation is:

$$\frac{x^2}{2} - \frac{y^2}{25} = 1$$

$$25x^2 - 2y^2 = 50$$

16. Here $a = 2$, so the form of the equation is $\dfrac{x^2}{4} + \dfrac{y^2}{b^2} = 1$. Substituting the point $(1,-2)$:

$$\frac{1^2}{4} + \frac{(-2)^2}{b^2} = 1$$
$$\frac{1}{4} + \frac{4}{b^2} = 1$$
$$\frac{4}{b^2} = \frac{3}{4}$$
$$b^2 = \frac{16}{3}$$

So the equation is:

$$\frac{x^2}{4} + \frac{y^2}{16/3} = 1$$
$$\frac{x^2}{4} + \frac{3y^2}{16} = 1$$
$$4x^2 + 3y^2 = 16$$

17. The form of the equation is $x^2 = 4py$. Substituting the point $(2,6)$:

$$x^2 = 4py$$
$$2^2 = 4p(6)$$
$$4p = \frac{2}{3}$$

So the equation is:

$$x^2 = \frac{2}{3}y$$
$$3x^2 = 2y$$

18. Here $a = 1$ and $c = \sqrt{10}$, so $b = 3$ and the equation is:

$$\frac{y^2}{1} - \frac{x^2}{9} = 1$$
$$9y^2 - x^2 = 9$$

19. The center is $(6,4)$, so $a = 3$ and $b = 1$. The equation is:

$$\frac{(x-6)^2}{1} + \frac{(y-4)^2}{9} = 1$$
$$9(x-6)^2 + (y-4)^2 = 9$$
$$9x^2 - 108x + 324 + y^2 - 8y + 16 = 9$$
$$9x^2 - 108x + y^2 - 8y + 331 = 0$$

20. Here $p = 2$, so the equation is:

$$(y+2)^2 = 4 \cdot 2(x-4)$$
$$y^2 + 4y + 4 = 8x - 32$$
$$y^2 + 4y - 8x + 36 = 0$$

21. The center is $(-5,-4)$, so $a = 1$, $c = 2$, and $b = \sqrt{3}$. The equation is:

$$\frac{(y+4)^2}{1} - \frac{(x+5)^2}{3} = 1$$
$$3(y+4)^2 - (x+5)^2 = 3$$
$$3y^2 + 24y + 48 - x^2 - 10x - 25 = 3$$
$$3y^2 + 24y - x^2 - 10x + 20 = 0$$

© 2013 Cengage Learning. All Rights Reserved. May not be scanned, copied or duplicated, or posted to a publicly accessible website, in whole or in part.

22.   The form of the equation is $(x+6)^2 = 4p(y+3)$. Substituting the point $(-5,-2)$:

$$(-5+6)^2 = 4p(-2+3)$$
$$4p = 1$$

So the equation is:

$$x^2 + 12x + 36 = y + 3$$
$$x^2 + 12x - y + 33 = 0$$

23.   The center is $(-5,0)$, so $b=2$ and $a=5$. The equation is:

$$\frac{(x+5)^2}{25} + \frac{y^2}{4} = 1$$
$$4(x+5)^2 + 25y^2 = 100$$
$$4x^2 + 40x + 100 + 25y^2 = 100$$
$$4x^2 + 40x + 25y^2 = 0$$

24.   The center is $(4,0)$, so $a=2$ and $b=4$. The equation is:

$$\frac{(x-4)^2}{4} - \frac{y^2}{16} = 1$$
$$4(x-4)^2 - y^2 = 16$$
$$4x^2 - 32x + 64 - y^2 = 16$$
$$4x^2 - 32x - y^2 + 48 = 0$$

25.   Solving the second equation for x yields $x = 4y - 17$. Substituting into the first equation:

$$(4y-17)^2 + y^2 = 17$$
$$16y^2 - 136y + 289 + y^2 = 17$$
$$17y^2 - 136y + 272 = 0$$
$$y^2 - 8y + 16 = 0$$
$$(y-4)^2 = 0$$
$$y = 4$$
$$x = -1$$

The solution set is $\{(-1,4)\}$.

26.   Solving the second equation for y yields $y = 3x - 8$. Substituting into the first equation:

$$x^2 - (3x-8)^2 = 8$$
$$x^2 - 9x^2 + 48x - 64 = 8$$
$$-8x^2 + 48x - 72 = 0$$
$$x^2 - 6x + 9 = 0$$
$$(x-3)^2 = 0$$
$$x = 3$$
$$y = 1$$

The solution set is $\{(3,1)\}$.

27.   Solving the first equation for y yields $y = x - 1$. Substituting into the second equation:

$$x - 1 = x^2 + 4x + 1$$
$$0 = x^2 + 3x + 2$$
$$0 = (x+2)(x+1)$$
$$x = -2, -1$$
$$y = -3, -2$$

The solution set is $\{(-1,-2),(-2,-3)\}$.

28. Multiply the first equation by 9:
$$36x^2 - 9y^2 = 144$$
$$9x^2 + 9y^2 = 16$$
Adding yields:
$$36x^2 - 9y^2 = 144$$
$$9x^2 + 9y^2 = 16$$
$$45x^2 = 160$$
$$x^2 = \frac{32}{9}$$
$$x = \pm\frac{4\sqrt{2}}{3}$$
Substituting to find y:
$$4\left(\frac{32}{9}\right) - y^2 = 16$$
$$y^2 = -\frac{16}{9}$$
$$y = \pm\frac{4}{3}i$$
The solution set is $\left\{\left(-\frac{4\sqrt{2}}{3}, -\frac{4}{3}i\right), \left(-\frac{4\sqrt{2}}{3}, \frac{4}{3}i\right), \left(\frac{4\sqrt{2}}{3}, -\frac{4}{3}i\right), \left(\frac{4\sqrt{2}}{3}, \frac{4}{3}i\right)\right\}$.

29. Multiply the first equation by -2:
$$-2x^2 - 4y^2 = -16$$
$$2x^2 + 3y^2 = 12$$
Adding yields:
$$-y^2 = -4$$
$$y^2 = 4$$
$$y = \pm 2$$
Substituting into the first equation:
$$x^2 + 2(4) = 8$$
$$x^2 = 0$$
$$x = 0$$
The solution set is $\left\{(0, -2), (0, 2)\right\}$.

30. Multiply the first equation by -1:
$$x^2 - y^2 = -1$$
$$4x^2 + y^2 = 4$$
Adding yields:
$$5x^2 = 3$$
$$x^2 = \frac{3}{5}$$
$$x = \pm\frac{\sqrt{15}}{5}$$
Substituting into the first equation:
$$y^2 - \frac{3}{5} = 1$$
$$y^2 = \frac{8}{5}$$
$$y = \pm\frac{2\sqrt{10}}{5}$$
The solution set is $\left\{\left(-\frac{\sqrt{15}}{5}, -\frac{2\sqrt{10}}{5}\right), \left(-\frac{\sqrt{15}}{5}, \frac{2\sqrt{10}}{5}\right), \left(\frac{\sqrt{15}}{5}, -\frac{2\sqrt{10}}{5}\right), \left(\frac{\sqrt{15}}{5}, \frac{2\sqrt{10}}{5}\right)\right\}$.

Chapter 8 Test

1. The vertex is $(0,0)$ and $p = -5$, so the focus is $(0,-5)$.

2. Completing the square:
$$y^2 - 4y - 8x - 20 = 0$$
$$y^2 - 4y + 4 = 8x + 20 + 4$$
$$(y-2)^2 = 8(x+3)$$
The vertex is $(-3,2)$.

3. The standard form of the equation is $y^2 = 12x$, so $p = 3$ and thus the directrix is $x = -3$.

4. Here $p = 6$, so the focus is $(6,0)$.

5. Completing the square:
$$x^2 + 4x - 12y - 8 = 0$$
$$x^2 + 4x + 4 = 12y + 12$$
$$(x+2)^2 = 12(y+1)$$
The vertex is $(-2,-1)$.

6. Here $p = -4$, so the directrix is $y = 4$.

7. The form of the equation is $y^2 = 4px$. Substituting the point $(-2,4)$:
$$(4)^2 = 4p(-2)$$
$$4p = -8$$
So the equation is $y^2 = -8x$, or $y^2 + 8x = 0$.

8. Here $p = -3$, so the equation is:
$$(x-3)^2 = 4(-3)(y-4)$$
$$x^2 - 6x + 9 = -12y + 48$$
$$x^2 - 6x + 12y - 39 = 0$$

9. The standard form is $\dfrac{x^2}{9} + \dfrac{y^2}{36} = 1$, so $a = 6$. The endpoints are $(0,6)$ and $(0,-6)$.

10. Completing the square:
$$x^2 - 4x + 9y^2 - 18y + 4 = 0$$
$$(x^2 - 4x + 4) + 9(y^2 - 2y + 1) = -4 + 4 + 9$$
$$(x-2)^2 + 9(y-1)^2 = 9$$
$$\frac{(x-2)^2}{9} + \frac{(y-1)^2}{1} = 1$$
Since $a = 3$, the length of the major axis is 6 units.

11. Completing the square:
$$9x^2 + 90x + 4y^2 - 8y + 193 = 0$$
$$9(x^2 + 10x + 25) + 4(y^2 - 2y + 1) = -193 + 225 + 4$$
$$9(x+5)^2 + 4(y-1)^2 = 36$$
$$\frac{(x+5)^2}{4} + \frac{(y-1)^2}{9} = 1$$
The center is $(-5,1)$ and $b = 2$, so the endpoints of the minor axis are $(-7,1)$ and $(-3,1)$.

12. The standard form is $\dfrac{x^2}{16} + \dfrac{y^2}{4} = 1$, so $a = 4$, $b = 2$, and $c = 2\sqrt{3}$. The foci are $\left(\pm 2\sqrt{3}, 0\right)$.

13. Completing the square:
$$3x^2 + 30x + y^2 - 16y + 79 = 0$$
$$3\left(x^2 + 10x + 25\right) + \left(y^2 - 16y + 64\right) = -79 + 75 + 64$$
$$3\left(x + 5\right)^2 + \left(y - 8\right)^2 = 60$$
$$\frac{\left(x + 5\right)^2}{60/3} + \frac{\left(y - 8\right)^2}{60} = 1$$
The center of the ellipse is $(-5,8)$.

14. Here $a = 10$ and $c = 8$, so $b = 6$ and the equation is:
$$\frac{x^2}{36} + \frac{y^2}{100} = 1$$
$$25x^2 + 9y^2 = 900$$

15. The center is $(6,-2)$, so $a = 4$ and $b = 2$. The equation is:
$$\frac{\left(x - 6\right)^2}{16} + \frac{\left(y + 2\right)^2}{4} = 1$$
$$\left(x - 6\right)^2 + 4\left(y + 2\right)^2 = 16$$
$$x^2 - 12x + 36 + 4y^2 + 16y + 16 = 16$$
$$x^2 - 12x + 4y^2 + 16y + 36 = 0$$

16. The standard form is $\dfrac{y^2}{8} - \dfrac{x^2}{32/9} = 1$. Thus the asymptotes are $y = \pm \dfrac{2\sqrt{2}}{\dfrac{4\sqrt{2}}{3}} x = \pm \dfrac{3}{2} x$.

17. Completing the square:
$$y^2 - 6y - 3x^2 - 6x - 3 = 0$$
$$\left(y^2 - 6y + 9\right) - 3\left(x^2 + 2x + 1\right) = 3 + 9 - 3$$
$$\left(y - 3\right)^2 - 3\left(x + 1\right)^2 = 9$$
$$\frac{\left(y - 3\right)^2}{9} - \frac{\left(x + 1\right)^2}{3} = 1$$
The center is $(-1,3)$ and $a = 3$, so the vertices are $(-1,0)$ and $(-1,6)$.

18. The standard form is $\dfrac{x^2}{4} - \dfrac{y^2}{5} = 1$, so $a = 2$, $b = \sqrt{5}$, and $c = 3$. Thus the foci are $(\pm 3,0)$.

19. Here $a = 6$ and $c = 4\sqrt{3}$, so $b = 2\sqrt{3}$. Thus the equation is:
$$\frac{x^2}{36} - \frac{y^2}{12} = 1$$
$$x^2 - 3y^2 = 36$$

20. The center is $(-1,4)$, so $a = 1$, $c = 3$, and $b = 2\sqrt{2}$. Thus the equation is:
$$\frac{\left(x + 1\right)^2}{1} - \frac{\left(y - 4\right)^2}{8} = 1$$
$$8\left(x + 1\right)^2 - \left(y - 4\right)^2 = 8$$
$$8x^2 + 16x + 8 - y^2 + 8y - 16 = 8$$
$$8x^2 + 16x - y^2 + 8y - 16 = 0$$

21. Subtracting the two equations yields:
$$y^2 + 4y = 8$$
$$y^2 + 4y - 8 = 0$$
$$y = \frac{-4 \pm \sqrt{16 + 32}}{2} = \frac{-4 \pm 4\sqrt{3}}{2} = -2 \pm 2\sqrt{3}$$
Thus the equation has two solutions.

22.   Solving the second equation for y yields $y = \dfrac{6}{x}$. Substituting into the first equation:

$$x^2 + 4\left(\frac{6}{x}\right)^2 = 25$$
$$x^4 + 144 = 25x^2$$
$$x^4 - 25x^2 + 144 = 0$$
$$\left(x^2 - 9\right)\left(x^2 - 16\right) = 0$$
$$x^2 = 9, 16$$
$$x = \pm 3, \pm 4$$
$$y = \pm 2, \pm \frac{3}{2}$$

The solution set is $\left\{ (3,2), (-3,-2), \left(4, \dfrac{3}{2}\right), \left(-4, -\dfrac{3}{2}\right) \right\}$.

23.   First complete the square:

$$y^2 + 4y + 8x - 4 = 0$$
$$y^2 + 4y + 4 = -8x + 4 + 4$$
$$(y + 2)^2 = -8(x - 1)$$

Graphing the parabola:

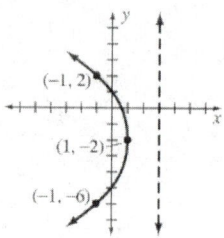

24.   First complete the square:

$$9x^2 - 36x + 4y^2 + 16y + 16 = 0$$
$$9\left(x^2 - 4x + 4\right) + 4\left(y^2 + 4y + 4\right) = -16 + 36 + 16$$
$$9(x - 2)^2 + 4(y + 2)^2 = 36$$
$$\frac{(x - 2)^2}{4} + \frac{(y + 2)^2}{9} = 1$$

Graphing the ellipse:

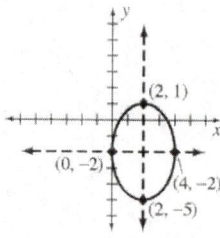

25.   First complete the square:

$$4x^2 + 24x - 3y^2 = 0$$
$$4\left(x^2 + 6x\right) - 3y^2 = 0$$
$$4\left(x^2 + 6x + 9\right) - 3y^2 = 0 + 36$$
$$4(x + 3)^2 - 3y^2 = 36$$
$$\frac{(x + 3)^2}{9} - \frac{y^2}{12} = 1$$

Graphing the hyperbola:

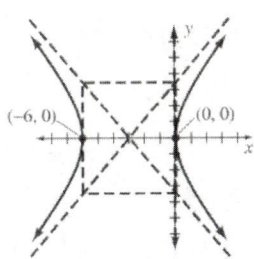

Chapters 0-8 Cumulative Review Problem Set

1. Evaluating: $\left(\dfrac{8}{27}\right)^{-2/3} = \left(\dfrac{27}{8}\right)^{2/3} = \left(\dfrac{3}{2}\right)^2 = \dfrac{9}{4}$

2. Evaluating: $\sqrt{\dfrac{16}{9}} = \dfrac{4}{3}$

3. Evaluating: $\sqrt[6]{\dfrac{1}{64}} = \dfrac{1}{2}$

4. Evaluating: $\left(2^{-1} - 3^{-2}\right)^{-1} = \left(\dfrac{1}{2} - \dfrac{1}{9}\right)^{-1} = \left(\dfrac{7}{18}\right)^{-1} = \dfrac{18}{7}$

5. Evaluating: $\left(2^{-3}\right)^{-2} = 2^6 = 64$

6. Evaluating: $9^{5/2} = 3^5 = 243$

7. Evaluating: $\log_4 64 = \log_4 4^3 = 3$

8. Evaluating: $\ln e^2 = 2\ln e = 2$

9. Evaluating: $\begin{vmatrix} 3 & -4 \\ 2 & -1 \end{vmatrix} = -3 + 8 = 5$

10. Adding 4 times row 3 to row 2, and 2 times row 3 to row 1:
$$\begin{vmatrix} 2 & 1 & -3 \\ 4 & 2 & -1 \\ -1 & -2 & 6 \end{vmatrix} = \begin{vmatrix} 0 & -3 & 9 \\ 0 & -6 & 23 \\ -1 & -2 & 6 \end{vmatrix} = -1\begin{vmatrix} -3 & 9 \\ -6 & 23 \end{vmatrix} = -(-69 + 54) = 15$$

11. Simplifying: $3(2x-1) - (3x+1) + 2(-x-1) = 6x - 3 - 3x - 1 - 2x - 2 = x - 6$

12. Simplifying: $(-2-x)(3+4x) = -6 - 3x - 8x - 4x^2 = -6 - 11x - 4x^2$

13. Simplifying: $(x-5)(3x^2 - 2x + 1) = 3x^3 - 2x^2 + x - 15x^2 + 10x - 5 = 3x^3 - 17x^2 + 11x - 5$

14. Simplifying:
$$\left(x^2 + x + 3\right)\left(2x^2 - x - 7\right) = 2x^4 - x^3 - 7x^2 + 2x^3 - x^2 - 7x + 6x^2 - 3x - 21 = 2x^4 + x^3 - 2x^2 - 10x - 21$$

15. Simplifying: $(2x-3)^3 = (2x-3)\left(4x^2 - 12x + 9\right) = 8x^3 - 24x^2 + 18x - 12x^2 + 36x - 27 = 8x^3 - 36x^2 + 54x - 27$

16. Simplifying: $\dfrac{-24x^3 y^2 + 36x^2 y^4}{4xy} = \dfrac{-24x^3 y^2}{4xy} + \dfrac{36x^2 y^4}{4xy} = -6x^2 y + 9xy^3$

17. Dividing using long division:

$$\begin{array}{r}
x^2 - 4x + 6 \\
3x+2\overline{)3x^3 - 10x^2 + 10x + 12} \\
\underline{3x^3 + 2x^2} \\
-12x^2 + 10x \\
\underline{-12x^2 - 8x} \\
18x + 12 \\
\underline{18x + 12} \\
0
\end{array}$$

The quotient is $x^2 - 4x + 6$.

18. Dividing using synthetic division:

$$2\overline{)\begin{array}{ccccc} 3 & -2 & -13 & 7 & 6 \\ & 6 & 8 & -10 & -6 \\ \hline 3 & 4 & -5 & -3 & 0 \end{array}}$$

The quotient is $3x^3 + 4x^2 - 5x - 3$.

19. Factoring: $-12 + 17x - 6x^2 = (-3 + 2x)(4 - 3x) = -(2x - 3)(3x - 4)$

20. Factoring: $x^2 - 3x + 2xy - 6y = x(x - 3) + 2y(x - 3) = (x - 3)(x + 2y)$

21. Factoring: $2x^3 + 6x^2 - 36x = 2x(x^2 + 3x - 18) = 2x(x + 6)(x - 3)$

22. This is not factorable using integers.

23. Factoring: $8x^3 + 27 = (2x + 3)(4x^2 - 6x + 9)$

24. Factoring: $4x^4 - 5x^2 + 1 = (4x^2 - 1)(x^2 - 1) = (2x + 1)(2x - 1)(x + 1)(x - 1)$

25. Simplifying: $\dfrac{96x^3y^4}{48x^2y^3} = 2xy$. Evaluating when $x = -6$ and $y = 4$: $2(-6)(4) = -48$

26. Simplifying: $\dfrac{x - 2}{4} - \dfrac{x + 9}{7} = \dfrac{7x - 14 - 4x - 36}{28} = \dfrac{3x - 50}{28}$. Evaluating when $x = 17$: $\dfrac{3(17) - 50}{28} = \dfrac{1}{28}$

27. Simplifying: $\dfrac{4a^2b}{9a^3b^2} \div \dfrac{8b}{3a} = \dfrac{4a^2b}{9a^3b^2} \cdot \dfrac{3a}{8b} = \dfrac{12a^3b}{72a^3b^3} = \dfrac{1}{6b^2}$. Evaluating when $a = 5$ and $b = -2$: $\dfrac{1}{6(-2)^2} = \dfrac{1}{24}$

28. Finding the product: $(5 + 2i)(4 - 3i) = 20 + 8i - 15i - 6i^2 = 26 - 7i$

29. Finding the product: $(-2 - i)(3 - 6i) = -6 - 3i + 12i + 6i^2 = -12 + 9i$

30. Finding the quotient: $\dfrac{2 - i}{3 + 4i} = \dfrac{2 - i}{3 + 4i} \cdot \dfrac{3 - 4i}{3 - 4i} = \dfrac{6 - 3i - 8i + 4i^2}{9 - 16i^2} = \dfrac{2 - 11i}{25} = \dfrac{2}{25} - \dfrac{11}{25}i$

31. Finding the product: $\sqrt{8}\left(\sqrt{6} - 2\sqrt{3}\right) = \sqrt{48} - 2\sqrt{24} = 4\sqrt{3} - 4\sqrt{6}$

32. Finding the product: $\left(2\sqrt{3} + 3\sqrt{5}\right)\left(2\sqrt{3} - 3\sqrt{5}\right) = 12 + 6\sqrt{15} - 6\sqrt{15} - 45 = -33$

33. Finding the quotient: $\dfrac{5\sqrt{2}}{3\sqrt{6}} = \dfrac{5\sqrt{2}}{3\sqrt{6}} \cdot \dfrac{\sqrt{6}}{\sqrt{6}} = \dfrac{5\sqrt{12}}{3\sqrt{36}} = \dfrac{10\sqrt{3}}{18} = \dfrac{5\sqrt{3}}{9}$

34. Finding the quotient: $\dfrac{\sqrt{5}}{\sqrt[3]{5}} = \dfrac{5^{1/2}}{5^{1/3}} = 5^{1/2 - 1/3} = 5^{1/6} = \sqrt[6]{5}$

35. Solving for y:

$$3x - 8y = -2$$
$$-8y = -3x - 2$$
$$y = \frac{3}{8}x + \frac{1}{4}$$

The slope is $\dfrac{3}{8}$.

36. Finding the point:

$$x = 4 + \frac{2}{3}(-6) = 4 - 4 = 0 \qquad\qquad y = -3 + \frac{2}{3}(9) = -3 + 6 = 3$$

The point is $(0, 3)$.

37. Finding the midpoint: $\left(\dfrac{-3 + 5}{2}, \dfrac{-4 + 10}{2}\right) = \left(\dfrac{2}{2}, \dfrac{6}{2}\right) = (1, 3)$

38. The vertical distance is: $0.04(4000) = 160$ feet

39. Finding the slope: $m = \dfrac{7-4}{-4+2} = -\dfrac{3}{2}$. Using the point-slope formula:

$$y - 4 = -\frac{3}{2}(x+2)$$
$$2y - 8 = -3x - 6$$
$$3x + 2y = 2$$

40. Using the slope-intercept formula:

$$y = \frac{5}{6}x - 3$$
$$6y = 5x - 18$$
$$5x - 6y = 18$$

41. Finding the slope:

$$-2x + 3y = -6$$
$$3y = 2x - 6$$
$$y = \frac{2}{3}x - 2$$

Using the point-slope formula:

$$y - 4 = \frac{2}{3}(x+1)$$
$$3y - 12 = 2x + 2$$
$$2x - 3y = -14$$

42. Finding the slope:

$$5x - 7y = 10$$
$$-7y = -5x + 10$$
$$y = \frac{5}{7}x - \frac{10}{7}$$

Using $m = -\dfrac{7}{5}$ in the point-slope formula:

$$y + 1 = -\frac{7}{5}(x+3)$$
$$5y + 5 = -7x - 21$$
$$7x + 5y = -26$$

43. Since $(x, -y)$ results in the same equation, the graph possesses x-axis symmetry.

44. Since $(x, -y)$, $(-x, y)$, and $(-x, -y)$ all result in the same equation, the graph possesses x-axis, y-axis, and origin symmetry.

45. Completing the square:

$$x^2 + 12x + y^2 - 8y + 42 = 0$$
$$\left(x^2 + 12x + 36\right) + \left(y^2 - 8y + 16\right) = -42 + 36 + 16$$
$$(x+6)^2 + (y-4)^2 = 10$$

The center is $(-6, 4)$.

46. Completing the square:

$$x^2 - 10x + y^2 + 6y + 18 = 0$$
$$\left(x^2 - 10x + 25\right) + \left(y^2 + 6y + 9\right) = -18 + 25 + 9$$
$$(x-5)^2 + (y+3)^2 = 16$$

The radius is 4.

47. The distance from $(-3, 4)$ to $(0, 0)$ is 5, so the equation is:

$$(x+3)^2 + (y-4)^2 = 5^2$$
$$x^2 + 6x + 9 + y^2 - 8y + 16 = 25$$
$$x^2 + 6x + y^2 - 8y = 0$$

48. The midpoint of this segment is the center of the circle, which is (5,1). Using the distance formula:

$$r = \sqrt{(6-5)^2 + (4-1)^2} = \sqrt{1+9} = \sqrt{10}$$

Thus the equation of the circle is:

$$(x-5)^2 + (y-1)^2 = 10$$
$$x^2 - 10x + 25 + y^2 - 2y + 1 = 10$$
$$x^2 - 10x + y^2 - 2y + 16 = 0$$

49. The standard form is $\dfrac{x^2}{14/9} - \dfrac{y^2}{14} = 1$, so $a = \dfrac{\sqrt{14}}{3}$ and $b = \sqrt{14}$. So the equations of the asymptotes are:

$$y = \pm \frac{\sqrt{14}}{\sqrt{14}/3} x = \pm 3x$$

50. Evaluating: $f(-3) = -(-3)^2 + 7(-3) - 4 = -34$

51. Finding the expression:

$$\frac{f(a+h) - f(a)}{h} = \frac{4(a+h)^2 - 2(a+h) - 7 - \left(4a^2 - 2a - 7\right)}{h}$$
$$= \frac{4a^2 + 8ah + 4h^2 - 2a - 2h - 7 - 4a^2 + 2a + 7}{h}$$
$$= \frac{8ah + 4h^2 - 2h}{h}$$
$$= 8a + 4h - 2$$

52. The denominator factors as $(x-2)^2$, so the domain is $(-\infty, 2) \cup (2, \infty)$.

53. Solving the inequality:

$$6 - x - x^2 \geq 0$$
$$(3+x)(2-x) \geq 0$$

Forming a sign chart:

The solution set is $[-3, 2]$.

54. Completing the square: $f(x) = 2\left(x^2 + 4x + 4\right) + 15 - 8 = 2(x+2)^2 + 7$. The vertex is $(-2, 7)$.

55. Completing the square: $f(x) = -\left(x^2 - 16x + 64\right) - 70 + 64 = -(x-8)^2 - 6$. The vertex is $(8, -6)$.

56. Finding the x-intercepts:

$$2x^2 - 19x + 9 = 0$$
$$(2x-1)(x-9) = 0$$
$$x = \frac{1}{2}, 9$$

57. Finding the compositions:

$$f(g(-1)) = f(2-3-6) = f(-7) = -21 - 2 = -23$$
$$g(f(3)) = g(9-2) = g(7) = 98 + 21 - 6 = 113$$

58. Finding the compositions:

$$f(g(x)) = f(-2x+5) = -3(-2x+5)^2 - (-2x+5) + 8 = -12x^2 + 60x - 75 + 2x - 5 + 8 = -12x^2 + 62x - 72$$
$$g(f(x)) = g(-3x^2 - x + 8) = -2\left(-3x^2 - x + 8\right) + 5 = 6x^2 + 2x - 11$$

59. Solving the equation:

$$-\frac{2}{3}y + 1 = x$$

$$-\frac{2}{3}y = x - 1$$

$$y = -\frac{3}{2}x + \frac{3}{2}$$

The inverse is $f^{-1}(x) = -\frac{3}{2}x + \frac{3}{2}$.

60. The variation equation is $y = Kxz$. Substituting $y = 24$, $x = 2$, and $z = -4$:

$$24 = K(2)(-4)$$
$$K = -3$$

So $y = -3xz$. Substituting $x = -6$ and $z = -5$: $y = -3(-6)(-5) = -90$

61. The variation equation is $y = \dfrac{Kx}{\sqrt{w}}$. Substituting $y = \dfrac{2}{3}$, $x = \dfrac{1}{2}$, and $w = 9$:

$$\frac{2}{3} = \frac{K\left(\frac{1}{2}\right)}{\sqrt{9}}$$

$$\frac{2}{3} = K\left(\frac{1}{6}\right)$$

$$K = 4$$

So $y = \dfrac{4x}{\sqrt{w}}$. Substituting $x = 8$ and $w = \dfrac{1}{4}$: $y = \dfrac{4(8)}{\sqrt{\frac{1}{4}}} = \dfrac{32}{\frac{1}{2}} = 64$

62. Using the function $N(t) = 750\left(\dfrac{1}{2}\right)^{t/13}$: $N(50) = 750\left(\dfrac{1}{2}\right)^{50/13} \approx 52.1$ milligrams

63. Using the function $A(t) = 30000(0.9)^t$: $A(3) = 30000(0.9)^3 = \$21{,}870$

64. Using the function $Q(t) = 150e^{-0.2t}$: $Q(12) = 150e^{-0.2(12)} \approx 13.6$ grams

65. Finding the compositions:

$$f\left(\frac{4-x}{3}\right) = -3\left(\frac{4-x}{3}\right) + 4 = -4 + x + 4 = x \qquad g(-3x+4) = \frac{4-(-3x+4)}{3} = \frac{4+3x-4}{3} = \frac{3x}{3} = x$$

So f and g are inverse functions.

66. Using synthetic division:

$$
\begin{array}{r|rrrrr}
-1 & 1 & -4 & -22 & 4 & 21 \\
   &   & -1 & 5 & 17 & -21 \\
\hline
   & 1 & -5 & -17 & 21 & 0
\end{array}
$$

Yes, it is a factor.

67. Using synthetic division:

$$
\begin{array}{r|rrrr}
4 & 2 & -3 & -4 & 6 \\
  &   & 8 & 20 & 64 \\
\hline
  & 2 & 5 & 16 & 70
\end{array}
$$

No, it is not a factor.

68. The possible rational roots are $\pm1, \pm2, \pm3, \pm6$. Using synthetic division:

$$
\begin{array}{r|rrrr}
1 & 1 & -2 & -5 & 6 \\
  &   & 1 & -1 & -6 \\
\hline
  & 1 & -1 & -6 & 0
\end{array}
$$

Therefore: $f(x) = (x-1)\left(x^2 - x - 6\right) = (x-1)(x-3)(x+2)$

The x-intercepts are 1, 3, and -2.

69. Using synthetic division:

$$1\overline{)\begin{array}{ccc} 1 & 2 & 1 \\ & 1 & 3 \\ \hline 1 & 3 & 4 \end{array}}$$

So $f(x) = x + 3 + \dfrac{4}{x-1}$. The oblique asymptote is $y = x + 3$.

70. Finding the decomposition:

$$\frac{x+10}{(2x-1)(x-4)} = \frac{A}{2x-1} + \frac{B}{x-4}$$
$$x + 10 = A(x-4) + B(2x-1)$$

Substituting $x = 1/2$ and $x = 4$:

$$\frac{21}{2} = A\left(-\frac{7}{2}\right) \qquad\qquad 14 = B(7)$$
$$A = -3 \qquad\qquad\qquad\qquad B = 2$$

The decomposition is $\dfrac{x+10}{(2x-1)(x-4)} = \dfrac{-3}{2x-1} + \dfrac{2}{x-4}$.

71. Computing: $2A - 3B = \begin{bmatrix} 4 & -2 \\ 6 & 8 \end{bmatrix} - \begin{bmatrix} 18 & -6 \\ -3 & 15 \end{bmatrix} = \begin{bmatrix} -14 & 4 \\ 9 & -7 \end{bmatrix}$

72. Computing each matrix:

$$AB = \begin{bmatrix} 1 & -2 \\ 3 & -4 \end{bmatrix}\begin{bmatrix} -3 & -4 \\ 2 & -1 \end{bmatrix} = \begin{bmatrix} -7 & -2 \\ -17 & -8 \end{bmatrix} \qquad BA = \begin{bmatrix} -3 & -4 \\ 2 & -1 \end{bmatrix}\begin{bmatrix} 1 & -2 \\ 3 & -4 \end{bmatrix} = \begin{bmatrix} -15 & 22 \\ -1 & 0 \end{bmatrix}$$

73. Form the augmented matrix: $\left[\begin{array}{cc|cc} 2 & 3 & 1 & 0 \\ 1 & 2 & 0 & 1 \end{array}\right]$

Switch row 1 and row 2: $\left[\begin{array}{cc|cc} 1 & 2 & 0 & 1 \\ 2 & 3 & 1 & 0 \end{array}\right]$

Add –2 times row 1 to row 2: $\left[\begin{array}{cc|cc} 1 & 2 & 0 & 1 \\ 0 & -1 & 1 & -2 \end{array}\right]$

Add 2 times row 2 to row 1, and multiply row 1 by –1: $\left[\begin{array}{cc|cc} 1 & 0 & 2 & -3 \\ 0 & 1 & -1 & 2 \end{array}\right]$

The inverse is $\begin{bmatrix} 2 & -3 \\ -1 & 2 \end{bmatrix}$.

74. a. This graph is an ellipse.   b. This graph is a parabola.
     c. This graph is a parabola.   d. This graph is a circle.
     e. This graph is a hyperbola.

75. Completing the square: 76. Completing the square:

$$4x^2 + 9y^2 - 16x - 18y - 11 = 0$$
$$4\left(x^2 - 4x + 4\right) + 9\left(y^2 - 2y + 1\right) = 11 + 16 + 9$$
$$4(x-2)^2 + 9(y-1)^2 = 36$$
$$\frac{(x-2)^2}{9} + \frac{(y-1)^2}{4} = 1$$

$$x^2 - 6x + y^2 + 2y + 1 = 0$$
$$\left(x^2 - 6x + 9\right) + \left(y^2 + 2y + 1\right) = -1 + 9 + 1$$
$$(x-3)^2 + (y+1)^2 = 9$$

Graphing the ellipse:

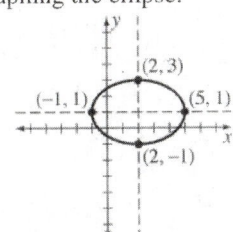

Graphing the circle:

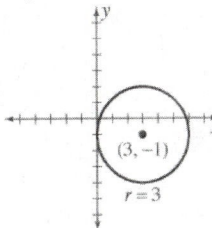

77.　Graphing the parabola $y^2 = x - 2$:

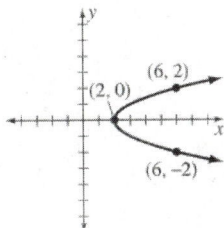

78.　Completing the square:

$$x^2 - y^2 - 2x + 4y - 12 = 0$$
$$\left(x^2 - 2x + 1\right) - \left(y^2 - 4y + 4\right) = 12 + 1 - 4$$
$$(x-1)^2 - (y-2)^2 = 9$$
$$\frac{(x-1)^2}{9} - \frac{(y-2)^2}{9} = 1$$

Graphing the hyperbola:

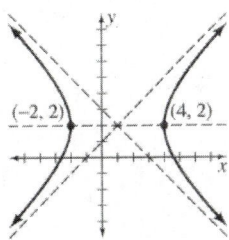

79.　Completing the square: $y = -2\left(x^2 - 6x\right) = -2\left(x^2 - 6x + 9\right) - 15 + 18 = -2(x-3)^2 + 3$

Graphing the parabola:

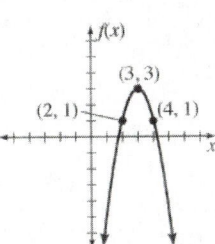

80.　Graphing the function $f(x) = \log_2(x+2)$:

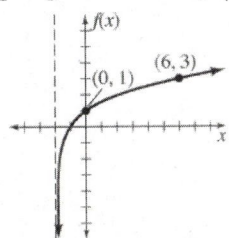

81.　Graphing the function $f(x) = -x(x+1)(x-3)$:

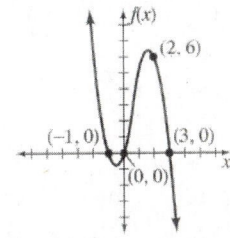

82.   Graphing the function $f(x) = 3|x-1| - 2$:

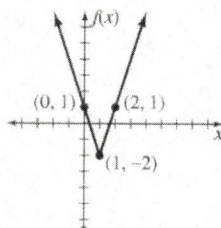

83.   Graphing the function $f(x) = -\sqrt{x+3} - 1$:

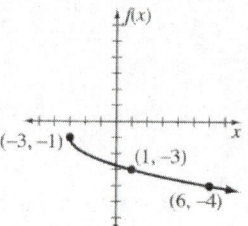

84.   Graphing the function $f(x) = e^x + 3$:

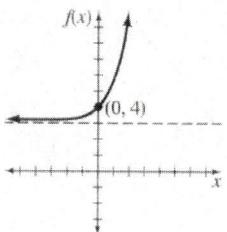

85.   Graphing the function $f(x) = -(x+2)^3 + 1$:

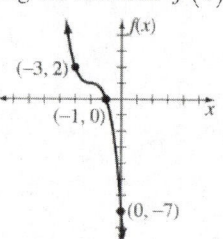

86.   Graphing the function $f(x) = -3x + 4$:

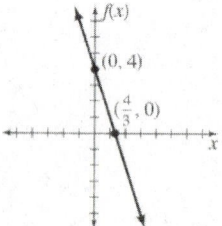

87.   Graphing the function $f(x) = (x+1)^2 + 1$:

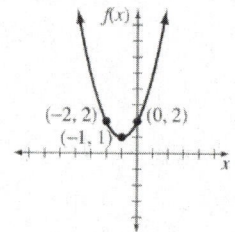

88.   Graphing the function $f(x) = (x-2)^4 + 1$:

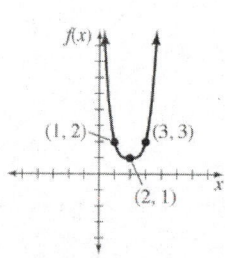

89.   The vertical asymptote is $x = 2$ and the horizontal asymptote is $y = 2$. Graphing the function $f(x) = \dfrac{2x}{x-2}$:

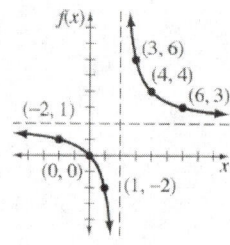

90. The vertical asymptote is $x = -1$. Using synthetic division:

$$
\begin{array}{r|rrr}
-1 & 1 & 0 & -3 \\
  &   & -1 & 1 \\
\hline
  & 1 & -1 & -2
\end{array}
$$

So $f(x) = x - 1 - \dfrac{2}{x+1}$. The oblique asymptote is $y = x - 1$. Graphing the function $f(x) = \dfrac{x^2 - 3}{x+1}$:

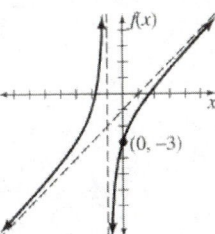

91. Solving the equation:

$$|3x - 1| = 8$$
$$3x - 1 = -8, 8$$
$$3x = -7, 9$$
$$x = -\frac{7}{3}, 3$$

The solution set is $\left\{ -\dfrac{7}{3}, 3 \right\}$.

92. Solving the equation:

$$(3x - 2)(2x + 1) = 0$$
$$x = -\frac{1}{2}, \frac{2}{3}$$

The solution set is $\left\{ -\dfrac{1}{2}, \dfrac{2}{3} \right\}$.

93. Solving the equation:

$$\sqrt{2x} + x = 4$$
$$\sqrt{2x} = 4 - x$$
$$2x = (4 - x)^2$$
$$2x = 16 - 8x + x^2$$
$$0 = x^2 - 10x + 16$$
$$0 = (x - 8)(x - 2)$$
$$x = 2 \qquad (x = 8 \text{ does not check})$$

The solution set is $\{2\}$.

94. Solving the inequality:

$$|2x + 6| \le 12$$
$$-12 \le 2x + 6 \le 12$$
$$-18 \le 2x \le 6$$
$$-9 \le x \le 3$$

The solution set is $[-9, 3]$.

95. Solving the equation:

$$(5x - 2)^2 = 18$$
$$5x - 2 = \pm 3\sqrt{2}$$
$$5x = 2 \pm 3\sqrt{2}$$
$$x = \frac{2 \pm 3\sqrt{2}}{5}$$

The solution set is $\left\{ \dfrac{2 \pm 3\sqrt{2}}{5} \right\}$.

96. Solving the equation:

$$2x^2 + x + 5 = 0$$
$$x = \frac{-1 \pm \sqrt{1 - 40}}{4} = \frac{-1 \pm \sqrt{-39}}{4} = \frac{-1 \pm i\sqrt{39}}{4}$$

The solution set is $\left\{ \dfrac{-1 \pm i\sqrt{39}}{4} \right\}$.

97. Multiply the first equation by 5 and the second equation by 2:
$$15x - 10y = 20$$
$$4x + 10y = 14$$
Adding yields:
$$19x = 34$$
$$x = \frac{34}{19}$$
Substituting into the second equation:
$$\frac{68}{19} + 5y = 7$$
$$5y = \frac{65}{19}$$
$$y = \frac{13}{19}$$
The solution set is $\left\{ \left(\frac{34}{19}, \frac{13}{19} \right) \right\}$.

98. Solving the equation:
$$(x+2)(5x-1) = 36$$
$$5x^2 + 9x - 2 = 36$$
$$5x^2 + 9x - 38 = 0$$
$$(5x+19)(x-2) = 0$$
$$x = -\frac{19}{5}, 2$$
The solution set is $\left\{ -\frac{19}{5}, 2 \right\}$.

99. Forming a sign chart:

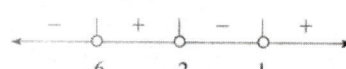

The solution set is $(-6, -2) \cup (1, \infty)$.

100. The possible rational roots are $\pm 1, \pm 2, \pm 3, \pm 4, \pm 6, \pm 12$. Using synthetic division:

$$
\begin{array}{r|rrrr}
1) & 1 & 7 & 4 & -12 \\
  &   & 1 & 8 & 12 \\
\hline
  & 1 & 8 & 12 & 0
\end{array}
$$

Factoring the equation:
$$(x-1)(x^2 + 8x + 12) = 0$$
$$(x-1)(x+6)(x+2) = 0$$
$$x = -6, -2, 1$$
The solution set is $\{-6, -2, 1\}$.

101. Form the augmented matrix: $\begin{bmatrix} 1 & -2 & 1 & | & 6 \\ -2 & 1 & -1 & | & -5 \\ 4 & 3 & 2 & | & 5 \end{bmatrix}$

Add 2 times row 1 to row 2, and –4 times row 1 to row 3: $\begin{bmatrix} 1 & -2 & 1 & | & 6 \\ 0 & -3 & 1 & | & 7 \\ 0 & 11 & -2 & | & -19 \end{bmatrix}$

Add 4 times row 2 to row 3: $\begin{bmatrix} 1 & -2 & 1 & | & 6 \\ 0 & -3 & 1 & | & 7 \\ 0 & -1 & 2 & | & 9 \end{bmatrix}$

Switch row 2 and row 3, and multiply row 2 by –1: $\begin{bmatrix} 1 & -2 & 1 & | & 6 \\ 0 & 1 & -2 & | & -9 \\ 0 & -3 & 1 & | & 7 \end{bmatrix}$

Add 3 times row 2 to row 3: $\begin{bmatrix} 1 & -2 & 1 & | & 6 \\ 0 & 1 & -2 & | & -9 \\ 0 & 0 & -5 & | & -20 \end{bmatrix}$

Divide row 3 by –5: $\begin{bmatrix} 1 & -2 & 1 & | & 6 \\ 0 & 1 & -2 & | & -9 \\ 0 & 0 & 1 & | & 4 \end{bmatrix}$

Add 2 times row 3 to row 2, and –1 times row 3 to row 1: $\begin{bmatrix} 1 & -2 & 0 & | & 2 \\ 0 & 1 & 0 & | & -1 \\ 0 & 0 & 1 & | & 4 \end{bmatrix}$

Add 2 times row 2 to row 1: $\begin{bmatrix} 1 & 0 & 0 & | & 0 \\ 0 & 1 & 0 & | & -1 \\ 0 & 0 & 1 & | & 4 \end{bmatrix}$

The solution set is $\{(0,-1,4)\}$.

102. Solving the equation:

$$3^{3x+1} = 9^{x-2}$$
$$3^{3x+1} = 3^{2x-4}$$
$$3x+1 = 2x-4$$
$$x = -5$$

The solution set is $\{-5\}$.

103. Solving the equation:

$$(4x+3)^2 = -16$$
$$4x+3 = \pm 4i$$
$$4x = -3 \pm 4i$$
$$x = \frac{-3 \pm 4i}{4}$$

The solution set is $\left\{ \dfrac{-3 \pm 4i}{4} \right\}$.

104. Solving the equation:

$$3x^2 - x - 3 = 0$$
$$x = \frac{1 \pm \sqrt{1 - 4(-9)}}{6} = \frac{1 \pm \sqrt{37}}{6}$$

The solution set is $\left\{ \dfrac{1 \pm \sqrt{37}}{6} \right\}$.

105. Factoring the inequality:

$$x^2 + 2x < 48$$
$$x^2 + 2x - 48 < 0$$
$$(x+8)(x-6) < 0$$

Forming a sign chart:

The solution set is $(-8, 6)$.

106. Substituting into the second equation:

$$2x + 5(3x - 4) = -9$$
$$2x + 15x - 20 = -9$$
$$17x = 11$$
$$x = \frac{11}{17}$$
$$y = 3\left(\frac{11}{17}\right) - 4 = -\frac{35}{17}$$

The solution set is $\left\{ \left(\frac{11}{17}, -\frac{35}{17} \right) \right\}$.

107. Solving the equation:

$$x^2 - 2x - 288 = 0$$
$$(x - 18)(x + 16) = 0$$
$$x = -16, 18$$

The solution set is $\{-16, 18\}$.

108. Solving the inequality:

$$|7x - 2| > 12$$

$$7x - 2 < -12 \quad \text{or} \quad 7x - 2 > 12$$
$$7x < -10 \qquad\qquad 7x > 14$$
$$x < -\frac{10}{7} \qquad\qquad x > 2$$

The solution set is $\left(-\infty, -\frac{10}{7} \right) \cup (2, \infty)$.

109. Solving the equation:

$$(2x + 3)(x - 4) = (3x - 1)(x - 4)$$
$$2x^2 - 5x - 12 = 3x^2 - 13x + 4$$
$$0 = x^2 - 8x + 16$$
$$0 = (x - 4)^2$$
$$x = 4$$

The solution set is $\{4\}$.

110. Solving the equation:

$$4^{x-2} = 12$$
$$\ln 4^{x-2} = \ln 12$$
$$(x - 2)\ln 4 = \ln 12$$
$$x \ln 4 - 2 \ln 4 = \ln 12$$
$$x \ln 4 = 2 \ln 4 + \ln 12$$
$$x = \frac{2 \ln 4 + \ln 12}{\ln 4} \approx 3.79$$

The solution set is $\{3.79\}$.

111. Solving the equation:

$$\log_2 (x - 1) + \log_2 (x + 4) = 1$$
$$\log_2 (x^2 + 3x - 4) = 1$$
$$x^2 + 3x - 4 = 2$$
$$x^2 + 3x - 6 = 0$$
$$x = \frac{-3 \pm \sqrt{9 + 24}}{2} = \frac{-3 + \sqrt{33}}{2}$$
$$\left(x = \frac{-3 - \sqrt{33}}{2} \text{ does not check} \right)$$

The solution set is $\left\{ \frac{-3 + \sqrt{33}}{2} \right\}$.

112. Solving the equation:
$$(6x-1)^2 = 45$$
$$6x - 1 = \pm 3\sqrt{5}$$
$$6x = 1 \pm 3\sqrt{5}$$
$$x = \frac{1 \pm 3\sqrt{5}}{6}$$
The solution set is $\left\{ \dfrac{1 \pm 3\sqrt{5}}{6} \right\}$.

113. Solving the inequality:
$$\frac{x-4}{x+2} + 3 \geq 0$$
$$\frac{x - 4 + 3x + 6}{x + 2} \geq 0$$
$$\frac{4x + 2}{x + 2} \geq 0$$
Forming a sign chart:

The solution set is $\left(-\infty, -2\right) \cup \left[-\dfrac{1}{2}, \infty\right)$.

114. Solving the equation:
$$x^{2/3} - 3x^{1/3} - 4 = 0$$
$$\left(x^{1/3} - 4\right)\left(x^{1/3} + 1\right) = 0$$
$$x^{1/3} = -1, 4$$
$$x = -1, 64$$
The solution set is $\left\{-1, 64\right\}$.

115. Let x represent the price per can of soda and y represent the price per bag of potato chips. The system of equations is:
$$6x + 3y = 8.37$$
$$8x + 5y = 12.45$$
Multiply the first equation by −4 and the second equation by 3:
$$-24x - 12y = -33.48$$
$$24x + 15y = 37.35$$
Adding yields:
$$3y = 3.87$$
$$y = 1.29$$
Substituting into the first equation:
$$6x + 3(1.29) = 8.37$$
$$6x = 4.50$$
$$x = 0.75$$
The soda costs $0.75 per can and the chips cost $1.29 per bag.

116. Solving the proportion:
$$\frac{1\frac{1}{4}}{2\frac{1}{2}} = \frac{3}{x}$$
$$\frac{5}{4}x = \frac{15}{2}$$
$$x = 6$$
Thus 6 cups of flour are needed.

117. Let n represent the number of nickels, d the number of dimes, and q the number of quarters. The system of equations is:

$$0.05n + 0.10d + 0.25q = 14$$
$$d = n + 5$$
$$q = 2n + 2$$

Substituting into the first equation:

$$0.05n + 0.10(n+5) + 0.25(2n+2) = 14$$
$$0.05n + 0.10n + 0.5 + 0.50n + 0.5 = 14$$
$$0.65n = 13$$
$$n = 20$$
$$d = 25$$
$$q = 42$$

Kaya has 20 nickels, 25 dimes, and 42 quarters.

118. Let x represent Sherry's age and $x + 4$ represent Tina's age. Their ages in 5 years are $x + 5$ and $x + 9$, so the equation is:

$$x + 5 + x + 9 = 48$$
$$2x + 14 = 48$$
$$2x = 34$$
$$x = 17$$

Sherry is 17 and Tina is 21.

119. Let x and $x + 72$ represent the two numbers. The equation is:

$$\frac{x+72}{x} = 6 + \frac{2}{x}$$
$$x + 72 = 6x + 2$$
$$70 = 5x$$
$$x = 14$$

The numbers are 14 and 86.

120. Let x represent the selling price. The equation is:

$$24 + 0.40x = x$$
$$24 = 0.6x$$
$$x = 40$$

She should charge $40 for the shirts.

121. Let x represent the amount to evaporate. The equation is:

$$0.10(30) = 0.20(30 - x)$$
$$3 = 6 - 0.2x$$
$$-3 = -0.2x$$
$$x = 15$$

Thus 15 gallons of water must be evaporated.

122. Let t represent the time Josh rode out, so $5\frac{5}{6} - t$ represents the time riding back. Setting the distances equal:

$$20t = 15\left(\frac{35}{6} - t\right)$$
$$120t = 15(35 - 6t)$$
$$8t = 35 - 6t$$
$$14t = 35$$
$$t = 2.5$$

The distance Josh rode out is: $20(2.5) = 50$ miles

123. Let r represent the required rate. The inequality is:

$$0.08(500) + r(500) > 100$$
$$40 + 500r > 100$$
$$500r > 60$$
$$r > 0.12$$

The rate must be higher than 12%.

124. Let T represent the time, l the length of road, and w the number of workers. The variation equation is $T = \dfrac{Kl}{w}$.

Substituting $w = 100$, $T = 4$, and $l = 2$:

$$4 = \frac{K(2)}{100}$$
$$K = 200$$

So $T = \dfrac{200l}{w}$. Substituting $w = 80$ and $l = 10$: $T = \dfrac{200(10)}{80} = 25$ weeks

125. Finding the time:

$$500e^{0.04t} = 2000$$
$$e^{0.04t} = 4$$
$$0.04t = \ln 4$$
$$t = \frac{\ln 4}{0.04} \approx 34.7 \text{ minutes}$$

Chapter 9
Sequences and Mathematical Induction

9.1 Arithmetic Sequences

1. The first five terms are: $-4, -1, 2, 5, 8$

3. The first five terms are: $2, 0, -2, -4, -6$

5. The first five terms are: $2, 11, 26, 47, 74$

7. The first five terms are: $0, 2, 6, 12, 20$

9. The first five terms are: $4, 8, 16, 32, 64$

11. Finding the terms:
$$a_{15} = -5(15) - 4 = -79$$
$$a_{30} = -5(30) - 4 = -154$$

13. Finding the terms:
$$a_{25} = (-1)^{26} = 1$$
$$a_{50} = (-1)^{51} = -1$$

15. The general term is: $a_n = 2n + 9$

17. The general term is: $a_n = -3n + 5$

19. The general term is: $a_n = \dfrac{n+2}{2} = \dfrac{1}{2}n + 1$

21. The general term is: $a_n = 4n - 2$

23. The general term is: $a_n = -3n$

25. The general term is $a_n = 5n - 2$. Finding the term: $a_{15} = 5(15) - 2 = 73$

27. The general term is $a_n = 11n + 4$. Finding the term: $a_{30} = 11(30) + 4 = 334$

29. The general term is $a_n = \dfrac{2}{3}n + \dfrac{1}{3}$. Finding the term: $a_{52} = \dfrac{2}{3}(52) + \dfrac{1}{3} = 35$

31. Forming the system of equations:
$$12 = a_1 + 5d$$
$$16 = a_1 + 9d$$
Subtracting yields:
$$4d = 4$$
$$d = 1$$
Substituting into the first equation:
$$12 = a_1 + 5$$
$$a_1 = 7$$

33. Forming the system of equations:
$$20 = a_1 + 2d$$
$$32 = a_1 + 6d$$
Subtracting yields:
$$4d = 12$$
$$d = 3$$
Substituting into the first equation:
$$20 = a_1 + 6$$
$$a_1 = 14$$
So $a_n = 14 + (n-1)3 = 3n + 11$. Finding the term: $a_{25} = 3(25) + 11 = 86$

35. Here $a_1 = 5$ and $a_{50} = 5 + 49(2) = 103$, so the sum is: $S_{50} = \dfrac{50(5 + 103)}{2} = 2700$

37. Here $a_1 = 2$ and $a_{40} = 2 + 39(4) = 158$, so the sum is: $S_{40} = \dfrac{40(2 + 158)}{2} = 3200$

39. Here $a_1 = 5$ and $a_{75} = 5 + 74(-3) = -217$, so the sum is: $S_{75} = \dfrac{75(5 - 217)}{2} = -7950$

41. Here $a_1 = \dfrac{1}{2}$ and $a_{50} = \dfrac{1}{2} + 49\left(\dfrac{1}{2}\right) = 25$, so the sum is: $S_{50} = \dfrac{50\left(\dfrac{1}{2} + 25\right)}{2} = 637\dfrac{1}{2}$

43. Here $a_1 = 1$ and $d = 4$, so:
$$197 = 1 + 4(n - 1)$$
$$196 = 4(n - 1)$$
$$n - 1 = 49$$
$$n = 50$$
Finding the sum: $S_{50} = \dfrac{50(1 + 197)}{2} = 4950$

45. Here $a_1 = 2$ and $d = 6$, so:
$$146 = 2 + 6(n - 1)$$
$$144 = 6(n - 1)$$
$$n - 1 = 24$$
$$n = 25$$
Finding the sum: $S_{25} = \dfrac{25(2 + 146)}{2} = 1850$

47. Here $a_1 = -7$ and $d = -3$, so:
$$-109 = -7 - 3(n - 1)$$
$$-102 = -3(n - 1)$$
$$n - 1 = 34$$
$$n = 35$$
Finding the sum: $S_{35} = \dfrac{35(-7 - 109)}{2} = -2030$

49. Here $a_1 = -5$ and $d = 2$, so:
$$119 = -5 + 2(n - 1)$$
$$124 = 2(n - 1)$$
$$n - 1 = 62$$
$$n = 63$$
Finding the sum: $S_{63} = \dfrac{63(-5 + 119)}{2} = 3591$

51. Here $a_1 = 1$ and $a_{200} = 1 + 199(2) = 399$, so: $S_{200} = \dfrac{200(1 + 399)}{2} = 40,000$

53. Here $a_1 = 18$ and $d = 2$, so:
$$482 = 18 + 2(n - 1)$$
$$464 = 2(n - 1)$$
$$n - 1 = 232$$
$$n = 233$$
Finding the sum: $S_{233} = \dfrac{233(18 + 482)}{2} = 58,250$

55. Here $a_1 = 1$ and $a_{30} = 5(30) - 4 = 146$, so: $S_{30} = \dfrac{30(1 + 146)}{2} = 2205$

57. Here $a_1 = -5$ and $a_{25} = -4(25) - 1 = -101$, so: $S_{25} = \dfrac{25(-5 - 101)}{2} = -1325$

59. Here $a_1 = 7$ and $a_{45} = 5(45) + 2 = 227$, so: $S_{45} = \dfrac{45(7 + 227)}{2} = 5265$

61. Here $a_1 = 2$ and $a_{30} = -2(30) + 4 = -56$, so: $S_{30} = \dfrac{30(2 - 56)}{2} = -810$

63. Here $a_1 = -7$, $a_3 = 3(3) - 10 = -1$, and $a_{32} = 3(32) - 10 = 86$, so:

$$S_{32} - S_3 = \frac{32(-7 + 86)}{2} - \frac{3(-7 - 1)}{2} = 1264 + 12 = 1276$$

65. Here $a_1 = 4$, $a_9 = 36$, and $a_{20} = 80$, so: $S_{20} - S_9 = \dfrac{20(4 + 80)}{2} - \dfrac{9(4 + 36)}{2} = 840 - 180 = 660$

67. Finding the sum: $\displaystyle\sum_{i=1}^{5} i^2 = 1 + 4 + 9 + 16 + 25 = 55$

69. Finding the sum: $\displaystyle\sum_{i=3}^{8} \left(2i^2 + i\right) = 21 + 36 + 55 + 78 + 105 + 136 = 431$

75. The terms are: 3, 3, 7, 7, 11, 11

77. The terms are: 4, 7, 10, 13, 17, 21

79. The terms are: 4, 12, 36, 108, 324, 972

81. The terms are: 1, 1, 2, 3, 5, 8

83. The terms are: 3, 1, 4, 9, 25, 256

9.2 Geometric Sequences

1. The general term is: $a_n = 3(2)^{n-1}$

3. The general term is: $a_n = 3(3)^{n-1} = 3^n$

5. The general term is: $a_n = \dfrac{1}{4}\left(\dfrac{1}{2}\right)^{n-1} = \left(\dfrac{1}{2}\right)^{n+1}$

7. The general term is: $a_n = 4(4)^{n-1} = 4^n$

9. The general term is: $a_n = 1(0.3)^{n-1} = (0.3)^{n-1}$

11. The general term is: $a_n = 1(-2)^{n-1} = (-2)^{n-1}$

13. Finding the term: $a_8 = \dfrac{1}{2}(2)^{8-1} = 2^6 = 64$

15. Finding the term: $a_9 = 729\left(\dfrac{1}{3}\right)^{9-1} = 729\left(\dfrac{1}{3}\right)^{8} = \dfrac{1}{9}$

17. Finding the term: $a_{10} = 1(-2)^{10-1} = (-2)^9 = -512$

19. Finding the term: $a_8 = \dfrac{1}{2}\left(\dfrac{1}{3}\right)^{8-1} = \dfrac{1}{2}\left(\dfrac{1}{3}\right)^{7} = \dfrac{1}{4374}$

21. Using the general term formula:

$$\frac{32}{3} = a_1(2)^4$$

$$\frac{32}{3} = 16a_1$$

$$a_1 = \frac{2}{3}$$

23. Using the general term formula:

$$12 = a_1 r^2$$

$$96 = a_1 r^5$$

Dividing the two equations:

$$8 = r^3$$

$$r = 2$$

25. Using the sum formula: $S_{10} = \dfrac{1(2)^{10} - 1}{2 - 1} = 2^{10} - 1 = 1023$

27. Using the sum formula: $S_9 = \dfrac{2(3)^9 - 2}{3 - 1} = \dfrac{39364}{2} = 19{,}682$

29. Using the sum formula: $S_8 = \dfrac{8\left(\dfrac{3}{2}\right)^8 - 8}{\dfrac{3}{2} - 1} = 394\dfrac{1}{16}$

31. Using the sum formula: $S_{10} = \dfrac{-4(-2)^{10} + 4}{-2 - 1} = \dfrac{-4092}{-3} = 1364$

33. Using the general term formula:

$$729 = 9(3)^{n-1}$$
$$(3)^{n-1} = 81$$
$$n - 1 = 4$$
$$n = 5$$

Finding the sum: $S_5 = \dfrac{9(3)^5 - 9}{3 - 1} = 1089$

35. Using the general term formula:

$$\frac{1}{512} = 4\left(\frac{1}{2}\right)^{n-1}$$
$$\left(\frac{1}{2}\right)^{n-1} = \frac{1}{2048}$$
$$n - 1 = 11$$
$$n = 12$$

Finding the sum: $S_{12} = \dfrac{4\left(\dfrac{1}{2}\right)^{12} - 4}{\dfrac{1}{2} - 1} = 7\dfrac{511}{512}$

37. Using the general term formula:

$$-729 = -1(-3)^{n-1}$$
$$(-3)^{n-1} = 729$$
$$n - 1 = 6$$
$$n = 7$$

Finding the sum: $S_7 = \dfrac{-1(-3)^7 + 1}{-3 - 1} = -547$

39. Using the sum formula: $S_9 = \dfrac{\dfrac{1}{4}(2)^9 - \dfrac{1}{4}}{2 - 1} = 127\dfrac{3}{4}$

41. Finding the sum: $\displaystyle\sum_{i=2}^{5}(-3)^{i+1} = -27 + 81 - 243 + 729 = 540$

43. Using the sum formula: $S_6 = \dfrac{\dfrac{3}{2}\left(\dfrac{1}{2}\right)^6 - \dfrac{3}{2}}{\dfrac{1}{2} - 1} = 2\dfrac{61}{64}$

45. Using the infinite sum formula: $S_\infty = \dfrac{2}{1 - \dfrac{1}{2}} = 4$

47. Using the infinite sum formula: $S_\infty = \dfrac{1}{1 - \dfrac{2}{3}} = 3$

49. Since $r = 2$, there is no infinite sum.

51. Using the infinite sum formula: $S_\infty = \dfrac{9}{1 + \dfrac{1}{3}} = \dfrac{27}{4}$

53. Using the infinite sum formula: $S_\infty = \dfrac{\dfrac{1}{2}}{1 - \dfrac{3}{4}} = 2$

55. Using the infinite sum formula: $S_\infty = \dfrac{8}{1 + \dfrac{1}{2}} = \dfrac{16}{3}$

57. Using the infinite sum formula: $0.\overline{3} = \dfrac{\dfrac{3}{10}}{1 - \dfrac{1}{10}} = \dfrac{3}{9} = \dfrac{1}{3}$

59. Using the infinite sum formula: $0.\overline{26} = \dfrac{\dfrac{26}{100}}{1 - \dfrac{1}{100}} = \dfrac{26}{99}$

61. Using the infinite sum formula: $0.\overline{123} = \dfrac{\dfrac{123}{1000}}{1 - \dfrac{1}{1000}} = \dfrac{123}{999} = \dfrac{41}{333}$

63. Using the infinite sum formula: $0.2\overline{6} = \dfrac{2}{10} + \dfrac{\dfrac{6}{100}}{1 - \dfrac{1}{10}} = \dfrac{2}{10} + \dfrac{6}{90} = \dfrac{1}{5} + \dfrac{1}{15} = \dfrac{4}{15}$

65. Using the infinite sum formula: $0.2\overline{14} = \dfrac{2}{10} + \dfrac{\dfrac{14}{1000}}{1 - \dfrac{1}{100}} = \dfrac{2}{10} + \dfrac{14}{990} = \dfrac{1}{5} + \dfrac{7}{495} = \dfrac{106}{495}$

67. Using the infinite sum formula: $2.\overline{3} = 2 + \dfrac{\dfrac{3}{10}}{1 - \dfrac{1}{10}} = 2 + \dfrac{1}{3} = \dfrac{7}{3}$

9.3 Another Look at Problem Solving

1. The sequence is \$19500, \$21200, \$22900, ... Finding the 21st term: $a_{21} = 19500 + 20(1700) = \$53,500$

3. The sequence is 15600, 16650, 17700, ... Finding the 16th term: $a_{16} = 15600 + 15(1050) = 31,350$ students

5. The sequence is 5000, 5500, 6050, .. Finding the 5th term: $a_5 = 5000(1.1)^4 \approx 7,321$ students

7. Finding the 8th term: $a_8 = 16000\left(\dfrac{1}{2}\right)^7 = 125$ liters

9. Finding the 7th term: $a_7 = 5832\left(\dfrac{2}{3}\right)^6 = 512$ gallons

11. Finding the sum: $S_{30} = \dfrac{30(1 + 30)}{2} = 465$ quarters $= \$116.25$

13. Finding the amount saved on the 15th day: $a_{15} = 1(2)^{14} = \$163.84$

    Finding the total amount saved: $S_{15} = \dfrac{1(2)^{15} - 1}{2 - 1} = \$327.67$

15. In the 11th second, the object falls: $a_{11} = 16 + 10(32) = 336$ feet

    Finding the total distance: $S_{11} = \dfrac{11(16 + 336)}{2} = 1,936$ feet

17. Finding the 7th term: $a_7 = 60\left(\dfrac{1}{2}\right)^6 = \dfrac{15}{16}$ gram

19. First find the total falls:

$$1458 + 1458\left(\frac{1}{3}\right) + 1458\left(\frac{1}{3}\right)^2 + 1458\left(\frac{1}{3}\right)^3 + 1458\left(\frac{1}{3}\right)^4 + 1458\left(\frac{1}{3}\right)^5 = \frac{1458\left(\frac{1}{3}\right)^6 - 1458}{\frac{1}{3} - 1} = 2184$$

Now find the total rebounds:

$$1458\left(\frac{1}{3}\right) + 1458\left(\frac{1}{3}\right)^2 + 1458\left(\frac{1}{3}\right)^3 + 1458\left(\frac{1}{3}\right)^4 + 1458\left(\frac{1}{3}\right)^5 = \frac{486\left(\frac{1}{3}\right)^5 - 486}{\frac{1}{3} - 1} = 726$$

Thus the total distance is $2184 + 726 = 2910$ feet.

21. The total number is: $S_{25} = \dfrac{25(1+25)}{2} = 325$ logs

23. Finding the eighth term: $a_8 = 1\left(\dfrac{2}{3}\right)^7 \approx 5.9\%$ remaining

25. The amount remaining is: $a_9 = 20\left(\dfrac{1}{2}\right)^8 = \dfrac{5}{64}$ gallon

9.4 Mathematical Induction

1. For $n = 1$: $S_1 = \dfrac{1 \bullet 2}{2} = 1$. Suppose that it is true for $n = k$: $1 + 2 + 3 + \ldots + k = \dfrac{k(k+1)}{2}$

   Now show it is true for $n = k+1$: $1 + 2 + 3 + \ldots + k + k + 1 = \dfrac{k(k+1)}{2} + (k+1) = \dfrac{k(k+1)}{2} + \dfrac{2(k+1)}{2} = \dfrac{(k+1)(k+2)}{2}$

   Thus $S_n = \dfrac{n(n+1)}{2}$ for all positive integers n.

3. For $n = 1$: $S_1 = \dfrac{1 \bullet 4}{2} = 2$. Suppose that it is true for $n = k$: $2 + 5 + 8 + \ldots + (3k-1) = \dfrac{k(3k+1)}{2}$

   Now show it is true for $n = k+1$:

   $$2 + 5 + 8 + \ldots + (3k-1) + (3k+2) = \dfrac{k(3k+1)}{2} + (3k+2) = \dfrac{k(3k+1)}{2} + \dfrac{6k+4}{2} = \dfrac{3k^2 + 7k + 4}{2} = \dfrac{(k+1)(3k+4)}{2}$$

   Thus $S_n = \dfrac{n(3n+1)}{2}$ for all positive integers n.

5. For $n = 1$: $S_1 = 2^1 = 2$. Suppose that it is true for $n = k$: $2 + 4 + 8 + \ldots + 2^k = 2(2^k - 1)$

   Now show it is true for $n = k+1$: $2 + 4 + 8 + \ldots + 2^k + 2^{k+1} = 2(2^k - 1) + 2^{k+1} = 2^{k+1} - 2 + 2^{k+1} = 2(2^{k+1} - 1)$

   Thus $S_n = 2(2^n - 1)$ for all positive integers n.

7.   For $n = 1$: $S_1 = \dfrac{1(2)(3)}{6} = 1$. Suppose that it is true for $n = k$: $1^2 + 2^2 + 3^2 + ... + k^2 = \dfrac{k(k+1)(2k+1)}{6}$

Now show it is true for $n = k + 1$:

$$
\begin{aligned}
1^2 + 2^2 + 3^2 + ... + k^2 + (k+1)^2 &= \frac{k(k+1)(2k+1)}{6} + (k+1)^2 \\
&= \frac{k(k+1)(2k+1) + 6(k+1)^2}{6} \\
&= \frac{(k+1)\left(2k^2 + k + 6k + 6\right)}{6} \\
&= \frac{(k+1)\left(2k^2 + 7k + 6\right)}{6} \\
&= \frac{(k+1)(k+2)(2k+3)}{6}
\end{aligned}
$$

Thus $S_n = \dfrac{n(n+1)(2n+1)}{6}$ for all positive integers n.

9.   For $n = 1$: $S_1 = \dfrac{1}{1+1} = \dfrac{1}{2}$. Suppose that it is true for $n = k$: $\dfrac{1}{2} + \dfrac{1}{6} + \dfrac{1}{12} + ... + \dfrac{1}{k(k+1)} = \dfrac{k}{k+1}$

Now show it is true for $n = k + 1$:

$$
\begin{aligned}
\frac{1}{2} + \frac{1}{6} + \frac{1}{12} + ... + \frac{1}{k(k+1)} + \frac{1}{(k+1)(k+2)} &= \frac{k}{k+1} + \frac{1}{(k+1)(k+2)} \\
&= \frac{k(k+2)+1}{(k+1)(k+2)} \\
&= \frac{k^2 + 2k + 1}{(k+1)(k+2)} \\
&= \frac{(k+1)^2}{(k+1)(k+2)} \\
&= \frac{k+1}{k+2}
\end{aligned}
$$

Thus $S_n = \dfrac{n}{n+1}$ for all positive integers n.

11.   For $n = 1$: $3^1 = 3 \geq 2 + 1$, so the statement is true. Suppose that it is true for $n = k$: $3^k \geq 2k + 1$

Now show that it is true for $n = k + 1$: $3^{k+1} = 3 \bullet 3^k \geq 3(2k+1) = 6k + 3 \geq 2k + 3$

Thus $3^n \geq 2n + 1$ for all positive integers n.

13.   For $n = 1$: $1^2 = 1 \geq 1$, so the statement is true. Suppose that it is true for $n = k$: $k^2 \geq k$

Now show that it is true for $n = k + 1$: $(k+1)^2 = k^2 + 2k + 1 \geq k + 2k + 1 = (k+1) + 2k \geq k + 1$

Thus $n^2 \geq n$ for all positive integers n.

15.   For $n = 1$: $4^1 - 1 = 3$, which is divisible by 3, so the statement is true.

Suppose that it is true for $n = k$: $4^k - 1 = 3a$, for some integer a.

Now show that it is true for $n = k + 1$: $4^{k+1} - 1 = 4 \bullet 4^k - 1 = 4(3a+1) - 1 = 12a + 4 - 1 = 12a + 3 = 3(4a+1)$

Thus $4^n - 1$ is divisible by 3 for all positive integers n.

17.   For $n = 1$: $6^1 - 1 = 5$, which is divisible by 5, so the statement is true.

Suppose that it is true for $n = k$: $6^k - 1 = 5a$, for some integer a.

Now show that it is true for $n = k + 1$: $6^{k+1} - 1 = 6 \bullet 6^k - 1 = 6(5a+1) - 1 = 30a + 6 - 1 = 30a + 5 = 5(6a+1)$

Thus $6^n - 1$ is divisible by 5 for all positive integers n.

19. For $n = 1$: $1^2 + 1 = 2$, which is divisible by 2, so the statement is true.

Suppose that it is true for $n = k$: $k^2 + k = 2a$, for some integer a.

Now show that it is true for $n = k + 1$:

$$(k+1)^2 + (k+1) = k^2 + 2k + 1 + k + 1 = k^2 + k + (2k + 2) = 2a + 2(k+1) = 2(a + k + 1)$$

Thus $n^2 + n$ is divisible by 2 for all positive integers n.

Chapter 9 Review Problem Set

1. The general term is: $a_n = 6n - 3$

2. The general term is: $a_n = 3^{n-2}$

3. The general term is: $a_n = 5(2)^n$

4. The general term is: $a_n = -3n + 8$

5. The general term is: $a_n = 2n - 7$

6. The general term is: $a_n = 3^{3-n}$

7. The general term is: $a_n = -(-2)^{n-1}$

8. The general term is: $a_n = 3n + 9$

9. The general term is: $a_n = \dfrac{n+1}{3}$

10. The general term is: $a_n = 4^{n-1}$

11. The general term is: $a_n = 4n - 3$. Finding the required term: $a_{19} = 4(19) - 3 = 73$

12. The general term is: $a_n = 4n - 6$. Finding the required term: $a_{28} = 4(28) - 6 = 106$

13. The general term is: $a_n = 8\left(\dfrac{1}{2}\right)^{n-1}$. Finding the required term: $a_9 = 8\left(\dfrac{1}{2}\right)^8 = \dfrac{1}{32}$

14. The general term is: $a_n = \dfrac{243}{32}\left(\dfrac{2}{3}\right)^{n-1}$. Finding the required term: $a_8 = \dfrac{243}{32}\left(\dfrac{2}{3}\right)^7 = \dfrac{4}{9}$

15. The general term is: $a_n = -3n + 10$. Finding the required term: $a_{34} = -3(34) + 10 = -92$

16. The general term is: $a_n = -32\left(-\dfrac{1}{2}\right)^{n-1}$. Finding the required term: $a_{10} = -32\left(-\dfrac{1}{2}\right)^9 = \dfrac{1}{16}$

17. Writing the equations:

$-19 = a_1 + 4d$

$-34 = a_1 + 7d$

Subtracting the two equations:

$3d = -15$

$d = -5$

18. Writing the equations:

$37 = a_1 + 7d$

$57 = a_1 + 12d$

Subtracting the two equations:

$5d = 20$

$d = 4$

Substituting:

$37 = a_1 + 28$

$a_1 = 9$

So $a_n = 9 + (n-1)4 = 4n + 5$. Finding the term: $a_{20} = 4(20) + 5 = 85$

19. Writing the equations:
$$5 = a_1 r^2$$
$$135 = a_1 r^5$$
Dividing yields:
$$r^3 = 27$$
$$r = 3$$
Substituting:
$$5 = a_1 (3)^2$$
$$a_1 = \frac{5}{9}$$

20. Writing the equations:
$$\frac{1}{2} = a_1 r$$
$$8 = a_1 r^5$$
Dividing yields:
$$r^4 = 16$$
$$r = \pm 2$$

21. Finding the sum: $S_9 = \dfrac{81\left(\dfrac{1}{3}\right)^9 - 81}{\dfrac{1}{3} - 1} = 121\dfrac{40}{81}$

22. Finding the 70th term: $a_{70} = -3 + 69(3) = 204$. Finding the sum: $S_{70} = \dfrac{70(-3 + 204)}{2} = 7035$

23. Finding the 75th term: $a_{75} = 5 + 74(-4) = -291$. Finding the sum: $S_{75} = \dfrac{75(5 - 291)}{2} = -10,725$

24. Finding the sum: $S_{10} = \dfrac{16\left(\dfrac{1}{2}\right)^{10} - 16}{\dfrac{1}{2} - 1} = 31\dfrac{31}{32}$

25. Finding the 95th term: $a_{95} = 7(95) + 1 = 666$. Finding the sum: $S_{95} = \dfrac{95(8 + 666)}{2} = 32,015$

26. First find the number of terms:
$$137 = 5 + 2(n - 1)$$
$$132 = 2(n - 1)$$
$$n - 1 = 66$$
$$n = 67$$
Finding the sum: $S_{67} = \dfrac{67(5 + 137)}{2} = 4,757$

27. First find the number of terms:
$$\frac{1}{64} = 64\left(\frac{1}{4}\right)^{n-1}$$
$$\left(\frac{1}{4}\right)^{n-1} = \frac{1}{4^6}$$
$$n - 1 = 6$$
$$n = 7$$
Finding the sum: $S_7 = \dfrac{64\left(\dfrac{1}{4}\right)^7 - 64}{\dfrac{1}{4} - 1} = 85\dfrac{21}{64}$

28. First find the number of terms:
$$384 = 8 + 2(n-1)$$
$$2(n-1) = 376$$
$$n - 1 = 188$$
$$n = 189$$

Finding the sum: $S_{189} = \dfrac{189(8+384)}{2} = 37,044$

29. First find the number of terms:
$$276 = 27 + 3(n-1)$$
$$3(n-1) = 249$$
$$n - 1 = 83$$
$$n = 84$$

Finding the sum: $S_{84} = \dfrac{84(27+276)}{2} = 12,726$

30. Finding the sum: $S_{45} = \dfrac{45(3-85)}{2} = -1845$

31. Finding the sum: $\displaystyle\sum_{i=1}^{5} i^3 = 1 + 8 + 27 + 64 + 125 = 225$

32. Finding the sum: $\displaystyle\sum_{i=1}^{8} 2^{8-i} = 128 + 64 + 32 + 16 + 8 + 4 + 2 + 1 = 255$

33. Finding the 3rd and 75th terms:
$$a_3 = 3(3) - 4 = 5 \qquad\qquad a_{75} = 3(75) - 4 = 221$$

Now finding the sum: $S_{75} - S_3 = \dfrac{75(-1+221)}{2} - \dfrac{3(-1+5)}{2} = 8250 - 6 = 8244$

34. Finding the sum: $S_\infty = \dfrac{64}{1-\dfrac{1}{4}} = 85\dfrac{1}{3}$

35. Finding the sum: $0.\overline{36} = \dfrac{\dfrac{36}{100}}{1-\dfrac{1}{100}} = \dfrac{36}{99} = \dfrac{4}{11}$

36. Finding the sum: $0.4\overline{5} = \dfrac{4}{10} + \dfrac{\dfrac{5}{100}}{1-\dfrac{1}{10}} = \dfrac{4}{10} + \dfrac{5}{90} = \dfrac{41}{90}$

37. Finding the 13th term: $a_{13} = 3750 + 12(-250) = \$750$

38. Finding the total saved: $S_{30} = \dfrac{30(1+30)}{2} = 465$ dimes . She saved \$46.50.

39. Finding the total saved: $S_{15} = \dfrac{1(2)^{15}-1}{2-1} = 32767$ dimes . Nancy saved \$3,276.70.

40. Finding the amount remaining: $a_7 = 61440\left(\dfrac{3}{4}\right)^6 = 10,935$ gallons

41. For $n = 1$: $5^1 = 5 > 4$, so the statement is true. Assume it is true for $n = k$: $5^k > 5k - 1$.
We show it is true for $n = k + 1$: $5^{k+1} = 5 \cdot 5^k > 5(5k-1) = 25k - 5 > 5k + 20 - 5 > 5k + 4$

Thus $5^n > 5n - 1$ for all positive integers n.

42. For $n = 1$: $1^3 - 1 + 3 = 3$, which is divisible by 3. Assume it is true for $n = k$: $k^3 - k + 3 = 3a$ for some integer a.
We show it is true for $n = k + 1$:
$$(k+1)^3 - (k+1) + 3 = k^3 + 3k^2 + 3k + 1 - k - 1 + 3 = (k^3 - k + 3) + (3k^2 + 3k) = 3a + 3(k^2 + k) = 3(a + k^2 + k)$$

Thus $n^3 - n + 3$ is divisible by 3 for all positive integers n.

43. For $n = 1$: $S_1 = \dfrac{1(1+3)}{4(1+1)(1+2)} = \dfrac{1}{6} = a_1$. Assume it is true for $n = k$: $\dfrac{1}{6} + \dfrac{1}{24} + \ldots \dfrac{1}{k(k+1)(k+2)} = \dfrac{k(k+3)}{4(k+1)(k+2)}$

We show it is true for $n = k + 1$:

$$\dfrac{1}{6} + \dfrac{1}{24} + \ldots \dfrac{1}{k(k+1)(k+2)} + \dfrac{1}{(k+1)(k+2)(k+3)} = \dfrac{k(k+3)}{4(k+1)(k+2)} + \dfrac{1}{(k+1)(k+2)(k+3)}$$

$$= \dfrac{k(k+3)^2 + 4}{4(k+1)(k+2)(k+3)}$$

$$= \dfrac{k^3 + 6k^2 + 9k + 4}{4(k+1)(k+2)(k+3)}$$

$$= \dfrac{(k+1)^2(k+4)}{4(k+1)(k+2)(k+3)}$$

$$= \dfrac{(k+1)(k+4)}{4(k+2)(k+3)}$$

Thus $S_n = \dfrac{n(n+3)}{4(n+1)(n+2)}$ for all positive integers n.

Chapter 9 Test

1. Finding the 15th term: $a_{15} = -(15)^2 - 1 = -226$

2. Finding the 5th term: $a_5 = 3(2)^4 = 48$

3. The general term is: $a_n = -5n + 2$

4. The general term is: $a_n = 5\left(\dfrac{1}{2}\right)^{n-1} = 5(2)^{1-n}$

5. The general term is: $a_n = 6n + 4$

6. The general term is: $a_n = 8\left(\dfrac{3}{2}\right)^{n-1}$. Finding the 7th term: $a_7 = 8\left(\dfrac{3}{2}\right)^6 = 91\dfrac{1}{8}$

7. The general term is: $a_n = 3n - 2$. Finding the 75th term: $a_{75} = 3(75) - 2 = 223$

8. Finding the number of terms:
$$243 = 7 + 4(n-1)$$
$$4(n-1) = 236$$
$$n - 1 = 59$$
$$n = 60$$

9. Finding the 40th term: $a_{40} = 1 + 3(39) = 118$. Now finding the sum: $S_{40} = \dfrac{40(1+118)}{2} = 2380$

10. Finding the sum: $S_8 = \dfrac{3(2)^8 - 3}{2 - 1} = 765$

11. Finding the 45th term: $a_{45} = 7(45) - 2 = 313$. Finding the sum: $S_{45} = \dfrac{45(5+313)}{2} = 7155$

12. Finding the sum: $S_{10} = \dfrac{6(2)^{10} - 6}{2 - 1} = 6138$

13. Finding the 150th term: $a_{150} = 2 + 149(2) = 300$. Finding the sum: $S_{150} = \dfrac{150(2+300)}{2} = 22,650$

14. Finding the number of terms:
$$193 = 11 + 2(n-1)$$
$$182 = 2(n-1)$$
$$n - 1 = 91$$
$$n = 92$$

Finding the sum: $S_{92} = \dfrac{92(11+193)}{2} = 9384$

15. Finding the 50th term: $a_{50} = 3(50) + 5 = 155$. Finding the sum: $S_{50} = \dfrac{50(8+155)}{2} = 4075$

16. Finding the sum: $S_{10} = \dfrac{1(-2)^{10} - 1}{-2 - 1} = -341$

17. Finding the sum: $S_{\infty} = \dfrac{3}{1 - \frac{1}{2}} = 6$

18. Finding the sum: $S_{\infty} = \dfrac{\frac{2}{9}}{1 - \frac{1}{3}} = \dfrac{1}{3}$

19. Finding the sum: $0.\overline{18} = \dfrac{\frac{18}{100}}{1 - \frac{1}{100}} = \dfrac{18}{99} = \dfrac{2}{11}$

20. Finding the sum: $0.2\overline{6} = \dfrac{2}{10} + \dfrac{\frac{6}{100}}{1 - \frac{1}{10}} = \dfrac{2}{10} + \dfrac{6}{90} = \dfrac{24}{90} = \dfrac{4}{15}$

21. Finding the 8th term: $a_8 = 49152 \left(\dfrac{1}{4}\right)^7 = 3$ liters

22. Finding the total amount: $S_{15} = \dfrac{0.10(2)^{14} - 0.10}{2 - 1} = \$1638.30$

23. The first year investment grows to: $a_1 = 350 + 350(0.06)(10) = \$560$

The tenth year investment grows to: $a_{10} = 350 + 350(0.06)(1) = \$371$

So the total of the investments is: $S_{10} = \dfrac{10(560 + 371)}{2} = \$4,655$

24. For $n = 1$: $S_1 = \dfrac{1(3-1)}{2} = 1$, which is true. Assume the sum is true for $n = k$: $1 + 4 + 7 + \ldots + (3k - 2) = \dfrac{k(3k-1)}{2}$

We must show it is true for $n = k + 1$:

$$1 + 4 + 7 + \ldots + (3k - 2) + (3k + 1) = \dfrac{k(3k-1)}{2} + (3k+1)$$
$$= \dfrac{k(3k-1) + 2(3k+1)}{2}$$
$$= \dfrac{3k^2 - k + 6k + 2}{2}$$
$$= \dfrac{3k^2 + 5k + 2}{2}$$
$$= \dfrac{(k+1)(3k+2)}{2}$$

So $S_n = \dfrac{n(3n-1)}{2}$ for all positive integers n.

25. For $n = 1$: $9^1 - 1 = 8$, so the result is true. Assume it is true for $n = k$: $9^k - 1 = 8a$, for some integer a.

We must show it is true for $n = k + 1$: $9^{k+1} - 1 = 9 \bullet 9^k - 1 = 9(8a + 1) - 1 = 72a + 9 - 1 = 72a + 8 = 8(9a + 1)$

So $9^n - 1$ is divisible by 8 for all positive integers n.

Appendix A
Binomial Theorem

1. Using the binomial theorem:

$$(x+y)^8 = \binom{8}{0}x^8 + \binom{8}{1}x^7y + \binom{8}{2}x^6y^2 + \binom{8}{3}x^5y^3 + \binom{8}{4}x^4y^4 + \binom{8}{5}x^3y^5 + \binom{8}{6}x^2y^6 + \binom{8}{7}xy^7 + \binom{8}{8}y^8$$

$$= x^8 + 8x^7y + 28x^6y^2 + 56x^5y^3 + 70x^4y^4 + 56x^3y^5 + 28x^2y^6 + 8xy^7 + y^8$$

3. Using the binomial theorem:

$$(x-y)^6 = \binom{6}{0}x^6 - \binom{6}{1}x^5y + \binom{6}{2}x^4y^2 - \binom{6}{3}x^3y^3 + \binom{6}{4}x^2y^4 - \binom{6}{5}xy^5 + \binom{6}{6}y^6$$

$$= x^6 - 6x^5y + 15x^4y^2 - 20x^3y^3 + 15x^2y^4 - 6xy^5 + y^6$$

5. Using the binomial theorem:

$$(a+2b)^4 = \binom{4}{0}a^4 + \binom{4}{1}a^3(2b) + \binom{4}{2}x^2(2b)^2 + \binom{4}{3}x(2b)^3 + \binom{4}{4}(2b)^4$$

$$= a^4 + 8a^3b + 24a^2b^2 + 32ab^3 + 16b^4$$

7. Using the binomial theorem:

$$(x-3y)^5 = \binom{5}{0}x^5 - \binom{5}{1}x^4(3y) + \binom{5}{2}x^3(3y)^2 - \binom{5}{3}x^2(3y)^3 + \binom{5}{4}x(3y)^4 - \binom{5}{5}(3y)^5$$

$$= x^5 - 15x^4y + 90x^3y^2 - 270x^2y^3 + 405xy^4 - 243y^5$$

9. Using the binomial theorem:

$$(2a-3b)^4 = \binom{4}{0}(2a)^4 - \binom{4}{1}(2a)^3(3b) + \binom{4}{2}(2a)^2(3b)^2 - \binom{4}{3}(2a)(3b)^3 + \binom{4}{4}(3b)^4$$

$$= 16a^4 - 96a^3b + 216a^2b^2 - 216ab^3 + 81b^4$$

11. Using the binomial theorem:

$$(x^2+y)^5 = \binom{5}{0}(x^2)^5 + \binom{5}{1}(x^2)^4y + \binom{5}{2}(x^2)^3y^2 + \binom{5}{3}(x^2)^2y^3 + \binom{5}{4}x^2y^4 + \binom{5}{5}y^5$$

$$= x^{10} + 5x^8y + 10x^6y^2 + 10x^4y^3 + 5x^2y^4 + y^5$$

13. Using the binomial theorem:

$$(2x^2-y^2)^4 = \binom{4}{0}(2x^2)^4 - \binom{4}{1}(2x^2)^3y^2 + \binom{4}{2}(2x^2)^2(y^2)^2 - \binom{4}{3}(2x^2)(y^2)^3 + \binom{4}{4}(y^2)^4$$

$$= 16x^8 - 32x^6y^2 + 24x^4y^4 - 8x^2y^6 + y^8$$

15. Using the binomial theorem:
$$(x+3)^6 = \binom{6}{0}x^6 + \binom{6}{1}x^5(3) + \binom{6}{2}x^4(3)^2 + \binom{6}{3}x^3(3)^3 + \binom{6}{4}x^2(3)^4 + \binom{6}{5}x(3)^5 + \binom{6}{6}(3)^6$$
$$= x^6 + 18x^5 + 135x^4 + 540x^3 + 1215x^2 + 1458x + 729$$

17. Using the binomial theorem:
$$(x-1)^9 = \binom{9}{0}x^9 - \binom{9}{1}x^8 + \binom{9}{2}x^7 - \binom{9}{3}x^6 + \binom{9}{4}x^5 - \binom{9}{5}x^4 + \binom{9}{6}x^3 - \binom{9}{7}x^2 + \binom{9}{8}x - \binom{9}{9}$$
$$= x^9 - 9x^8 + 36x^7 - 84x^6 + 126x^5 - 126x^4 + 84x^3 - 36x^2 + 9x - 1$$

19. Using the binomial theorem:
$$\left(1+\frac{1}{n}\right)^4 = \binom{4}{0} + \binom{4}{1}\left(\frac{1}{n}\right) + \binom{4}{2}\left(\frac{1}{n}\right)^2 + \binom{4}{3}\left(\frac{1}{n}\right)^3 + \binom{4}{4}\left(\frac{1}{n}\right)^4 = 1 + \frac{4}{n} + \frac{6}{n^2} + \frac{4}{n^3} + \frac{1}{n^4}$$

21. Using the binomial theorem:
$$\left(a-\frac{1}{n}\right)^6 = \binom{6}{0}a^6 - \binom{6}{1}a^5\left(\frac{1}{n}\right) + \binom{6}{2}a^4\left(\frac{1}{n}\right)^2 - \binom{6}{3}a^3\left(\frac{1}{n}\right)^3 + \binom{6}{4}a^2\left(\frac{1}{n}\right)^4 - \binom{6}{5}a\left(\frac{1}{n}\right)^5 + \binom{6}{6}\left(\frac{1}{n}\right)^6$$
$$= a^6 - \frac{6a^5}{n} + \frac{15a^4}{n^2} - \frac{20a^3}{n^3} + \frac{15a^2}{n^4} - \frac{6a}{n^5} + \frac{1}{n^6}$$

23. Using the binomial theorem:
$$(1+\sqrt{2})^4 = \binom{4}{0} + \binom{4}{1}(\sqrt{2}) + \binom{4}{2}(\sqrt{2})^2 + \binom{4}{3}(\sqrt{2})^3 + \binom{4}{4}(\sqrt{2})^4 = 1 + 4\sqrt{2} + 12 + 8\sqrt{2} + 4 = 17 + 12\sqrt{2}$$

25. Using the binomial theorem:
$$(3-\sqrt{2})^5 = \binom{5}{0}(3)^5 - \binom{5}{1}(3)^4(\sqrt{2}) + \binom{5}{2}(3)^3(\sqrt{2})^2 - \binom{5}{3}(3)^2(\sqrt{2})^3 + \binom{5}{4}(3)(\sqrt{2})^4 - \binom{5}{5}(\sqrt{2})^5$$
$$= 243 - 405\sqrt{2} + 540 - 180\sqrt{2} + 60 - 4\sqrt{2}$$
$$= 843 - 589\sqrt{2}$$

27. Using the binomial theorem, the first four terms are:
$$\binom{12}{0}x^{12} + \binom{12}{1}x^{11}y + \binom{12}{2}x^{10}y^2 + \binom{12}{3}x^9y^3 = x^{12} + 12x^{11}y + 66x^{10}y^2 + 220x^9y^3$$

29. Using the binomial theorem, the first four terms are:
$$\binom{20}{0}x^{20} - \binom{20}{1}x^{19}y + \binom{20}{2}x^{18}y^2 - \binom{20}{3}x^{17}y^3 = x^{20} - 20x^{19}y + 190x^{18}y^2 - 1140x^{17}y^3$$

31. Using the binomial theorem, the first four terms are:
$$\binom{14}{0}(x^2)^{14} - \binom{14}{1}(x^2)^{13}(2y^3) + \binom{14}{2}(x^2)^{12}(2y^3)^2 - \binom{14}{3}(x^2)^{11}(2y^3)^3$$
$$= x^{28} - 28x^{26}y^3 + 364x^{24}y^6 - 2912x^{22}y^9$$

33. Using the binomial theorem, the first four terms are:
$$\binom{9}{0}a^9 + \binom{9}{1}a^8\left(\frac{1}{n}\right) + \binom{9}{2}a^7\left(\frac{1}{n}\right)^2 + \binom{9}{3}a^6\left(\frac{1}{n}\right)^3 = a^9 + \frac{9a^8}{n} + \frac{36a^7}{n^2} + \frac{84a^6}{n^3}$$

35. Using the binomial theorem, the first four terms are:
$$\binom{10}{0}(-x)^{10} + \binom{10}{1}(-x)^9(2y) + \binom{10}{2}(-x)^8(2y)^2 + \binom{10}{3}(-x)^7(2y)^3 = x^{10} - 20x^9y + 180x^8y^2 - 960x^7y^3$$

37. The fourth term is: $\binom{8}{3}x^5y^3 = 56x^5y^3$

39. The fifth term is: $\binom{9}{4}x^5(-y)^4 = 126x^5y^4$

41. The sixth term is: $\begin{pmatrix} 7 \\ 5 \end{pmatrix}(3a)^2 b^5 = 189a^2 b^5$

43. The eighth term is: $\begin{pmatrix} 10 \\ 7 \end{pmatrix}(x^2)^3 (y^3)^7 = 120x^6 y^{21}$

45. The seventh term is: $\begin{pmatrix} 15 \\ 6 \end{pmatrix}(1)^9 \left(-\frac{1}{n}\right)^6 = \frac{5005}{n^6}$